Nanotechnology for Sustainable Food Packaging

Nanotechnology for Sustainable Food Packaging

Edited by

Dr. C. Anandharamakrishnan
Director, CSIR–National Institute for Interdisciplinary Science and Technology (NIIST)
Ministry of Science and Technology–Government of India, Industrial Estate PO
Thiruvananthapuram 695019
Kerala
India

Dr. Jeyan A. Moses
Assistant Professor, Computational Modeling and Nanoscale Processing Unit,
Department of Food Process Engineering
National Institute of Food Technology, Entrepreneurship and Management,
Thanjavur (NIFTEM-T)
Thanjavur 613005
Tamil Nadu
India

Dr. M. Maria Leena
Assitant Professor, Department of Biotechnology
Faculty of Engineering and Technology
SRM Institute of Science and Technology (SRMIST)
Tiruchirappalli 621105
Tamil Nadu
India

WILEY

This edition first published 2025.

Registered Offices
John Wiley & Sons, Inc., 111 River Street, Hoboken, NJ 07030, USA
John Wiley & Sons Ltd, New Era House, 8 Oldlands Way, Bognor Regis, West Sussex, PO22 9NQ, UK

For details of our global editorial offices, customer services, and more information about Wiley products visit us at www.wiley.com.

The manufacturer's authorized representative according to the EU General Product Safety Regulation is Wiley-VCH GmbH, Boschstr. 12, 69469 Weinheim, Germany, e-mail: Product_Safety@wiley.com.

Wiley also publishes its books in a variety of electronic formats and by print-on-demand. Some content that appears in standard print versions of this book may not be available in other formats.

Library of Congress Cataloging-in-Publication Data
Names: Anandharamakrishnan, C., editor. | Moses, Jeyan A., editor. | Leena, Maria, editor.
Title: Nanotechnology for sustainable food packaging / edited by C. Anandharamakrishnan, Jeyan A. Moses, Maria Leena.
Description: Hoboken, NJ : Wiley, 2025. | Includes index.
Identifiers: LCCN 2024039489 (print) | LCCN 2024039490 (ebook) | ISBN 9781119875123 (hardback) | ISBN 9781119875130 (adobe pdf) | ISBN 9781119875147 (epub)
Subjects: LCSH: Food–Packaging. | Nanotechnology. | Sustainable development.
Classification: LCC TP374 .N39 2025 (print) | LCC TP374 (ebook) | DDC 664/.09–dc23/eng/20241031
LC record available at https://lccn.loc.gov/2024039489
LC ebook record available at https://lccn.loc.gov/2024039490

Cover design: Wiley
Cover image: © Dimitri Tymchenko/Shutterstock

Set in 9.5/12.5pt STIXTwoText by Straive, Pondicherry, India
Printed and bound by CPI Group (UK) Ltd, Croydon, CR0 4YY

C9781119875123_150325

Contents

List of Contributors

C. Anandharamakrishnan
CSIR—National Institute for Interdisciplinary Science and Technology (NIIST)
Ministry of Science and Technology—Government of India, Industrial Estate PO
Thiruvananthapuram, Kerala
India

Betty Del Carmen Jarma Arroyo
Consumer Science Department
Universidade Federal Rural de Pernambuco
Recife
Brazil

J. Cyril Robinson Azariah
Department of Nanotechnology
Institute of Electronics and Communication Engineering
Saveetha School of Engineering, Saveetha Institute of Medical and Technical Sciences
Saveetha University
Chennai, Tamil Nadu
India

Adity Bahndral
Department of Food Technology and Nutrition, School of Agriculture
Lovely Professional University
Phagwara, Punjab
India

Vimala S.K. Bharathi
Department of Biosystems Engineering
University of Manitoba
Winnipeg, Manitoba
Canada

President's Office
University of Lethbridge
Lethbridge, Alberta
Canada

Kanishka Bhunia
Agricultural and Food Engineering Department
Indian Institute of Technology Kharagpur
Kharagpur, West Bengal
India

Divyajyoti Biswal
Department of Chemical Engineering
Visvesvaraya National Institute of Technology (VNIT)
Nagpur, Maharashtra
India

Luana S.C. Carnaval
Centre for Sustainable Packaging and Bioproducts (CSPB)
School of Food Science and Environmental Health
Technological University Dublin
Dublin
Ireland

Sustainability and Health Institute (SHI)
Technological University Dublin
Dublin
Ireland

Miguel Angelo Cerqueira
International Iberian Nanotechnology Laboratory
Braga
Portugal

Pintu Choudhary
Department of Food Technology
Chaudhary Bansal Lal Government Polytechnic
Bhiwani, Haryana
India

Maria José Costa
International Iberian Nanotechnology Laboratory
Braga
Portugal

Centre of Biological Engineering
University of Minho
Braga
Portugal

Shweta Deotale
Department of Chemical Engineering
Visvesvaraya National Institute of Technology (VNIT)
Nagpur, Maharashtra
India

Lisa-Marie Dietz
Sustainable Packaging Institute SPI,
Albstadt-Sigmaringen University,
Sigmaringen
Germany

Anns Annie Gigi
Agro-Processing and Technology Division
CSIR-National Institute for Interdisciplinary Science and Technology
Thiruvananthapuram
India

Academy of Scientific and Innovative Research (AcSIR)
Ghaziabad
India

Hamid
Department of Food Technology and Nutrition, School of Agriculture
Lovely Professional University
Phagwara, Punjab
India

Amit K. Jaiswal
School of Food Science and Environmental Health
Technological University Dublin
Dublin
Ireland

Sustainability and Health Institute (SHI)
Technological University Dublin
Dublin
Ireland

Swarna Jaiswal
Centre for Sustainable Packaging and Bioproducts (CSPB)
School of Food Science and Environmental Health
Technological University Dublin
Dublin
Ireland

Sustainability and Health Institute (SHI)
Technological University Dublin
Dublin
Ireland

Digvir S. Jayas
Department of Biosystems Engineering
University of Manitoba
Winnipeg, Manitoba
Canada

President's Office
University of Lethbridge
Lethbridge, Alberta
Canada

Vidushi Kapoor
CSIR—National Institute for Interdisciplinary Science and Technology (NIIST)
Ministry of Science and Technology—Government of India
Industrial Estate PO
Thiruvananthapuram, Kerala
India

Navjot Kaur
Department of Food Technology and Nutrition, School of Agriculture
Lovely Professional University
Phagwara, Punjab
India

Rahul Kumar
Agricultural and Food Engineering Department
Indian Institute of Technology Kharagpur
Kharagpur, West Bengal
India

L. Mahalakshmi
CSIR—National Institute for Interdisciplinary Science and Technology (NIIST)
Ministry of Science and Technology—Government of India
Industrial Estate PO
Thiruvananthapuram, Kerala
India

M. Maria Leena
Department of Biotechnology
Faculty of Engineering and Technology
SRM Institute of Science and Technology (SRMIST)
Tiruchirappalli
Tamil Nadu
India

Sachin A. Mandavgane
Department of Chemical Engineering
Visvesvaraya National Institute of Technology (VNIT)
Nagpur, Maharashtra
India

Enayde de Almeida Melo
Consumer Science Department
Universidade Federal Rural de Pernambuco
Recife
Brazil

Jeyan A. Moses
Computational Modeling and Nanoscale Processing Unit
Department of Food Process Engineering
National Institute of Food Technology Entrepreneurship and Management Thanjavur (NIFTEM-T)
Thanjavur
Tamil Nadu
India

Vanmathi Mugasundari
Computational Modeling and Nanoscale Processing Unit,
Department of Food Process Engineering
National Institute of Food Technology, Entrepreneurship and Management, Thanjavur (NIFTEM-T)
Thanjavur
Tamil Nadu
India

Ishita Neogi
CSIR—National Institute for Interdisciplinary Science and Technology (NIIST)
Ministry of Science and Technology—Government of India
Industrial Estate PO
Thiruvananthapuram, Kerala
India

Sundus Nida
Department of Food Technology, Faculty of Engineering and Technology
JAIN (Deemed-to-be University)
Ramanagara, Karnataka
India

Gopinath Packirisamy
Centre for Nanotechnology
Indian Institute of Technology
Roorkee, Uttarakhand
India

Department of Biosciences and Bioengineering
Indian Institute of Technology
Roorkee, Uttarakhand
India

Lorenzo Pastrana
International Iberian Nanotechnology Laboratory
Braga
Portugal

Kalpani Y. Perera
Centre for Sustainable Packaging and Bioproducts (CSPB)
School of Food Science and Environmental Health, Faculty of Science, Technological University Dublin
Dublin
Ireland

Sustainability and Health Institute (SHI)
Technological University Dublin
Dublin
Ireland

Prasanth K.S. Pillai
Department of Food Science and Nutrition
University of Minnesota
St. Paul, MN
United States

K.V. Ragavan
Agro-Processing and Technology Division
CSIR-National Institute for Interdisciplinary Science and Technology
Thiruvananthapuram
India

Academy of Scientific and Innovative Research (AcSIR)
Ghaziabad
India

Vijayakumar Raja
CSIR—National Institute for Interdisciplinary Science and Technology (NIIST)
Ministry of Science and Technology—Government of India
Industrial Estate PO
Thiruvananthapuram, Kerala
India

Corina L. Reichert
Sustainable Packaging Institute SPI, Faculty of Life Sciences
Albstadt-Sigmaringen University, Sigmaringen
Germany

P. Santhoshkumar
Computational Modeling and Nanoscale Processing Unit
Department of Food Process Engineering
National Institute of Food Technology, Entrepreneurship and Management—Thanjavur (NIFTEM-T)
Thanjavur, Tamil Nadu
India

Andrelina Maria Pinheiro Santos
Department of Chemical Engineering, Bioprocess Laboratory
Universidade Federal de Pernambuco, Recife
Brazil

Markus Schmid
Sustainable Packaging Institute SPI,
Albstadt-Sigmaringen University,
Sigmaringen
Germany

Rafeeya Shams
Department of Food Technology and
Nutrition, School of Agriculture
Lovely Professional University
Phagwara, Punjab
India

Akanksha Shetty
Department of Food Technology, Faculty
of Engineering and Technology
JAIN (Deemed-to-be University)
Ramanagara, Karnataka
India

Dravin Pratap Singh
Centre for Nanotechnology
Indian Institute of Technology
Roorkee, Uttarakhand
India

Ila Singh
Centre for Nanotechnology
Indian Institute of Technology
Roorkee, Uttarakhand
India

Singam Suranjoy Singh
Department of Food Science
University of Guelph
Guelph, Ontario
Canada

Victor Souza
International Iberian Nanotechnology
Laboratory
Braga
Portugal

Rishab Subramaniam
Department of Food Technology, Faculty
of Engineering and Technology,
JAIN (Deemed-to-be University)
Ramanagara, Karnataka
India

Sungeetha
Department of Nanotechnology
Institute of Electronics and
Communication Engineering, Saveetha
School of Engineering, Saveetha Institute
of Medical and Technical Sciences,
Saveetha University
Chennai, Tamil Nadu
India

Ashley George Thomas
Department of Medical Biotechnology and
Integrative Physiology
Institute of Biotechnology, Saveetha
School of Engineering, Saveetha Institute
of Medical and Technical Sciences,
Saveetha University
Chennai, Tamil Nadu
India

B.S. Unnikrishnan
Centre for Nanotechnology
Indian Institute of Technology
Roorkee, Uttarakhand
India

Devika Yadav
CSIR—National Institute for
Interdisciplinary Science and Technology
(NIIST)
Ministry of Science and Technology—
Government of India
Thiruvananthapuram, Kerala
India

Preface

The packaging component holds a prominent space in the food industry. Over the years, several innovations and advances have been made, and recent endeavors focus on developing sustainable packaging solutions. Among the various approaches to achieving this target, copious research focus and commercial interest have been diverted to nanotechnology-driven concepts, which is the motivation for this book. To achieve this, the book has excellent contributions and expert insights from leading voices in the field, from across different countries.

The introductory chapters present an overview of nanotechnology and its applications in the food industry, the emerging trends in food packaging, and the scope of nanotechnology. This is followed by chapters highlighting different material compositions such as polymer-based nanocomposites and Inorganic and metal oxide nanomaterials. The next set of chapters present interesting applications such as edible coatings, multilayer flexible films for bio-based food packaging, nanoencapsulation and nanodelivery through packaging, active packaging, and intelligent packaging. With information on concepts, approaches, and advances, this section is followed by chapters showcasing the integration of quality detection and sensing applications, and emerging application fields such as the significance of additive manufacturing technology and nanoscale surface modification. The last application-focused chapter is on how nanotechnology is expanding in the packaging space for special/space foods. The concluding chapters explain concepts of life cycle analysis, migratory effects, safety, and concerns of nanofood packaging, global regulatory frameworks for nanomaterials in food packaging, and a strong focus on sustainability and the approaches thereof.

We are confident that this book will be a valuable resource for researchers and industry, who aim to develop sustainable food packaging solutions in the fascinating field of nanotechnology. This will be of significance as the shift to sustainable packaging technologies gains momentum.

1

Nanotechnology in the Food Industry

P. Santhoshkumar and Jeyan A. Moses

Computational Modeling and Nanoscale Processing Unit, Department of Food Process Engineering, National Institute of Food Technology Entrepreneurship and Management, Thanjavur (NIFTEM-T), Thanjavur, Tamil Nadu - 613005, India

1.1 Introduction

In the broadest sense, nanotechnology deals with the manipulation of matter on a near-atomic scale, aiming to develop new structures, materials, or devices. Given its uniqueness in terms of material functionalities to beneficial properties and capabilities, nano-based concepts have gained enormous applications across different sectors, ranging from medicine to energy to environment to agriculture to electronics to drug delivery to material sciences (Anandharamakrishnan, 2021; Maria et al., 2019). This also explains that nano-based concepts were in use centuries before and these are documented in different other literature sources. Irrespective of the application, one concept that is unarguably considered as the deciding factor for the level of changes a material can undergo when present in nanoscale vs. bulk counterpart is the significant variation in surface-to-volume ratio (Perumal et al., 2022). Now this can have prominent effects on a range of properties, spanning from thermal to physical to electrical to absorption to reaction to several others. In addition, for many applications, the impact of quantum confinement on the properties thereof has been well documented.

Particularly in the past couple of decades, copious works have been taken up and the field has gained considerable research and commercial interest. Just as every other technology, nanotechnology has also been associated with both merits and limitations (Maria et al., 2019). This explains the need for a judicious evaluation of the technology in a broad perspective, alongside both near and long-term impact. On a different note, nano-based concepts have been recognized with several Noble Prizes for discoveries ranging from characterization techniques to applications. Nano-based materials have been classified under various categories and scientists have used different scales to explain their classification. For a simple understanding, such classifications are done based on the number of dimensions (e.g., 0D, 1D, 2D, or 3D materials), shapes (e.g., spheres, shells, rods, etc.), and even based on composition or source (e.g., nature-derived, nature identical, engineered, etc.) (Anandharamakrishnan, 2021).

Nanotechnology for Sustainable Food Packaging, First Edition. Edited by C. Anandharamakrishnan, Jeyan A. Moses, and M. Maria Leena.

Several technologies have emerged in the food industry and some of them have transformed our existing approaches in handling, manufacturing, sensing, quality evaluation, processing, and other aspects (Anandharamakrishnan, 2021). Similarly, nanotechnology has also found diverse applications in the food sector and the focus of this chapter is to present such applications, highlighting examples of research work done during the recent past. Having said this, it is important to note that the following approach is logical and valid for all such applications:

1) Material fabrication/synthesis/preparation
2) Material characterization
3) In-depth assessment of the functionalities and identification of the application range
4) Application
5) Assessment of merits and limits
6) Iterations for technology modification/improvement
7) Final optimized process
8) Detailed impact assessments, validation, and documentation
9) Commercialization

Here, typically, (1) involves either physical, chemical, biological, or hybrid approaches for the production of nanomaterials or for structuring at the nanoscale or for allied applications. Then, in (2) different techniques and instrumental approaches are used to characterize the developed material (Preethi et al., 2019). This could be in terms of surface morphology and topography assessments, particle size, surface composition, or other measurements. Following suitable studies as in (3), the application range can be assessed. In the lab scale, once an application (4) is made, the pros and cons are identified (5), and any further improvement required is done (6). A final optimized process with a set process and product variables is then developed (7) and this is subjected to a detailed assessment in terms of various aspects (8), before being commercialized (9). A popular example is the synthesis and characterization of metal nanoparticles, and the assessment of functionalities to meet different applications (Lakshmanan et al., 2021). It is important to note that for specific requirements and applications, the order listed here may have minor variations.

1.2 Nanotechnology in the Food Industry

Along the food supply chain, nano-based concepts can be used at any or multiple points (Figure 1.1) (Anandharamakrishnan, 2021). Typical applications are detained in this section. In a broad sense, when nano-based approaches are involved, the product is categorized as "nanofood."

1.2.1 Food Production

Nanotechnology has found several applications at the production site. Popularly, several innovations have come forth in aspects of seed quality improvement, fertilizers, pesticides, soil quality improvement, growth stimulators, and an array of sensing

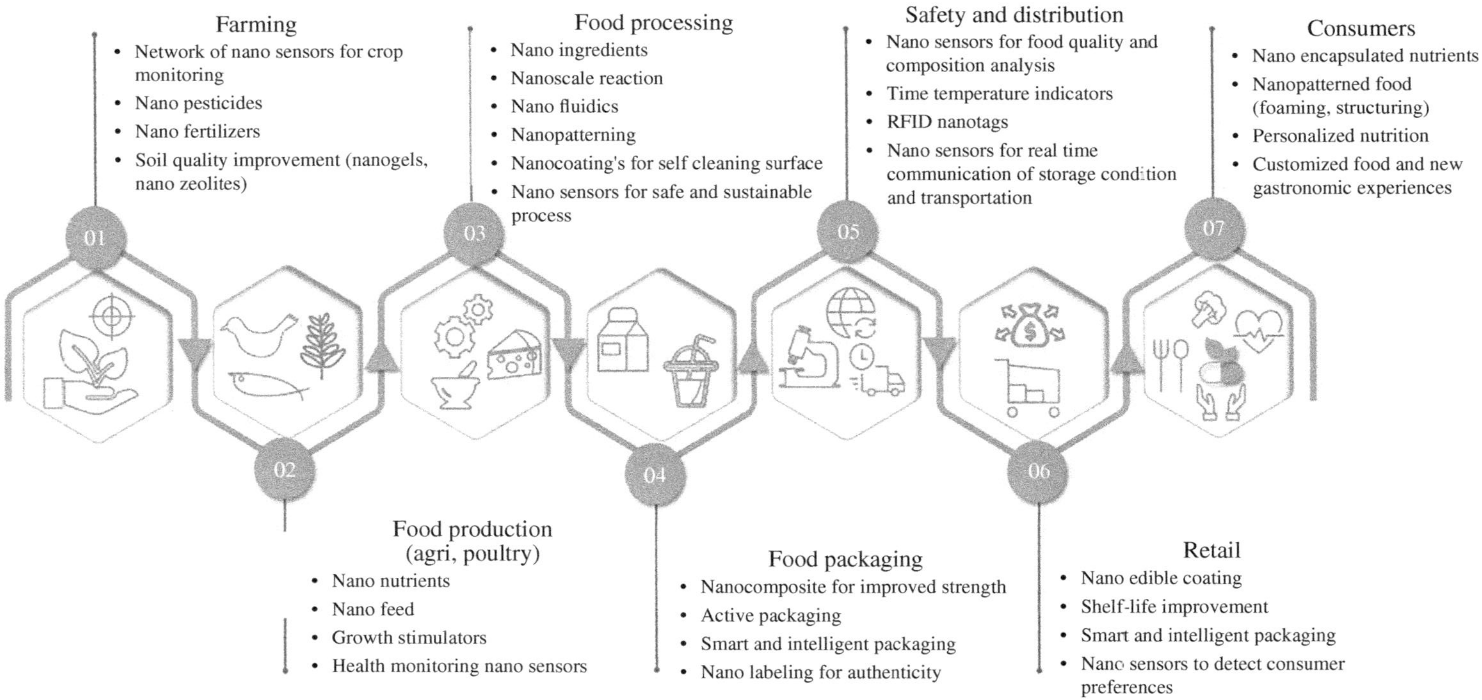

Figure 1.1 Potential applications of nanotechnology in Agro-Food Sector (*Source:* Anandharamakrishnan (2021)/with permission of Elsevier)

devices to monitor soil and water quality. There has been a keen focus on crop yield improvement and targeted destruction of weeds or insect pests. Also, nanosensing devices are valued for their speed, specificity, and accuracy. Nano-based concepts are also being explored to improve feed quality; nutrient-dense feeds can prove beneficial for fish, poultry, and animals. Overall, such aspects are better reviewed by other researchers.

1.2.2 Food Processing

Taking roots from applications in different other sectors such as chemical engineering, biomedical engineering, mechanical engineering, etc., the food industry has also benefitted from nano-based concepts. There are also nature-based nanostructures that researchers have taken ideas from. A classic example is the beautifully structured milk casein micelles. Accordingly, engineered nanostructures have been fabricated for different applications (Preethi et al., 2019), including the delivery of nutraceutical ingredients (Choudhary et al., 2021; Jayan et al., 2019; Leena et al., 2022; Mahalakshmi et al., 2020; Selvakumar et al., 2019). Certainly, a vast majority of such applications are still at the lab level. Some of them are being explored currently in pilot-scale trials and a couple of them have been commercialized. Some interesting applications include the improvement of reaction rates and kinetics, nanofluidics for different applications, nanopatterning and nanostructuring of foods and beverages, even for improved sensorial perception, and self-cleaning equipment, including nanocoatings, and machine parts (Deotale et al., 2020; Joseph et al., 2024; Logesh et al., 2021). In such examples, either the food or the equipment used or the process uses nanotechnology concepts. One recent example is the usage of an electrospun nanomembrane for the filtration of coconut neera, wherein the authors reported a substantial improvement in the product's quality (Leena et al., 2021). Electrospun nanomembrane are potential material for ensuring quality and safety of beverages (Leena et al., 2020b) and delivery of nutraceutical ingredients (Kavimughil et al., 2023). Similarly, nanomaterials can be used for disinfection applications in the food and water sector (Murugesan et al., 2019). Innovations in the past and recent advances have been reviewed and reported by different research groups (Anandharamakrishnan and Parthasarathi, 2019).

1.2.3 Food Additives

Closely linked to food processing, several works have demonstrated the capabilities of using nanoscale ingredients and additives in foods for improved characteristics and specific functionalities. For instance, nano additives can provide intense flavors, stable colors, improved efficiencies against microorganisms, better water dispersibility of targeted compounds, and improved functional characteristics. Another interesting application is the scope of using nano-sized salt to lower dietary intake of sodium (Figure 1.2) (Vinitha et al., 2021). Researchers used an electrospraying approach and demonstrated its scope in a study conducted in comparison with conventional and spray-dried micron-sized salts (Vinitha et al., 2021). Such applications have been reviewed, consolidated, and reported

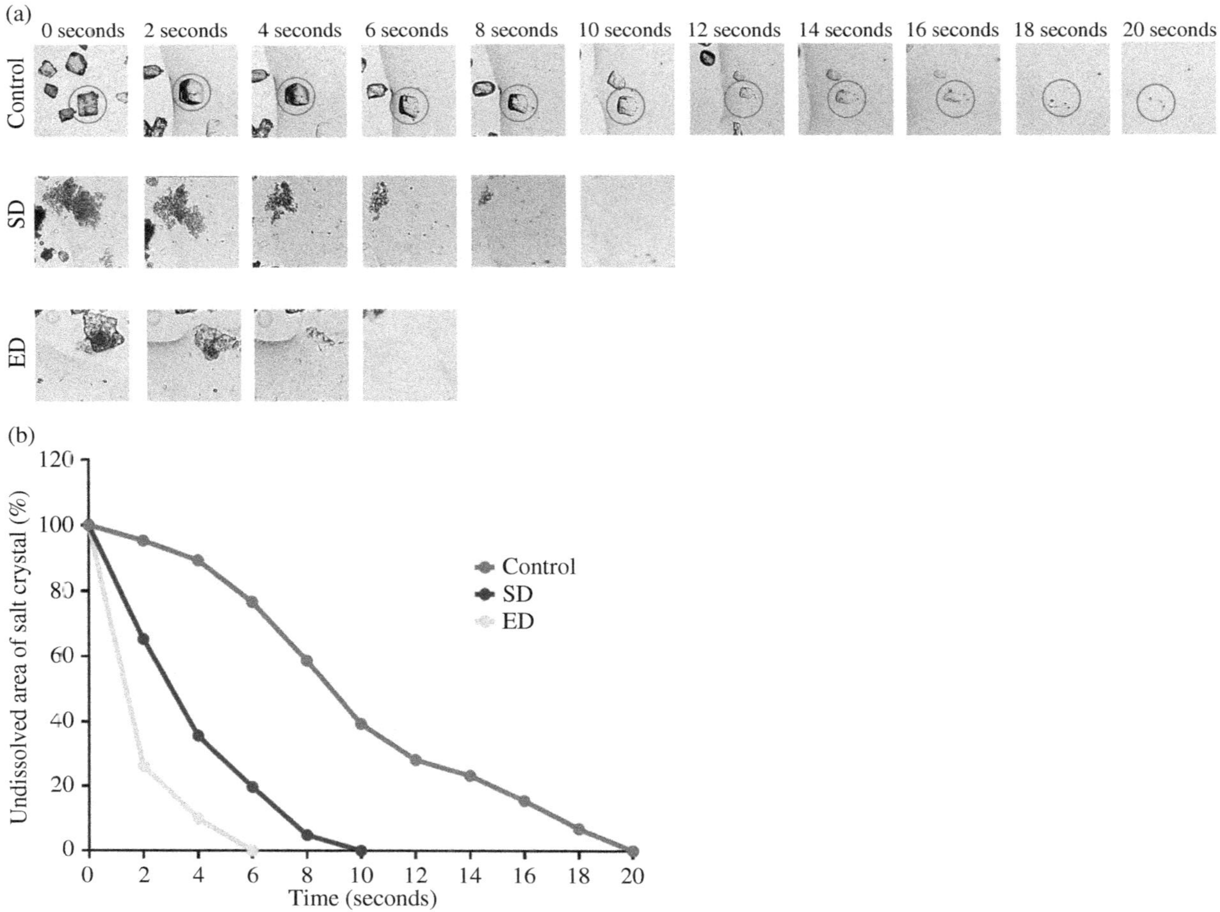

Figure 1.2 Dissolution kinetics of control, spray-dried, and electrosprayed salts in artificial saliva. (a) Video microscopy images of the dissolution sequence at every 2 seconds (*Source:* Vinitha et al. (2021)/with permission of Elsevier). (b) Dissolution time profile of control, SD and ED salts calculated from equivalent exposure area (*Source:* Vinitha et al. (2021)/with permission of Elsevier)

(Leena et al., 2020a; Mahalakshmi et al., 2024). Further, possible toxicity and detrimental effects are also being studied extensively.

1.2.4 Food Packaging and Retailing

Nanotechnology has a vast range of applications in the field of food packaging. The relevance for nanotechnology in the food packaging sector from 2000 to 2025 is shown in Figure 1.3. Popular keyword examples link with the development of packaging materials with nano-based concepts includes edible and other coatings, nanocomposites, food quality sensing systems, active and intelligent packaging, RFID nanotags, systems for real-time monitoring and communication of product status, etc. (Bharathi et al., 2019; Kalpana et al., 2019; Preethi et al., 2020). For instance, nanostructured edible coatings can also support in providing improved bioaccessibility of encapsulated ingredients while controlling moisture loss and retaining the color of minimally processed fruits (Figure 1.4) (Leena et al., 2020c). Similarly, nanofiber-based coatings have been explored as an innovative approach to improving the quality of minimally processed fruits/vegetables (Bharathi et al., 2020). Accordingly, the selection of materials, processes involved, characterization studies, application studies, performance, and validation go hand-in-hand (Theagarajan et al., 2021). Given the significance of such concepts, upcoming chapters of this book, detail such ideas, presenting the state-of-the-art. Studies must expand on the impact of using nano-based concepts in food packaging on their reuse, recycling, and disposal.

1.2.5 Food Safety and Quality

Nano-based sensing and biosensing systems are also finding an increasing number of applications. Owing to their unique properties, they can prove to be excellent alternatives and complementors for several conventional testing methodologies (Manoj et al., 2021). Copious studies have explored the feasibility of using nanomaterials and nanofabrication approaches in developing systems for the quality and safety assessment of solid and liquid foods. An example of a nanomaterial-based approach is the food waste-derived carbon dots that have been used for the detection of melamine (Murugesan et al., 2023). In a similar approach, heavy metals and metal ions can also be detected in water and different liquid foods (Murugesan et al., 2020). Similarly, for nanofabrication, a popular approach is the electrohydrodynamic process of electrospraying/electrospinning to develop natural bioactive-based nanosensors for food quality analysis (Amal Nath et al., 2022). Conventionally, the approach involves the usage of needle(s) and recently, particularly for higher process outputs, needleless electrospinning is considered (Raja et al., 2024). From a different perspective, studying the crystal structure can also help in assessing the quality and safety of foods. For example, researchers used a powder X-ray diffraction approach to screen for different adulterants in foodstuffs (Paranthaman et al., 2021a, 2021b). Overall, these are extensively reviewed in different articles (Shanthamma et al., 2022; Ukkunda et al., 2024). Nanozymes are also finding newer

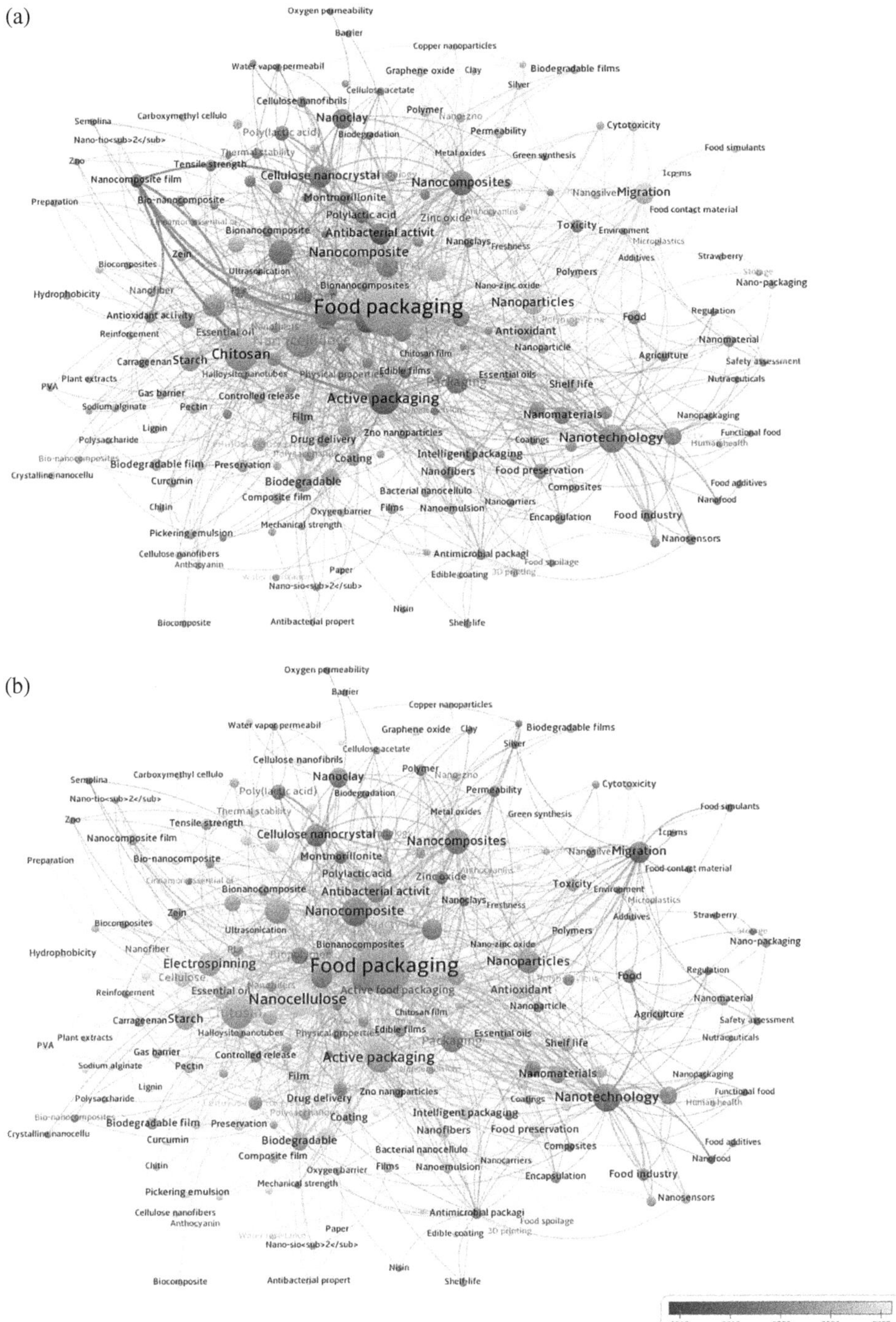

Figure 1.3 Keywords relevant to nanotechnology in the food packaging sector: trends in (a) 2000 to (b) 2025

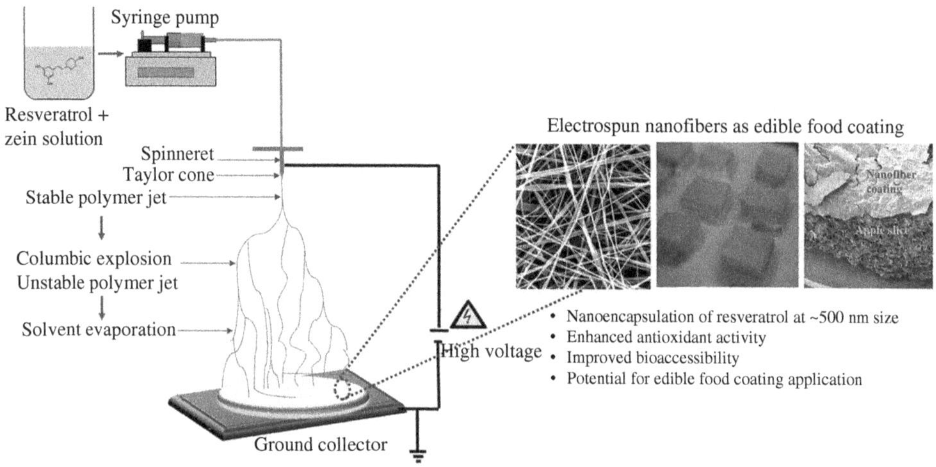

Figure 1.4 Edible coating with resveratrol-loaded electrospun zein nanofibers to control moisture loss and retain the color of minimally processed fruits (*Source:* Maria Leena et al. (2020)/with permission of Elsevier)

application ranges in food quality assessment (Payal et al., 2021). One aspect that must be emphasized is the cost economics involved.

1.2.6 Food and the Consumer

The ultimate decider for a food product is consumer perception and preference. Accordingly, several approaches have evolved to target this requirement. However, this remains dynamic and changes over the years have been phenomenal. Without technological innovations, food products do not survive in the market. One promising approach is the scope of using nano-based concepts to fill the rising gap between consumer demands and available products. For instance, foods can be structured/nano-patterned for a better gastronomic experience (Figure 1.5) (Deotale et al., 2023; Subhasri et al., 2022). Further, it is also convenient to personalize nutrient and nutraceutical ingredients through nanoencapsulation and other approaches. Particularly, with nanoencapsulation it is possible to achieve enhanced stability and better biological performance of bioactives (Leena et al., 2020a; Mahalakshmi et al., 2022). Such findings are also well-documented (Subhasri et al., 2023). Further, studies have also demonstrated the superior performance of nanosystems as compared with their micro-counterparts (Mahalakshmi et al., 2020, 2023). Apart from conventional in vitro systems, studies have also considered the use of dynamic *in vitro* systems to better understand such effects (Jayan et al., 2019). Researchers have also integrated nanoencapsulation strategies with novel food manufacturing approaches such as 3D printing for the fabrication of customized and personalized nutrient/nutraceutical delivery vehicles (Leena et al., 2022).

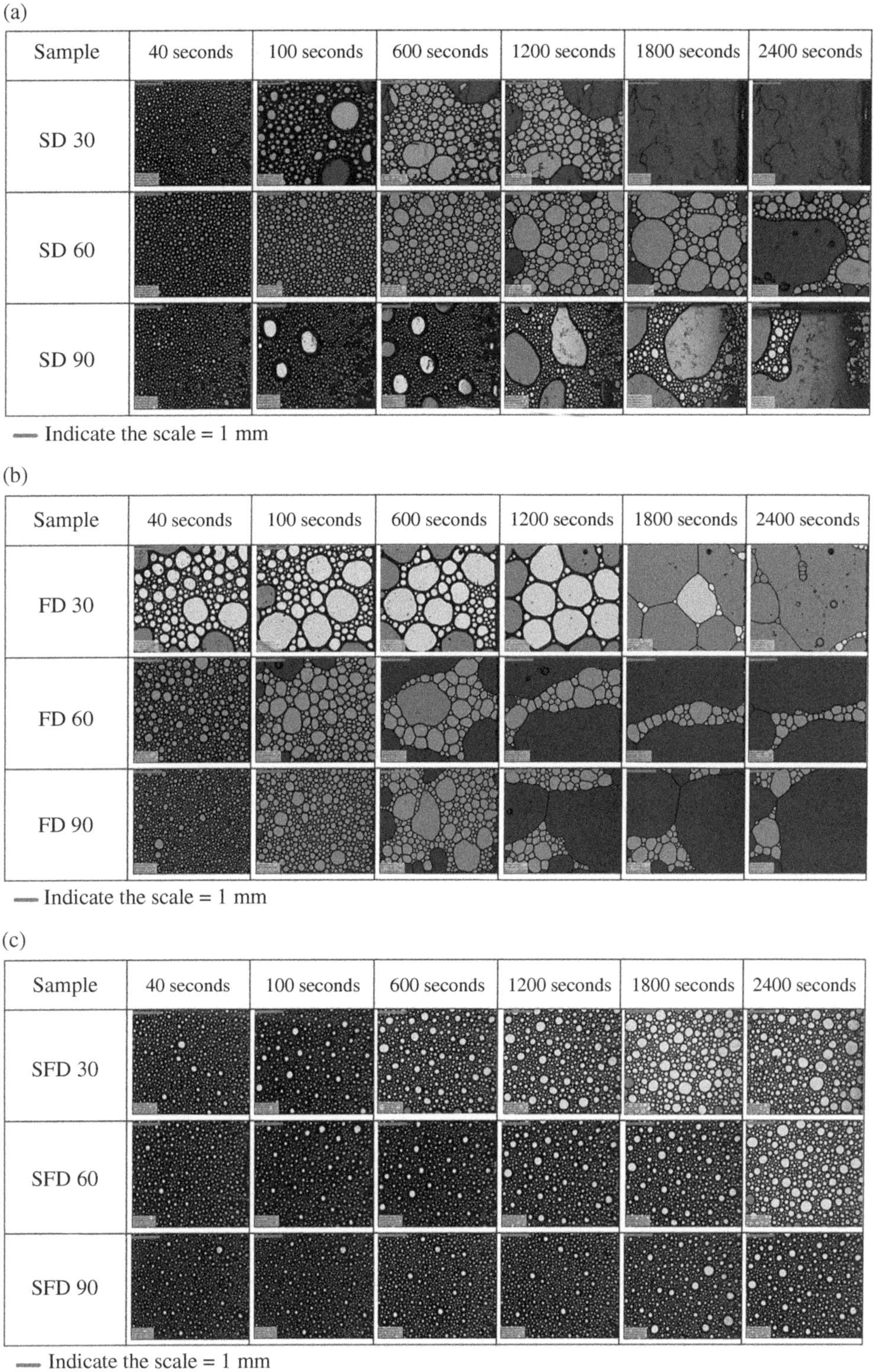

Figure 1.5 Effect of temperature on foaming and bubble structure of coffee solutions prepared from coffee obtained by (a) spray drying (SD), (b) freeze drying (FD), and (c) spray-freeze drying (SFD) techniques (SD 30, SD 60, SD 90 = SD powder dissolved in water at 30, 60, and 90 °C, respectively; FD 30, FD 60, FD 90 = FD powder dissolved in water at 30, 60, and 90 °C, respectively; SFD 30, SFD 60, SFD 90 = SFD powder dissolved in water at 30, 60, and 90 °C, respectively) (*Source:* Deotale et al. (2020)/with permission of Springer Nature)

1.3 Conclusion

Certainly, the field of nanotechnology offers promising scope for the food industry. This chapter has detailed recent advances in the field. Nonetheless, there is a huge requirement in addressing the multiple challenges and limitations of the field. One major aspect is the possible ill effects of this advancing field, particularly for food and allied applications. Such studies can establish better confidence and consumer acceptance of this emerging field, also supporting the creation of laws and regulations to govern the field. Overall, the applications are ample, and subsequent chapters of this book will focus exclusively on the food packaging sector.

References

Amal Nath, V., Vijayakumar, R., Leena, M. M., Moses, J. A., & Anandharamakrishnan, C. (2022). Co-electrospun-electrosprayed ethyl cellulose-gelatin nanocomposite pH-sensitive membrane for food quality applications. *Food Chemistry*, *394*, 133420. https://doi.org/10.1016/j.foodchem.2022.133420

Anandharamakrishnan, C. (2021). Trends and Impact of Nanotechnology in Agro-Food Sector. In K. Knoerzer, & K. B. T.-I. F. P. T. Muthukumarappan (Eds.), *Innovative Food Processing Technologies A Comprehensive Review* (pp. 523–531). Elsevier. https://doi.org/10.1016/B978-0-12-815781-7.00005-6

Anandharamakrishnan, C., & Parthasarathi, S. (Eds.). (2019). *Food nanotechnology: principles and applications.* CRC Press.

Bharathi, S. K. V., Leena, M. M., Moses, J. A., & Anandharamakrishnan, C. (2020). Nanofibre-based bilayer biopolymer films: enhancement of antioxidant activity and potential for food packaging application. *International Journal of Food Science & Technology*, *55*(4), 1477–1484. https://doi.org/10.1111/ijfs.14492

Bharathi, S. V., Rohini, B., Moses, J. A., & Anandharamakrishnan, C. (2019). Nanocomposite for Food Packaging. In C. Anandharamakrishnan, & S. Parthasarathi (Eds.), *Food Nanotechnology* (pp. 275–307). CRC Press.

Choudhary, P., Dutta, S., Moses, J. A., & Anandharamakrishnan, C. (2021). Nanoliposomal encapsulation of chia oil for sustained delivery of α-linolenic acid. *International Journal of Food Science & Technology*, *56*(9), 4206–4214. https://doi.org/10.1111/ijfs.15064

Deotale, S. M., Dutta, S., Moses, J. A., & Anandharamakrishnan, C. (2020). Stability of instant coffee foam by nanobubbles using spray-freeze drying technique. *Food and Bioprocess Technology*, *13*(11), 1866–1877. https://doi.org/10.1007/s11947-020-02526-6

Deotale, S. M., Dutta, S., Moses, J. A., & Anandharamakrishnan, C. (2023). Foaming and defoaming—concepts and their significance in food and allied industries: a review. *Discover Chemical Engineering*, *3*(1), 9. https://doi.org/10.1007/s43938-023-00025-6

Jayan, H., Maria Leena, M., Sivakama Sundari, S. K., Moses, J. A., & Anandharamakrishnan, C. (2019). Improvement of bioavailability for resveratrol through encapsulation in zein using electrospraying technique. *Journal of Functional Foods*, *57*, 417–424. https://doi.org/10.1016/J.JFF.2019.04.007

Joseph, S. M., Drishya, C., Perumal, A. B., Leena, M. M., Moses, J. A., & Anandharamakrishnan, C. (2024). Micro- and Nano-Fluidics in Food Engineering. In C. Anandharamakrishnan, & J. A. M. (Eds.), *Emerging Technologies for the Food Industry* (pp. 189–222). Apple Academic Press. https://doi.org/10.1201/9781003413660-7

Kalpana, S., Priyadarshini, S. R., Maria Leena, M., Moses, J. A., & Anandharamakrishnan, C. (2019). Intelligent packaging: trends and applications in food systems. *Trends in Food Science & Technology*, *93*, 145–157. https://doi.org/10.1016/j.tifs.2019.09.008

Kavimughil, M., Maria Leena, M., Moses, J. A., & Anandharamakrishnan, C. (2023). Electrospining and Electrospraying Technologies for Nutraceutical Delivery System Development. In C. Anandharamakrishnan, & P. B. T.-I. A. of F. F. Subramanian Ingredients and Nutraceuticals (Eds.), *Industrial Application of Functional Foods, Ingredients and Nutraceuticals Extraction, Processing and Formulation of Bioactive Compounds* (pp. 279–323). Academic Press. https://doi.org/10.1016/B978-0-12-824312-1.00009-1

Lakshmanan, M., Ragul, D., Michael, M. L., Moses, J. A., & Anandharamakrishanan., C. (2021). Characterization of silver nanoparticles synthesized using *Ocimum basilicum* seed extract. *Letters in Applied NanoBioScience*, *11*(2), 3411–3420. https://doi.org/10.33263/LIANBS112.34113420

Leena, M. M., Anukiruthika, T., Moses, J. A., & Anandharamakrishnan, C. (2022). Co-delivery of curcumin and resveratrol through electrosprayed core-shell nanoparticles in 3D printed hydrogel. *Food Hydrocolloids*, *124*(Part A), 107200.

Leena, M. M., Bharathi, S. K. V., Moses, J. A., & Anandharamakrishnan, C. (2020b). Potential Applications of Nanofibers in Beverage Industry. In A. M. Grumezescu, & A. M. B. T.-N. in the B. I. Holban (Eds.), *Nanoengineering in the Beverage Industry* (pp. 333–368). Academic Press. https://doi.org/10.1016/B978-0-12-816677-2.00011-9

Leena, M. M., Mahalakshmi, L., Moses, J. A., & Anandharamakrishnan, C. (2020a). Nanoencapsulation of Nutraceutical Ingredients. In K. Pal, I. Banerjee, P. Sarkar, D. Kim, W.-P. Deng, N. K. Dubey, & K. B. T.-B.-B. F. Majumder (Eds.), *Biopolymer-Based Formulations Biomedical and Food Applications* (pp. 311–352). Elsevier. https://doi.org/10.1016/B978-0-12-816897-4.00014-X

Leena, M. M., Yoha, K. S., Moses, J. A., & Anandharamakrishnan, C. (2020c). Edible coating with resveratrol loaded electrospun zein nanofibers with enhanced bioaccessibility. *Food Bioscience*, *36*, 100669. https://doi.org/10.1016/j.fbio.2020.100669

Leena, M. M., Yoha, K. S., Moses, J. A., & Anandharamakrishnan, C. (2021). Electrospun nanofibrous membrane for filtration of coconut neera. *Nanotechnology for Environmental Engineering*, *6*(24), 1–10. https://doi.org/10.1007/s41204-021-00116-1

Logesh, D., Vallikkadan, M. S., Leena, M. M., Moses, J. A., & Anandharamakrishnan, C. (2021). Advances in microfluidic systems for the delivery of nutraceutical ingredients. *Trends in Food Science & Technology*, *116*, 501–524. https://doi.org/10.1016/j.tifs.2021.07.011

Mahalakshmi, L., Choudhary, P., Moses, J. A., & Anandharamakrishnan, C. (2023). Emulsion electrospraying and spray drying of whey protein nano and microparticles with curcumin. *Food Hydrocolloids for Health*, *3*, 100122. https://doi.org/10.1016/j.fhfh.2023.100122

Mahalakshmi, L., Leena, M. M., Moses, J. A., & Anandharamakrishnan, C. (2020). Micro- and nano-encapsulation of β-carotene in zein protein: size-dependent release and absorption behavior. *Food & Function*, *11*(2), 1647–1660. https://doi.org/10.1039/C9FO02088H

Mahalakshmi, L., Priyanga, S., Gover, A., Moses, J. A., & Anandharamakrishnan, C. (2024). Nanotechnology for the Food Industry. In C. Anandharamakrishnan & Jeyan Arthur Moses (eds.) *Emerging Technologies for the Food Industry* (pp. 321–358). Apple Academic Press. https://doi.org/10.1201/9781003413592-8

Mahalakshmi, L., Yoha, K. S., Moses, J. A., & Anandharamakrishnan, C. (2022). Nano Delivery Systems for Food Bioactives. In K. P., A. S., P. S., N. B., & V. J. (Eds.), *Food, Medical, and Environmental Applications of Nanomaterials* (pp. 205–230). Elsevier. https://doi.org/10.1016/B978-0-12-822858-6.00008-X

Manoj, D., Shanmugasundaram, S., & Anandharamakrishnan, C. (2021). Nanosensing and nanobiosensing: concepts, methods, and applications for quality evaluation of liquid foods. *Food Control, 126*, 108017. https://doi.org/10.1016/j.foodcont.2021.108017

Maria, L., Moses, J. A., & Anandharamakrishnan, C. (2019). Ethical and Regulatory Issues in Applications of Nanotechnology in Food. In S. P. C. Anandharamakrishnan (Ed.), *Food Nanotechnology Principles and Applications* (1st Edition, pp. 67–92). CRC Press. https://www.taylorfrancis.com/chapters/edit/10.1201/9781315153872-5/ethical-regulatory-issues-applications-nanotechnology-food-leena-maria-moses-anandharamakrishnan

Murugesan, P., Libiya, N., Moses, J. A., & Anandharamakrishnan, C. (2023). Fluorescence resonance energy transfer-based sensor with silver-conjugated orange peel waste-derived carbon dots for melamine detection. *Carbon Letters, 33*(7), 2335–2348. https://doi.org/10.1007/s42823-023-00563-7

Murugesan, P., Moses, J. A., & Anandharamakrishnan, C. (2019). Photocatalytic disinfection efficiency of 2D structure graphitic carbon nitride-based nanocomposites: a review. *Journal of Materials Science, 54*(19), 12206–12235. https://doi.org/10.1007/s10853-019-03695-2

Murugesan, P., Moses, J. A., & Anandharamakrishnan, C. (2020). One step synthesis of fluorescent carbon dots from neera for the detection of silver ions. *Spectroscopy Letters, 53*(6), 407–415. https://doi.org/10.1080/00387010.2020.1764589

Paranthaman, R., Moses, J. A., & Anandharamakrishnan, C. (2021a). Development and validation of a screening method for simultaneous detection of $KBrO_3$ and KIO_3 in baking ingredients and additives using powder XRD. *Journal of Food Composition and Analysis, 102*, 104007. https://doi.org/10.1016/j.jfca.2021.104007

Paranthaman, R., Moses, J. A., & Anandharamakrishnan, C. (2021b). Development of a method for qualitative detection of lead chromate adulteration in turmeric powder using X-ray powder diffraction. *Food Control, 126*, 107992. https://doi.org/10.1016/j.foodcont.2021.107992

Payal, A., Krishnamoorthy, S., Elumalai, A., Moses, J. A., & Anandharamakrishnan, C. (2021). A review on recent developments and applications of nanozymes in food safety and quality analysis. *Food Analytical Methods, 14*(8), 1537–1558. https://doi.org/10.1007/s12161-021-01983-9

Perumal, A. B., Nambiar, R. B., Moses, J. A., & Anandharamakrishnan, C. (2022). Nanocellulose: recent trends and applications in the food industry. *Food Hydrocolloids, 127*, 107484. https://doi.org/10.1016/j.foodhyd.2022.107484

Preethi, R., Maria Leena, M., Moses, J. A., & Anandharamakrishnan, C. (2020). Biopolymer Nanocomposites and Its Application in Food Processing. In S. Ahmed, & W. Ali (Eds.),

Green Nanomaterials. Advanced Structured Materials (pp. 283–317). Springer, Singapore. https://doi.org/10.1007/978-981-15-3560-4_12

Preethi, R., Maria, L., Moses, J. A., & Anandharamakrishnan, C. (2019). Fabrication of Nanomaterials. In S. P. C. Anandharamakrishnan (Ed.), *Food Nanotechnology Principles and Applications* (1st Edition, pp. 95–124). CRC Press.

Raja, V., Mahalakshmi, L., Leena, M. M., Moses, J. A., & Anandharamakrishnan, C. (2024). Needleless electrospinning: concepts and applications in the food industry. *Food Engineering Reviews*, *16*(2), 252–269. https://doi.org/10.1007/s12393-023-09362-2

Selvakumar, S., Janakiraman, A. B., Michael, M. L., Jeyan Arthur, M., & Chinnaswamy, A. (2019). Formulation and characterization of β-carotene loaded solid lipid nanoparticles. *Journal of Food Processing and Preservation*, *43*(12), e14212. https://doi.org/10.1111/jfpp.14212

Shanthamma, S., Leena, M. M., Moses, J. A., & Anandharamakrishnan, C. (2022). Potential Applications of Nanosensors in the Food Supply Chain. In P. J., R. B., M. H. N., K. K., & J. S. (Eds.), *Food Engineering Innovations Across the Food Supply Chain* (pp. 369–388). Elsevier. https://doi.org/10.1016/B978-0-12-821292-9.00004-2

Subhasri, D., Dutta, S., Leena, M. M., Moses, J. A., & Anandharamakrishnan, C. (2022). Gastronomy: an extended platform for customized nutrition. *Future Foods*, *5*, 100147. https://doi.org/10.1016/j.fufo.2022.100147

Subhasri, D., Leena, M. M., Moses, J. A., & Anandharamakrishnan, C. (2023). Factors affecting the fate of nanoencapsulates post administration. *Critical Reviews in Food Science and Nutrition*, 1–25. https://doi.org/10.1080/10408398.2023.2245462

Theagarajan, R., Krishnamoorthy, S., Moses, J. A., & Anandharamakrishnan, C. (2021). Advanced Applications of Green Materials in Food Applications. In S. A. (E.) (Ed.), *Applications of Advanced Green Materials* (pp. 1–31). Elsevier. https://doi.org/10.1016/B978-0-12-820484-9.00001-5

Theagarajan, R., Krishnamoorthy, S., Moses, J. A., & Anandharamakrishnan, C. (2021). Advanced Applications of Green Materials in Food Applications. In S. A. (E.) (Ed.), *Applications of Advanced Green Materials* (pp. 1–31). Elsevier. https://doi.org/10.1016/B978-0-12-820484-9.00001-5

Ukkunda, N. S., Santhoshkumar, P., Paranthaman, R., & Moses, J. A. (2024). X-ray diffraction and its emerging applications in the food industry. Critical Reviews in Food Science and Nutrition, 1–16. https://doi.org/10.1080/10408398.2024.2395487

Vinitha, K., Leena, M. M., Moses, J. A., & Anandharamakrishnan, C. (2021). Size—dependent enhancement in salt perception: spraying approaches to reduce sodium content in foods. *Powder Technology*, *378*, 237–245. https://doi.org/10.1016/j.powtec.2020.09.079

2

Trends in Food Packaging and the Scope of Nanotechnology

Dravin Pratap Singh[1], *Ila Singh*[1], *B. S. Unnikrishnan*[1], *and Gopinath Packirisamy*[1,2]

[1] *Centre for Nanotechnology, Indian Institute of Technology, Roorkee, Uttarakhand, India*
[2] *Department of Biosciences and Bioengineering, Indian Institute of Technology, Roorkee, Uttarakhand, India*

2.1 Introduction

Food packaging is the process of enclosing or safeguarding food items for the purposes of transportation, storage, sale, and use. Due to its role in preserving food and preventing contamination, packaging is a crucial component of the food industry. Additionally, it aids in extending the shelf life of food items, improving their accessibility and convenience for consumers (Nile et al., 2020). With the development of new and cutting-edge materials, technologies, and processes, food packaging has experienced a tremendous transformation over time. These innovations have paved the way for creating more practical and effective packaging options to safeguard food items better, increase their shelf life, and raise their quality. Food packaging has existed since the dawn of time and has changed as new materials and technologies have emerged. Ancient civilizations used a wide range of organic materials for food storage, including gourds, shells, animal hides, intestines, bones, plant leaves, stems, and bark. These components, which were widely accessible, offered a straightforward and efficient method of carrying and storing food (Teixeira-Costa and Andrade, 2021); for instance, the Egyptians transported and stored food and other goods in amphorae, baskets, and clay jars. In addition to storing and transporting food, the Ancient Greeks and Romans employed clay jars for various purposes. Food was frequently kept in wooden barrels during the Middle Ages or waxed or sealed with clay. The invention of tin cans in the 16th century changed food storage and preservation by keeping food fresher and longer. Due to their capacity to create a safe and airtight seal, glass jars and bottles gained popularity in the 19th century (Twede, 2002). In the early 20th century, the invention of aluminum foil and waxed paper significantly enhanced food packaging by permitting longer-term preservation by wrapping and sealing from clay jars and baskets to tin cans, glass jars and bottles, aluminum foil, and waxed paper.

Nanotechnology for Sustainable Food Packaging, First Edition. Edited by C. Anandharamakrishnan, Jeyan A. Moses, and M. Maria Leena.

The Industrial Revolution brought significant advancements to the food packaging industry, as it led to the mass production of tin cans and glass jars, making packaging more affordable and accessible. The development of paper and cardboard packaging also helped to reduce costs, as they were cheaper and easier to produce (Lamberti and Escher, 2007). The mid-20th century saw the growth of convenience products and packaging. Prepackaged food products became trendy, as they were convenient and easy to prepare. Plastic packaging became popular as it was more lightweight and cost-effective than glass and tin cans. The invention of new technologies, such as vacuum sealing, also assisted in extending the shelf-life of food products and making them easier to store. In the last few decades, there has been a rising trend toward more sustainable packaging solutions in response to increased awareness of environmental issues (Geyer et al., 2017). Several food companies have switched to reusable materials, such as paper and cardboard, as well as bioplastics and biodegradable materials, to lessen their environmental effect. Overall, the trend toward more sustainable and innovative packaging solutions is likely to continue as consumers and businesses become more environmentally conscious (Martins et al., 2019). The food packaging industry is continually evolving and adapting to meet the needs of a growing consumer base (Ravichandran, 2010). Packaging producers are capitalizing on new technologies such as "nanotechnology" and "nanomaterials" to create more efficient, cost-effective, and sustainable packaging solutions. For example, many companies are investing in "active" packaging, which uses nanomaterials that absorb or release gases and odors to help safeguard the food. Such technology helps reduce food wastage and increases the shelf-life of food products (Debiagi et al., 2014). Corporations also invest in "smart" packaging, which uses nanosensors and other technologies to detect microbial contamination and spoilage (Drago et al., 2020). A significant influence on the food industry has come from the food packaging sector. Food items benefit from the packaging by having a longer shelf life, which increases their usability and convenience for consumers.

Additionally, it helps to prevent food contamination and reduction in food waste. This is crucial for perishable foods like dairy, meat, and fruit. Presently, nanotechnology has emerged as a promising technology for food packaging; using the nanotechnological approach makes it possible to make convenient, sustainable food packaging with enhanced food safety features. Overall, packaging based on nanotechnology has the potential to be more advantageous than conventional packaging in terms of safety, shelf life, and environmental effects (Li et al., 2020). To ensure the responsible and safe application of nanotechnology in food packaging, additional research is required.

2.2 Role of Nanotechnology in Food Packaging

2.2.1 Classification of Food Packaging

The food packaging materials can be classified as depicted in Figure 2.1. Materials like glass and metals are generally used for juices, beverages, and milk products.

Type I soda-lime glasses contain elements like silica, borate, and metal oxides, while type II borosilicate glass contains sulfur and metal oxides. Paper-based packaging materials are generally used for dairy products, oil-rich food items, and bakery products. To increase

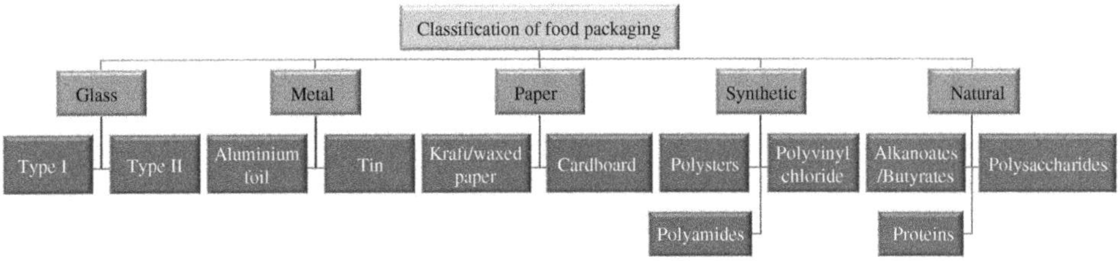

Figure 2.1 Classification of food packaging

the shelf-life and transportation, food items are stored in synthetic packaging materials. Recently, research has focused on the development of biodegradable natural polymers as food packaging material, which is found to be cost-effective, biodegradable, eco-friendly, and edible (Verma et al., 2021).

2.2.2 Effect of Food Packaging on the Shelf-Life of Food

Food packaging is an essential part of the food industry, used to maintain the quality and safety of food products during transportation, storage, and delivery. Proper packaging can help prevent deterioration, preserve flavor and texture, ensure food safety, and prevent contamination from external sources. However, the efficacy of food packaging is dependent on several factors, including the type of packaging material, the packaging size, and the storage conditions (Hemalatha et al., 2017). The use of cutting-edge materials and technology for food packaging has gained popularity in recent years to extend the shelf life of food products. One of these is using nanotechnology-based packaging, which can offer better barrier qualities, antimicrobial activity, and intelligent packaging. Increased food safety, decreased food waste, and greater sustainability are possible advantages of these cutting-edge packaging materials and technologies (Primožič et al., 2021). Various elements, such as microbial growth, oxidation, moisture content, and physical damage, impact the shelf-life of food products. Food items can become spoiled and contaminated due to microbial development (including bacteria, yeasts, molds, and viruses), while oxidation can impact its color, flavor, and nutritional value. The presence of moisture can influence the growth of microorganisms and the rate of chemical reactions, whilst physical harm can cause bruising or other types of damage that increase the risk of spoiling (Siegrist et al., 2007). The shelf-life of food products is significantly influenced by the type of food packaging material used and the packaging style. Due to their affordability, adaptability, and convenience, plastic packaging materials like polyethylene terephthalate (PET) and polypropylene (PP) are frequently utilized for packaging food goods (GIACIN, 1980; van Willige et al., 2002). However, due to the poor barrier qualities of these materials, oxygen, and moisture may enter, causing food to deteriorate and perish. Metal packaging materials, such as aluminum and tinplate, are commonly used in packaging canned food products. Still, they can be subject to corrosion and can impact the taste and quality of food (Sultana et al., 2022). Glass packaging materials are suitable for packaging food items that require superior protection against oxygen and moisture since they have exceptional barrier qualities and are inert (Gallucci et al., 2021; Srivastava et al., 2007). Although paper-based packaging materials are recyclable and biodegradable, their usefulness for some food products may be hampered by their poor barrier qualities and susceptibility to moisture and grease (Dai et al., 2021; Hakovirta et al., 2015). The design of the packaging, particularly the type of seal, closure, and label, can also affect how long food products will stay fresh (Bou-Mitri et al., 2021). The shelf life of food products can be increased by preventing the entry of air and moisture with a tight seal and closure. Antimicrobial agents, moisture absorbers, and oxygen scavengers can all be used to extend the shelf life of food goods (Gould, 1996). The shelf-life of food products can also be impacted by storage conditions like temperature, humidity, and light. Food items can degrade more quickly when exposed to high temperatures and humidity, while flavor and color changes can result from light exposure (Forsido et al., 2021). No specific ideal packaging material can fulfill all the requirements for improving barrier characteristics, shelf-life extension, biodegradability, and biocompatibility.

Nevertheless, several packing materials have been created that satisfy some of these requirements, e.g., chitosan, cellulose-based materials, polyhydroxyalkanoates, bio-based plastics, and edible coatings (Ahmed et al., 2022). We will further discuss in detail about these packaging materials in the coming sections of this chapter. While each of these materials has benefits and drawbacks, they may be able to address some of the issues that come with using conventional packaging materials. However, further research is required to create packaging materials that can fulfill these requirements for improving barrier characteristics, shelf-life extension, biodegradability, and biocompatibility.

2.2.3 Role of Packaging Material from a Health Perspective

The major concerns with most of the packaging material is the migration of molecules from the packaging material to the food as well as the rate of degradation. Concerning health issues, the use of recyclable materials, including paper can create a great impact on health due to the absorption of dyes or volatile chemicals from these materials into the food items. Although many materials are available in the market, the inability to mitigate the health concerns raised due to the overuse of nonbiodegradable packaging material welcomes most bio-composite-based materials. Research on the migrants of molecules from packaging material to the food should be undertaken using sensor technology in the future to avoid these upcoming health issues like nano plastics in body fluids. The dyes from the packaging materials transported to food items at high temperatures are considered to be a reason for lung and bladder cancer. The presence of excess levels of chemicals, including diisobutyl phthalate, in baby food packaging material was reported earlier. Another concern is the presence of wax in packaging material to provide a water barrier. These can migrate to the food items and can cause endocrine disruption (Deshwal et al., 2019).

2.3 Types of Nanomaterials Deployed in Food Packaging

Nanomaterials are being employed more frequently to increase the features of food packaging—such as extending its shelf life, expanding its barrier capabilities, and minimizing food waste. Several kinds of nanomaterials are used in food packaging to improve their properties and are categorized into nanoscale polymers, nanoscale metals, carbon

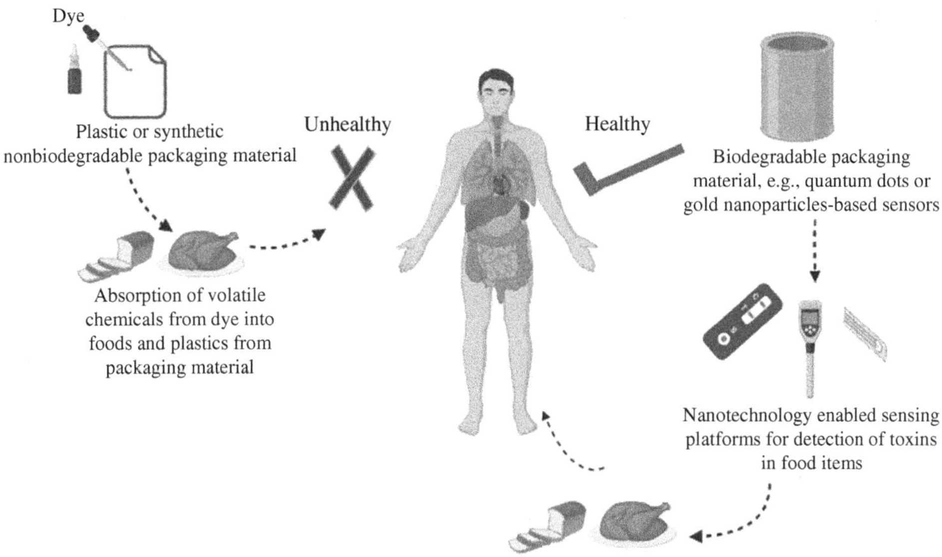

Figure 2.2 Synthetic food packaging materials contain dyes and chemicals that get absorbed via food items, while biodegradable packaging materials containing nanoparticles can sense toxins in food to provide safe, healthy food (*Source:* Created with BioRender.com)

nanotubes and carbon dots, nanoscale clays, and nanocomposites (Figures 2.2 and 2.3). Let us study each of these materials in further detail.

2.3.1 Nanoscale Polymers

Nanoscale polymers are materials consisting of polymer chains at a nanoscale. Due to their distinctive qualities, these materials are appropriate for various applications, including food packaging. Nanoscale polymers are used to improve the qualities of food packaging, such as extending shelf life, improving barrier properties, boosting mechanical strength, and improving sensory and visual appeal (Silvestre et al., 2011). They can be incorporated into packaging materials in various ways, including coatings, films, and nanoparticles. The use of nanoscale polymers in food packaging is described here with some illustrations. (i) Coatings: Nanoscale polymers can be utilized to make coatings that act as a barrier against oxygen and moisture, preventing food from spoiling (Bang and Kim, 2012). For instance, a work by Xu et al. (2017) led to the development of a graphene oxide-chitosan (GO-CS)-TiO_2 coating, which considerably enhanced the barrier qualities of a film against oxygen and water vapor. In another study, researchers created a nanocoating using chitosan and AgNPs that successfully extended the shelf life of strawberries by preventing the formation of spoiling germs (Shankar et al., 2021). (ii) Films: Nanoscale polymers can also make packaging films with enhanced mechanical and barrier qualities. In the work by Nuruddin et al. (2021), for instance, cellulose nanocrystals (CNCs) were used to create polyvinyl alcohol (PVA) film, which improved the mechanical properties of the film, such as its tensile strength and modulus. To preserve

(a)

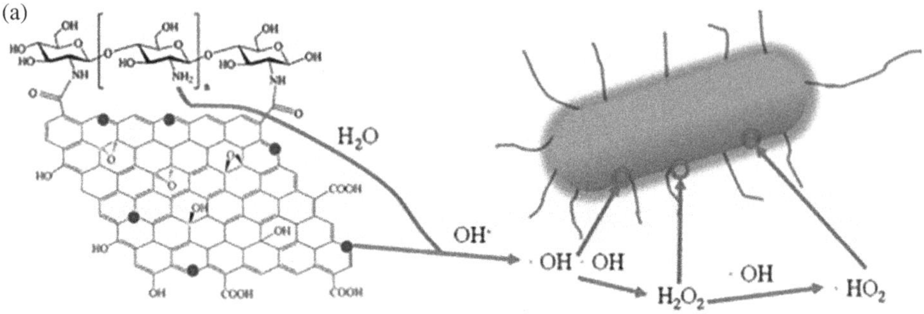

(b)

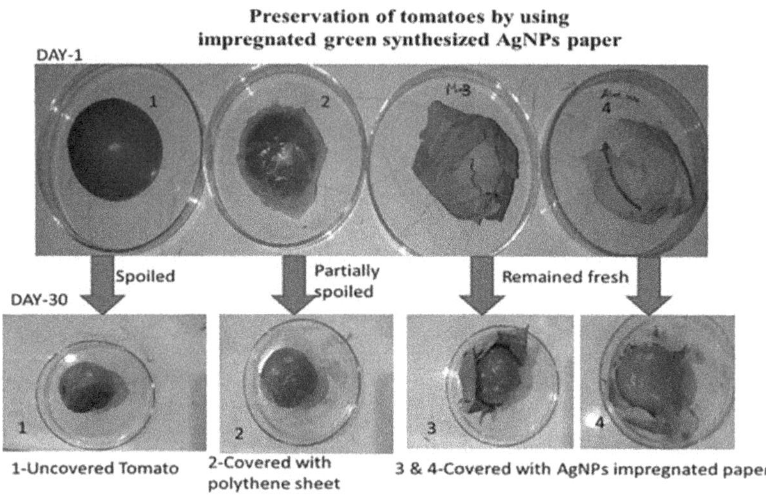

Preservation of tomatoes for about 30 days

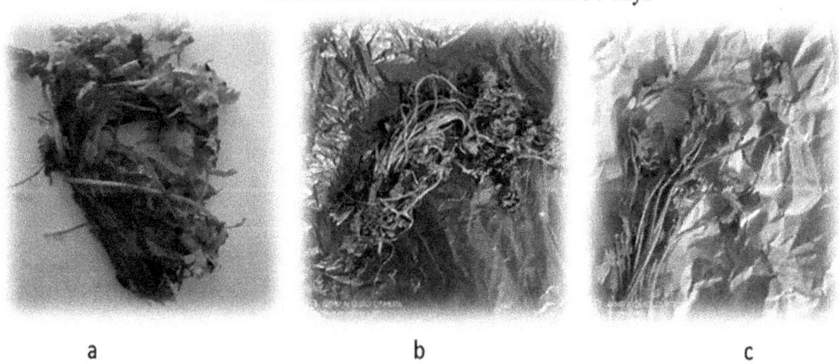

Preservation of corainder leaf for about 15 days: **a**-Stored at Room temperature; **b**-Wrapped in polythene sheet; **c**-Wrapped in AgNPs coated packing material

Figure 2.3 Types of nanomaterials deployed in food packaging. (a) Antibacterial Graphene oxide@Chitosan@TiO_2 (Reprinted from "The graphene oxide and chitosan biopolymer loads TiO_2 for antibacterial and preservative research," 221, pp. 267–277, Xu et al., Food Chemistry, Copyright (2017), with permission from Elsevier) (*Source:* Xu et al. (2017)/with permission of Elsevier) (b) Preservation of perishable food items using silver nanoparticles (AgNPs) coated packing material (*Source:* Kumari et al. (2021)/Journal of Pure and Applied Microbiology/CC BY 4.0) (c) Mechanism of Nylon12 clay (*Source:* Gill et al. (2020)/John Wiley & Sons) (d) Nanocomposites for food packaging applications (*Source:* Perera et al. (2022)/with permission of Elsevier)

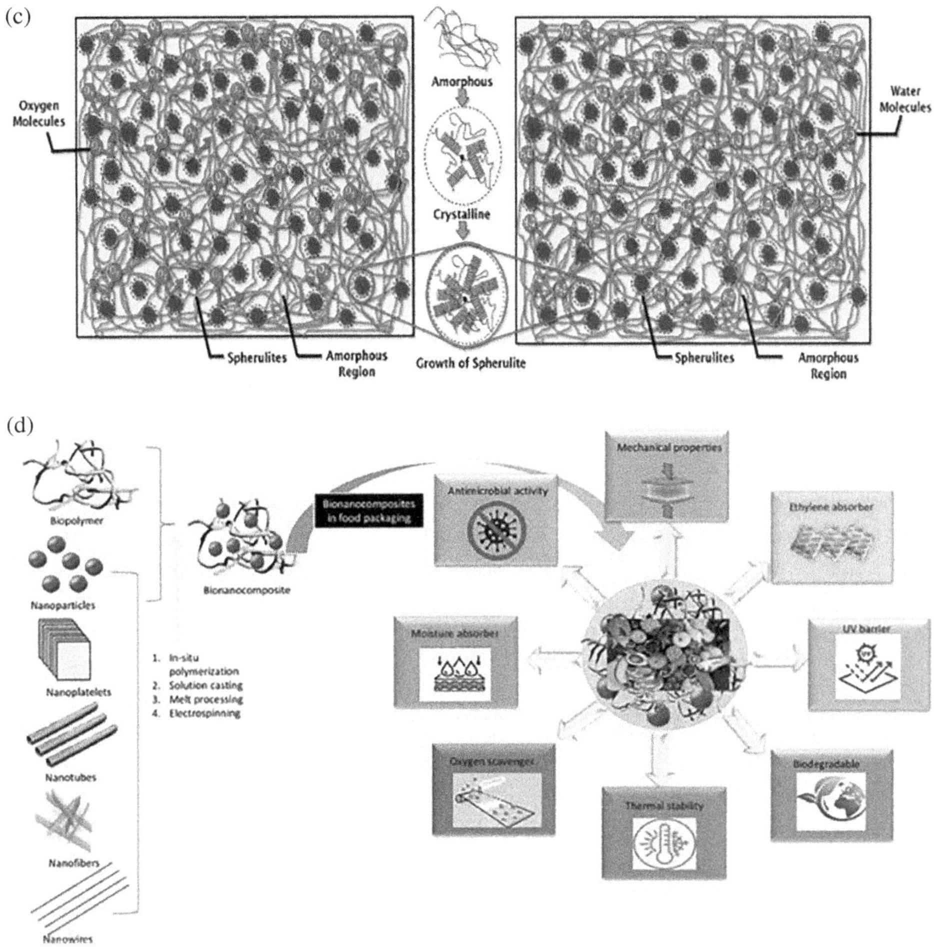

Figure 2.3 (Continued)

Schizothorax prenati fillets, Xin et al. (2020) created a zein and potato starch coating and strengthened it with chitosan nanoparticles and curcumin. Additionally, the film reduced the rate at which lipids in food degraded, extending the product's shelf life. (iii) Nanoparticles: Nanoscale polymers can be exploited as nanoparticles to give food packaging antibacterial qualities, improved barrier properties, and enhanced sensory and aesthetic appeal. For instance, a study by (Xing et al., 2019) created an AgNP coating that improved the mechanical and barrier properties of the packaging material while simultaneously demonstrating high antibacterial activity against *Escherichia coli* and *Staphylococcus aureus*; similarly, in another study, researchers (Chougule et al., 2021) created a polyethylene (PE) film with AgNPs that showed potent antibacterial effects. (iv) Active Packaging: In a study (Surendhiran et al., 2022) scientists created a nanocomposite film using chitosan loaded with turmeric essential oil and magnetic silica nanocomposites to stop the growth of bacteria that cause animal food to decay. Additionally, the film prevented lipid oxidation and preserved the meat's quality while it was being stored.

2.3.2 Nanoscale Metals

Engineered metals with diameters between one and one hundred nanometers are known as nanoscale metals (nm) because of their distinctive characteristics, such as their high surface area, superior thermal and mechanical stability, and potent antibacterial activity; nanoscale metals have emerged as interesting materials for use in food packaging (Bajpai et al., 2018). Listed below are a few benefits for nanoscale metals in food packaging. (i) Due to their potent antibacterial action, AgNPs are frequently employed in food packaging. AgNPs have been added to polyethylene films to prevent spoilage bacteria and increase the shelf-life of fresh food, such as strawberries and tomatoes (Kumari et al., 2021). To stop microbial development and preserve freshness, AgNPs have also been utilized in commercial packaging materials or added to food packaging (Mackevica et al., 2016) (ii) Zinc oxide nanoparticles: Due to their antibacterial properties, zinc oxide nanoparticles are helpful tool in the fight against bacterial and fungal growth in food packaging materials. For instance, zinc oxide nanoparticles have been added to enhance the barrier qualities and antibacterial activity of chitosan films (Souza et al., 2021). These are just a few of the numerous applications for metal nanoparticles that can be added to food packaging materials to enhance their antibacterial and antioxidant activity, barrier qualities, and overall performance. To make sure that the advantages of these materials outweigh any potential concerns to human health or the environment, it is crucial to keep exploring their effectiveness and safety.

2.3.3 Carbon Nanotubes and Carbon Dots

Nanoparticles in bio-composite can provide more benefits compared to naked composites. Organic nanoparticles, including carbon nanotubes and dots, are carbon nanomaterials with high thermal stability, mechanical and barrier properties used for food packaging applications. Recently, multiwalled carbon nanotubes (MWCNTs) and zinc oxide nanoparticles doped PVA composites were fabricated and used as antibacterial packaging that had excellent food preservation ability. The films made were useful for preserving fruits, vegetables, meat, and packaging frozen foods (Boro and Moholkar, 2022). Another research team has synthesized MWCNTs from waste plastic bottles and made a composite with gelatin and garlic microparticles for antibacterial activity. This film can avoid the interaction of food with MWCNTs and solve the problem of migration. The use of carbon dots can improve the water barrier properties of food packaging materials (Pattanshetti et al., 2020). Researchers at Nanjing Forestry University have reported cyanobacteria-based carbon quantum dots/PVA/nanocellulose composite films that possess UV-protection, flexibility, and water barrier properties. Certain carbon dots can provide mechanical strength to the packaging material (Xu et al., 2020). Although many research reports have been published so far, the rate of migration, toxicity, and degradability of carbon nanomaterials are still under study. Further research is underway to explore the potential benefits of these organic nanoparticles in food packaging materials.

2.3.4 Nanoscale Clays

Nanoscale clays exhibit low permeability and can be used as an excellent additive in packaging materials. These nanoscale clays provide permeability and more hydration for packaging materials. $ZnFe_2O_4$/kaolinite nanocomposite was doped in polystyrene films, which exhibited considerable structural deformation, and the rate of degradation was proportional to the doping percentage of clay composites (Saleem et al., 2022). Similarly, organo-modified montmorillonite along with alginate, were used as bio-based films for food packaging. These films were reinforced with chitosan to provide a high impact on the enhancement of the film. The nanoscale clay-based packaging material exhibits a high biodegradability rate and water retention effect. The barrier performance of packaging composites against water vapor and air can be enhanced using clay-based composites (El Bourakadi et al., 2022). The nanocomposites prepared via a twin-screw extruder generated Nylon12 clay nanocomposites with BHX-kaolin clay and BLX-kaolin clay. These materials provide a barrier to air due to the hydrogen bonding existing between the polymeric matrix. The polymers undergo a transition from amorphous/crystalline, and the growth of individual spherulite acts as a barrier (Gill et al., 2020). Thus, these polymeric composites serve as an outstanding food packaging material.

2.3.5 Nanocomposites

Composites with polymer matrix with fillers at a nanoscale range are found to be cost-effective, biodegradable, and possess barrier properties. Organic and inorganic fillers were used in various nanocomposites, which can provide versatile applications in the food packaging industry. The market value of these products can be increased by using cheaper fillers, which focuses on sustainable development. Inorganic fillers, including AgNPs, copper nanoparticles, gold nanoparticles, sulfur nanoparticles, metal oxide nanoparticles, zeolites, and nanoclays, impart properties like antibacterial, mechanical barrier, thermal stability, tensile strength, flexibility, transparency, antioxidant activity, and oxygen barrier properties.

Organic fillers like nanocellulose and polysaccharide nanoparticles provide barrier properties and edibility for the packaging material (Perera et al., 2022). The amalgamation of both organic and inorganic fillers to produce nanohybrid composites can tackle issues currently faced by the sustainable food packaging market.

2.4 Categorization of Food Packaging

There are numerous ways that nanotechnology has been applied to improve the functionality of different food packaging materials (Figure 2.4). Food packaging categorization on food type is illustrated here with some examples.

2.4.1 Packaging for Horticulture Products

Horticultural items, such as fruits and vegetables, are frequently wrapped in modified atmosphere packaging (MAP) to preserve their freshness and increase their shelf life (MAP is a type of packaging that alters the atmosphere inside the package to assist in extending the shelf life of food (Church, 1994). It usually lowers the oxygen level and increases the carbon dioxide level to delay respiration and impede microbial growth. MAP can be produced by employing an array of packaging materials, such as polyethylene (PE), polypropylene (PP), and biodegradable films); by enhancing their barrier qualities and antibacterial activity, nanotechnology can improve MAP films' performance (Geeson, 1989).

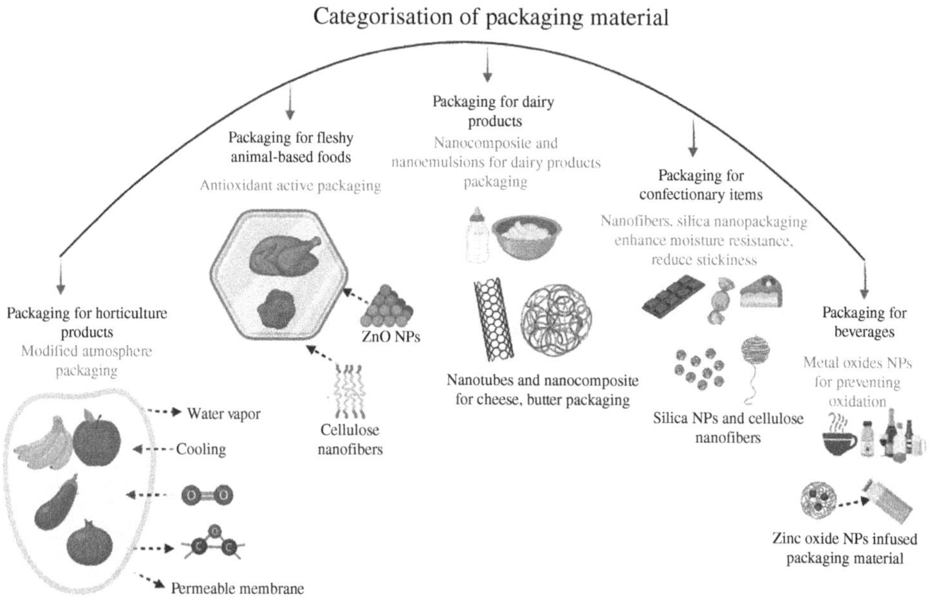

Figure 2.4 Classification of packaging materials into different categories based on their applications (*Source:* created with BioRender.com)

For instance, clay mineral nanoparticles, such as montmorillonite, have been added to polylactide films to enhance their gas barrier qualities and lengthen the shelf life of fresh goods (Svagan et al., 2012). Carbon nanotubes have been employed in horticulture food packaging to strengthen the material's mechanical strength and provide gas barrier qualities. For instance, carbon nanotubes were used to create a fresh-cut fruit packaging material that decreased the rate of respiration and increased the product's shelf life (Rezić et al., 2017). Using lignocellulose nanofibers from agri-food wastes is a novel eco-friendly packaging technique for freshly cut lettuce. It has been utilized in horticulture food packaging to give mechanical strength, flexibility, and gas barrier qualities (Bascón-Villegas et al., 2022). The orange's nutritional value, sensory acceptance, weight loss, and respiration rate were all significantly improved by the nanocomposite layer made of nano clay and Carnauba wax (Motamedi et al., 2018). In a study to extend the shelf life of freshly cut vegetables during postharvest storage, cellulose nanofibrillated covering was created and applied; nanocellulose's hydrophilic-hydrogel properties enabled leaves to retain moisture, freshly cut fruits and vegetables have a longer shelf life due to the packaging ability to decrease their rate of respiration and possess superior barrier characteristics (Pacaphol et al., 2019).

In another study, researchers created a nanoemulsions-based covering for mangoes by mixing cinnamon essential oil into a chitosan matrix. The mangoes' shelf life was increased by the coating's ability to stop the growth of germs and fungi (Yu et al., 2021). To extend the shelf life of orange fruits, an alginate-chitosan-based coating that is enriched with ZnO nanoparticles has been created (*Citrus sinensis L.*); this study is the first to examine the overall effects of coatings made of sodium alginate and chitosan with nano-zinc oxide on the quality assessment (pH, weight loss, firmness, total soluble solids, total acidity, antioxidant, ascorbic acid, respiration rate, decay rate, antioxidant enzymes, and microbial population) of orange fruit (*Citrus sinensis*). It was discovered that the coating was sufficient to increase the shelf life of oranges by up to 20 days at 4 °C (Dulta et al., 2022). In fruit preservation, lipid-based edible nanocoating has received considerable research, and a candelilla wax and tarbush bioactive-based nanocoating is one illustration of such a coating. It has been tested to see how well it preserves the quality of apples after harvest. In a study, the scientists coated apples and then kept them at room condition for up to four weeks and in the refrigerator for eight weeks. Comparing coated to uncoated apples, a substantial difference was seen in the weight loss and respiration rate. The covering preserved the fruit's general quality, color, and firmness while it was being stored. The fruit is covered by a thin layer of lipid-based covering, which prevents rotting by reducing water loss and gas exchange; candelilla wax provides the essential lipid structure, while the tarbush, a bioactive compound, prevents the fruit from getting oxidized (De León-Zapata et al., 2018). The fact that lipid-based coatings are typically regarded as safe for consumption and can increase the fruit's nutritional content is one benefit of employing them. These coatings are a feasible choice for large-scale fruit preservation since they are simple to apply using spray or dipping methods. These illustrations show how packaging based on nanotechnology can extend the shelf life and increase the safety of horticulture products.

2.4.2 Packaging for Fleshy Animal-Based Food

Packaging that can insulate meat and poultry products from bacterial infection and spoilage is crucial. Materials with an antibacterial effect that stops the growth of bacteria and other microbes can be produced using nanotechnology (Lamri et al., 2021). For example, an active packaging material for chicken was created utilizing a mix of zinc oxide nanoparticles, cellulose nanofibers as a reinforcement filler and gelatin. In addition to extending the chicken's shelf life by lowering the deterioration rate, the packaging was found to have good antibacterial efficacy (Ahmadi et al., 2020). Similarly, in another study, a nanocomposite film-based active packaging material for pork was created utilizing a combination of poly (lactic acid) (PLA)/PVA and zinc oxide nanoparticles (Duan et al., 2022). Starch-based edible packaging has gained a lot of interest in recent years. Different forms of starch, including maize starch, potato starch, tapioca starch, and others, can be used to make starch-based packaging. Starch-based packaging has several benefits for the packaging of fleshy animal-based food, e.g., starch has good barrier qualities that can aid in lowering the exchange of gases and moisture between the environment and the meat (Cheng et al., 2021). To extend the shelf life of the food product, starch-based packaging can also be employed as a carrier for active substances like antioxidants or antibacterial agents. Another benefit of packaging made of starch is that it can be decomposed and degraded naturally, so it may be discarded without harming the environment. This is crucial in the context of foods made from fleshy animals since packaging waste can be a serious environmental issue. Packaging made of starch has several drawbacks as well. It is prone to moisture and can become weaker mechanically in a moist environment. This may restrict its use in some circumstances, especially when the food product has a high moisture content. To make starch-based packaging more suited for a larger variety of food goods, ongoing research is aimed at enhancing its mechanical characteristics (Sapper and Chiralt, 2018).

2.4.3 Packaging for Dairy Products

Milk, cheese, and other dairy products need packaging to keep them safe from infection and spoiling. The use of nanoemulsions for milk and cream is one illustration of dairy product packaging based on nanotechnology. Nanoemulsions are surfactant-stabilized dispersions of nanoscale droplets of one liquid (oil) in another immiscible liquid (water) and are thermodynamically stable mixtures. Nanoemulsions can be used to make a barrier shielding dairy products from moisture absorption, oxygen exposure, and microbial growth (Aswathanarayan and Vittal, 2019). Nanoemulsions, according to research, can add up to 14 days to the shelf life of milk and cream. Another illustration is the utilization of nanocoating in the packaging of cheese. Nanocoating is very thin coatings of nanomaterials that are applied to the surface of cheese to act as a barrier against pollutants like moisture and oxygen. For instance, to preserve Cheddar cheese, a study was conducted by (Jafarzadeh et al., 2021); they created an edible film incorporating whey protein isolate (WPI) nanofibers (WPNFs) and carvacrol (CA). The results showed that *S. aureus, Salmonella enteritidis, Listeria monocytogenes,* and *E. coli* were all susceptible to the antimicrobial effects of the edible films. The antioxidant activity was also enhanced. They consequently asserted that the coatings created by WPNF might improve the quality of fresh-cut Cheddar cheese over the course of storage. For the packaging of dairy goods, including yogurt, cream cheese, and butter, nanocomposites have also been created. Polymer matrices reinforced by nanoscale particles, fibers, or tubes are known as nanocomposites. They are perfect for dairy product packaging because they can offer superior mechanical qualities, good barrier properties, and improved thermal stability. In the Journal of Agricultural and Food Chemistry, for instance, researchers found that a nanocomposite film comprised of PLA/zinc oxide deposited halloysite nanotubes can greatly increase packaging's barrier qualities and increase the shelf life of the product (Boro and Moholkar, 2022). Yogurt is one of the dairy products that need proper packaging to maintain quality and increase shelf life. Packaging made of PLA and silanized nano cellulose composite film has been investigated as a possible dairy products packaging solution. The barrier qualities of packaging materials can be enhanced by coatings made of nano-cellulose. The key elements that cause the yogurt to spoil are the migration of oxygen and moisture, which these coverings can stop.

Additionally, nano-cellulose coatings can increase the package material's mechanical strength and offer a smooth surface that is simple to print on. The study investigated how barrier qualities and mechanical strength of PLA films were affected by nano-cellulose. According to the findings, the addition of nano-cellulose enhanced PLA films' barrier qualities and mechanical strength. Nano-cellulose-based coatings for dairy product packaging provide some potential benefits, but there are also disadvantages that must be considered. The high expense of making nano-cellulose is one of the key disadvantages. Although there is little proof to support this, there are also worries regarding the possible toxicity of nano-cellulose (Jin et al., 2020).

2.4.4 Packaging for Confectionery Items

Packaging that can shield confectionery products from moisture and keep them from sticking together is necessary for items like chocolates and sweets. Packaging materials with better surface features and barrier qualities can be made using nanotechnology. By managing the rate of moisture absorption and lowering oxidation, packaging for confectionery items can aid in maintaining texture and freshness. Packaging films have been enhanced with nanomaterials to give moisture barriers, and antioxidant qualities, such as silica nanoparticles and cellulose nanofibers, have been used to improve moisture resistance and reduce the stickiness of the packaging materials for confectionery items. The capacity of green tea extract applied to the laminating adhesive in sealable multilayer films, which were not in direct touch with the packaged food, to scavenge free radicals (peanuts and cereals covered with chocolate). The authors claim that this packaging solution prevented food from oxidizing for 16 months (Carrizo et al., 2016). It can prolong the shelf life of chocolate goods by stopping the growth of bacteria and fungi. Another illustration is the usage of nanoclays to improve the barrier qualities of the packaging for sweets. According to a study, adding nano clays to PLA films can enhance their water vapor barrier qualities, preventing moisture loss and preserving the texture and quality of confectionery goods like candies and chocolates. In a study researchers produced edible coatings made of whey protein concentrate (WPC) and used them to cover sweets made of jaggery. The coating increased the surface smoothness and texture of the candies while dramatically reducing moisture absorption. According to the study, coatings made of whey protein may be an adequate substitute for synthetic packaging materials for confectionery products. However, two of their key downsides are whey protein-based coatings' low water resistance and propensity to dissolve in the presence of moisture. Therefore, additional study is required to improve coatings' formulation and processing parameters based on whey protein for confectionery applications (Chand et al., 2014).

2.4.5 Packaging for Beverages

Materials that can act as a barrier against oxygen, light, and moisture are needed for beverage packaging, such as those used for juices and soft drinks. These elements may oxidize the beverage, cause it to deteriorate, and change its flavor and quality. Several methods have been used to improve the performance of beverage packaging materials using nanotechnology: (i) Oxygen barrier properties: Oxygen can cause oxidation, which can cause beverages to deteriorate and have unpleasant flavors. Materials for beverage packaging have been infused with nanoparticles of metals and metal oxides, such as aluminum oxide, titanium dioxide, and zinc oxide, to enhance their oxygen barrier qualities (Frounchi and Dourbash, 2009). (ii) Light barrier properties: Exposure to light can lead to the deterioration of a beverage's color, flavor, and nutritional content. To enhance their light-barrier capabilities and shield the beverage from UV radiation, zinc oxide nanoparticles have been added to beverage packaging materials (Hari et al., 2021), e.g., to prevent beer from being exposed to light, which can cause it to deteriorate and acquire an unpleasant taste, some makers are coating glass bottles with nanoparticles. (iii) Antimicrobial qualities: Microbiological contamination can lead to beverage deterioration and compromise its safety. Silver and copper nanoparticles have been added to packaging materials for beverages to give antibacterial activity against germs that might lead to deterioration. (iv) Self-cleaning qualities: During handling and storage, the materials used to package beverages might become contaminated with dirt and bacteria. Polydimethylsiloxane (PDMS) or paraffin wax emulsion with water-based SiO_2 nanoparticle dispersion achieves a water contact angle of 145° and a sliding angle of 1.5° to provide polyester fabric with extremely high-water repellency and has been integrated into beverage packaging materials to give self-cleaning capabilities, which can aid to preserve the quality and hygienic of the packaging (Ruzi et al., 2022).

2.5 Effects of Food Packaging Materials on the Environment

Food packaging is crucial for safeguarding and keeping food, yet it may also harm the environment. The following are some ways that food packaging materials may harm the environment: (i) Waste production: Plastic packaging materials can produce a lot of waste that ends up in landfills, the ocean, and other ecosystems. Particularly, plastic packaging can take hundreds of years to decompose, which causes pollution and environmental damage. (ii) Energy use: Producing, moving, and disposing of food packaging materials all need energy, which adds to global warming and greenhouse gas emissions. For instance, producing aluminum cans for beverage packaging is an energy-intensive operation that contributes significantly to atmospheric carbon dioxide emissions. (iii) Depletion of resources: Oil, gas, and water are among the natural resources needed to produce food packaging materials. Overuse of these resources can result in environmental harm, including waterway pollution and the loss of wildlife habitat, as well as resource depletion. (iv) Chemical pollution: Some materials used in food packaging include chemicals that can seep into the food supply or the environment, endangering both people and wildlife's health. For instance, food containers made of polystyrene foam may emit styrene, which may be carcinogenic, into the environment and food. At the same time, some plastics contain the chemical bisphenol A (BPA), which has been linked to health issues like cancer and developmental difficulties. (v) Land use: The manufacture of food packaging materials may potentially have an adverse effect on ecosystems and biodiversity by causing deforestation and other types of land use change; (vi) Microplastics: Food packaging, mainly plastic packaging, can degrade into minute particles known as microplastics that could pose health concerns to people if consumed by marine life and make their way into the food chain; (vii) Regulation and oversight: The use of nanotechnology in food packaging is not adequately governed by explicit legislation. Because of this, determining the potential environmental dangers connected to these products and ensuring they are being utilized safely and responsibly can be challenging (Nilsen-Nygaard et al., 2021).

Overall, even while waste can be reduced and shelf life can be extended through nanotechnology-based food packaging, it is crucial to weigh the potential environmental dangers involved with these materials carefully. The type of material used, the methods utilized for manufacture and disposal, and the efficiency of recycling and waste management systems all impact how environmentally friendly food packaging materials are. It is crucial to use more ecologically friendly and sustainable packaging materials, encourage recycling, and make waste reduction efforts to lessen the damaging effects of food packaging on the environment (Ramos et al., 2018).

2.6 Future Scope of Nanotechnology in Food Packaging

The future scope section talks about the potential for more research and development into food packaging using nanotechnology and new trends that are starting to emerge in this field. One recent development in the food industry is the creation of "smart" food packaging that can track and detect changes in food quality and safety in real time. Another trend is using sustainable and biodegradable nanoparticles in food packaging, which can lessen waste and increase environmental sustainability. The part also examines the possibilities for using nanotechnology to create active, intelligent packaging systems that can communicate with food and the environment to offer further functionality. Advanced techniques may have advantages, but several issues must be resolved before they can be used effectively and practically. One of the biggest obstacles is the complexity of the food matrix, which might obstruct measurements and analyses of nanomaterials.

Furthermore, the general use of modern techniques is constrained by their high cost and technical complexity. It is also difficult to compare the findings of various studies because there are no defined techniques for determining the presence of nanomaterials in food. The possible negative impacts of employing nanoparticles in food packaging on both human health and the environment must be taken into consideration. It is crucial to ascertain the toxicological characteristics of nanomaterials, both when used as an ingredient in food and when ingested. Further investigation is required to determine the toxicity and biotransformation of nanomaterials in the human body and the possibility of their build-up in organs and tissues. Regularly holding special meetings to discuss the most recent advancements and future uses of nanotechnology in the food sector is crucial for promoting its use in food packaging. These gatherings can bring together academics, business leaders, and representatives of the regulatory community to discuss the advantages and disadvantages of using nanotechnology in food packaging. These gatherings can also serve as a venue for networking and teamwork, which can create creative food packaging solutions (Murugan et al., 2023; Sahoo et al., 2021).

2.7 The Commercial Value of Food Packaging in the Market

The world food packaging market was expected to rise at a compound annual growth rate of 5.1% during the 2021–2028 period in terms of US dollars. The market size was determined as USD 323.81 billion in 2021, which was projected to be US$478.18 billion in 2028, attributed to the increased demand and growth of packaging materials in the market. The encounter with COVID-19 had a huge impact on the increased global market value of packaged products. Consumer demand has risen as eating habits and changing lifestyles have changed. To protect against contamination, increase shelf-life, potion control, and convenience, packaged food was preferred. Similarly, Indian food packaging was projected to be 29.88% of the compound annual growth rate during the 2021–2028 period. The increased population, especially the need for packed food in the busy world, makes the packaging industry more demandable. Among packaging materials, plastics are more preferable to the food market. The use of plastic cling film was increased in packing fruits, vegetables, fish, dairy, and meat products. However, the need for biodegradable and intelligent packaging material is required to meet the increasing demand for food storage materials. The emergence of nanotechnology in the food packaging industry also wakens sustainable food packaging that can upsurge the market.

2.8 Patents Related to Food Packaging

2.8.1 Safety Concerns and Judicial Aspects of Food Packaging

Concerns have been expressed about the potential consequences to human health and the environment because of the use of nanotechnology in food packaging. One of the main worries is the potential toxicity of nanomaterials, which can result in harmful health effects such as inflammation, oxidative stress, and genotoxicity. Moreover, due to nanomaterials' small size, they may accumulate in organs and tissues, which could have long-term consequences on health. Also, there are worries regarding the potential migration of nanomaterials from food packaging into food products, which could result in consumer ingestion. Nanoparticles may migrate due to their interaction with the food matrix, such as when they are released due to mechanical stress or diffuse through packaging. Nanomaterials' toxicity depends on their physicochemical characteristics, such as size, shape, surface charge, and composition (Fiordaliso et al., 2022). The study advises adopting alternative testing techniques, such as *in vitro* and *in silico* models, to lessen the number of animal studies and claims that the toxicity of nanomaterials can be assessed using standardized toxicity assays. Using barrier coatings, such as polymer coatings, to stop the diffusion of nanoparticles through the packaging material is one way to limit the migration of nanomaterials from food packaging. The study also suggests using analytical methods like atomic force microscopy (AFM) and electron microscopy to detect nanoparticles that have migrated from food packaging. It is crucial to develop sophisticated methods for characterizing and assessing the properties of nanomaterials to properly comprehend the potential advantages and concerns related to their application in food packaging. The accuracy, sensitivity, and cost-effectiveness of current technologies, including transmission electron microscopy (TEM), AFM, and dynamic light scattering (DLS), are constrained. Thus, more sophisticated techniques are required to precisely analyze the characteristics of nanomaterials and evaluate their biotransformation in food, such as X-ray photoelectron spectroscopy (XPS), Raman spectroscopy, and high-resolution TEM (HRTEM) (Buchman et al., 2019; Gupta et al., 2020).

Many nations are still working on defining rules and guidelines for the safe and responsible use of nanomaterials in food packaging; therefore, the regulatory environment for nanotechnology-based food packaging is still changing. The US Food and Drug Administration (FDA) has published guidelines for applying nanotechnology to food and food packaging, highlighting the necessity of safety evaluations and labeling regulations for goods using nanoparticles (Hamburg, 2012). Similarly to, the European Union has set rules for the use of nanoparticles in food packaging, necessitating that producers provide proof that their goods are safe before they can be sold (Coles and Frewer, 2013). Guidelines and regulations for using nanotechnology in food packaging have been developed in several other nations, including Australia and Canada. According to a study that was published in the journal Nanoethics, the regulatory environment for food packaging based on nanotechnology is complicated and fragmented, with several nations having their own rules and regulations. According to the article, international regulatory harmonization is necessary to maintain consistency in the safety evaluation and regulation of nanomaterials used in food packaging. Another study shows that the potential for exposure and the material's danger should be considered when assessing the risk of nanomaterials in food packaging. This study was published in the journal *Regulatory Toxicology and Pharmacology*. The article suggests using a tier system for risk assessment, beginning with a quick risk evaluation and progressing to a thorough evaluation of the material's toxicity and probable exposure. To fulfil consumer needs and regulatory concerns, nanoproducts must be categorized and given names. Nanomaterials must be classified and named in a way that considers the distinctive characteristics that set them apart from their bulk equivalents. The potential dangers and advantages of nano products to human health and the environment must be taken into account when classifying and labeling them (Tirumala et al., 2021). Regulatory bodies can guarantee that nanoproducts are appropriately labeled and that consumers are aware of their possible risks and advantages by developing standardized methods for the categorization and naming of nanomaterials. The major patents involved in the protection of food packaging materials are tabulated in Table 2.1' under section 2.8.

Table 2.1 Patents associated with food packaging materials

Patent Number	DWPI Title	Title	Applicant and Publication Date
WO2009142739A1	Coating composition for forming a coating on a paper-based substrate, e.g., pizza boxes, pop corn, and pet food bags comprises nanofiller having specific aspect ratio; binder; and a crosslinking agent	Paper coating composition having nanofiller (Wan Qichun, Zhang Huiqing; 2009, 2009)	Wan Qichun, Zhang Huiqing; 2009
CN115339209A	Nonstick milk yogurt cup used in food packaging material, comprises base layer	Nonstaining yoghurt cup and preparation process thereof (QU Rui-long, 2022, 2022)	QU Rui-long, 2022
KR2469257B1	Coating film useful as coating agent for food and medicine, comprises an iron-tannic acid coating layer comprising iron and tannic acid, and a pullulan coating layer containing pullulan laminated on iron-tannic acid coating layer	Coating layer comprising pullulan and manufacturing method thereof (Univ Ewha Ind Collaboration Found and (EWHA-C); 2018, KR2469257B1)	Univ Ewha Ind Collaboration Found (EWHA-C); 2018
US20220112381A1	Antimicrobial coated substrate for killing microbes in food, comprises substrate with one or more surfaces and titanium dioxide coating on one of surfaces of substrate, where coating has rhombohedral microstructure, that has exhibited high antimicrobial activity, and high mechanical stability	Antimicrobial coating for food contact surfaces (Hunt Heather, Columbia, MO, US, 2022)	Hunt Heather, Columbia, MO, US

(Continued)

Table 2.1 (Continued)

Patent Number	DWPI Title	Title	Applicant and Publication Date
KR2497422B1	Composite coating film for self-cleaning food packaging material, comprises, e.g., self-cleaning porous nanocoating layer with chitin-based nanofiber layers and inorganic particle layers alternately stacked on surface of film substrate	Self-cleaning composite coating film (Korea Research Institute of Chemical, 2023)	Korea Research Institute of Chemical Technology, 34114, KR
WO2000053413A1	Moldings and coatings used for packaging food and cosmetics comprise antimicrobial agent(s) and polymeric material	Antimicrobial plastics (Sarangapani Shanthaus; 2000, 2000)	Sarangapani Shanthaus; 2000
EP2967093A1	Composition, useful for preventing or mitigating pre- and/or postharvest damage and weight loss, drip loss, and UV damage of plant part, comprises a cellulose nanomaterial, an inorganic salt component, and a crosslinking agent	Nano-cellulose edible coatings and uses thereof (Univ Oregon Stateus; 2016, 2016)	Univ Oregon Stateus; 2016
KR2022102768A	Manufacture of cellulose nanofiber-coated thin film used as barrier film for food packaging, involves preparing cellulose nanofibers, coating aqueous composite solution on polymer film, hot air drying, and thermally pressing dried coating film	Method for manufacturing cellulose nanofiber-coated thin film and film manufactured by the same (Youlchon Chemical Co Ltdkr; 2022, 2022)	Youlchon Chemical Co Ltdkr; 2022
US20190364916A1	Coating composition used to coat plant part, which is tropical fruit comprising banana, papaya, avocado, melon, or mango, comprises cellulose nanomaterial and an emulsion system comprising fatty acid and surfactant	Food product coatings (Univ Oregon State (UORE-C); 2019, 2024)	Univ Oregon State (UORE-C); 2019
MX2015002794A	Edible coating used for increasing the shelf life of the food, based on nanocapsules dispersion comprises nanocapsules dispersion is formed by nanocapsules, and coating is applied to freshly cut fruits and vegetables	Edible coatings based on nanometric dispersions charged with functional ingredients and its use to increase shelf life. (GUERRERO DAVID QUINTANAR, MX, 2016)	Guerrero David Quintanar, MX

2.9 Conclusion

The use of nanotechnology is becoming more and more significant as the world of food packaging continues to change. Food packaging can be made more effective and economical by leveraging the special qualities of nanoparticles, which will promote customer safety, extend product shelf life, and even reduce waste. Moreover, new packaging materials that are robust and durable but lightweight can be created using nanotechnology. Despite the many potential benefits, using nanoparticles in food packaging raises concerns regarding their influence on human health and the environment. More research must be done to completely comprehend the possible hazards and advantages of using nanotechnology in food packaging. To ensure the sustainable and safe use of nanomaterials in food packaging, adopting the proper laws is crucial. In the future, nanotechnology will likely play a significant role in developing cutting-edge approaches to food packaging. Nonetheless, it is critical to proceed cautiously and thoroughly weigh the advantages and disadvantages of these materials. Nanotechnology can be used to improve food safety, lower waste, and advance sustainability in the food packaging sector with continued research and regulatory monitoring.

Acknowledgment

Gopinath Packirisamy is thankful to the Department of Biotechnology, Government of India for the TATA Innovation fellowship. Dravin Pratap Singh is grateful to the Ministry of Education (MoE) for the fellowship. Unnikrishnan BS acknowledges the Science and Engineering Research Board (SERB), New Delhi, India, for the National Post Doctoral Fellowship (N-PDF). The authors are thankful to the Centre for Nanotechnology and Department of Biosciences and Bioengineering, IIT Roorkee, for providing the required infrastructure and various facilities to execute the research work.

References

Ahmadi, A., Ahmadi, P., & Ehsani, A. (2020). Development of an active packaging system containing zinc oxide nanoparticles for the extension of chicken fillet shelf life. *Food Science & Nutrition*, *8*(10), 5461–5473. https://doi.org/10.1002/fsn3.1812

Ahmed, S., Sameen, D. E., Lu, R., Li, R., Dai, J., Qin, W., & Liu, Y. (2022). Research progress on antimicrobial materials for food packaging. *Critical Reviews in Food Science and Nutrition*, *62*(11), 3088–3102. https://doi.org/10.1080/10408398.2020.1863327

Amenta, Valeria, Karin Aschberger, Maria Arena, Hans Bouwmeester, Filipa Botelho Moniz, Puck Brandhoff, Stefania Gottardo, Hans J. P. Marvin, Agnieszka Mech, Laia Quiros Pesudo, Hubert Rauscher, Reinhilde Schoonjans, Maria Vittoria Vettori, Stefan Weigel, and Ruud J. Peters. 2015. "Regulatory Aspects of Nanotechnology in the Agri/Feed/Food Sector in EU and Non-EU Countries." *Regulatory Toxicology and Pharmacology* 73(1):463–76. doi: https://doi.org/10.1016/j.yrtph.2015.06.016.

Aswathanarayan, J. B., & Vittal, R. R. (2019). Nanoemulsions and their potential applications in food industry. *Frontiers in Sustainable Food Systems*, *3*, 95. https://doi.org/10.3389/fsufs.2019.00095

Bajpai, V. K., Kamle, M., Shukla, S., Mahato, D. K., Chandra, P., Hwang, S. K., Kumar, P., Huh, Y. S., & Han, Y.-K. (2018). Prospects of using nanotechnology for food preservation, safety, and security. *Journal of Food and Drug Analysis*, *26*(4), 1201–1214. https://doi.org/10.1016/j.jfda.2018.06.011

Bang, G., & Kim, S. W. (2012). Biodegradable poly(lactic acid)-based hybrid coating materials for food packaging films with gas barrier properties. *Journal of Industrial and Engineering Chemistry*, *18*(3), 1063–1068. https://doi.org/10.1016/j.jiec.2011.12.004

Bascón-Villegas, I., Pereira, M., Espinosa, E., Sánchez-Gutiérrez, M., Rodríguez, A., & Pérez-Rodríguez, F. (2022). A new eco-friendly packaging system incorporating lignocellulose nanofibres from agri-food residues applied to fresh-cut lettuce. *Journal of Cleaner Production*, *372*, 133597. https://doi.org/10.1016/j.jclepro.2022.133597

Boro, U., & Moholkar, V. S. (2022). Antimicrobial bionanocomposites of poly(lactic acid)/ZnO deposited halloysite nanotubes for potential food packaging applications. *Materials Today Communications*, *33*, 104787. https://doi.org/10.1016/j.mtcomm.2022.104787

Bou-Mitri, C., Abdessater, M., Zgheib, H., & Akiki, Z. (2021). Food packaging design and consumer perception of the product quality, safety, healthiness and preference. *Nutrition & Food Science*, *51*(1), 71–86. https://doi.org/10.1108/NFS-02-2020-0039

Buchman, J. T., Hudson-Smith, N. V., Landy, K. M., & Haynes, C. L. (2019). Understanding nanoparticle toxicity mechanisms to inform redesign strategies to reduce environmental impact. *Accounts of Chemical Research*, *52*(6), 1632–1642. https://doi.org/10.1021/acs.accounts.9b00053

Carrizo, D., Taborda, G., Nerín, C., & Bosetti, O. (2016). Extension of shelf life of two fatty foods using a new antioxidant multilayer packaging containing green tea extract. *Innovative Food Science & Emerging Technologies*, *33*, 534–541. https://doi.org/10.1016/j.ifset.2015.10.018

Chand, K., Verma, A. K., Kumar, A., & Shahi, N. C. (2014). Effect of edible coating on quality parameters of jaggery during storage. *Sugar Tech*, *16*(1), 80–85. https://doi.org/10.1007/s12355-013-0244-7

Cheng, H., Chen, L., McClements, D. J., Yang, T., Zhang, Z., Ren, F., Miao, M., Tian, Y., & Jin, Z. (2021). Starch-based biodegradable packaging materials: a review of their preparation, characterization and diverse applications in the food industry. *Trends in Food Science & Technology*, *114*, 70–82. https://doi.org/10.1016/j.tifs.2021.05.017

Chougule, S. S., Gurme, S. T., Jadhav, J. P., Dongale, T. D., & Tiwari, A. P. (2021). Low density polyethylene films incorporated with biosynthesised silver nanoparticles using *Moringa oleifera* plant extract for antimicrobial, food packaging, and photocatalytic degradation applications. *Journal of Plant Biochemistry and Biotechnology*, *30*(1), 208–214. https://doi.org/10.1007/s13562-020-00584-7

Chu, X., Zhang, P., Shi, S., Liu, Y., Feng, W., Zhou, N., Li, J., & Shen, J. (2023). Self-cleaning composite coating film. *Engineering Aspects*, *658*, 130621. https://doi.org/10.1016/j.colsurfa.2022.130621

Church, N. (1994). Developments in modified-atmosphere packaging and related technologies. *Trends in Food Science & Technology*, *5*(11), 345–352. https://doi.org/10.1016/0924-2244(94)90211-9

Coles, D., & Frewer, L. J. (2013). Nanotechnology applied to European food production—a review of ethical and regulatory issues. *Trends in Food Science & Technology*, *34*(1), 32–43. https://doi.org/10.1016/j.tifs.2013.08.006

Dai, L., Xi, X., Li, X., Li, W., Du, Y., Lv, Y., Wang, W., & Ni, Y. (2021). Self-assembled all-polysaccharide hydrogel film for versatile paper-based food packaging. *Carbohydrate Polymers*, *271*, 118425. https://doi.org/10.1016/j.carbpol.2021.118425

D'Arcy, N. (2001), Antimicrobials in plastics: A global review. Plastics, *Additives and Compounding*, *3*(12), 12–15. https://doi.org/10.1016/S1464-391X(01)80328-7

De León-Zapata, M. A., Ventura-Sobrevilla, J. M., Salinas-Jasso, T. A., Flores-Gallegos, A. C., Rodríguez-Herrera, R., Pastrana-Castro, L., Rua-Rodríguez, M. L., & Aguilar, C. N. (2018). Changes of the shelf life of candelilla wax/tarbush bioactive based-nanocoated apples at industrial level conditions. *Scientia Horticulturae*, *231*, 43–48. https://doi.org/10.1016/j.scienta.2017.12.005

Debiagi, F., Kobayashi, R. K. T., Nakazato, G., Panagio, L. A., & Mali, S. (2014). Biodegradable active packaging based on cassava bagasse, polyvinyl alcohol and essential oils. *Industrial Crops and Products*, *52*, 664–670. https://doi.org/10.1016/j.indcrop.2013.11.032

Deshwal, G. K., Panjagari, N. R., & Alam, T. (2019). An overview of paper and paper based food packaging materials: health safety and environmental concerns. *Journal of Food Science and Technology*, *56*(10), 4391–4403. https://doi.org/10.1007/s13197-019-03950-z

Dominguez, E. T., Nguyen, P. H., Hunt, H. K., & Mustapha, A. (2019). Antimicrobial coatings for food contact surfaces: Legal framework, mechanical properties, and potential applications. *Comprehensive Reviews in Food Science and Food Safety*, *18*(6), 1852–1858. https://doi.org/10.1111/1541-4337.12502

Drago, E., Campardelli, R., Pettinato, M., & Perego, P. (2020). Innovations in smart packaging concepts for food: an extensive review. *Foods*, *9*(11), 1628. https://doi.org/10.3390/foods9111628

Duan, L., Yan, F., Zhang, L., Liu, B., Zhang, Y., Tian, X., Liu, Z., Wang, X., Wang, S., Tian, J., Bao, H., & Liu, T. (2022). ZnO@polyvinyl alcohol/poly(lactic acid) nanocomposite films for the extended shelf life of pork by efficient antibacterial adhesion. *ACS Omega*, *7*(49), 44657–44669. https://doi.org/10.1021/acsomega.2c03016

Dulta, K., Koşarsoy Ağçeli, G., Thakur, A., Singh, S., Chauhan, P., & Chauhan, P. K. (2022). Development of alginate-chitosan based coating enriched with ZnO nanoparticles for increasing the shelf life of orange fruits (*Citrus sinensis* L.). *Journal of Polymers and the Environment*, *30*(8), 3293–3306. https://doi.org/10.1007/s10924-022-02411-7

El Bourakadi, K., El Kacem Qaiss, A., & Bouhfid, R. (2022). Bio-films based on alginate/modified clay through spray drying: mechanical, rheological, morphological, and transport properties for potential use as active food packaging. *International Journal of Biological Macromolecules*, *210*, 663–668. https://doi.org/10.1016/j.ijbiomac.2022.04.222

Fiordaliso, F., Bigini, P., Salmona, M., & Diomede, L. (2022). Toxicological impact of titanium dioxide nanoparticles and food-grade titanium dioxide (E171) on human and environmental health. *Environmental Science: Nano*, *9*(4), 1199–1211. https://doi.org/10.1039/D1EN00833A

Forsido, S. F., Welelaw, E., Belachew, T., & Hensel, O. (2021). Effects of storage temperature and packaging material on physico-chemical, microbial and sensory properties and shelf life of extruded composite baby food flour. *Heliyon*, *7*(4), e06821. https://doi.org/10.1016/j.heliyon.2021.e06821

Frounchi, M., & Dourbash, A. (2009). Oxygen barrier properties of poly(ethylene terephthalate) nanocomposite films. *Macromolecular Materials and Engineering*, *294*(1), 68–74. https://doi.org/10.1002/mame.200800238

Gallucci, T., Lagioia, G., Piccinno, P., Lacalamita, A., Pontrandolfo, A., & Paiano, A. (2021). Environmental performance scenarios in the production of hollow glass containers for food packaging: an LCA approach. *The International Journal of Life Cycle Assessment*, *26*(4), 785–798. https://doi.org/10.1007/s11367-020-01797-7

Geeson, J. D. (1989). Modified atmosphere packaging of fruits and vegetables. *Acta Horticulturae*, *258*, 143–150. https://doi.org/10.17660/ActaHortic.1989.258.14

Geyer, R., Jambeck, J. R., & Law, K. L. (2017). Production, use, and fate of all plastics ever made. *Science Advances*, *3*(7). https://doi.org/10.1126/sciadv.1700782

Giacin, J. R. (1980). Evaluation of plastics packaging materials for food packaging applications: food safety considerations. *Journal of Food Safety*, *2*(4), 257–276. https://doi.org/10.1111/j.1745-4565.1980.tb00402.x

Gill, Y. Q., Abid, U., & Song, M. (2020). High performance Nylon12/clay nanocomposites for potential packaging applications. *Journal of Applied Polymer Science*, *137*(41). https://doi.org/10.1002/app.49247

Gould, G. W. (1996). Methods for preservation and extension of shelf life. *International Journal of Food Microbiology*, *33*(1), 51–64. https://doi.org/10.1016/0168-1605(96)01133-6

Goswami, S., Sharma, A., Anjana, & Bhattacharya, M. Chapter 2 – Pullulan films and natural compounds: Applications and perspectives. *Biopolymer-Based Nano Films*, *4*, 19–41. https://doi.org/10.1016/B978-0-12-823381-8.00006-5

Gupta, A., Kumar, S., & Kumar, V. (2020). Challenges for Assessing Toxicity of Nanomaterials. In M. Ince (Ed.), *Biochemical Toxicology—Heavy Metals a/nd Nanomaterials*. IntechOpen. https://doi.org/10.5772/intechopen.89601

Hakovirta, M., Aksoy, B., & Hakovirta, J. (2015). Self-assembled micro-structured sensors for food safety in paper based food packaging. *Materials Science and Engineering: C*, *53*, 331–335. https://doi.org/10.1016/j.msec.2015.04.020

Hamburg, M. A. (2012). FDA's approach to regulation of products of nanotechnology. *Science*, *336*(6079), 299–300. https://doi.org/10.1126/science.1205441

Hari, K. D., Garcia, C. V., Shin, G.-H., & Kim, J.-T. (2021). Improvement of the UV barrier and antibacterial properties of crosslinked pectin/zinc oxide bionanocomposite films. *Polymers*, *13*(15), 2403. https://doi.org/10.3390/polym13152403

Hemalatha, T., UmaMaheswari, T., Senthil, R., Krithiga, G., & Anbukkarasi, K. (2017). Efficacy of chitosan films with basil essential oil: perspectives in food packaging. *Journal of Food Measurement and Characterization*, *11*(4), 2160–2170. https://doi.org/10.1007/s11694-017-9601-7

Jafarzadeh, S., Salehabadi, A., Mohammadi Nafchi, A., Oladzadabbasabadi, N., & Jafari, S. M. (2021). Cheese packaging by edible coatings and biodegradable nanocomposites; improvement in shelf life, physicochemical and sensory properties. *Trends in Food Science & Technology*, *116*, 218–231. https://doi.org/10.1016/j.tifs.2021.07.021

Jin, K., Tang, Y., Zhu, X., & Zhou, Y. (2020). Polylactic acid based biocomposite films reinforced with silanized nanocrystalline cellulose. *International Journal of Biological Macromolecules*, *162*, 1109–1117. https://doi.org/10.1016/j.ijbiomac.2020.06.201

Jurić, M., Bandić, L. M., Carullo, D., & Jurić, S. (2024). Technological advancements in edible coatings: Emerging trends and applications in sustainable food preservation. *Food Bioscience*, *58*, 103835. https://doi.org/10.1016/j.fbio.2024.103835

Kumari, S. C., Padma, P. N., & Anuradha, K. (2021). Green silver nanoparticles embedded in cellulosic network for fresh food packaging. *Journal of Pure and Applied Microbiology*, *15*(3), 1236–1244. https://doi.org/10.22207/JPAM.15.3.13

Lamberti, M., & Escher, F. (2007). Aluminium foil as a food packaging material in comparison with other materials. *Food Reviews International*, *23*(4), 407–433. https://doi.org/10.1080/87559120701593830

Lamri, M., Bhattacharya, T., Boukid, F., Chentir, I., Dib, A. L., Das, D., Djenane, D., & Gagaoua, M. (2021). Nanotechnology as a processing and packaging tool to improve meat quality and safety. *Foods*, *10*(11), 2633. https://doi.org/10.3390/foods10112633

Leena Maria, J.A. Moses, C. Anandharamakrishnan, "Ethical and Regulatory Issues in Applications of Nanotechnology in Food", in Food Nanotechnology: Principles and Applications (1st ed.)., Ch–5, pp 67–92, 2019, CRC Press

Li, Y., Chu, F., Côté, J.-F., Coelho, L. C., & Chu, C. (2020). The multi-plant perishable food production routing with packaging consideration. *International Journal of Production Economics*, *221*, 107472. https://doi.org/10.1016/j.ijpe.2019.08.007

Ma, H., Burger, C., Hsiao, B. S., & Chu, B. (2014). Fabrication and characterization of cellulose nanofiber based thin-film nanofibrous composite membranes. *Journal of Membrane Science*, *454*, 272–282. https://doi.org/10.1016/j.memsci.2013.11.055

Mackevica, A., Olsson, M. E., & Hansen, S. F. (2016). Silver nanoparticle release from commercially available plastic food containers into food simulants. *Journal of Nanoparticle Research*, *18*(1), 5. https://doi.org/10.1007/s11051-015-3313-x

Marrani, Daniela. 2013. "Nanotechnologies and Novel Foods in European Law." *NanoEthics* 7(3):177–88. doi: 10.1007/s11569-013-0176-4.

Martins, V. G., Romani, V. P., Martins, P. C., & da Silva Filipini, G. (2019). Innovative Packaging that Saves Food. In C. M. Galanakis (Ed.), *Saving Food* (pp. 171–202). Elsevier. https://doi.org/10.1016/B978-0-12-815357-4.00006-7

Motamedi, E., Nasiri, J., Malidarreh, T. R., Kalantari, S., Naghavi, M. R., & Safari, M. (2018). Performance of carnauba wax-nanoclay emulsion coatings on postharvest quality of "Valencia" orange fruit. *Scientia Horticulturae*, *240*, 170–178. https://doi.org/10.1016/j.scienta.2018.06.002

Murugan, S., Karuppiah, V., Thangavel, K., & Panneerselvam, S. (2023). Applications of Nanotechnology in Food Sector: Boons and Banes. In A. Sharma, P. S. Vijayakumar, P. K. Prabhakar, & R. Kumar (Eds.), *Nanotechnology Applications for Food Safety and Quality Monitoring* (pp. 473–492). Elsevier. https://doi.org/10.1016/B978-0-323-85791-8.00009-4

Nile, S. H., Baskar, V., Selvaraj, D., Nile, A., Xiao, J., & Kai, G. (2020). Nanotechnologies in food science: applications, recent trends, and future perspectives. *Nano-Micro Letters*, *12*(1), 45. https://doi.org/10.1007/s40820-020-0383-9

Nilsen-Nygaard, J., Fernández, E. N., Radusin, T., Rotabakk, B. T., Sarfraz, J., Sharmin, N., Sivertsvik, M., Sone, I., & Pettersen, M. K. (2021). Current status of biobased and biodegradable food packaging materials: impact on food quality and effect of innovative processing technologies. *Comprehensive Reviews in Food Science and Food Safety*, *20*(2), 1333–1380. https://doi.org/10.1111/1541-4337.12715

Njuguna, J., Ansari, F., Sachse, S., Rodriguez, V. M., Siqqique, S., & Zhu, H. (2021). 1 – Nanomaterials, nanofillers, and nanocomposites: types and properties. *Health and Environmental Safety of Nanomaterials*, *23*, 3–37. https://doi.org/10.1016/B978-0-12-820505-1.00011-0

Nuruddin, M., Chowdhury, R. A., Szeto, R., Howarter, J. A., Erk, K. A., Szczepanski, C. R., & Youngblood, J. P. (2021). Structure–property relationship of cellulose nanocrystal–polyvinyl alcohol thin films for high barrier coating applications. *ACS Applied Materials & Interfaces*, *13*(10), 12472–12482. https://doi.org/10.1021/acsami.0c21525

Pacaphol, K., Seraypheap, K., & Aht-Ong, D. (2019). Development and application of nanofibrillated cellulose coating for shelf life extension of fresh-cut vegetable during postharvest storage. *Carbohydrate Polymers*, *224*, 115167. https://doi.org/10.1016/j.carbpol.2019.115167

Pattanshetti, A., Pradeep, N., Chaitra, V., & Uma, V. (2020). Synthesis of multi-walled carbon nanotubes (MWCNTs) from plastic waste & analysis of garlic coated gelatin/MWCNTs nanocomposite films as food packaging material. *SN Applied Sciences*, *2*(4), 730. https://doi.org/10.1007/s42452-020-2442-8

Perera, K. Y., Jaiswal, S., & Jaiswal, A. K. (2022). A review on nanomaterials and nanohybrids based bio-nanocomposites for food packaging. *Food Chemistry*, *376*, 131912. https://doi.org/10.1016/j.foodchem.2021.131912

Primožič, M., Knez, Ž., & Leitgeb, M. (2021). (Bio)nanotechnology in food science—food packaging. *Nanomaterials*, *11*(2), 292. https://doi.org/10.3390/nano11020292

Ramos, Ó. L., Pereira, R. N., Cerqueira, M. A., Martins, J. R., Teixeira, J. A., Malcata, F. X., & Vicente, A. A. (2018). Bio-Based Nanocomposites for Food Packaging and Their Effect in Food Quality and Safety. In *Food Packaging and Preservation* (pp. 271–306). Elsevier. https://doi.org/10.1016/B978-0-12-811516-9.00008-7

Ravichandran, R. (2010). Nanotechnology applications in food and food processing: innovative green approaches, opportunities and uncertainties for global market. *International Journal of Green Nanotechnology: Physics and Chemistry*, *1*(2), P72–P96. https://doi.org/10.1080/19430871003684440

Rezić, I., Haramina, T., & Rezić, T. (2017). Metal Nanoparticles and Carbon Nanotubes—Perfect Antimicrobial Nano-fillers in Polymer-Based Food Packaging Materials. In *Food Packaging* (pp. 497–532). Elsevier. https://doi.org/10.1016/B978-0-12-804302-8.00015-7

Roubert, François, Marie-Gabrielle Beuzelin-Ollivier, Margarethe Hofmann-Amtenbrink, Heinrich Hofmann, and Alessandra Hool. 2016. "'Nanostandardization' in Action: Implementing Standardization Processes in a Multidisciplinary Nanoparticle-Based Research and Development Project." *NanoEthics* 10(1):41–62. doi: 10.1007/s11569-015-0248-8.

Ruzi, M., Celik, N., & Onses, M. S. (2022). Superhydrophobic coatings for food packaging applications: a review. *Food Packaging and Shelf Life*, *32*, 100823. https://doi.org/10.1016/j.fpsl.2022.100823

Sahoo, M., Vishwakarma, S., Panigrahi, C., & Kumar, J. (2021). Nanotechnology: current applications and future scope in food. *Food Frontiers*, *2*(1), 3–22. https://doi.org/10.1002/fft2.58

Saleem, W., Nazir, R., Chaudhry, M. N., Saleem, M., & Abidi, S. H. (2022). Packaging material-based polystyrene $ZnFe_2O_4$/clay nanocomposite: preparation, characterization and degradation studies. *Applied Nanoscience*, *12*(5), 1459–1473. https://doi.org/10.1007/s13204-021-02316-2

Sapper, M., & Chiralt, A. (2018). Starch-based coatings for preservation of fruits and vegetables. *Coatings*, *8*(5), 152. https://doi.org/10.3390/coatings8050152

Shankar, S., Khodaei, D., & Lacroix, M. (2021). Effect of chitosan/essential oils/silver nanoparticles composite films packaging and gamma irradiation on shelf life of strawberries. *Food Hydrocolloids*, *117*, 106750. https://doi.org/10.1016/j.foodhyd.2021.106750

Siegrist, M., Cousin, M.-E., Kastenholz, H., & Wiek, A. (2007). Public acceptance of nanotechnology foods and food packaging: the influence of affect and trust. *Appetite*, *49*(2), 459–466. https://doi.org/10.1016/j.appet.2007.03.002

Silvestre, C., Duraccio, D., & Cimmino, S. (2011). Food packaging based on polymer nanomaterials. *Progress in Polymer Science*, *36*(12), 1766–1782. https://doi.org/10.1016/j.progpolymsci.2011.02.003

Souza, V., Alves, M., Santos, C., Ribeiro, I., Rodrigues, C., Coelhoso, I., & Fernando, A. (2021). Biodegradable hitosan films with ZnO nanoparticles synthesized using food industry by-products—production and characterization. *Coatings*, *11*(6), 646. https://doi.org/10.3390/coatings11060646

Srivastava, A., Akoh, C. C., Yi, W., Fischer, J., & Krewer, G. (2007). Effect of storage conditions on the biological activity of phenolic compounds of blueberry extract packed in glass bottles. *Journal of Agricultural and Food Chemistry*, *55*(7), 2705–2713. https://doi.org/10.1021/jf062914w

Sultana, A., Kathuria, A., & Gaikwad, K. K. (2022). Metal–organic frameworks for active food packaging. A review. *Environmental Chemistry Letters*, *20*(2), 1479–1495. https://doi.org/10.1007/s10311-022-01387-z

Surendhiran, D., Roy, V. C., Park, J.-S., & Chun, B.-S. (2022). Fabrication of chitosan-based food packaging film impregnated with turmeric essential oil (TEO)-loaded magnetic-silica nanocomposites for surimi preservation. *International Journal of Biological Macromolecules*, *203*, 650–660. https://doi.org/10.1016/j.ijbiomac.2022.01.178

Svagan, A. J., Åkesson, A., Cárdenas, M., Bulut, S., Knudsen, J. C., Risbo, J., & Plackett, D. (2012). Transparent films based on PLA and montmorillonite with tunable oxygen barrier properties. *Biomacromolecules*, *13*(2), 397–405. https://doi.org/10.1021/bm201438m

Teixeira-Costa, B. E., & Andrade, C. T. (2021). Natural polymers used in edible food packaging—history, function and application trends as a sustainable alternative to synthetic plastic. *Polysaccharides*, *3*(1), 32–58. https://doi.org/10.3390/polysaccharides3010002

Tirumala, M. G., Anchi, P., Raja, S., Rachamalla, M., & Godugu, C. (2021). Novel methods and approaches for safety evaluation of nanoparticle formulations: a focus towards in vitro models and adverse outcome pathways. *Frontiers in Pharmacology*, *12*, 612659. https://doi.org/10.3389/fphar.2021.612659

Traxler, I., Laske, S., & Fischer, J. (2022). Closed-loop recycling of polypropylene: A case study on mechanical recycling of separately collected yogurt cups in Austria. *Resources, Conservation and Recycling*, *205*, 107537. https://doi.org/10.1016/j.resconrec.2024.107537

Twede, D. (2002). The packaging technology and science of ancient transport amphoras. *Packaging Technology and Science*, *15*(4), 181–195. https://doi.org/10.1002/pts.597

Verma, M. K., Shakya, S., Kumar, P., Madhavi, J., Murugaiyan, J., & Rao, M. V. R. (2021). Trends in packaging material for food products: historical background, current scenario, and future prospects. *Journal of Food Science and Technology*, *58*(11), 4069–4082. https://doi.org/10.1007/s13197-021-04964-2

van Willige, R. W. G., Linssen, J. P. H., Meinders, M. B. J., van der Stege, H. J., & Voragen, A. G. J. (2002). Influence of flavour absorption on oxygen permeation through LDPE, PP, PC and PET plastics food packaging. *Food Additives and Contaminants*, *19*(3), 303–313. https://doi.org/10.1080/02652030110081146

Xin, S., Xiao, L., Dong, X., Li, X., Wang, Y., Hu, X., Sameen, D. E., Qin, W., & Zhu, B. (2020). Preparation of chitosan/curcumin nanoparticles based zein and potato starch composite films for Schizothorax prenati fillet preservation. *International Journal of Biological Macromolecules*, *164*, 211–221. https://doi.org/10.1016/j.ijbiomac.2020.07.082

Xing, Y., Li, W., Wang, Q., Li, X., Xu, Q., Guo, X., Bi, X., Liu, X., Shui, Y., Lin, H., & Yang, H. (2019). Antimicrobial nanoparticles incorporated in edible coatings and films for the preservation of fruits and vegetables. *Molecules*, *24*(9), 1695. https://doi.org/10.3390/molecules24091695

Xu, L., Li, Y., Gao, S., Niu, Y., Liu, H., Mei, C., Cai, J., & Xu, C. (2020). Preparation and properties of cyanobacteria-based carbon quantum dots/polyvinyl alcohol/nanocellulose composite. *Polymers*, *12*(5), 1143. https://doi.org/10.3390/polym12051143

Xu, W., Xie, W., Huang, X., Chen, X., Huang, N., Wang, X., & Liu, J. (2017). The graphene oxide and chitosan biopolymer loads TiO_2 for antibacterial and preservative research. *Food Chemistry*, *221*, 267–277. https://doi.org/10.1016/j.foodchem.2016.10.054

Yaashikaa, P. R., Kamalesh, R., Senthil Kumar, P., Saravanan, A., Vijayasri, K., & Rangasamy, G. (2023). Recent advances in edible coatings and their application in food packaging. *Food Research International*, *173*, 113366. https://doi.org/10.1016/j.foodres.2023.113366

Yu, K., Xu, J., Zhou, L., Zou, L., & Liu, W. (2021). Effect of chitosan coatings with cinnamon essential oil on postharvest quality of mangoes. *Foods*, *10*(12), 3003. https://doi.org/10.3390/foods10123003

Zambrano-Zaragoza, M. L., González-Reza, R., Mendoza-Muñoz, N., Miranda-Linares, V., Bernal-Couoh, T. F., Mendoza-Elvira, S., & Quintanar-Guerrero, D. (2016). Edible coatings based on nanometric dispersions charged with functional ingredients and it's use to increase shelf life. *International Journal of Molecular Sciences*, *19*(3), 705. http://dx.doi.org/10.3390/ijms19030705

3

Polymer-Based Nanocomposites in Food Packaging

Vijayakumar Raja[1], L. Mahalakshmi[1], M. Maria Leena[2], and C. Anandharamakrishnan[1]

[1] *CSIR–National Institute for Interdisciplinary Science and Technology (NIIST), Ministry of Science and Technology–Government of India, Industrial Estate PO, Thiruvananthapuram, Kerala - 695019, India*
[2] *Department of Biotechnology, Faculty of Engineering and Technology, SRM Institute of Science and Technology (SRMIST), Tiruchirappalli, Tamil Nadu - 621105, India*

3.1 Introduction

Polymeric materials are employed in modified forms in food packaging systems, such as multilayer packaging materials, to boost mechanical strength and gas/water vapor barrier properties, hence increasing the shelf life of packaged products. However, the multilayer packaging materials developed and used for food packaging raise concerns regarding their recyclability and difficulty in the removal of synthetic adhesives used for their preparation (Aznar et al., 2011). Thus, novel packaging approaches where nanotechnological aspects are included aiming at improving the mechanical, barrier properties, and achieving simple and lightweight packaging systems. The polymeric nanocomposites play a key role in achieving the above-mentioned objectives.

Polymer-based nanocomposites are a type of multicomponent solid materials that consist of polymers (biodegradable or nonbiodegradable synthetic polymers) reinforced with inorganic molecules or solids that possess various mechanical and structural properties. On a nanocomposite matrix, the polymer is the continuous phase, while the reinforcement agent is a dispersed phase. The reinforcement agent is distributed throughout the polymeric structure, which provides better structural and barrier properties than conventional polymers. The use of fillers improves the functionality and applicability of polymers while preserving the characteristic flexibility of polymer materials and their manufacturing conditions (Winey and Vaia, 2007). Based on the morphology of reinforcement fillers, these are classified as particulate nanocomposites, elongated particle nanocomposites, and layered nanocomposites. The major challenge in preparing a polymer nanocomposite system is to distribute the nanoparticles homogeneously in the polymer phase. The distribution of nanofillers is majorly affected by the degree of nanoparticle separation, based on which the nanocomposites may be categorized as conventional composites, exfoliated, and intercalated nanocomposites (Bousmina, 2006).

Nanotechnology for Sustainable Food Packaging, First Edition. Edited by C. Anandharamakrishnan, Jeyan A. Moses, and M. Maria Leena.

The polymer/filler interface and the type of polymers and fillers used primarily influence the characteristics of the generated nanocomposites. Polymeric nanocomposite materials prepared with suitable fillers possess improved resistance against microbial growth (Omerović et al., 2021). The homogeneity of the nanofiller dispersion also controls the movement of gas/water vapor through the polymer, resulting in increasing the effective diffusion path length (Checchetto et al., 2014), by increasing the tortuosity. The permeability is also dependent on the size of the filler particle (Dougnac et al., 2010), the filler shape/aspect ratio (Mandal et al., 2024), and the alignment of nanofillers (Zarshad et al., 2022) in the polymer. When the fillers with a high aspect ratio are added to a polymeric matrix, a higher degree of reinforcement effect is achieved, as the fillers possess an increased specific surface area (Martone et al., 2011).

The consistent distribution of nanoparticles creates a huge interface of the polymer/filler, modifying the mobility, mechanical, and thermal characteristics of the materials (Idumah and Obele, 2021). Additionally, the interface region between each reinforcement nanoparticle results in the percolation of an interfacial network which majorly influences the properties of the nanocomposites developed (Idumah, 2023). When the filler material of reduced particle size is incorporated into the polymer, the individual filler particles are taken closer to each other, which results in the overlapping of interfacial layers and affects the quality of the nanocomposite. Thus, the characteristics of polymers/nanofillers and their impact on the properties of the nanocomposites play a vital role in achieving a suitable nanocomposite film for food packaging. This chapter discusses an outline of the production of polymer nanocomposites, the types of materials used, and their applications in food packaging substantiated by the existing literature and market trends.

3.2 Types of Polymers

Various polymers such as polyethylene (PE) (Youssef et al., 2023), polyethylene terephthalate (PET) (Sanchez-Garcia et al., 2007), polypropylene (PP) (Manikantan and Varadharaju, 2011), polyvinyl chloride (PVC) (Erlantz et al., 2017), poly-butylene adipate-co-terephthalate (PBAT) (Venkatesan and Rajeswari, 2017), and polystyrene (PS) (Ibrahim et al., 2020), are used for the production of novel polymer nanocomposite structures. The suitability of the polymers for nanocomposite production is determined by considering the melt flow index (MFI) of the polymers, which represents their flow behavior. The appropriate polymer material for food applications is selected based on the intrinsic characteristics of the polymer and its intended use. Apart from the synthetic nonbiodegradable polymers, various biopolymers have also been studied for the preparation of nanocomposites.

3.3 Types of Fillers

Various inorganic and organic fillers with a minimum of one dimension are used for the fabrication of nanocomposites. Three types of nanofiller materials are classified based on the dimensions (in the nanometer range) and morphological differences.

Crystal structures (unidimensional)
Elongated structures (two-dimensional; nanotubes, and nanowhiskers)
Isodimensional nanoparticles (three-dimensional; spherical silica nanoparticles, semiconductor nanoclusters, etc.)

When the morphology of the filler becomes complex, their dispersion in the polymer is constrained because they tend to form agglomerates due to interparticle interaction. In such cases, nanofiller materials subjected to surface modification strategies were used for nanocomposite production. The addition of surface-modified nanofillers tends to increase the interaction at the polymer/filler interface, thus reducing the formation of filler agglomerates. This results in increasing the hydrophobicity of the fillers, which improves the interfacial adhesion through chemical binding or chain entanglement that helps in loosening the filler agglomerates formed during the composite formation (Rong et al., 2006).

3.4 Polymer Nanocomposites – Preparation Methods

There are numerous approaches for the production of nanocomposites, where solution casting is the generally used method due to its simplicity of production. It is a direct dispersion approach in which the fillers are dissolved in a suitable solvent by agitation/sonication (Sheng et al., 2017). The filler solution is combined with the polymer solution. The nanocomposites are formed by casting, where the solvent is removed via evaporation or distillation (Silva et al., 2018).

In situ polymerization is the most effective approach to creating nanocomposites with uniformly distributed filler molecules. In this approach, the fillers are blended with the prepolymer or monomer material to initiate the chemical bond formation and to improve the polymer-filler interaction, followed by the initiation of polymerization using heat or radiation (Mao and Wang, 2020) (Figure 3.1). *In situ* intercalative polymerization, (Namazi et al., 2012), solution intercalation, chemical reduction (Daoush and Hong, 2013), and melt intercalation (Lepoittevin et al., 2003) are all examples of intercalative polymerization used for nanocomposite production. In the solution intercalation method, the polymer solution and the nanoparticles are blended and polymer chains result in intercalation and displacement of the solvent within the nanoparticle interlayer. The melt intercalation process entails heating the polymer and filler blend over the polymer's softening point. Biopolymer chains are disseminated in layered silicates during annealing, resulting in two types of structures: intercalations and exfoliations. The extent to which the polymer penetrates the layered silicate determines the development of either shape. This method is useful for creating nano-sized particles with increased impact strength, thermal stability, and tensile strength.

Electrospinning is an emerging approach to producing nanocomposites with a uniform distribution of particles (Schmatz et al., 2019). In this method, the liquid for spinning is filled and fed through a syringe feeder at a predetermined flow rate. Immobilized charges on the liquid droplet surface tend to produce a continuous jet. The electrospinning technique is used to generate micro- and nanofibers of different morphologies (Cui et al., 2023). Reinforcement of nanoparticles/fillers has proven to be effective in the electrospinning process as reported by researchers due to its ability to control nanofiber's compositional, structural, and functional

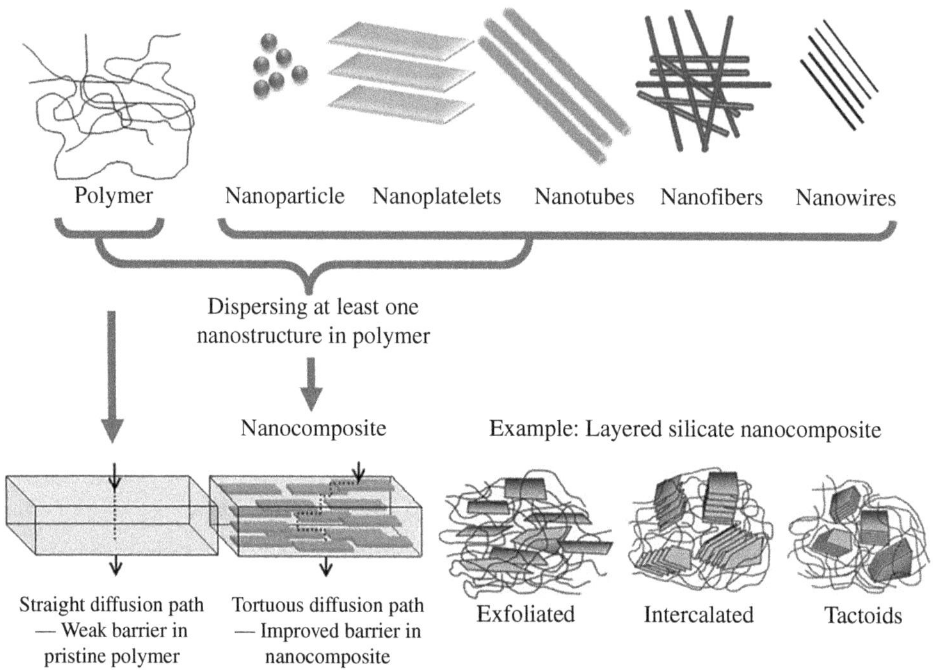

Figure 3.1 Preparation of nanocomposites with improved barrier and mechanical properties. *Source:* Mihindukulasuriya and Lim (2014)/with permission of Elsevier.

properties. The polymer nanocomposites prepared using the electrospinning approach were found to undergo shear flow-based deformation called nanonecking as a result of overlapping of stress fields due to the occurrence of nanopores in the nanocomposite films (Kim et al., 2005). Another novel nanocomposite preparation approach was also established by using PLA-based nanofibers (by electrospinning) in which the phycocyanin/polyvinyl alcohol was reinforced using the electrospraying approach.

3.5 Mechanism of Reinforcement

In the solution casting method, the polymer is mixed in a solvent, which is then added with the nanofiller materials such as graphene, etc. In this approach, it is necessary to note the solvent compatibility with nanofiller and polymer. The mixing of the polymer and reinforcement filler is achieved by either agitation (simple/shear) or ultrasonication. This aids in the adsorption of the nanofiller onto the polymer surface. The solvent is then evaporated to ensure the formation of nanocomposite with a homogeneous distribution of fillers.

The polymer is melted by heating, and the filler is added directly to the molten polymer in the melt mixing approach, where mixing is achieved using internal mixers or extruders. Temperature, mixer/extrusion screw configuration, feeding, and mixing rates are the prime influencing factors that affect the dispersion of filler. Various other melt processing methods such as extrusion processing, injection molding, and compression molding are also used for the production of nanocomposites.

In the *in situ* polymerization process, nanofillers are mixed with the monomers in their native form, and then polymerization is initiated with the application of heat or irradiation. Once the reaction is initiated, the nanofillers get delaminated, and a final composite is formed with the polymerization of the monomer. Layer-by-layer assembling is another approach to producing multifunctional nanocomposites. In this method, the assembly of nanocomposites is achieved by the adsorption of oppositely charged materials through the application of various interactive mechanisms. This approach is useful in controlling the interfacial characteristics and surface and is thus used to produce customized nanocomposite assemblies for various uses (Zhang et al., 2019).

Some studies have also explored hybrid approaches by employing combinations of techniques to prepare high-performing composites with improved reinforcement efficiencies. In some cases, various approaches, such as electric field/magnetic field (Shi et al., 2018; Wu et al., 2018) applications, are also used for nanocomposite production to induce the alignment of the nanofiller within the polymeric matrix.

3.6 Common Polymer Nanocomposites Used in Food-Packaging Systems

3.6.1 Polymer/Carbon Nanotube-Based Nanocomposites

Carbon nanotubes (CNTs) are used as nanofillers to reconstruct various materials, especially polymers with increased mechanical strength, due to their superior mechanical and thermal characteristics. The incorporation of CNTs can raise the thermal conductivity of the polymeric system, as the thermal conductivity of CNTs ranges between 2000 and 6000 W/mK (Karaipekli et al., 2017). However, ensuring proper distribution and orientation of CNTs in the polymeric structure is a challenge, as CNTs generate stable agglomerates when introduced to the polymer through the Van der Waals interaction (Sahoo et al., 2010). Functionalization of CNTs is an approach to prevent this bundle formation and to ensure the compatibility between the polymers and CNTs, which helps in improving the performance of the composites. Numerous studies have developed Polymer/CNT-based nanocomposites with control over surface morphology (Cebeci et al., 2009). The tensile strength of the multiwalled CNTs incorporated nanocomposite films increased by 116% when compared to the prepared simple polymeric films (Wen et al., 2022). Polylactic acid nanocomposites were prepared in another study using functionalized CNTs wrapped using polydopamine via melt blending method. The prepared nanocomposite has shown better properties with 3%wt of multiwalled CNTs where the hardness was increased by 79% and Young's modulus was increased by 113% as compared to pristine polylactic acid films (Yakdoumi et al., 2022).

3.6.2 Polymer/Graphene Nanocomposites

Graphene is another type of filler material employed in the production of food packaging films and coating ingredients. Derivatives of graphene such as graphene oxide, graphene nanoplatelets (Lin et al., 2018), etc. are used in nanocomposites as they exhibit higher electrical conductivity, transparency, low density, and high aspect ratio.

In addition, graphene/graphite-incorporated polymeric nanocomposites have displayed improved water vapor permeation inhibition properties. In a study conducted by Mittal and Chaudhry (2015), the melt mixing approach is employed for the production of thermally reduced graphene-incorporated nanocomposites. The high-density poly ethylene (HDPE) nanocomposite incorporated with graphene reported an increase in the tensile (>50%) and storage modulus with an increase in the filler concentration in the nanocomposite. Low-Density Poly Ethylene (LLDPE) polymer was found to have the highest improvement (about 100%) in tensile modulus, break stress (60%), and tensile stress (30%). The molecular impermeability attributes of graphene nanosheets and the interactions of graphene nanosheets with the polymeric molecules are reported as major influencing factors for the improved barrier properties, even at lower concentration levels of graphene (Allahbakhsh, 2017). Some studies have also investigated the production of graphene nanocomposites via a solution-casting approach. For example, Manikandan et al. (2020) used a solution casting approach to produce polyhydroxy butyrate-based graphene nanocomposites with graphene nanoplatelet loading of 0.7 wt%, which has been shown to increase the melting point, thermal stability, and mechanical strength of the polymer. The incorporation of graphene nanoplatelet also resulted in controlling the UV and visible light penetration through the packaging material. Thus, a 4-fold increase in the shelf life of the food products was reported. Graphene oxide was incorporated into polycaprolactone polymer in another study, which has shown an increase in the bactericidal effect (Malik, 2022). Complex architectures were also achieved using graphene derivatives with improved barrier properties. For example, Goh et al. (2016) have fabricated a composite film with sandwich architecture using impermeable reduced graphene oxide in its core, where polymeric film serves as an outer encapsulation. The complex structure of the film resulted in about 87% reduction of its water vapor permeability. The film has also been reported to raise (~8-fold) the shelf life of moisture and oxygen-sensitive food materials.

3.6.3 Polymer/Nanoclay-Based Nanocomposites

Nanoclay reinforcement is one of the widely employed approaches for polymer nanocomposite production, as clay materials can increase the dispersion tendency, and improve physical characteristics. The intercalation of polymers in nanoclay interlayer levels has been investigated in numerous studies recently. Montmorillonite is the generally used nanoclay, which contains two tetrahedral silica sheets attached to an edge-shared octahedral alumina sheet. In a study, the polycaprolactone-based film was developed using organo nanoclay (0.4 wt%) with chitosan and oleic acid as plasticizers. The presence of organo nanoclay significantly enhanced the antimicrobial activity of the polycaprolactone film against *Candida albicans* (Cesur et al., 2018). Another study has recently explored the grafting of nanoclay material with essential oils, which is incorporated into LLDPE to produce nanocomposite films. The films have shown strong antimicrobial activity with enhanced mechanical characteristics (Tornuk et al., 2018). The application of clay in nanocomposites reportedly increases the mobility distance of gas molecules which results in reducing the oxygen permeability of the pure polymeric films (Choi et al., 2011). In addition to the barrier properties, nanoclay materials also possess ethylene-scavenging activity, due to which numerous studies have developed polymer nanocomposite-based active packaging materials for ethylene-sensitive food materials (Tas et al., 2017; Wang et al., 2023).

3.6.4 Polymer/Inorganic Nanoparticle Nanocomposites

Apart from clay minerals and CNTs, several other inorganic nanomaterials such as semiconductors, metals (Ag, Au, etc.), and metal oxides (TiO_2, Al_2O_3, Fe_2O_3, SiO_2, etc.) are also being used as nanofillers due to their nanoscale size (Kango et al., 2013). In a study, the addition of 5 wt% ZnO nanofiller caused a uniform filler dispersion in the polymer. The addition of ZnO also decreased the packing material's diffusion coefficient, oxygen permeability, and water absorption capacity. ZnO nanoparticles also improved the weatherability of the polymer (Seo et al., 2011), as they efficiently absorb the UV light at wavelength below 350 nm. Thus, these nanocomposites can be used in UV-protective packaging materials (Lizundia et al., 2016). PLA-based nanocomposites developed by the addition of inorganic metal/metallic oxide, silica-based and graphite nanoparticles reveal significant changes in their physical properties, and a decline in water vapor and oxygen permeability properties in comparison with the pristine PLA films (Mulla et al., 2021). The inorganic or metal nanoparticles also result in increasing the optical properties of the nanocomposite membranes developed with increased transparency (Caseri, 2008). Another study used gold nanoparticles as reinforcement in a polyvinyl alcohol polymer matrix and reported an increased formation of the inhibition zone that prevents microbial contamination, thus improving the shelf life of the foods (Chowdhury et al., 2020). Silver nanoparticles are another type of efficient nanofiller that provides superior mechanical and antimicrobial properties to the nanocomposites (Mathew et al., 2019).

3.7 Applications in Food Packaging

Polymer nanocomposites have gained significant attention in the field of food packaging owing to their capability to prolong the shelf life of packaged goods through increased barrier and antibacterial characteristics. A study has reported that nanocomposite packaging films are capable of minimizing water loss in vegetables and could retain water during storage (Wen et al., 2022). In the same study, the chicken meat wrapped in ZnO-loaded PVA nanocomposite films demonstrated the prevention of microbial growth for up to 36 hours at refrigerated conditions. Few studies have also studied the use of biopolymers as capping agents to stabilize the nanofillers by reducing the development of agglomerates (Lomate et al., 2018; Lü et al., 2003).

Polymer nanocomposites are useful in the fabrication of ultrafast sensors with a higher sensitivity. Moustafa et al. (2021) have developed a rapid humidity measurement sensor using polyvinyl chloride/graphene oxide nanocomposites. In these smart sensors, hydrophilic oxygen reactive groups absorb water from the atmosphere, increasing the material's electrical conductivity. The change in electrical conductivity is recorded and correlated with the absorbed humidity. This is useful in the intelligent packaging of food systems.

For the packaging of high-moisture foods, food trays made up of polymers should be modified to improve their water barrier properties. In a study, carbon-based nanofillers (multiwalled CNTs, graphene oxide, and carbon blacks) were used in the HDPE polymer, and food tray packaging material was formed by the twin-screw extrusion technique (Gill et al., 2020). In addition, various nanocomposites with ethylene scavenging activity were developed and used in fruit and vegetable packaging, which also have reportedly delayed the ripening of fruits packed in nanocomposite packaging systems (Hu et al., 2011).

3.8 Properties of Polymer Nanocomposites

The reinforcing agents introduced into polymeric films not only increase their mechanical properties but also have a substantial effect on their physical and structural characteristics. Several metal oxide or synthetic reinforcement agents have been shown to have potential antimicrobial activity, thus improving the shelf life of foods when used in nanocomposite systems. Researchers are also investigating the role of reinforcement fillers in controlling the release of active compounds. This section discusses the effect of nanofillers on various quality attributes of polymer nanocomposites.

3.8.1 Gas and Water Vapor Permeability

The incorporation of fillers has the main focus of decreasing the transmission rate of water vapor and gas through the matrix (Castillo et al., 2015). The water vapor permeability properties of several polymers, such as HDPE, LDPE, and PP, that possess good film-forming properties can be improved by forming their nanocomposites.

The addition of fillers results in a physical blockage to water molecule transportation by forming a complex route. This is influenced by the concentration of nanofillers and the chemical interactions between the polymer matrix and nanofillers. The oxygen and water vapor permeability of various polymers and their nanocomposites are presented in Table 3.1. When the nanofiller concentration increases above a threshold level, the water and oxygen permeability again tend to increase as shown in Figure 3.2. This may have been due to the excessive settlement of nano filler in the crystallization regions of the polymer, thus instigating the rupture of the formed structure. Incorporation of graphene oxide nanoparticles (0.35 vol%) into poly-(butylene adipate-co-terephthalate) film has decreased water vapor permeability (36% reduction in H_2O permeability as compared to PBAT film) due to its higher water resistance as a result of its crystalline structure and hydrophobic character and increase in the tortuosity. Thus, a balance in the nanoparticle distribution and polymeric structure is important to obtain an improved barrier performance.

3.8.2 Mechanical Properties

Various mechanical properties such as tensile strength, elongation at break, Young's modulus, etc. are the determining factors representing the impact of fillers or reinforcement agents' incorporation on the mechanical properties. These characteristics are derived from the stress–strain curves in the standard texture profile analysis technique. Most polymeric materials improve their tensile strength and other mechanical characteristics with the addition of reinforcement agents. The mechanical properties of nanocomposites vary based on the type, concentration of the nanofiller, and the molecular interaction in the polymer/filler interface (Shankar and Rhim, 2016). Nanocomposites made of polymeric materials and loaded with nanofillers are flexible in nature and can withstand large levels of applied forces (Murali et al., 2019). The use of nanofibers as reinforcement has been reported to possess high tensile and puncture strength (Carvalho et al., 2020). Generally, three-dimensional fillers show a higher degree of tensile roughness, as they contribute to a

Table 3.1 Mechanical, thermal, and barrier properties of polymer nanocomposites.

Composite Configuration	Method of Preparation	Mechanical Properties	Permeability	Thermal Properties	Reference
Low-Density Poly Ethylene (LDPE)/ Cu-nanoparticles (3 wt%)	Solvent evaporation and doctor blade film applicator (for thin film formation)	16 MPa—tensile strength	Water Vapor Permeability (WVP) – 0.55×10^{10} g/s mPa	Melting temperature – 127.31 °C	Lomate et al. (2018)
High-Density Poly Ethylene (HDPE)/carbon nanotubes (4.06 wt%)	Melt mixing and compression molding	26.44 MPa – Tensile strength, 1063.83 MPa – Young's modulus, 9.28% – elongation at break	–	Decomposition temperature – 479 °C	Evgin et al. (2019)
Polypropylene/Ag decorated multiwalled Carbon Nanotube (0.2 phr)	Melt blending	56% elongation at break	Oxygen Transmission Rate (OTR) – 14 : 7 × 10^{-4} cm^3/m^2 d Pa	Melting temperature – 164.2 °C (less than pure PP film 167.5 °C)	Wen et al. (2020)
Poly-(butylene adipate-co-terephthalate) with ZnO	Solution casting	Tensile strength 45.1 MPa (for pure PBAT [37.9 MPa])	1052.4 cc/m^2 per 24 h OTR for pure PBAT, which decreases with increase in ZnO concentration	Degradation temperature for composite 428.90 °C (for pure PBAT 412 °C)	Venkatesan and Rajeswari (2017)
Polypropylene (PP) and montmorillonite (1 wt%)	Melt intercalation in twin-screw extruder	Minimum difference in mechanical properties of pure PP and nanocomposite due to the presence of partly immobilized polymer	Water Vapor Permeability (WVP) – 320 μm g/m^2 dia for pure PP and 301 μm g/m^2 dia for nanocomposite	Melting temperature – 168 °C (for pure PP 169 °C)	Zehetmeyer et al. (2012)

Phr, parts per hundred resins; LDPE, low-density poly ethylene; HDPE, high-density poly ethylene; WVP, water vapor permeability; OTR, oxygen transmission rate.

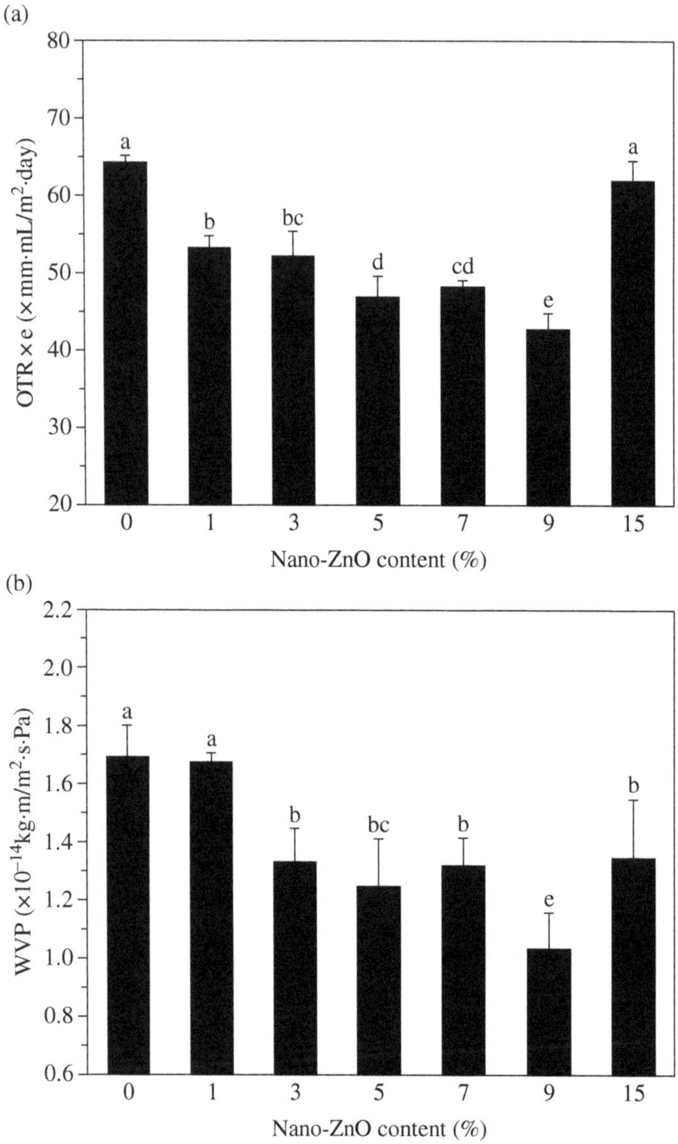

Figure 3.2 Effect of ZnO concentration on oxygen and water vapor permeability of PLA/ZnO nanocomposite films. *Source:* Tang et al. (2020))/CC BY 4.0/MDPI.

uniform dispersion of filler and improved interlocking of the filler and polymer as reported by Gill et al. (2020).

3.8.3 Melt Flow Index

The MFI represents the flow behavior of the polymeric systems. MFI is measured to study the impact of filler addition on the flow characteristics of the molten polymer as well as to determine the processing conditions suitable for industrial production using the extrusion approach (Tayfun et al., 2016). The use of filler material reportedly affects the MFI of the polymer nanocomposites. MFI of the polymer nanocomposites depends on the viscoelastic characteristics of the matrix and the properties of fillers (structure and distribution) (Zhu and Xanthos, 2004). In general, the MFI decreased with nanofiller addition. In a study conducted by Tayfun et al. (2016), the inclusion of unmodified CNTs was reported to decrease the MFI of polyurethane. This is due to the increased adhesion between CNT particles that resulted in the development of huge agglomerates, which resisted the flow of the polymer. Similar results of a decrease in the MFI due to the flow resistance offered by the nanofiller have been reported in other studies (Sahu and Rama Sreekanth, 2022; Zago et al., 2019). Whereas, the modified fillers reportedly amplified the MFI of the composite, which is due to the increased surface interactions between the polymeric and dispersed phases that resulted in a homogeneous dispersion of the added filler. Another study has reported an increase in MFI during the incorporation of clay as the filler due to a decline in the cross-linking of polymeric chains (Khalaj et al., 2016).

3.8.4 Microstructure

The levels of nanoparticle (nanofiller) incorporation affect the uniformity of nanoparticle distribution in the formed film. The study that formed the incorporation of thermally exfoliated graphene oxide in the polyurethane polymer film has shown that the film was homogeneous and smooth until the addition of 1% of the nanoparticle. Where the high concentration of incorporated nanoparticles led to a lack of affinity between the nanofillers and the polymer. As a result, the aggregation of nanoparticles and inhomogeneity in surface morphology occurs. The orientation of the polymer composite structure varies with the processing methods, which affects the morphology and structure of prepared nanocomposites. For example, when polymer composites are produced using simple solution mixing or through an *in situ* polymerization approach, the produced nanocomposites are randomly oriented exfoliated single or multiple-layer platelets, whereas the composites developed by the melt mixing process produce more aligned and intercalated or stacked composite structure. Processing methods can induce the structural orientation of the dispersed platelet polymer composite and favor the reinforcement but may potentially increase the percolation threshold. In the injection molding method, the platelets may be positioned more randomly close to the interior portion of the sample, where the platelets are arranged parallel to the surface. In contrast, thin solution-cast films or compression-molded films may have aligned platelets throughout their cross-section. Figure 3.3a shows the more exfoliated state of dispersion of functionalized graphene sheet/polyurethane composites produced by a simple solution mixing process (Potts et al., 2011). Figure 3.3b shows the graphene oxide nanosheets/poly-(butylene adipate-terephthalate) nanocomposite material developed for food packaging. The addition of graphene oxide nanosheets resulted in the formation of individual exfoliated fractions without the agglomerate formation, and graphene oxide nanosheets were also observed throughout the film in a wrinkled paper type of structure (Ren et al., 2017). Duan et al. (2013) reported that the poly(lactic acid) nanocomposite containing nanoclay showed an even distribution and uniform alignment (in flow direction) of the clay nanoparticles when the molten polymer mixture was applied

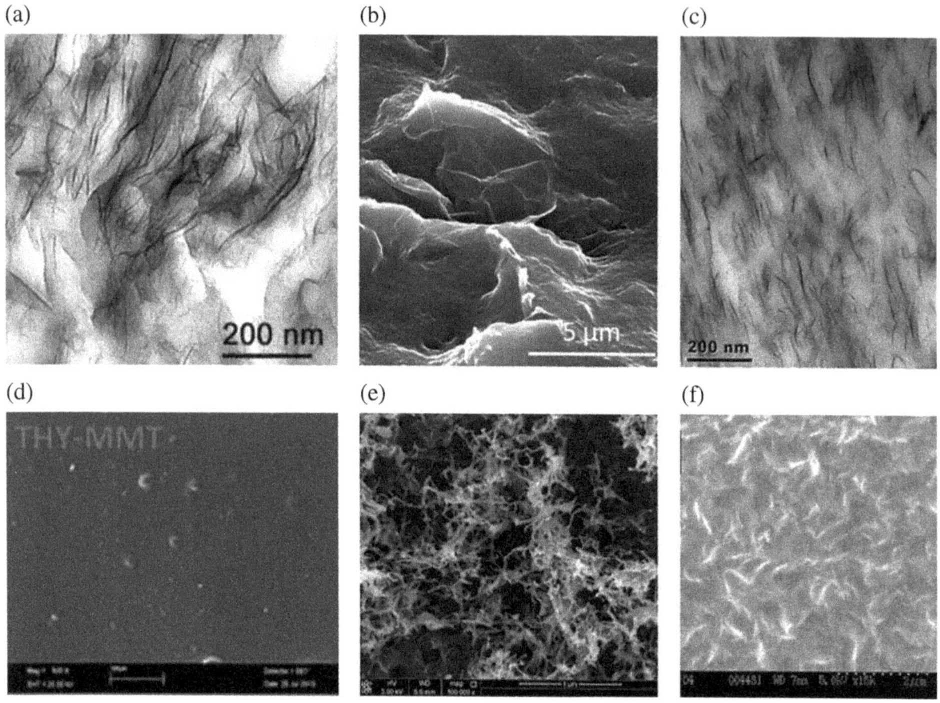

Figure 3.3 Morphology of (a) polyurethane nanocomposite prepared with thermally exfoliated graphene oxide. *Source:* Potts et al. (2011))/with permission of Elsevier, (b) graphene oxide nanosheet incorporated PBAT film. *Source:* Ren et al. (2017))/with permission of Elsevier, (c) PLA nanocomposites with nanoclay (3 wt%). *Source:* Duan et al. (2013))/with permission of Elsevier, (d) thymol-nanoclay-incorporated LLDPE film. *Source:*(Tornuk et al. (2018))/with permission of Elsevier, (e) carbon nanotube-incorporated LDPE-based nanocomposite film. *Source:* Majeed et al. (2018))/with permission of Elsevier, and (f) multiwalled CNT incorporated PE film. *Source:* (McNally et al. (2005))/with permission of Elsevier.

with mechanical shear between the two rotors in the torque rheometer (Figure 3.3c). The filler alignment was also retained when the molten mixture was compressed for the formation of sheets.

3.8.5 Antibacterial Properties

Nanoparticles exhibiting antibacterial properties have been explored to be incorporated in packaging films and proven effective against food spoilage bacteria. The direct integration of metal nanoparticles into the nanocomposite films poses health issues due to their potential transfer from the package to the food and associated toxicity. Surface modification or functionalization of inorganic nanoparticles is an approach to increase the stability and antimicrobial attributes of the nanocomposite matrix (Hajizadeh et al., 2020). In a study conducted recently, ZnO-reinforced chitosan nanocomposite film has shown intrinsic antimicrobial properties for the storage of fresh meat, when used as a primary packaging system (Souza et al., 2020). Some of the polymers have a characteristic antimicrobial potential themselves, which can be further enhanced using inorganic fillers with antimicrobial activity. For instance, chitosan film possessed antimicrobial activity due to its interaction with negatively charged microbes. The addition of ZnO nanofiller improved its potential antimicrobial activity (Rodrigues et al., 2020). Another study has incorporated colloidal silver nanoparticles into LDPE films by melt blending approach, which has shown higher levels of antibacterial activity against various strains such as *Escherichia coli, Candida albicans,* and *Staphylococcus aureus* (Jokar et al., 2012). This results from the interaction of negatively charged phosphorus/sulfur ions present in bio-macromolecular compounds (such as proteins/nucleic acids) with positively charged nanoparticles, which can cause structural alterations and rupture of the bacterial cell walls and membranes, resulting in metabolic process disruption and cell death (Butkus et al., 2003; Rhim et al., 2014). Apart from the metal oxides, the nanocomposites prepared with the incorporation of essential oils have also shown to be beneficial in terms of their antimicrobial activity. The reinforcement agents in these studies often hinder the release of active compounds over time (Souza et al., 2019). Metal nanoparticles also have proven the potential to treat bacteria, including both Gram-positive and Gram-negative strains, with a less-toxic impact. In a study, poly(lactic acid) film was embedded with ZnO nanoparticles and essential oils to reduce the lipid peroxidation of packed fish meat. The essential oils have shown improved antimicrobial and antioxidant activities, and aid in controlling the transport and migration of Zn^{2+} ions into the packed meat (Heydari-Majd et al., 2019).

3.8.6 Thermal Properties

The reinforcement fillers increase the glass transition temperature of the nanocomposites. In addition, the incorporation of nanofibers has reportedly increased the thermal resistance of the polymer, because of the increasing crystalline structure results. This causes a rise in the temperature necessary for thermal decomposition (Abral et al., 2020). Similar results of enhanced thermal stability were also reported in other studies (Agrawal et al., 2011). Apart from thermal decomposition kinetics, studies also have revealed thermal stability at high temperatures in terms of mass loss. The mass loss was lesser for pure polymer films as compared to the nanocomposites (Wu et al., 2019). At higher thermal ranges, mass loss occurs due to depolymerization, and evaporation of water and volatile molecules. The preparation of nanocomposites results in lowering the mass loss, which is indicative of improved thermal stability (Safaei et al., 2019).

3.9 Environmental Impact

Though nanofiller-incorporated composite polymers can result in increasing barrier properties and provide higher shelf life, the increased production and consumption raise the possibilities of human and environmental exposure to the nanomaterials. The release of nanoparticles into the medium takes place either through passive diffusion, desorption, and further dissolution or through the partial degradation of these polymers (Duncan and Pillai, 2015). The entangled structures however get released only when the complete degradation of the polymers takes place (Kotsilkov et al., 2018). In a study, UV and O_3 degradation of the polymer nanocomposite films was studied at accelerated weathering conditions and it was reported that the release of nanoclay was taking place at a slower phase during the first 130 hours of degradation. This was observed to be increasing upon the complete degradation of the film. The presence of nanoclay is also reported to increase the UV-assisted oxidation in LDPE (Han et al., 2018). The release of the nanofillers from the polymer nanocomposites during their exposure to light, UV irradiation, moisture, mechanical force, aging, and other parameters. The carbon nanotube nanofiller causes potential cytotoxicity effects on human cells even during skin contact. On the other hand, upon oral exposure, the ingested nanoparticles also have the potential of being readily absorbed, and translocated among different organs and cause toxic effects. This requires further investigation to develop novel approaches for producing nanocomposites with a lower risk of potential toxicity when exposed to all of these factors.

3.10 Conclusion

The development of nanocomposites with polymers and other nanomaterials as reinforcement agents is a unique approach to improving the quality of food packaging systems. Reduced gas and water vapor transmission rates of the nanofiller-incorporated polymer nanocomposites indicate their suitability for food packaging with an enhanced shelf storage stability of various fresh and processed food systems. The reinforcement fillers commonly employed in the nanocomposites also result in property enhancements even at lower concentrations. This also reduces the weight of the nanocomposites while significantly improving their mechanical and thermal properties. Utilization of nanofillers of various dimensions for nanocomposite production is also associated with antimicrobial effects, oxidation control, ripening prevention via ethylene scavenging, etc. In addition, the lack of regulation for polymer nanocomposite films that are to be in food contact, the release and migration of nanofillers upon partial and complete degradation, and their interaction during weathering are to be explored to ensure their environmental impact as well. In addition, more studies are required on the optimization of production strategies, determining their potential for shelf-life extension, and investigating the feasibility of the nanocomposite production technologies for producing biocompatible nanocomposites for food applications.

References

Abral, H., Ariksa, J., Mahardika, M., Handayani, D., Aminah, I., Sandrawati, N., Sapuan, S. M., & Ilyas, R. A. (2020). Highly transparent and antimicrobial PVA based bionanocomposites reinforced by ginger nanofiber. *Polymer Testing*, *81*, 106186. https://doi.org/10.1016/J.POLYMERTESTING.2019.106186

Agrawal, S., Patidar, D., & Saxena, N. S. (2011). Glass transition temperature and thermal stability of ZnS/PMMA nanocomposites. *Phase Transitions*, *84*, 888–900. https://doi.org/10.1080/01411594.2011.563152

Allahbakhsh, A. (2017). High barrier graphene/polymer nanocomposite films. *Food packaging*, 699–737. https://doi.org/10.1016/B978-0-12-804302-8.00020-0

Aznar, M., Vera, P., Canellas, E., Nerín, C., Mercea, P., & Störmer, A. (2011). Composition of the adhesives used in food packaging multilayer materials and migration studies from packaging to food. *Journal of Materials Chemistry*, *21*, 4358–4370. https://doi.org/10.1039/C0JM04136J

Bousmina, M. (2006). Study of intercalation and exfoliation processes in polymer nanocomposites. *Macromolecules*, *39*, 4259–4263. https://doi.org/10.1021/MA052647F

Butkus, M. A., Edling, L., & Labare, M. P. (2003). The efficacy of silver as a bactericidal agent: advantages, limitations and considerations for future use. *Journal of Water Supply: Research and Technology–AQUA*, *52*, 407–416. https://doi.org/10.2166/AQUA.2003.0037

Carvalho, R. A., de Oliveira, A. C. S., Santos, T. A., Dias, M. V., Yoshida, M. I., & Borges, S. V. (2020). WPI and Cellulose Nanofibres Bio-nanocomposites: Effect of Thyme Essential Oil on the Morphological, Mechanical, Barrier and Optical Properties. *Journal of Polymers and the Environment*, *28*, 231–241. https://doi.org/10.1007/S10924-019-01598-6

Caseri, W. (2008). Inorganic nanoparticles as optically effective additives for polymers. *Chemical Engineering Communications*, *196*, 549–572. https://doi.org/10.1080/00986440802483954

Castillo, L. A., López, O. V., Ghilardi, J., Villar, M. A., Barbosa, S. E., & García, M. A. (2015). Thermoplastic starch/talc bionanocomposites. Influence of particle morphology on final properties. *Food Hydrocolloid*, *51*, 432–440. https://doi.org/10.1016/J.FOODHYD.2015.05.030

Cebeci, H., de Villoria, R. G., Hart, A. J., & Wardle, B. L. (2009). Multifunctional properties of high volume fraction aligned carbon nanotube polymer composites with controlled morphology. *Composites Science and Technology*, *69*, 2649–2656. https://doi.org/10.1016/J.COMPSCITECH.2009.08.006

Cesur, S., Köroğlu, C., & Yalçın, H. T. (2018). Antimicrobial and biodegradable food packaging applications of polycaprolactone/organo nanoclay/chitosan polymeric composite films. *Journal of Vinyl and Additive Technology*, *24*, 376–387. https://doi.org/10.1002/VNL.21607

Checchetto, R., Miotello, A., Nicolais, L., & Carotenuto, G. (2014). Gas transport through nanocomposite membrane composed by polyethylene with dispersed graphite nanoplatelets. *Journal of Membrane Science*, *463*, 196–204. https://doi.org/10.1016/J.MEMSCI.2014.03.065

Choi, R. N., Cheigh, C. I., Lee, S. Y., & Chung, M. S. (2011). Preparation and Properties of Polypropylene/Clay Nanocomposites for Food Packaging. *Journal of Food Science*, *76*, N62–N67. https://doi.org/10.1111/J.1750-3841.2011.02351.X

Chowdhury, S., Teoh, Y. L., Ong, K. M., Rafflisman Zaidi, N. S., & Mah, S. K. (2020). Poly(vinyl) alcohol crosslinked composite packaging film containing gold nanoparticles on shelf life extension of banana. *Food Packaging and Shelf Life*, *24*, 100463. https://doi.org/10.1016/J.FPSL.2020.100463

Cui, J., Yu, X., Shen, Y., Sun, B., Guo, W., Liu, M., Chen, Y., Wang, L., Zhou, X., Shafiq, M., Mo, X., 2023. Electrospinning Inorganic Nanomaterials to Fabricate Bionanocomposites for Soft and Hard Tissue Repair. *Nanomaterials*. 2023, Vol. 13, Page 204. https://doi.org/10.3390/NANO13010204

Daoush, W. M., & Hong, S. H. (2013). Synthesis of multi-walled carbon nanotube/silver nanocomposite powders by chemical reduction in aqueous solution. *Journal of Experimental Nanoscience*, *8*, 742–751. https://doi.org/10.1080/17458080.2011.604959

Dougnac, V. N., Alamillo, R., Peoples, B. C., & Quijada, R. (2010). Effect of particle diameter on the permeability of polypropylene/SiO2 nanocomposites. *Polymer*, *51*, 2918–2926. https://doi.org/10.1016/J.POLYMER.2010.02.014

Duan, Z., Thomas, N. L., & Huang, W. (2013). Water vapour permeability of poly(lactic acid) nanocomposites. *Journal of Membrane Science*, *445*, 112–118. https://doi.org/10.1016/J.MEMSCI.2013.06.008

Duncan, T. V., & Pillai, K. (2015). Release of engineered nanomaterials from polymer nanocomposites: Diffusion, dissolution, and desorption. *ACS Applied Materials & Interfaces*, *7*, 2–19. https://doi.org/10.1021/AM5062745

Erlantz, L., Ruiz-Rubio, L., Vilas, J. L., & León, L. M. (2017). Properties of UV protective films of poly(vinyl-chloride)/TiO2 nanocomposites for food packaging. *Polymer Bulletin*, *74*, 1387–1404. https://doi.org/10.1007/S00289-016-1782-4

Evgin, T., Turgut, A., Šlouf, M., Špitalsky, Z., Micusik, M., Sarikanat, M., Nogellova, Z., Novak, I., & Omastová, M. (2019). Morphological, electrical, mechanical and thermal properties of high-density polyethylene/multiwall carbon nanotube nanocomposites: effect of aspect ratio. *Materials Research Express*, *6*, 095079. DOI: 10.1088/2053-1591/AB11A6

Gill, Y. Q., Jin, J., & Song, M. (2020). Comparative study of carbon-based nanofillers for improving the properties of HDPE for potential applications in food tray packaging. *Polymers and Polymer composites*, *28*, 562–571. https://doi.org/10.1177/0967391119892091

Goh, K., Heising, J. K., Yuan, Y., Karahan, H. E., Wei, L., Zhai, S., Koh, J. X., Htin, N. M., Zhang, F., Wang, R., Fane, A. G., Dekker, M., Dehghani, F., & Chen, Y. (2016). Sandwich-Architectured Poly(lactic acid)-Graphene Composite Food Packaging Films. *ACS Applied Materials & Interfaces*, *8*, 9994–10004. https://doi.org/10.1021/ACSAMI.6B02498

Hajizadeh, H., Peighambardoust, S. J., Peighambardoust, S. H., & Peressini, D. (2020). Physical, mechanical, and antibacterial characteristics of bio-nanocomposite films loaded with Ag-modified SiO2 and TiO2 nanoparticles. *Journal of Food Science*, *85*, 1193–1202. https://doi.org/10.1111/1750-3841.15079

Han, C., Zhao, A., Varughese, E., & Sahle-Demessie, E. (2018). Evaluating weathering of food packaging polyethylene-nano-clay composites: Release of nanoparticles and their impacts. *NanoImpact*, *9*, 61–71. https://doi.org/10.1016/J.IMPACT.2017.10.005

Heydari-Majd, M., Ghanbarzadeh, B., Shahidi-Noghabi, M., Najafi, M. A., & Hosseini, M. (2019). A new active nanocomposite film based on PLA/ZnO nanoparticle/essential oils for the preservation of refrigerated Otolithes ruber fillets. *Food Packaging and Shelf Life*, *19*, 94–103. https://doi.org/10.1016/J.FPSL.2018.12.002

Hu, Q., Fang, Y., Yang, Y., Ma, N., & Zhao, L. (2011). Effect of nanocomposite-based packaging on postharvest quality of ethylene-treated kiwifruit (Actinidia deliciosa) during cold storage. *Food Research International*, *44*, 1589–1596. https://doi.org/10.1016/J.FOODRES.2011.04.018

Ibrahim, S., El-Naggar, M. E., Youssef, A. M., & Abdel-Aziz, M. S. (2020). Functionalization of Polystyrene Nanocomposite with Excellent Antimicrobial Efficiency for Food Packaging Application. *Journal of Cluster Science*, *31*, 1371–1382. https://doi.org/10.1007/S10876-019-01748-9

Idumah, C. I. (2023). Influence of interfacial engineering on properties of polymeric nanoarchitectures. *Polymer-Plastics Technology and Materials, 62*, 1844–1877. https://doi.org/10.1080/25740881.2023.2240393

Idumah, C. I., & Obele, C. M. (2021). Understanding interfacial influence on properties of polymer nanocomposites. *Surfaces and Interfaces, 22*, 100879. https://doi.org/10.1016/J.SURFIN.2020.100879

Jokar, M., Abdul Rahman, R., Ibrahim, N. A., Abdullah, L. C., & Tan, C. P. (2012). Melt Production and Antimicrobial Efficiency of Low-Density Polyethylene (LDPE)-Silver Nanocomposite Film. *Food and Bioprocess Technology, 5*, 719–728. https://doi.org/10.1007/S11947-010-0329-1

Kango, S., Kalia, S., Celli, A., Njuguna, J., Habibi, Y., & Kumar, R. (2013). Surface modification of inorganic nanoparticles for development of organic–inorganic nanocomposites—A review. *Progress in Polymer Science, 38*, 1232–1261. https://doi.org/10.1016/J.PROGPOLYMSCI.2013.02.003

Karaipekli, A., Biçer, A., Sarı, A., & Tyagi, V. V. (2017). Thermal characteristics of expanded perlite/paraffin composite phase change material with enhanced thermal conductivity using carbon nanotubes. *Energy Conversion and Management, 134*, 373–381. https://doi.org/10.1016/J.ENCONMAN.2016.12.053

Khalaj, M. J., Ahmadi, H., Lesankhosh, R., & Khalaj, G. (2016). Study of physical and mechanical properties of polypropylene nanocomposites for food packaging application: Nano-clay modified with iron nanoparticles. *Trends in Food Science and Technology, 51*, 41–48. https://doi.org/10.1016/J.TIFS.2016.03.007

Kim, G. M., Lach, R., Michler, G. H., & Chang, Y. W. (2005). The Mechanical Deformation Process of Electrospun Polymer Nanocomposite Fibers. *Macromolecular Rapid Communications, 26*, 728–733. https://doi.org/10.1002/MARC.200500023

Kotsilkov, S., Ivanov, E., Vitanov, N.K., 2018. Release of Graphene and Carbon Nanotubes from Biodegradable Poly(Lactic Acid) Films during Degradation and Combustion: Risk Associated with the End-of-Life of Nanocomposite Food Packaging Materials. *Materials*. 2018, Vol. 11, Page 2346. https://doi.org/10.3390/MA11122346

Lepoittevin, B., Pantoustier, N., Devalckenaere, M., Alexandre, M., Calberg, C., Jérôme, R., Henrist, C., Rulmont, A., & Dubois, P. (2003). Polymer/layered silicate nanocomposites by combined intercalative polymerization and melt intercalation: a masterbatch process. *Polymer, 44*, 2033–2040. https://doi.org/10.1016/S0032-3861(03)00076-4

Lin, S., Anwer, M. A. S., Zhou, Y., Sinha, A., Carson, L., & Naguib, H. E. (2018). Evaluation of the thermal, mechanical and dynamic mechanical characteristics of modified graphite nanoplatelets and graphene oxide high-density polyethylene composites. *Composites Part B: Engineering, 132*, 61–68. https://doi.org/10.1016/J.COMPOSITESB.2017.08.010

Lizundia, E., Ruiz-Rubio, L., Vilas, J. L., & Luis, M. L. (2016). Poly(l-lactide)/zno nanocomposites as efficient UV-shielding coatings for packaging applications. *Journal of Applied Polymer Science, 133*. https://doi.org/10.1002/APP.42426

Lomate, G. B., Dandi, B., & Mishra, S. (2018). Development of antimicrobial LDPE/Cu nanocomposite food packaging film for extended shelf life of peda. *Food Packaging and Shelf Life, 16*, 211–219. https://doi.org/10.1016/J.FPSL.2018.04.001

Lü, C., Cui, Z., Wang, Y., Li, Z., Guan, C., Yang, B., & Shen, J. (2003). Preparation and characterization of ZnS–polymer nanocomposite films with high refractive index. *Journal of Materials Chemistry, 13*, 2189–2195. https://doi.org/10.1039/B304154A

Majeed, K., Al Ali AlMaadeed, M., & Zagho, M. M. (2018). Comparison of the effect of carbon, halloysite and titania nanotubes on the mechanical and thermal properties of LDPE based nanocomposite films. *Chinese Journal of Chemical Engineering*, *26*, 428–435. https://doi.org/10.1016/J.CJCHE.2017.09.017

Malik, N. (2022). Thermally exfoliated graphene oxide reinforced polycaprolactone-based bactericidal nanocomposites for food packaging applications. *Materials Technology*, *37*, 345–354. https://doi.org/10.1080/10667857.2020.1842150

Mandal, S., Roy, D., Mukhopadhyay, K., Dwivedi, M., & Joshi, M. (2024). Mechanistic insight into the role of the aspect ratio of nanofillers in the gas barrier properties of polymer nanocomposite thin films. *RSC Applied Interfaces*. https://doi.org/10.1039/D3LF00214D

Manikandan, N. A., Pakshirajan, K., & Pugazhenthi, G. (2020). Preparation and characterization of environmentally safe and highly biodegradable microbial polyhydroxybutyrate (PHB) based graphene nanocomposites for potential food packaging applications. *International Journal of Biological Macromolecules*, *154*, 866–877. https://doi.org/10.1016/J.IJBIOMAC.2020.03.084

Manikantan, M. R., & Varadharaju, N. (2011). Preparation and properties of polypropylene-based nanocomposite films for food packaging. *Packaging Technology and Science*, *24*, 191–209. https://doi.org/10.1002/PTS.925

Mao, H. N., & Wang, X. G. (2020). Use of in-situ polymerization in the preparation of graphene / polymer nanocomposites. *New Carbon Materials*, *35*, 336–343. https://doi.org/10.1016/S1872-5805(20)60493-0

Martone, A., Faiella, G., Antonucci, V., Giordano, M., & Zarrelli, M. (2011). The effect of the aspect ratio of carbon nanotubes on their effective reinforcement modulus in an epoxy matrix. *Composites Science and Technology*, *71*, 1117–1123. https://doi.org/10.1016/J.COMPSCITECH.2011.04.002

Mathew, S., Snigdha, S., Mathew, J., & Radhakrishnan, E. K. (2019). Biodegradable and active nanocomposite pouches reinforced with silver nanoparticles for improved packaging of chicken sausages. *Food Packaging and Shelf Life*, *19*, 155–166. https://doi.org/10.1016/J.FPSL.2018.12.009

McNally, T., Pötschke, P., Halley, P., Murphy, M., Martin, D., Bell, S. E. J., Brennan, G. P., Bein, D., Lemoine, P., & Quinn, J. P. (2005). Polyethylene multiwalled carbon nanotube composites. *Polymer*, *46*, 8222–8232. https://doi.org/10.1016/J.POLYMER.2005.06.094

Mihindukulasuriya, S. D. F., & Lim, L. T. (2014). Nanotechnology development in food packaging: A review. *Trends in Food Science and Technology*, *40*, 149–167. https://doi.org/10.1016/J.TIFS.2014.09.009

Mittal, V., & Chaudhry, A. U. (2015). Polymer – graphene nanocomposites: effect of polymer matrix and filler amount on properties. *Macromolecular Materials and Engineering*, *300*, 510–521. https://doi.org/10.1002/MAME.201400392

Moustafa, H., Morsy, M., Ateia, M. A., & Abdel-Haleem, F. M. (2021). Ultrafast response humidity sensors based on polyvinyl chloride/graphene oxide nanocomposites for intelligent food packaging. *Sensors and Actuators A: Physical*, *331*, 112918. https://doi.org/10.1016/J.SNA.2021.112918

Mulla, M.Z., Rahman, M.R.T., Marcos, B., Tiwari, B., Pathania, S., 2021. Poly Lactic Acid (PLA) Nanocomposites: Effect of Inorganic Nanoparticles Reinforcement on Its Performance and

Food Packaging Applications. *Molecules*. 2021, Vol. 26, Page 1967. https://doi.org/10.3390/MOLECULES26071967

Murali, S., Kumar, S., Koh, J., Seena, S., Singh, P., Ramalho, A., & Sobral, A. J. F. N. (2019). Bio-based chitosan/gelatin/Ag@ZnO bionanocomposites: synthesis and mechanical and antibacterial properties. *Cellulose*, *26*, 5347–5361. https://doi.org/10.1007/S10570-019-02457-2

Namazi, H., Dadkhah, A., & Mosadegh, M. (2012). New Biopolymer Nanocomposite of Starch-Graft Polystyrene/Montmorillonite Clay Prepared Through Emulsion Polymerization Method. *Journal of Polymers and the Environment*, *20*, 794–800. https://doi.org/10.1007/S10924-012-0496-4

Omerović, N., Djisalov, M., Živojević, K., Mladenović, M., Vunduk, J., Milenković, I., Knežević, N., Gadjanski, I., & Vidić, J. (2021). Antimicrobial nanoparticles and biodegradable polymer composites for active food packaging applications. *Comprehensive Reviews in Food Science and Food Safety*, *20*, 2428–2454. https://doi.org/10.1111/1541-4337.12727

Potts, J. R., Dreyer, D. R., Bielawski, C. W., & Ruoff, R. S. (2011). Graphene-based polymer nanocomposites. *Polymer*, *52*, 5–25. https://doi.org/10.1016/J.POLYMER.2010.11.042

Ren, P. G., Liu, X. H., Ren, F., Zhong, G. J., Ji, X., & Xu, L. (2017). Biodegradable graphene oxide nanosheets/poly-(butylene adipate-co-terephthalate) nanocomposite film with enhanced gas and water vapor barrier properties. *Polymer Testing*, *58*, 173–180. https://doi.org/10.1016/J.POLYMERTESTING.2016.12.022

Rhim, J. W., Wang, L. F., Lee, Y., & Hong, S. I. (2014). Preparation and characterization of bio-nanocomposite films of agar and silver nanoparticles: Laser ablation method. *Carbohydrate Polymers*, *103*, 456–465. https://doi.org/10.1016/J.CARBPOL.2013.12.075

Rodrigues, C., de Mello, J. M. M., Dalcanton, F., Macuvele, D. L. P., Padoin, N., Fiori, M. A., Soares, C., & Riella, H. G. (2020). Mechanical, Thermal and Antimicrobial Properties of Chitosan-Based-Nanocomposite with Potential Applications for Food Packaging. *Journal of Polymers and the Environment*, *28*, 1216–1236. https://doi.org/10.1007/S10924-020-01678-Y

Rong, M. Z., Zhang, M. Q., & Ruan, W. H. (2006). Surface modification of nanoscale fillers for improving properties of polymer nanocomposites: a review. *Materials science and technology*, *22*, 787–796. https://doi.org/10.1179/174328406X101247

Safaei, M., Taran, M., & Imani, M. M. (2019). Preparation, structural characterization, thermal properties and antifungal activity of alginate-CuO bionanocomposite. *Materials Science and Engineering: C*, *101*, 323–329. https://doi.org/10.1016/J.MSEC.2019.03.108

Sahoo, N. G., Rana, S., Cho, J. W., Li, L., & Chan, S. H. (2010). Polymer nanocomposites based on functionalized carbon nanotubes. *Progress in Polymer Science*, *35*, 837–867. https://doi.org/10.1016/J.PROGPOLYMSCI.2010.03.002

Sahu, S. K., & Rama Sreekanth, P. S. (2022). Mechanical, thermal and rheological properties of thermoplastic polymer nanocomposite reinforced with nanodiamond, carbon nanotube and graphite nanoplatelets. *Advances in Materials and Processing Technologies*, *8*, 2086–2096. https://doi.org/10.1080/2374068X.2022.2034309

Sanchez-Garcia, M. D., Gimenez, E., & Lagaron, J. M. (2007). Novel PET Nanocomposites of Interest in Food Packaging Applications and Comparative Barrier Performance with Biopolyester Nanocomposites. *Journal of Plastic Film & Sheeting*, *23*, 133–148. https://doi.org/10.1177/8756087907083590

Schmatz, D. A., Costa, J. A. V., & de Morais, M. G. (2019). A novel nanocomposite for food packaging developed by electrospinning and electrospraying. *Food Packaging and Shelf Life*, *20*, 100314. https://doi.org/10.1016/J.FPSL.2019.100314

Seo, J., Jeon, G., Jang, E. S., Bahadar Khan, S., & Han, H. (2011). Preparation and properties of poly(propylene carbonate) and nanosized ZnO composite films for packaging applications. *Journal of Applied Polymer Science*, *122*, 1101–1108. https://doi.org/10.1002/APP.34248

Shankar, S., & Rhim, J. W. (2016). Polymer Nanocomposites for Food Packaging Applications. *Functional and physical properties of polymer nanocomposites*, 29–55. https://doi.org/10.1002/9781118542316.CH3

Sheng, S. K., Amin, K. A. M., Hong, L. L., Hassan, M. F., & Ismail, M. (2017). Investigation of Morphological, Structural and Electrical Properties of Cds/ PMMA Nanocomposite Film Prepared by Solution Casting Method. *International Journal of Electrochemical Science*, *12*, 10023–10031. https://doi.org/10.20964/2017.11.75

Shi, Y. D., Yu, H. O., Li, J., Tan, Y. J., Chen, Y. F., Wang, M., Wu, H., & Guo, S. (2018). Low magnetic field-induced alignment of nickel particles in segregated poly(l-lactide)/poly(ε-caprolactone)/multi-walled carbon nanotube nanocomposites: Towards remarkable and tunable conductive anisotropy. *Chemical engineering journal*, *347*, 472–482. https://doi.org/10.1016/J.CEJ.2018.04.147

Silva, M., Alves, N. M., & Paiva, M. C. (2018). Graphene-polymer nanocomposites for biomedical applications. *Polymers for Advanced Technologies*, *29*, 687–700. 10.1002/PAT.4164

Souza, V. G. L., Rodrigues, C., Ferreira, L., Pires, J. R. A., Duarte, M. P., Coelhoso, I., & Fernando, A. L. (2019). In vitro bioactivity of novel chitosan bionanocomposites incorporated with different essential oils. *Industrial Crops and Products*, *140*, 111563. https://doi.org/10.1016/J.INDCROP.2019.111563

Souza, V.G.L., Rodrigues, C., Valente, S., Pimenta, C., Pires, J.R.A., Alves, M.M., Santos, C.F., Coelhoso, I.M., Fernando, A.L., 2020. Eco-Friendly ZnO/Chitosan Bionanocomposites Films for Packaging of Fresh Poultry Meat. *Coatings* 2020, Vol. 10, Page 110. https://doi.org/10.3390/COATINGS10020110

Tang, Z., Fan, F., Chu, Z., Fan, C., Qin, Y., 2020. Barrier Properties and Characterizations of Poly(lactic Acid)/ZnO Nanocomposites. *Molecules*. 2020, Vol. 25, Page 1310. https://doi.org/10.3390/MOLECULES25061310

Tas, C. E., Hendessi, S., Baysal, M., Unal, S., Cebeci, F. C., Menceloglu, Y. Z., & Unal, H. (2017). Halloysite Nanotubes/Polyethylene Nanocomposites for Active Food Packaging Materials with Ethylene Scavenging and Gas Barrier Properties. *Food and Bioprocess Technology*, *10*, 789–798. https://doi.org/10.1007/S11947-017-1860-0

Tayfun, U., Kanbur, Y., Abacı, U., Güney, H. Y., & Bayramlı, E. (2016). Mechanical, electrical, and melt flow properties of polyurethane elastomer/surface-modified carbon nanotube composites. *Journal of Composite Materials*, *51*, 1987–1996. https://doi.org/10.1177/0021998316666158

Tornuk, F., Sagdic, O., Hancer, M., & Yetim, H. (2018). Development of LLDPE based active nanocomposite films with nanoclays impregnated with volatile compounds. *Food Research International*, *107*, 337–345. https://doi.org/10.1016/J.FOODRES.2018.02.036

Venkatesan, R., & Rajeswari, N. (2017). ZnO/PBAT nanocomposite films: Investigation on the mechanical and biological activity for food packaging. *Polymers for Advanced Technologies*, *28*, 20–27. https://doi.org/10.1002/PAT.3847

Wang, K., Wu, L., Li, Y., & Li, H. (2023). Preparation and characterization of chitosan/halloysite nanotubes composite film with ethylene scavenging and gas resistance for active food packaging. *Journal of Food Safety*, *43*, e13027. https://doi.org/10.1111/JFS.13027

Wen, Y. H., De Guzman, M. R., Lin, X., Chen, J. C., Gao, C., Wu, C. S., Wang, Z. H., Huang, D., Du, J., Yu, Y. Q., Zhao, W., Suen, M. C., & Tsou, C. H. (2020). Antibacterial Nanocomposites of Polypropylene Modified with Silver-Decorated Multiwalled Carbon Nanotubes. *Nano*, 15. https://doi.org/10.1142/S179329202050112X

Wen, Y. H., Tsou, C. H., de Guzman, M. R., Huang, D., Yu, Y. Q., Gao, C., Zhang, X. M., Du, J., Zheng, Y. T., Zhu, H., & Wang, Z. H. (2022). Antibacterial nanocomposite films of poly(vinyl alcohol) modified with zinc oxide-doped multiwalled carbon nanotubes as food packaging. *Polymer Bulletin*, *79*, 3847–3866. https://doi.org/10.1007/S00289-021-03666-1

Winey, K. I., & Vaia, R. A. (2007). Polymer Nanocomposites. *MRS Bulletin*, *32*, 314–322. https://doi.org/10.1557/MRS2007.229

Wu, C., Li, Y., Du, Y., Wang, L., Tong, C., Hu, Y., Pang, J., & Yan, Z. (2019). Preparation and characterization of konjac glucomannan-based bionanocomposite film for active food packaging. *Food Hydrocolloids*, *89*, 682–690. https://doi.org/10.1016/J.FOODHYD.2018.11.001

Wu, S., Peng, S., Wang, C.H., 2018. Multifunctional Polymer Nanocomposites Reinforced by Aligned Carbon Nanomaterials. *Polymers*. 2018, Vol. 10, Page 542. https://doi.org/10.3390/POLYM10050542

Yakdoumi, F. Z., Hadj-Hamou, A. S., Rahoui, N., Rahman, M. M., & Abetz, V. (2022). Polylactic acid nanocomposites containing functionalized multiwalled carbon nanotubes as antimicrobial packaging materials. *International Journal of Biological Macromolecules*, *213*, 55–69. https://doi.org/10.1016/J.IJBIOMAC.2022.05.142

Youssef, A. M., Abd El-Aziz, M. E., & Morsi, S. M. M. (2023). Development and evaluation of antimicrobial LDPE/TiO2 nanocomposites for food packaging applications. *Polymer Bulletin*, *80*, 5417–5431. https://doi.org/10.1007/S00289-022-04346-4

Zago, D. M., Jikan, S. S., Badarulzaman, N. A., Nuhu, A. H., & Bano, N. (2019). Investigation on structural, mechanical and melt flow properties of in situ polypropylene/kaolin nanocomposite. *AIP Conference Proceedings*, *2068*. https://doi.org/10.1063/1.5089403/823966

Zarshad, S., Naghib, S. M., Zare, Y., & Rhee, K. Y. (2022). A simple model for gas barrier performance of polymer nanocomposites considering filler alignment angle and diffusion direction. *Composites Science and Technology*, *230*, 109397. https://doi.org/10.1016/J.COMPSCITECH.2022.109397

Zehetmeyer, G., Soares, R. M. D., Brandelli, A., Mauler, R. S., & Oliveira, R. V. B. (2012). Evaluation of polypropylene/montmorillonite nanocomposites as food packaging material. *Polymer Bulletin*, *68*, 2199–2217. https://doi.org/10.1007/S00289-012-0722-1

Zhang, X., Xu, Y., Zhang, X., Wu, H., Shen, J., Chen, R., Xiong, Y., Li, J., & Guo, S. (2019). Progress on the layer-by-layer assembly of multilayered polymer composites: Strategy, structural control and applications. *Progress in Polymer Science*, *89*, 76–107. https://doi.org/10.1016/J.PROGPOLYMSCI.2018.10.002

Zhu, L., & Xanthos, M. (2004). Effects of process conditions and mixing protocols on structure of extruded polypropylene nanocomposites. *Journal of Applied Polymer Science*, *93*, 1891–1899. https://doi.org/10.1002/APP.20658

4

Inorganic and Metal Oxide Nanomaterials in Food Packaging

Akanksha Shetty, Rishab Subramaniam, and Sundus Nida

Department of Food Technology, Faculty of Engineering and Technology, JAIN (Deemed-to-be University), Ramanagara, Karnataka, India

4.1 Introduction

Food packaging is crucial in today's world because it safeguards food products when they are transported over large distances, preserves their quality, and offers consumers convenience and information (Garcia et al., 2018). The primary purpose of food packaging is to protect food products from unfavorable circumstances (such as spoilage caused by microorganisms, chemical contaminants, oxygen, moisture, light, and external forces) during storage and transportation as well as to increase their shelf life (Bumbudsanpharoke and Ko, 2015). Each stage of the food production process requires packaging, yet the primary problem with traditional food packaging materials is their permeable nature. There are no materials for packaging that completely withstand water vapor and air gases. To safeguard and monitor the quality of packaged foods, participants in the food supply chain also look for innovative, economical, ecologically friendly, and resourceful food packaging methods. This is made possible by a commitment to food safety, quality, and traceability. As a result, there are several important variables that drive the exploration of new food packaging materials. Researchers in the food industry have been encouraged to take advantage of the strong and green biodegradable nanocomposites, which present new opportunities and challenges for the development of nanomaterials in the food industry. These factors include rising demands for high food quality and safety as well as concerns over environmentally sustainable development. Nearly every area of science and technology has been transformed by nanotechnology, and copper nanoparticles can be utilized in materials that come into contact with food (Cabedo and Gamez-Pérez, 2018). Materials created with one or more exterior dimensions of less than 100 nm are known as nanomaterials. The nanoscale has a size range of 1–100 nm. Based on their properties, sizes, and structures, nanomaterials can be classified into various categories (Adeyeye, 2019) due to their large surface-area-to-volume ratio and distinctive chemical and physical properties, inorganic compounds with nanoscale present potent antibacterial activity at low doses. Additionally, some of

Nanotechnology for Sustainable Food Packaging, First Edition. Edited by C. Anandharamakrishnan, Jeyan A. Moses, and M. Maria Leena.

them are thought to be harmless and even include mineral elements that are necessary for humans. They are also more stable in harsh situations, such as high temperatures and pressures (Jagadish et al., 2018). Nanoparticles, nanofibers, and nanoplates are frequently used nano-packaged materials. In general, there are two types of nanomaterials used in food packaging: inorganic and organic compounds. Phenols, halogenated substances, quaternary ammonium salts, as well as recently highly valued natural materials like chitosan and chitin, are some examples of the first group. Metals, metal oxides, and metals linked to phosphates can be noted when it comes to the latter group. Regarding the inorganic agents, certain of the oxidized nanoparticles, such as TiO_2, ZnO, MgO, and CaO, have gained a lot of interest because they withstand the rigorous processing conditions and exert potent biocidal effects on pathogens that cause food-borne illness (Cabedo and Gamez-Pérez, 2018; Huang et al., 2018). The advantages of packaging with these nanoparticle-containing nanocomposites include a decrease in the need for preservatives and a faster rate of responses to limit microbiological development. Overall, there are three primary uses for nanocomposite materials in food packaging that have been reported in the last three years, i.e., improved, intelligent, and active packaging (Huang et al., 2018). Even after the package has been opened, nanoparticles could be used as carriers to add enzymes, antioxidants, antibrowning agents, tastes, and other bioactive elements to food. This would increase the food's shelf life (Adeyeye, 2019).

4.2 Types and Functions of Nanomaterials in Food Packaging

4.2.1 Titanium Dioxide Nanoparticles

TiO_2 research has been a major topic since Fujishima and Honda addressed its unique properties in 1972. Titanium oxide, also referred to as titanium (IV) oxide as well as titania, is the naturally occurring form of oxidized titanium. TiO_2 polymorphs include anatase, rutile, and brookite (Hoseinnejad et al., 2018). TiO_2 nanoparticles can be synthesized in rutile, anatase, or brookite forms, each with unique properties and reactivities based on their band gap (Garcia et al., 2018). Titanium dioxide nanoparticles (TiO_2-nanoparticles) are beneficial metal oxide nanomaterials with thermostability but also inert and are one of the most studied materials. The material can also alter the characteristics of biodegradable films. Furthermore, this material has numerous advantages, including being inexpensive, non-toxic, and photo-stable (Huang et al., 2018). The American Food and Drug Administration has approved the use of these nanoparticles in food, drugs, cosmetics, and surfaces in contact with foods (Hoseinnejad et al., 2018).

It is gaining popularity as a superior photocatalyst material for energy and environmental applications like water and air purification, antimicrobial, self-cleaning surfaces, and water splitting. Numerous research findings have demonstrated that TiO_2 nanoparticles can be utilized in a variety of fields due to their abundance, durability, and low cost; however, they are not classified as generally recognized safe (GRAS) materials by the Food and Drug Administration (FDA) and other organizations because they may end up causing genotoxicity as well as cytotoxicity to human cells in toxicology studies

(Hoseinnejad et al., 2018; Jagadish et al., 2018). However, the European Food Safety Authority (EFSA) stated in 2016 that more studies on TiO_2 safety are required. The EFSA has deemed TiO_2 unsafe to be used as a food additive even though they cannot rule out the possibility of TiO_2 genotoxicity or accumulation in the body after ingestion (Jagadish et al., 2018). Even so, TiO_2 nanoparticles are not prohibited from being used in the food industry. The most common TiO_2 synthesis method is sol–gel processing (Garcia et al., 2018).

TiO_2 colloids produced by the sol–gel method and blended with pectin to frame aerogels have shown promise for use in food packaging. When *Lactobacillus acidophilus, Saccharomyces cerevisiae*, and *Escherichia coli* were incubated with TiO_2/Pt nanoparticles during photoelectrochemical oxidation, their growth was completely inhibited (Jagadish et al., 2018). TiO_2, on the other hand, is thermodynamically unstable, agglomerates, and thus is hard to separate from a treated solution. Because TiO_2 photocatalyst will only be active at UV levels that are harmful to human cells, visible irradiation could solve this problem (Huang et al., 2018).

Different nanocomposites incorporated with TiO_2, along with their distinctive characteristics, are shown in Table 4.1 (Jafarzadeh et al., 2020). Doping TiO_2 or trying to form nanocomposites are two methods. Thus, the antibacterial property of visible-light-irradiated nitrogen as well as carbon-doped TiO_2 against several microbial was investigated, with

Table 4.1 Different nanocomposites incorporated with TiO_2, along with their distinctive characteristics

Host Polymer/TiO_2	Method of Preparation	Properties/Activity	Target Microbial Site	Reference
Chitosan	Solvent	Antioxidant ethylene scavenging antimicrobial pH-sensitive	*Escherichia coli* *Salmonella* *Staphylococcus aureus* *Listeria monocytogenes*	Zhang *et al.* (2017)
High density polyethylene (HDPE)	Blown film extrusion	Photocatalytic microbiological	*Pseudomonas* *E. coli* Lactic acid bacteria	Gumiero *et al.* (2013)
Guar gum	Solvent casting	UV, light, oxygen barrier	*L. monocytogenes, Salmonella enterica sv. Typhimurium*	Arfat *et al.* (2017)
Soybean polysaccharides	Solvent casting	Antimicrobial	*S. aureus* *Pseudomonas aeruginosa* *E. coli*	Salarbashi et al. (2018)
Wheat gluten/ nanocellulose	Solvent casting	Antimicrobial	*Saccharomyces cerevisiae* G -ve bacteria *E. coli* G +ve bacteria *S. aureus*	El-Wakil *et al.* (2015)
Pectin/TiO_2	Solgel	Insulation antimicrobial	*E. coli*	Nešić *et al.* (2018)

Source: Jafarzadeh et al. (2020)/with permission of Elsevier

nitrogen doping showing better bactericidal activity against microorganisms like *Listeria monocytogenes, Shigella flexneri, Vibrio parahaemolyticus, Staphylococcus aureus, Streptococcus pyogenes,* as well as *Acinetobacter baumannii* (Jagadish et al., 2018).

TiO_2 nanoparticles have photocatalytic antibacterial effects, producing reactive oxygen species (ROS) in the presence of UV light and thus killing bacteria through peroxidation in the cell surface and oxidation of DNA and proteins (Garcia et al., 2018). The anatase form has an elevated photocatalytic performance and can turn active after being exposed to near-UV light. It is also inexpensive and nontoxic. UV-A light exposure has the greatest antimicrobial effect. TiO_2's photocatalytic activity helps make it a possibly beneficial material for sanitizing food contact surfaces and decontaminating water.

Furthermore, this property might be useful in the development of antimicrobial active packaging. Such packaging can be made from a variety of polymeric matrices, including conventional plastics and biopolymers. TiO_2 nanoparticles (2–5 wt.%) were mixed with ethylene-vinyl alcohol to create self-sterilizing nanocomposites that were effective against Gram-positive as well as Gram-negative bacteria in both in vitro and actual packaging for cut lettuce, oriented-polypropylene (PP) films coated of TiO_2 nanoparticles found to inhibit the development of *E. coli* after exposure to black light. Under fluorescent light, low-density polyethylene (LDPE) films containing TiO_2 nanoparticles inhibited *Pseudomonas spp.* and yeasts, as well as the effect was improved by UV-A light exposure (Huang et al., 2018). Furthermore, TiO_2 incorporation into polymers produces nanocomposites that can prevent UV light and oxygen. UV light blocking is desirable because it not only protects the packaged food but also delays the aging as well as deterioration of the nanocomposite. TiO_2 nanotubes (0.05–0.1% w/v) were mixed with chitosan to create UV-blocking nanocomposite films that have been resistant to *E. coli, S. aureus,* and *Salmonella enterica* (Jagadish et al., 2018).

TiO_2 can also be used in a synergistic way to develop active packaging with enhanced properties; for example, a report described a high-density polyethylene (HDPE)-TiO_2-marigold extract nanocomposite for maintaining the stability of soybean oil through the antioxidant effect of the carotenoid component; because carotenoids degrade easily when exposed to light, the UV blocking effect of TiO_2 preserved their effectiveness.

Whey protein edible films with TiO_2 nanoparticles demonstrated excellent mechanical as well as moisture obstruction characteristics, along with antimicrobial influence. Other foods such as strawberries, cheese, and shrimp have already been mentioned to benefit from similar nanocomposite films in terms of decay and quality preservation. Aside from the improved mechanical as well as antibacterial properties of TiO_2-containing nanocomposites, a study found that such a nanomaterial could even prevent the movement of plastic additives such as ethylene glycol from packaging into food (Garcia et al., 2018).

4.2.2 ZnO Nanoparticles

Zinc oxide (ZnO) has sparked widespread interest due to its superior properties, particularly because of the successful development of nanomaterials. ZnO nanoparticles have a high surface-to-volume ratio, a highly crystalline structure, enhanced mechanical properties, good thermal conductivity, as well as excellent optical absorption in the UV range, which is advantageous for bacterial interactions (Jagadish et al., 2018).

Given their high luminescent effectiveness with a huge exciton binding energy (60 meV) as well as a wide band gap, significant research of ZnO—nanoparticles has indeed been stimulated just on generation of nanoparticles utilizing various methods of synthesis and on their future applications (3.36 eV) (Huang et al., 2018).

The FDA has classified ZnO as a GRAS material, permitting its use in the food and drug industries, particularly as an antifungal and antibacterial agent. ZnO nanoparticles are toxic to a wide range of bacteria. The hydrothermal method was used to create cluster-like ZnO nanoparticles that were grown on a Polydopamine (PDA)-polyethylene terephthalate (PET) substrate (Huang et al., 2018). The rise of *Gluconobacter cerinus* was slowed by destroying the bacterial cell membrane, while Ultraviolet protection capacity increased approximately 500-fold (Jagadish et al., 2018). This technique was also used to make ZnO nanorods. ZnO nanoparticles can be utilized as an antimicrobial substance because of their low cost, toxicity, and UV-blocking properties, which give them an advantage over silver nanoparticles (Hoseinnejad et al., 2018). ZnO nanoparticles are generally produced through mechanochemical processing, but they can also be produced through the sol–gel technique as well as spray pyrolysis. Like TiO_2, the antimicrobial effect of nano-ZnO is due to a photocatalytic reaction, but unlike TiO_2, it also releases ions (Zn^{+2}) that damage and kill bacterial cells, unlike TiO_2, and the antimicrobial activity does not need to be triggered by UV exposure (Garcia et al., 2018).

Because of their antimicrobial activity, these nanoparticles are of great interest, particularly given the increasing demand for alternative approaches generating novel antibiotics that can control disease outbreaks. ZnO is a type of zinc derivative that can be used alone or in combination with other compounds, such as gelatin, to keep foods from spoiling. Among metal oxides, ZnO inhibits the growth of microorganisms, including *E. coli, S. aureus, Salmonella spp, Campylobacter jejuni,* and *L. monocytogenes* (Hoseinnejad et al., 2018).

ZnO-coated polyvinyl chloride (PVC) films inhibited the growth of both *E. coli* and *S. aureus*, though the latter was more effective. Furthermore, nanocomposites containing nano-ZnO showed enhanced mechanical as well as oxygen barrier properties. Nano-ZnO (0.5 wt.%) was incorporated into HDPE and altered by the coupling agent KH550, likely to result in nanocomposite films with improved UV-blocking barrier, as well as mechanical characteristics, in addition to antimicrobial properties, particularly against *S. aureus* (Garcia et al., 2018).

Furthermore, research documented the development of antimicrobial glass by encasing with ZnO nanoparticles, achieving an 89% lowering in *E. coli* numbers (Huang et al., 2018). The development of biodegradable packaging is appealing because it would result in less waste and pollution while reducing reliance on oil; however, one drawback of bioplastics is their high water permeability. As a result, nano-ZnO has been added to bioplastics to create nanomaterials with enhanced properties (Garcia et al., 2018). In addition, a study found that ZnO nanocomposites are effective against yeasts and molds that spoil bread.

4.2.3 Silica Nanoparticles

Silicon is the largest plentiful solid element on the planet. It is commonly found in the forms of SiO_2 as well as silicate. Because of the increased surface area, SiO_2 exhibits excellent antimicrobial behavior at the nanoscale; additionally, these nanoparticles significantly reduce microbial cell attachment to surfaces as well as their proliferation (Huang et al., 2018). Antimicrobial behavior associated with silver-silica nanostructure is more effective compared to common materials, including silver zeolite as well as silver nitrate (Egger et al., 2009).

Nano-SiO_2 is added to plastics to improve mechanical properties as well as thermal stability Ethylene-vinyl alcohol (EVOH)-SiO_2 nanocomposites containing 5% nano-SiO_2 were found to have significantly improved water as well as gas barrier properties as well as mechanical strength when compared to the standard polymer. Nano-SiO_2 (0.09 wt.%) modified with ethylene/vinyl acetate (EVA) and blended with PP resulted in nanocomposites with increased tensile strength and decreased gas permeability, as well as nanocomposites that adsorbed fewer ink solvents than just the polymer composite, which could be useful for laminated food packaging because the PP layer is used for printing (Garcia et al., 2018).

Nitric oxide-releasing silica nanoparticles, when applied to biofilms of (*S. aureus, Pseudomonas aeruginosa, Staphylococcus epidermidis, E. coli,* as well as *Candida albicans*) show that nearly 99% decay of the microorganisms (Hetrick et al., 2009). Integration of $SiO_2/TiO_2/Ag$ nanoparticles integrated into LDPE produced a multifunctional nanocomposite capable of regulating CO_2 and O_2 levels, scavenging ethylene, and inhibiting microbial growth, thereby prolonging the shelf life of packaged mushrooms, implying that incorporating nanomaterials could be a great approach for nano-packaging development (Garcia et al., 2018).

Another method for creating SiO_2-containing nanocomposites is to coat polymers. The sol–gel process was used to create a PLA-based coating implementing nano-SiO_2 utilizing tetraethoxysilane as well as 3-isocyanatopropyltriethoxysilane as both a precursor as well as coupling agent, respectively; the membrane could be added to PLA films to create eco-friendly packaging with gas barrier properties approximately to 70% higher than pure PLA films while remaining transparent (Cabedo and Gamez-Pérez, 2018; Garcia et al., 2018). There have been few studies on SiO_2 nanocomposite applications. Nonetheless, one study found that LDPE-SiO_2 packaging for chilled white shrimp increased shelf life and kept the shrimp fresh for eight days, implying that this nanomaterial can also be embedded into polymers to create active packaging with antimicrobial as well as enzyme-inhibiting properties (Wang et al., 2013).

4.2.4 Nanoclay and Silicate Nanoparticles

At the moment, nanoclay has a global market value of approximately 70%, making the most commercial production applied nanomaterial. Clay minerals are classified according to their structures, chemistries, and sources. These materials are classified into four major classes based on their layered structures: chlorite, montmorillonite (MMT)/smectite, illite, and kaolinite. Among these clays, MMT is among the most frequently utilized in the production of nanocomposites. Because of its high surface reactivity as well as surface area, layered clay has the greatest acceptability. As a result, numerous studies have shown that natural biopolymer-layered silicate nanostructures drastically enhanced packaging properties (Huang et al., 2018). Despite this, organoclays as nanoparticles in food packaging have received less attention than other nano-encapsulation systems.

4.2.5 Silver-Based Nanoparticles

Silver nanoparticles are newly developed biocidal agents that can be used to reduce infection rates. They have been used in some commercial products due to their high thermal resistance, ease of use, as well as broad range of antimicrobial performance (Hoseinnejad et al., 2018). So far, silver-based nanoparticles have taken an interesting significance in all types of nanoparticles established and characterized because of their intrinsic characteristic of antimicrobial properties also in solid-state samples, and have thus been used as bacteriostatic agents since ancient times. Ag nanoparticles, silver nanoclusters (Ag NCs), and silver-based alloy materials are examples of silver-based nanoparticles (Huang et al., 2018). To date, some companies, such as DurethamVR and KodakVR, use silver nanoparticles in their own packaging polymer to increase shelf-life and inhibit the growth of microorganisms in perishable products, particularly foodstuff, despite their maximum release in food medium of around 0.01 mg/kg (Huang et al., 2018). The antimicrobial action of silver nanoparticles is primarily determined by their size, concentration, target microbe species, composition, and shape (Cabedo and Gamez-Pérez, 2018). Because of their exceedingly large specific surface area, these nanoparticles are more active in interacting with bacterial cell membranes. Also, because their sizes were much less than 10 nm, triangular as well as truncated morphological characteristics are known to be extremely reactive (Hoseinnejad et al., 2018). According to recent researchers, the antimicrobial activity of silver nanoparticles below 10 nm is primarily related to the nanoparticle itself; nevertheless, as the size is increased, silver ions become the dominant antimicrobial agent (Durán et al., 2016). In addition, silver nanoparticles can enter microbial cells more easily than silver ions. Silver nanoparticles cause cell membrane destruction as well as ROS reactions, which eventually damage the cells' DNA (Jagadish et al., 2018).

4.2.6 Other Nanoparticles

Other metal oxides, including Cu_2O nanoparticles, MgO nanoparticles, Fe_3O_4 nanoparticles, $FeMnO_3$, as well as α-Fe_2O_3 nanoparticles alone or in nanocomposites, have shown an increased potential to be utilized as antimicrobial agents used in food packaging. Nanocomposites consisting of metal-doped metal oxides as well as mixed metal oxides, such as Ag/ZnO/CuO in small amounts or CuO/montmorillonite nanocomposite fully integrated in chitosan film, have demonstrated high antimicrobial activity (Jagadish et al., 2018).

4.2.6.1 Copper Oxide Nanoparticles

Copper is appealing for active packaging because it has broad antimicrobial properties. Cu-nanoparticles as well as copper oxide nanoparticles, are the most common types of copper-based nanoparticles (CuO-nanoparticles) (Huang et al., 2018). According to most studies done on CuO nanoparticles, this substance is one of the most thoroughly researched metal oxide nanoparticles. Copper does have benefits for use as a biocidal agent because it is inexpensive and shares properties with noble metals like silver as well as gold. In terms of the steady advancement of nanotechnology, nanoparticles of copper oxide have sparked considerable interest due to their impact on a diverse range of microorganisms (Table 4.2) (Hoseinnejad et al., 2018).

Table 4.2 Antimicrobial activity of copper nanoparticles toward various microorganisms

Polymer/Cu-NPs	Antimicrobial Activity Toward	Reference
Low-density polyethylene (LDPE)	*Coliforms*	Beigmohammadi *et al.* (2016)
Polylactic acid	*Pseudomonas spp.*	Longano *et al.* (2012)
Sodium alginate and cellulose nanowhisker	*S. aureus* *E. coli* *Salmonella sp.* *Candida albicans Trichoderma sp*	Saravanakumar *et al.* (2020)
Alginate	*E. coli* *L. monocytogenes*	Roy and Rhim (2020)
Agar, alginate, carrageenan, chitosan, and CMC films	*E. coli* *L. monocytogenes*	Shankar and Rhim (2018)
Chitosan nanofiber	Gram-positive and Gram-negative bacteria	Almasi et al. (2018)

Copper ions operate by freely accepting as well as donating electrons, exhibiting high redox potential as well as the capacity to disrupt microbial cell components, ultimately killing them. After a month of refrigerated storage, nano CuO was added to LDPE to create a nanocomposite with antibacterial effects; the material was tested for cheese packaging and resulted in a decrease of 4.21 log CFU/g of coliforms (Garcia et al., 2018).

Because of its antimicrobial activity, this particle can be used to inhibit the development of bacteria, viruses, and fungi. Because of their large surface area, the nano-sized CuO-nanoparticles have been capable of interfacing with the cell membrane, resulting in an enhanced antimicrobial effect. As a result, CuO-nanoparticles' antibacterial activity has been broadly applied in the fields of materials for food packaging polymer nanocomposites and water purification (Huang et al., 2018).

4.2.6.2 Selenium Nanoparticles

Selenium is indeed a micronutrient that can be found in both human and animal bodies. The average level of selenium managed to be found in adults is around 10–20 mg, as well as the daily consumption is around 40 μg (with a maximum content of around 300 μg) per day. An excess of such a component (administration of 3200–6700 μg Se/d) leads to toxicity in humans, and the excess selenium load is excreted through urination. Because of its impressive biological effects, including antibacterial, antiviral, as well as antioxidant properties, selenium has found its way into the field of nanotechnology (Hoseinnejad et al., 2018).

4.2.6.3 Calcium Oxide Nanoparticles

Scallop shells contain calcium oxide (CaO), a natural antimicrobial agent. CaO consumption as a food ingredient is currently not regulated; however, EU regulations prohibit its use in certain products, including butter, honey, fermented milk, and so on (Hoseinnejad et al., 2018). Antimicrobial agents can also be made from CaO nanoparticles.

4.3 Inorganic and Metal Oxide Nanomaterials Applied in Food Packaging

It is possible to produce nanomaterials using a "top-down" or "bottom-up" strategy. Breaking down bigger particles into nano-sized materials is known as "top-down." The top-down methodology employs milling and homogenization as processes. However, "bottom-up" manufacturing involves assembling simpler molecules by processes like self-assembly, crystallization, layer-by-layer deposition, solvent extraction/evaporation, and biomass reactions (Joshi et al., 2019). When added to the polymer matrix as reinforcement, nanoparticles' reduced permeability to gases and water causes a longer, more complicated diffusion path, which improves the barrier qualities compared to the clean polymer. Additionally, the number of sites in the polymer chain that could interact with water molecules decreases as bonds between the polymer and the nanoparticles form. The molecular mobility and relaxation of the polymer chains are altered by the high aspect ratio and uniform dispersion of a nanomaterial in the polymer matrix, enhancing the material's mechanical and thermal resistance. Additionally, the strengthening of the nanocomposites may be influenced by the bonding between the polymer chains and nanoparticles (Adeyeye, 2019). Nano-sized materials are used extensively in the field of consumer goods due to the variety of their qualities. They enhance the functionality of packaging materials by giving them better barrier qualities, optical and recycling capabilities, greater flexibility, temperature and flame resistance, and durability (Anirudhan et al., 2018). Nano-sized materials are widely utilized in food packaging as antimicrobials to lower microbial deterioration of packaged foods. Additionally, the barrier properties of the packages are enhanced. The longer storage life of packaged food products is a result of the enhanced functionality of packaging materials. By incorporating nanoparticles into their polymer matrix, food packaging materials can improve their barrier performance and stop gases from entering and moving out of the food (Cushen et al., 2012; Rovera et al., 2020). Polymers include polyamides, nylons, polyolefins, ethylene-vinyl acetate copolymer, polystyrene, epoxy resins, polyurethane, polyvinyl chloride, and polyethylene terephthalate all contain nanosized components. Silver, gold, zinc oxide, silica, titanium dioxide, alumina, and iron oxides are among the frequently used metals and metal oxides in the nanoscale, along with carbon-based materials (Adeyeye, 2019).

In an active packaging system, to protect the product from unfavorable environmental conditions, nanomaterials are employed to interact directly with the fish and fish products or the surroundings of fish and fish products. Due to their characteristics, silver nanoparticles and coatings are used as antimicrobial materials. The usage of other nanomaterials as oxygen or ultraviolet scavengers is a possibility (Adeyeye, 2019). As intelligent packaging materials, nanomaterials are used to notify customers of microbiological contamination, chemical contamination, or the presence of gases indicating contamination (Cushen et al., 2012). The packaging polymers polyamides (PA), nylons (PO), polyolefins (PO), polystyrene (PS), epoxy resins (ER), polyurethane (PU), PVC, and PET all contain nanomaterials (PET). Nanomaterials are now employed in fish and the packaging of fish-related products. They are added to metals and metal oxides such as silver, gold, zinc oxide, silica, titanium dioxide, alumina, and iron oxides, among others. Chemical sensors are now being used by food firms, including Nestle, Cadbury, and Monoprix Supermarket, to detect color changes in their products (Adeyeye, 2019).

Due to their large surface-area-to-volume ratio and distinctive chemical and physical properties, inorganic compounds with nanoscale present potent antibacterial activity at low doses. Additionally, some of them are thought to be harmless and even include mineral elements that are necessary for humans. They are also more stable in extreme circumstances, such as high temperatures and pressures. Metal nanoparticles (MNPs) and metal oxide nanoparticles, such as Ag, Cu, TiO_2, and ZnO, make up the majority of inorganic antibacterial substances. ZnO is a food additive that is now accepted as generally recognized as safe material by the Food and Drug Administration. An inorganic substance called ZnO is frequently utilized in daily life. One viable alternative to several antibacterial techniques could be the application of metal and metal oxide nanoparticles. Chemical disinfectants, coating-based applications, and food preparation processes may all benefit from the antimicrobial nanoparticles (Jagadish et al., 2018). These nanoparticles can improve the tensile strength, gas barrier, and UV barrier properties of the nanocomposites in addition to being employed as photocatalysts with antibacterial and ethylene-scavenging properties (Figure 4.1) (Nikolic et al., 2021). Fresh products tend to lose water and are easily ruined by the action of bacteria and enzymes; thus, these characteristics would be helpful for packaging them (Garcia et al., 2018).

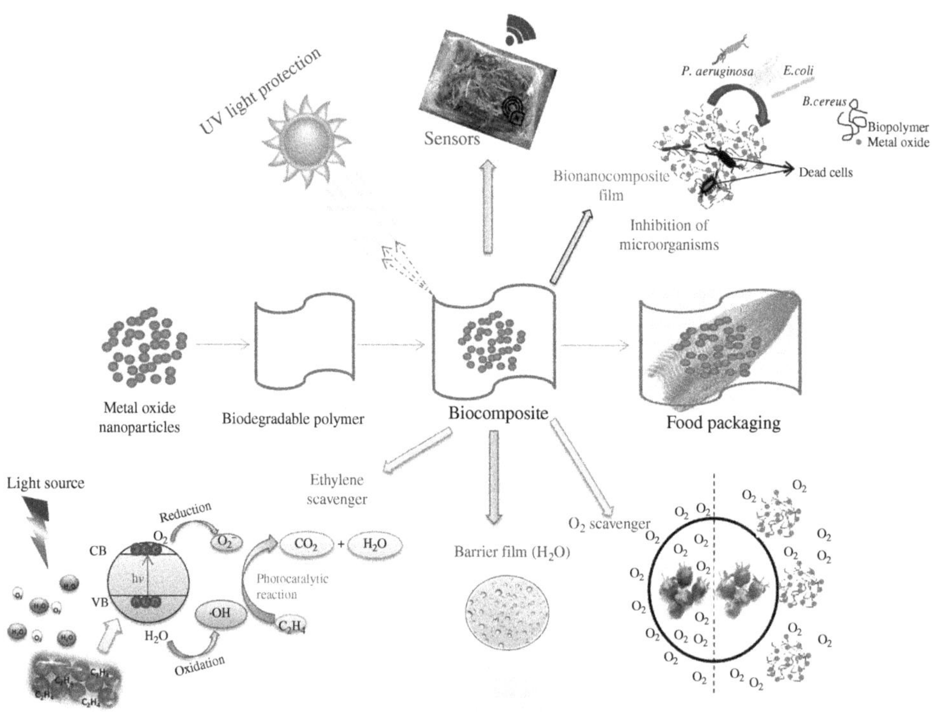

Figure 4.1 A visual illustration depicting various functions of metal oxide nanoparticles in smart food packaging (*Source:* Nikolic et al. (2021)/with permission of Elsevier)

4.3.1 Inorganic-Based Nanocomposites

4.3.1.1 Titanium Dioxide (TiO_2)

Because of its bright white coloring, TiO_2 is frequently utilized as a pigment in a variety of items, including foods and cosmetics. TiO_2 nanoparticles can be produced in rutile, anatase, or brookite form, each of which has unique characteristics and reactivities based on its distinctive band gap. TiO_2 has the potential to be a beneficial material for cleaning food contact surfaces and decontaminating water due to its photocatalytic activity. Additionally, the development of active packaging with antibacterial qualities may benefit from this capability. Such packaging can be created using a variety of polymeric matrices, including traditional plastics and biopolymers. EVOH was combined with TiO_2 nanoparticles (2–5 wt.%) to create nanocomposites with self-sterilizing capabilities that were efficient against both Gram-positive and Gram-negative bacteria. TiO_2 nanoparticle-infused edible whey protein sheets showed excellent mechanical properties, moisture barrier qualities and antimicrobial effects (Garcia et al., 2018).

4.3.1.2 Zinc Oxide (ZnO)

White powdery zinc oxide (ZnO) is frequently used in sunscreens and the food sector. Because of their economical rate, low toxicity, and UV-blocking capabilities, ZnO nanoparticles outperform silver nanoparticles as an antibacterial material. Similar to TiO_2, ZnO has been mixed into or coated over synthetic and biological polymers to create nanocomposites with antibacterial characteristics. Biodegradable packaging is developed due to less waste and eco-friendly. However, one drawback of bioplastics is their high water permeability. In order to create nanocomposites with better characteristics, nano-ZnO has also been added to bioplastics (Garcia et al., 2018).

4.3.1.3 Nanoclay

Durethan® KU2-2601 is a commercially available nanoclay composite made from engineered plastic polyamide 6 (nylon 6) and nano clay. Due to the clay nanoparticles' dispersion throughout the polymer matrix, which provides good features of gas and moisture barrier, strength, toughness, abrasion, and chemical resistance, it may be used in a variety of packaging fields, from common foodstuffs to medical industry (Bumbudsanpharoke and Ko, 2015).

4.3.1.4 Silver

Metallic silver atom clusters known as silver (Ag) nanoparticles were created primarily for antibacterial and sterilizing applications (Bumbudsanpharoke and Ko, 2015). Silver is one of the nanomaterials included in transition metals, having more electrical and thermal conductivity. It is produced through physical and chemical methodologies. Due to their bactericidal and catalytic properties, silver nanoparticles (Ag nanoparticles) are used as a potential food and beverage protector. For protecting food and beverages, Ag nanoparticles are used as antibacterial agents (Adeyeye, 2019). Silver is very toxic to a wide range of microbes (Honarvar et al., 2016). When included in food packaging, Numerous methods could be used to incorporate Ag nanoparticles into polymeric polymers for food packaging. Porous substances like zeolite may be used to incorporate or deposit silver ions, which are then applied to polymers (Adeyeye, 2019).

4.3.1.5 Copper

Copper nanoparticles (Cu nanoparticles) and copper oxide nanoparticles are the two primary types of copper-based nanoparticles (CuO-nanoparticles). Its key characteristic is antibacterial action, which enables this substance to be utilized to inhibit the growth of bacteria, viruses, and fungi. Due to their vast surface area, the nano-sized CuO-nanoparticles were able to interact with the cell membrane and thus displayed a stronger antibacterial activity. CuO-nanoparticles have seen extensive usage in the fields of chemical engineering, food science, and biomedicine. They are employed as gas sensors, catalysts, water purifiers, polymer reinforcing agents, semiconductors, magnetic storage devices, solar cell fields, emission devices, and other things (Huang et al., 2018). Due to their numerous toxic effects, copper and copper oxide nanoparticles can also prevent the growth of germs on polymer composites (Cabedo and Gamez-Pérez, 2018). Due to its low cost and similar qualities to noble metals like silver and gold, copper has some advantages when used as a biocidal agents. Due to their effects on a variety of microorganisms, copper oxide nanoparticles have garnered significant interest in relation to the steady advancement of nanotechnology (Hoseinnejad et al., 2018).

4.4 Properties of Inorganic and Metal/Metal Oxide Nanoparticles

The rise of antibiotic-resistant pathogens has evolved into a serious public health concern, prompting numerous studies to be published in order to improve current antimicrobial therapies. More than 70% of bacterial infections have been known to be resistant to at least one of the antibiotics commonly used to treat the infection. The advancement of new and efficient antimicrobial agents appears to be critical. Metals with antimicrobial activity, including Ag, Cu, Au, Ti, and Zn, each with different properties, potencies, and activity spectra, have been known and used for centuries (Cabedo and Gamez-Pérez, 2018).

Numerous kinds of oxide nanoparticles have gotten a lot of attention because of their potential antimicrobial properties. Antimicrobial activity has been found in metal oxide nanoparticles, including Ag_2O, Si, TiO_2, CuO, CaO, ZnO, as well as MgO (Wang et al., 2013). MNPs suppressed several microbial species in vitro, according to research. Furthermore, the morphological and physicochemical properties of nanometals have been shown to influence their antimicrobial activities. Small nanoparticles are known to have the most potent bactericidal effect (Cabedo and Gamez-Pérez, 2018). The positive surface charge of MNPs allows them to bind to the negatively charged surfaces of bacteria, potentially increasing their bactericidal effect.

4.4.1 Antiviral Activity of Metal Oxide Nanoparticles

One of the most important routes for virus spread is through contaminated surfaces. Some metal oxide nanoparticles' antiviral activity has sparked interest in the production of customer-protective packaging. CuO, ZnO, TiO_2, and $LaxMnO_3$ have demonstrated virucidal activity against enveloped viruses like Influenza A, yellow fever virus, and respiratory virus, as well as nonenveloped viruses like rhinovirus-2. Because surfaces coated with nanoparticles were more effective against enveloped viruses than nonenveloped viruses, it was hypothesized that the primary mechanism involved ROS generation. ROS effectively damages the outer lipid envelope while having a lesser impact on protein capsids. Another theory would be that metal oxide nanoparticles protect virus entry into human cells. Lately, ZnO nanoparticles have been discovered to target the SARS-CoV-2 ACE2 receptor, which is a crucial protein that allows the virus to enter host cells (Jagadish et al., 2018).

4.4.2 Antimicrobial Mechanisms of Metal Oxide Nanoparticles

Microbe contamination of food means reducing quality, shortening product shelf life, as well as raises the risk of food-borne illness to consumers while also being the leading cause of food waste. Because contamination occurs primarily on the surface of food that comes into contact with the package, the development of antimicrobial materials opens the door to new methods for dealing with the primary concerns of microbial contamination. Some metal and metal oxide nanomaterials or nanocomposite materials have already been utilized in the sector of food packaging due to their antimicrobial properties. It is possible to produce new and enhanced antimicrobial systems by regulating the size, morphological characteristics, surface, as well as interfacial properties of antimicrobial nanoparticles (Wang et al., 2013).

The antimicrobial activity of nanoparticles can be attributed to direct interaction with microbial cells (trying to disrupt the transmembrane transfer of electrons, trying to disrupt the cell envelope); oxidization of cell components; as well as production of secondary products (i.e., reactive oxygen compounds or disintegrated heavy metal ions), which causes cell damage (Figure 4.2) (Dash et al., 2022; Wang et al., 2013). Nanoparticles'

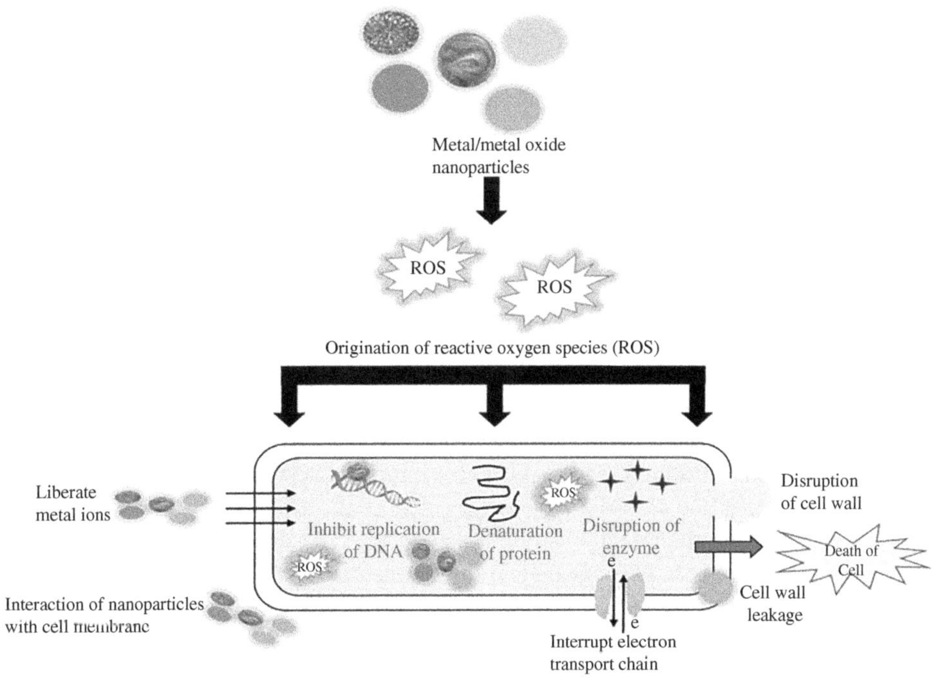

Figure 4.2 Mechanism of action of metal and metal oxide nanoparticles on bacterial cells (*Source:* Dash et al. (2022)/MDPI/CC BY 4.0)

antimicrobial activity can manifest as inhibition of growth, antimicrobial compounds, or antimicrobial as well as antifouling surfaces. The nanoparticles can be introduced into direct contact with the food for an immediate effect (e.g., via a package coating) or incorporated within the package's polymer, slowly migrating to the food over time (Cabedo and Gamez-Pérez, 2018).

The use of nanomaterials that have demonstrated effective antibacterial effectiveness in vitro in food packaging requires further verification since the food composition and structure may affect nanoparticle antibacterial activity. ZnO nanoparticles were shown to interact directly with *C. jejuni* cells, destabilize the membrane, and penetrate the bacterial cell, inducing oxidative stress. Higher concentrations and smaller nanoparticle sizes produce more ROS, resulting in increased antibacterial efficiency. For antibacterial activity measurements on plastics as well as other nonporous surfaces, the authorized ISO 22196:2011 is used (Jagadish et al., 2018). As previously stated, active nanoparticles in the polymer matrices are only those that migrate from film to product or those found on the film surface that come in contact with the food product.

4.4.3 Toxicity of Metal Oxide Nanoparticles

Human beings could be exposed to nanoparticles dissolute from food packaging film directly through food or indirectly through inhalation of particles. As a result, testing the significant cytotoxicity of nano-enforcers being used in active packaging is critical. Cell viability after exposure to nanoparticles in a buffer or cell culture medium has been used most commonly to assess nanoparticle cytotoxicity. Metal oxide nanoparticles have been shown in various mammalian cell lines to decrease cell viability, induce membrane lipid peroxidation, and damage DNA (Bumbudsanpharoke and Ko, 2015). The cytotoxic pattern differs depending on the metal oxide and cell type, and it is dose as well as time dependent. Smaller nanoparticles are generally more active and may be internalized by cells more quickly than larger ones (Jagadish et al., 2018).

The surface of nanoparticles greatly influences their biocompatibility. Based on their surface chemistry, ingested nanoparticles could either stimulate or suppress immune responses. The cytotoxicity of ZnO nanoparticles on human immune cells has been found to be related to the nanoparticles' intracellular solubility in Zn^{2+}-ions. Different anions have a significant impact on nanoparticle suspension stability and metal ion release from nanoparticles. TiO_2 and ZnO nanoparticles' pro-oxidative and pro-inflammatory effects were reduced by using a medium containing anion, including chloride as well as phosphate (Bumbudsanpharoke and Ko, 2015). Cytotoxicity was found to be concentration-dependent, as macrophages were able to neutralize the harmful effect of Mg-ZnO nanoparticles at concentration levels less than 1 mg/ml, whereas higher concentrations disrupted membranes and caused cell death (Jagadish et al., 2018).

Several studies indicate that nanoparticles such as Ag, TiO_2, and ZnO have an effect on the microbiota, as evidenced by a change in the *Firmicutes/Bacteroidetes* ratio, a decrease in the abundance of *Lactobacillus* strains, and a boost in the abundance of proteobacteria (Bumbudsanpharoke and Ko, 2015). Indeed, the negative effects of nanoparticles may resemble microbiome shifts seen in bowel disease, colorectal cancer, or obesity, where gut dysbiosis is a key pathogenic factor (Jagadish et al., 2018).

Furthermore, recent evidence suggests that ZnO nanoparticle-induced disruption of the microbiota-gut-brain axis could lead to neurobehavioral disorder by affecting gut microbiota (Wang et al., 2013). According to the extensive advancement of active packaging, test methods must be standardized as well as validated, both positive and negative limits must be identified, and cytotoxicity data must be harmonized (Jagadish et al., 2018). Indeed, there is insufficient information available about the safety risks of nanoparticles found in consumer products.

4.5 Legislation and Regulatory Aspects

A conference titled "Nanotechnologies in the Food and Agriculture Sectors: Potential Food Safety Implications" was organized in 2009 by the World Health Organization (WHO) and the Food and Agriculture Organization of the United Nations (FAO). In addition to Australia/New Zealand, Brazil, Canada, China, the European Union, Indonesia, Japan, Malaysia, Mexico, the Republic of Korea, South Africa, Switzerland, and the United States, experts from 13 other nations also participated in this meeting. They all expressed the same worries about the lack of knowledge regarding the impacts of nanomaterial uses on human health and demanded that their safety be given early consideration. The meeting gave attendees information on both established and new uses for nanotechnology, as well as what is known about their possible effects on food safety, potential dangers, and our current ability to evaluate those risks. It was anticipated that a medium-term strategy for additional work would be created to precisely identify and anticipate dangers. The meeting included "Nano-enabled food contact materials (FCMs) and packaging" as a topic. The FAO and the WHO expressed their intention to give national governments direction and counsel on food safety-related matters. After the meeting, the participants came to the consensus that nanotechnology offers significant advantages and potential for the creation of novel products for the treatment of water, agriculture, food processing, packaging, and preservation. The Codex Alimentarius Commission could be able to fill this gap in the definition of terminology used in the context of food in this worldwide discussion, according to FAO/WHO (2013). A draft paper titled "State of the art on the efforts and actions relevant to risk assessment and risk management of nanotechnologies in the food and agriculture sectors" was made available for comments by the FAO and the WHO in 2012 (Takeuchi et al., 2014). This paper's objective was to follow up with each nation on the advice of the prior expert meeting in 2009 regarding their approach to regulating food safety (Bumbudsanpharoke and Ko, 2015).

In the United States, the FDA has not yet defined "nanomaterial," "nanotechnology," or "nanoscale." The various draft guidance documents the Agency has issued to date on the topic reiterate the FDA's view that nanotechnology represents an "evolving state of the science" and that a single definition of nanotechnology and its related terms is not appropriate for regulatory purposes until more can be learned about the interaction of nanomaterials with biological systems.

The European Union (EU) has adopted a more rigid strategy. A nanomaterial, according to the European Commission (EC), is "a natural, incidental, or manufactured material containing particles, in an unbound state, as an aggregate, or as an agglomerate, and where, for 50% or more of the particles in the number size distribution, one or more external dimensions is in the size range 1–100 nm." A definition of nanotechnology has been adopted in Canada. The working definition used by this Agency includes particles with sizes between 1 and 100 nm, but it also includes substances with lesser or larger sizes if they have one or more nanoscale characteristics (Clark, 2012).

4.6 Applications in Food Systems

4.6.1 Intelligent Packaging—Application of Metal Oxide Nanoparticles in Food Safety Sensors

Intelligent packaging does not interact with food but instead observes the situation of the packaged food and informs on food quality degradation through the use of indicators (labels) and sensors, as well as enabling traceability through the use of unique codes as well as tags such as bar codes, Radio Frequency Identification (RFID) tags, smart tags, or Near Field Communication (NFC) tags (Bumbudsanpharoke and Ko, 2015). Time temperature, crispness but also gas leakage indicators, but also relative humidity sensors are among the environmental factors supervised within or outside the packaging. Freshness indicators, which are typically color-changing labels on the container/packaging, display the pH change or characteristic gases released throughout food spoilage as detected by sensors inside the packaging. Recent studies have also concentrated on multifunctional pH dependent color changing intelligent packaging made up of a biopolymer (chitosan, starch, etc.), metal oxide (ZnO, TiO_2), and pH-sensitive component (Jagadish et al., 2018).

TiO_2 nanoparticles are commonly used in UV-activated oxygen indicators. Advances in printed and flexible electronics, as well as the advancement of advanced bio-nanocomposite materials, have resulted in numerous advances in intelligent packaging (Garcia et al., 2018). Metal oxides were extensively studied and used as sensors to detect components for a wide variety of gases such as CO_2, NH_3, H_2S, and H_2O, but also dimethylamine as well as trimethylamine released throughout food spoilage (Jagadish et al., 2018). Recent research includes the development of a Ni-SnO_2 sensor to detect the presence of ethylene in apple fruit quality assessment that used a simple sol–gel spin coating method (Cabedo and Gamez-Pérez, 2018).

Niobium doping of TiO_2 nanotubes led to good selectivity and the ability to detect dimethylamine at low concentrations (5–50 ppm) (Jagadish et al., 2018). Gelatin-based nanocomposite films implementing ZnO nanoparticles demonstrated great prospects like a relative humidity sensing layer in food packaging at room temperature (Huang et al., 2018).

4.6.2 Applications of Inorganic Nano-Additives, as a Function of Their Structure

4.6.2.1 Titanium Dioxide Nanoparticles

The utilization of iso-dimensional nanoparticles as nano additives is primarily targeted at interacting with food or providing new functionalities to the package rather than modifying the real characteristics of the polymeric package (Wang et al., 2013). As a result, nano additives have typically been utilized to enhance the package with active or intelligent characteristics, including antimicrobial and optical activity, agent encapsulation and controlled release, and oxygen scavengers. As a result, the majority of inorganic isodimensional nanoparticles are metals or oxides. Some metals have been shown to be efficient as antimicrobials and otherwise oxygen scavengers, whereas oxides have demonstrated an effective UV radiation-blocking effect as well as an antimicrobial effect (Wang et al., 2013). There is also polyhedral oligomeric silsesquioxane (a special organic-inorganic hybrid nano additive) that was documented as a reinforcing agent for applications such as packaging; however, its use in food packaging applications is limited to a few scientific papers.

4.6.2.2 1D Nano-Additives

Incorporating nanofibers into a polymer matrix is an efficient method for enhancing mechanical characteristics (in aspects of stiffness as well as stress to failure) without sacrificing final material transparency (Wang et al., 2013). Carbon nanotubes (CNTs), as well as tubular (halloysite) as well as fibrillary (sepiolite/palygorskite) nanoclays are the most commonly used nanofibers in food packaging nanocomposites (Liao et al., 2019).

4.6.2.3 Nanolayers

The most widely used type of nanocomposites are those that use nanolayers as nano additives. The great feature predicted from moreover of 2D nano additives is a rise in barrier properties in addition to the strength properties common to nanofibers (Wang et al., 2013). Because of this, layered nanocomposites are among the most exciting for the food packaging industry. Different layered nanoparticles have been used as nanoparticles in polymer layered nanomaterials in recent years; however, the clay minerals (so-called nanoclays) as nanoparticles have been the most studied extensively and technologically relevant due to its abundance, reduced margin, major property improvements accomplished, and relatively simple processability (Huang et al., 2018; Wang et al., 2013). In terms of popularity and volume of examples, the carbon-based nanolayers graphene and graphene oxide rank second. The synthetic layered clays (including laponite or synthetic mica), as well as layered double hydroxides, are the most intriguing (LDH) (Huang et al., 2018).

4.7 Conclusion and Future Trends

Recent advances in science and technology have pushed the use of nanomaterials in food packaging due to their potential in the food business. Due to an increase in food productivity and a decrease in food waste and losses, the use of nanomaterials in food packaging will give long-term economic benefits to nations worldwide. These benefits include solutions to a number of issues with food packaging, storage, and distribution. Additionally, this will facilitate the distribution of wiser, healthier food to the people. Inorganic nanoparticles are known to be among the most widely employed antibacterial materials in several industries, particularly the food packaging industry. It has been shown that the antibacterial activity of the mentioned inorganic nanoparticles can protect against a wide range of microorganisms, including food-borne diseases. Since they have a high efficacy at extremely low levels and may be suitable substitutes for typical chemical antibacterial materials, certain antimicrobial agents, such as titanium oxide, silver, zinc oxide, copper, gold, selenium, etc., are proper agents. Additionally, nanotechnology offers a variety of options for affordable, environmentally friendly, biodegradable, and renewable packaging materials, which are gaining popularity and acceptance as a means of addressing the problems of environmental pollution and food shortages by ensuring that food is available to everyone. However, the food sector is still developing applications of nanomaterials as effective nanocomponents of food packaging. Researchers from all over the world are focusing on "Smart Packaging." Research is being done to create polymer nanocomposite food packaging films as well as antigen-specific biomarkers suited for food packaging. Given the prominent applications of nanotechnology, there are numerous complex considerations regarding whether and how nanomaterials should be regulated. This intricacy is clearly visible in regulatory agencies' attempts to develop practical definitions for nanomaterials, a job that has proven challenging due to the wide range of materials that these definitions may possibly cover. Without clear direction at the outset on figuring out whether a substance qualifies as a "nanomaterial" for regulatory purposes, adhering to the regulatory criteria applicable to nanomaterials internationally would be more difficult. In conclusion, if properly controlled, nanomaterials in food will have a significant positive impact on sustainability and also have positive health and environmental effects.

References

Adeyeye, S. A. O. (2019). Food packaging and nanotechnology: safeguarding consumer health and safety. *Nutrition and Food Science*, *49*(6), 1164–1179. https://doi.org/10.1108/NFS-01-2019-0020

Almasi, H., Jafarzadeh, P., & Mehryar, L. (2018). Fabrication of novel nanohybrids by impregnation of CuO nanoparticles into bacterial cellulose and chitosan nanofibers: Characterization, antimicrobial and release properties. *Carbohydrate Polymers*, *186*, 273–281. https://doi.org/10.1016/j.carbpol.2018.01.067

Anirudhan, T. S., Athira, V. S., & Chithra Sekhar, V. (2018). Electrochemical sensing and nano molar level detection of Bisphenol-A with molecularly imprinted polymer tailored on multiwalled carbon nanotubes. *Polymer*, *146*, 312–320. https://doi.org/10.1016/j.polymer.2018.05.052

Arfat, Y. A., Ejaz, M., Jacob, H.,x & Ahmed, J. (2017). Deciphering the potential of guar gum/Ag-Cu nanocomposite films as an active food packaging material. *Carbohydrate Polymers*, *157*, 65–71. https://doi.org/10.1016/J.CARBPOL.2016.09.069

Beigmohammadi, F., Peighambardoust, S. H., Hesari, J., Azadmard-Damirchi, S., Peighambardoust, S. J., & Khosrowshahi, N. K. (2016). Antibacterial properties of LDPE nanocomposite films in packaging of UF cheese. *LWT - Food Science and Technology*, *65*, 106–111. https://doi.org/10.1016/j.lwt.2015.07.059

Bumbudsanpharoke, N., & Ko, S. (2015). Nano-food packaging: an overview of market, migration research, and safety regulations. *Journal of Food Science*, *80*(5), R910–R923. https://doi.org/10.1111/1750-3841.12861

Cabedo, L., & Gamez-Pérez, J. (2018). Inorganic-based Nanostructures and Their Use in Food Packaging. In M. Â. P. R. Cerqueira, J. M. Lagaron, & A. A. M. de Oliveira Soares Vicente (Eds.), *Nanomaterials for Food Packaging: Materials, Processing Technologies, and Safety Issues* (pp. 13–45), Elsevier. https://doi.org/10.1016/B978-0-323-51271-8.00002-4

Clark, M. N. (2012). Nanotechnology Regulation of Food Packaging. *Packaging Law.*

Cushen, M., Kerry, J., Morris, M., Cruz-Romero, M., & Cummins, E. (2012). Nanotechnologies in the food industry—recent developments, risks and regulation. *Trends in Food Science & Technology, 24*(1), 30–46. https://doi.org/10.1016/j.tifs.2011.10.006

Dash, K. K., Deka, P., Bangar, S. P., Chaudhary, V., Trif, M., & Rusu, A. (2022). Applications of inorganic nanoparticles in food packaging: a comprehensive review. *Polymers, 14*(3), 521.

Durán, N., Durán, M., de Jesus, M. B., Seabra, A. B., Fávaro, W. J., & Nakazato, G. (2016). Silver nanoparticles: a new view on mechanistic aspects on antimicrobial activity. *Nanomedicine: Nanotechnology, Biology, and Medicine, 12*(3), 789–799. https://doi.org/10.1016/j.nano.2015.11.016

Egger, S., Lehmann, R. P., Height, M. J., Loessner, M. J., & Schuppler, M. (2009). Antimicrobial properties of a novel silver-silica nanocomposite material. *Applied and Environmental Microbiology, 75*(9), 2973–2976. https://doi.org/10.1128/AEM.01658-08

El-Wakil, N. A., Hassan, E. A., Abou-Zeid, R. E., & Dufresne, A. (2015). Development of wheat gluten/nanocellulose/titanium dioxide nanocomposites for active food packaging. *Carbohydrate Polymers, 124*, 337–346. https://doi.org/10.1016/J.CARBPOL.2015.01.076

Garcia, C. V., Shin, G. H., & Kim, J. T. (2018). Metal oxide-based nanocomposites in food packaging: applications, migration, and regulations. *Trends in Food Science & Technology, 82*, 21–31. https://doi.org/10.1016/j.tifs.2018.09.021

Gumiero, M., Peressini, D., Pizzariello, A., Sensidoni, A., Iacumin, L., Comi, G., & Toniolo, R. (2013). Effect of TiO_2 photocatalytic activity in a HDPE-based food packaging on the structural and microbiological stability of a short-ripened cheese. *Food Chemistry, 138*, 1633–1640. https://doi.org/10.1016/J.FOODCHEM.2012.10.139

Hetrick, E. M., Shin, J. H., Paul, H. S., & Schoenfisch, M. H. (2009). Anti-biofilm efficacy of nitric oxide-releasing silica nanoparticles. *Biomaterials, 30*(14), 2782–2789. https://doi.org/10.1016/j.biomaterials.2009.01.052

Honarvar, Z., Hadian, Z., & Mashayekh, M. (2016). Nanocomposites in food packaging applications and their risk assessment for health. *Electronic Physician, 8*(6), 2531–2538. https://doi.org/10.19082/2531

Hoseinnejad, M., Jafari, S. M., & Katouzian, I. (2018). Inorganic and metal nanoparticles and their antimicrobial activity in food packaging applications. *Critical Reviews in Microbiology, 44*(2), 161–181.

Huang, Y., Mei, L., Chen, X., & Wang, Q. (2018). Recent developments in food packaging based on nanomaterials. *Nanomaterials, 8*(10), 1–29. https://doi.org/10.3390/nano8100830

Jafarzadeh, S., Salehabadi, A., & Jafari, S. M. (2020). Metal Nanoparticles as Antimicrobial Agents in Food Packaging. In S. M. Jafari (Ed.), *Handbook of Food Nanotechnology: Applications and Approaches*. INC. https://doi.org/10.1016/B978-0-12-815866-1.00010-8

Jagadish, K., Shiralgi, Y., Chandrashekar, B. N., Dhananjaya, B. L., & Srikantaswamy, S. (2018). Ecofriendly Synthesis of Metal/Metal Oxide Nanoparticles and Their Application in Food Packaging and Food Preservation. In: *Impact of Nanoscience in the Food Industry*. Edited by Alexandru Mihai Grumezescu and Alina Maria Holban, Academic Press, 197–216. https://doi.org/10.1016/B978-0-12-811441-4.00008-X

Joshi, K., Chandra, A., Jain, K., & Talegaonkar, S. (2019). Nanocrystalization: an emerging technology to enhance the bioavailability of poorly soluble drugs. *Pharmaceutical Nanotechnology, 7*(4), 259–278. https://doi.org/10.2174/2211738507666190405182524

Liao, Y., Zhang, R., & Qian, J. (2019). Printed electronics based on inorganic conductive nanomaterials and their applications in intelligent food packaging. *RSC Advances*, *9*(50), 29154–29172. https://doi.org/10.1039/c9ra05954g

Longano, D., Ditaranto, N., Cioffi, N. et al. (2012). Analytical characterization of laser-generated copper nanoparticles for antibacterial composite food packaging. *Analytical and Bioanalytical Chemistry*, *403*, 1179–1186. https://doi.org/10.1007/s00216-011-5689-5

Nesić, A., Gordić, M., Davidović, S., Radovanović, Z., Nedeljković, J., Smirnova, I., & Gurikov, P. (2018). Pectin-based nanocomposite aerogels for potential insulated food packaging application. *Carbohydrate Polymers*, *195*, 128–135. https://doi.org/10.1016/J.CARBPOL.2018.04.076

Nikolic, M. V., Vasiljevic, Z. Z., Auger, S., & Vidic, J. (2021). Metal oxide nanoparticles for safe active and intelligent food packaging. *Trends in Food Science & Technology*, *116*, 655–668. https://doi.org/10.1016/j.tifs.2021.08.019

Rovera, C., Ghaani, M., & Farris, S. (2020). Nano-inspired oxygen barrier coatings for food packaging applications: an overview. *Trends in Food Science and Technology*, *97*, 210–220. https://doi.org/10.1016/j.tifs.2020.01.024

Roy, S., & Rhim, J. W. (2020). Effect of CuS reinforcement on the mechanical, water vapor barrier, UV-light barrier, and antibacterial properties of alginate-based composite films. *International Journal of Biological Macromolecules*, *164*, 37–44. https://doi.org/10.1016/j.ijbiomac.2020.07.092

Salarbashi, D., Tafaghodi, M. & Bazzaz, B.S.F. (2018). Soluble soybean polysaccharide/TiO_2 bionanocomposite film for food application. *Carbohydrate Polymers*, *186*, 384–393. https://doi.org/https://doi.org/10.1016/J.CARBPOL.2017.12.081

Saravanakumar, K., Sathiyaseelan, A., Mariadoss, A. V. A., Xiaowen, H., & Wang, M. H. (2020). Physical and bioactivities of biopolymeric films incorporated with cellulose, sodium alginate and copper oxide nanoparticles for food packaging application. *International Journal of Biological Macromolecules*, *153*, 207–214. https://doi.org/10.1016/j.ijbiomac.2020.02.250

Shankar, S., & Rhim, J. W. (2018). Antimicrobial wrapping paper coated with a ternary blend of carbohydrates (alginate, carboxymethyl cellulose, carrageenan) and grapefruit seed extract. *Carbohydrate Polymers*, *196*, 92–101. https://doi.org/10.1016/j.carbpol.2018.04.128

Takeuchi, M. T., Kojima, M., & Luetzow, M. (2014). State of the art on the initiatives and activities relevant to risk assessment and risk management of nanotechnologies in the food and agriculture sectors. *Food Research International*, *64*, 976–981. https://doi.org/10.1016/j.foodres.2014.03.022

Wang, H., Du, L. J., Song, Z. M., & Chen, X. X. (2013). Progress in the characterization and safety evaluation of engineered inorganic nanomaterials in food. *Nanomedicine*, *8*(12), 2007–2025. https://doi.org/10.2217/nnm.13.176

Zhang, X., Xiao, G., Wang, Y., Zhao, Y., Su, H., & Tan, T. (2017). Preparation of chitosan-TiO_2 composite film with efficient antimicrobial activities under visible light for food packaging applications. *Carbohydrate Polymers*, *169*, 101–107. https://doi.org/10.1016/j.carbpol.2017.03.073

5

Edible Coatings: Concept, Applications and Toxicological Aspects

Luana S. C. Carnaval[1,2,*], *Andrelina Maria Pinheiro Santos*[3,*], *Betty Del Carmen Jarma Arroyo*[4,*], *Enayde de Almeida Melo*[4,*], *Amit K. Jaiswal*[1,2], *and Swarna Jaiswal*[1,2]

[1] *Centre for Sustainable Packaging and Bioproducts (CSPB), School of Food Science and Environmental Health, Technological University Dublin, Dublin, Ireland*
[2] *Sustainability and Health Research Institute (SHRI), Technological University Dublin, Dublin, Ireland*
[3] *Department of Chemical Engineering, Bioprocess Laboratory, Universidade Federal de Pernambuco, Recife, Brazil*
[4] *Consumer Science Department, Universidade Federal Rural de Pernambuco, Recife, Brazil*

5.1 Introduction

Edible coatings have shown to be a very promising technology in terms of increasing food shelf life, extending of postharvest life of various fresh produce, maintaining or improving the quality of food products, meeting consumer demand for safe/healthy foods, reliability and convenience, besides meets sustainability by reducing the use of nonbiodegradable packaging materials. This type of coating has been applied especially to fruit due to its ability to minimize moisture loss, respiration rates, ethylene production, and ripening. In addition, some of these materials make fruits more attractive to the consumer by adding shine to them. Edible coatings have also been studied for other food products, especially as carriers of food additives such as antimicrobial agents, antioxidants, antibrowning agents, flavorings, dyes, and nutrients (Kumarihami et al., 2022).

These materials can be prepared in two ways: as coatings applied directly on the food surface and then representing an edible primary packaging, or as films being molded first into a sheet and then used as wrapping materials. Depending on the method of preparation, the raw materials or the desired function, edible films, and coatings find different applications such as individual packaging, at an interface between foods, as an inner layer of another packaging, and as a carrier for functional compounds. Although the edible film/coating concept is not new, current research in this field has focused on developing active packaging that provides extra functionality beyond mechanical and barrier protection (Maan et al., 2021).

---

* All authors contributed equally

Nanotechnology for Sustainable Food Packaging, First Edition. Edited by C. Anandharamakrishnan, Jeyan A. Moses, and M. Maria Leena.

Different active coating systems offer extra protection to foods, being developed with the application of oxygen and moisture scavengers, CO_2 emitters, antioxidant and antimicrobial agents, and others. Furthermore, the emerging use of nanotechnology, such as nanoparticles, has received particular interest in the food-packaging field due to their high specific surface area, providing better reinforcement and important active properties to the matrix (Jafarzadeh et al., 2021a). For example, Arroyo et al. (2020) developed an active edible coating based on alginate and chitosan added with ZnO nanoparticles, and this coating was able to extend the shelf life of guavas up to 20 days compared to 7 days for uncoated ones. Resende et al. (2018) developed edible coatings based on chitosan/cellulose nanofibril and identified that high concentrations of nanofibril improved the postharvest quality of strawberry fruits by minimizing oxygen diffusion, decreasing respiration, and delaying oxidation.

This chapter will give an overview of edible coating; raw materials/base materials used in edible coatings matrices, methods of coating; essential properties of bio-nanocomposite coatings; active and intelligent coatings; coatings as substance releasers or carriers and their application; toxicological, safety, and migration aspects; besides future trends, challenges, and potential commercial applications. All these topics will highlight nanotechnological approaches applied to them.

5.2 Edible Coating Concept, Nanotechnology, and Raw Materials Applied to These Matrices

Edible films and coatings are described as thin-layered consumable materials that work as barriers against moisture, gases, mechanical stress, etc. (Maan et al., 2021; Maringgal et al., 2020). In addition to outstanding features such as biocompatibility, edibility, and a wide range of applications, these materials can ensure the quality and safety of products by antimicrobial and antioxidant actions preventing oxidative deterioration, contamination, and spoilage, besides facilitating transportation, storage, and display (Maan et al., 2021). These edible materials are one of the most studied outlooks to enhance food quality and shelf life. Depending on their application, they have to fulfill numerous requirements, such as balanced gas permeability properties for a normal CO_2/O_2 exchange in fruits and vegetables, limited water vapor permeability to inhibit moisture escape in fresh commodities, and antimicrobial activity, especially for fresh and fresh-cut food. In addition to variable adhesion requests, for example, fruits and vegetables with naturally waxed hydrophobic surfaces or fresh-cut fruits and vegetables with hydrophilic hard-to-adhere surfaces (Arnon-Rips and Poverenov, 2018).

Nanocomposites have been widely explored as new alternatives for coating materials, and the nature of the polymeric materials and nanoparticles governs the final properties of the nanocomposites. Nanoscaled particles in nanocomposite technologies have achieved great interest in both academia and industries due to their unique physical, chemical, optical, antimicrobial, and biological properties, bringing relevant alternatives for the enhancement of biopolymer properties (Jafarzadeh et al., 2021a). The recent application of nanotechnology for coatings has demonstrated its efficacy in enhancing food quality and storage life. Gardesh et al. (2016) found that the climacteric respiration peak was significantly suppressed by 0.5% nano-chitosan coatings, which were able to slow down the natural ripening process, too.

In addition to the numerous methods used to obtain edible coatings, there is also a wide variety of raw materials used to form the matrices of these materials (Arnon-Rips and Poverenov, 2018). Edible films/coatings can be prepared based on polysaccharides, proteins, lipids, and their combinations, in addition to some additives used to achieve extra properties or improve an existing one. Some examples of natural polysaccharides applied in the development of these materials are starch, cellulose, pectin, gums, chitosan, agar, and alginate. Among the proteins are gelatin, corn zein, and whey protein. Lipid-based materials have also been used in the coating, for example, carnauba wax, bees wax, paraffin, and glycerides (Maan et al., 2021).

Polysaccharides are highly available, do not cause allergenic reactions, generally dissolve in water, and often have satisfactory mechanical properties. The chemical structure defined for each monomer unit is an advantage of polysaccharides as it allows better prediction and control of coating properties. However, the low moisture barrier is a challenge for the development of polysaccharide-based materials (Arnon-Rips and Poverenov, 2018).

Proteins offer a wide variety of physical and mechanical properties that can meet the demands of different types of food. Water-soluble proteins tend to form ionic interactions or hydrogen bonds. High gas permeability, good mechanical properties, and low moisture barrier are some characteristics of protein-based coatings. Furthermore, proteins with low water solubility generally tend to have lower vapor permeability. However, a major drawback is that the use of proteins is limited due to the considerable risk of causing allergenic reactions (Arnon-Rips and Poverenov, 2018).

Lipids, the first edible coatings, have been widely applied as coatings on meat, poultry, and seafood. They are hydrophobic and, therefore, have good moisture barrier properties. In addition, lipid coatings can add shine to foods, making them more attractive. In contrast, lipids have poor mechanical properties and low adhesion to foods with hydrophilic surfaces. Considering the advantages and disadvantages of each type of material, the concept of composite has received great attention in recent years. A composite is defined as the combination of two or more components, from different classes or not, which aim to improve the properties of edible coatings, thus managing to combine the advantages offered by different materials (Arnon-Rips and Poverenov, 2018).

5.3 Methods of Coating

In the development of a successful coating, the adhesion of this material to the food is an important factor and depends on some properties of the formulated coating, such as viscosity, density, and surface tension (Md Nor and Ding, 2020), besides properties of the food products (Suhag et al., 2020). In addition, the effectiveness of edible coatings also depends on the method used for their application. Several techniques are currently used, such as dipping, spraying, layer-by-layer, vacuum impregnation, fluidized-bed, foaming, and panning. The choice of technique will depend on the nature of the food to be coated, the surface attributes, and the main purpose of the coating. Physical parameters of foods, such as shape, size, and desired coating thickness, influence the selection of coating system and method. Therefore, different methods of applying edible coatings on foods have been studied to improve quality and safety, as well as to extend shelf life (Suhag et al., 2020).

The new generation of edible coatings has been developed over the years through the application of nanotechnology systems, for example, nanocomposite edible films, nanoencapsulation of active compounds aiming to control their release, nanolaminates to create multilayered systems, etc. (Andrade et al., 2012). Figure 5.1 illustrates some examples of the general nanosystems in edible food packaging and their mechanisms.

5.3.1 Dipping

The dipping method, the most common coating technique for food, is described as the process of immersing a food sample in a coating-forming solution. It is a simple and low-cost technique. The coating application by this method is carried out in three stages: (i) immersion and dwelling, (ii) deposition, and (iii) evaporation of solvents. In the first step, the food product is immersed in the coating solution, and the permanence ensures an adequate amount of solution, as well as a good interaction between food and the coating matrix. Excess liquid on the food surface is drained and removed by deposition, and this

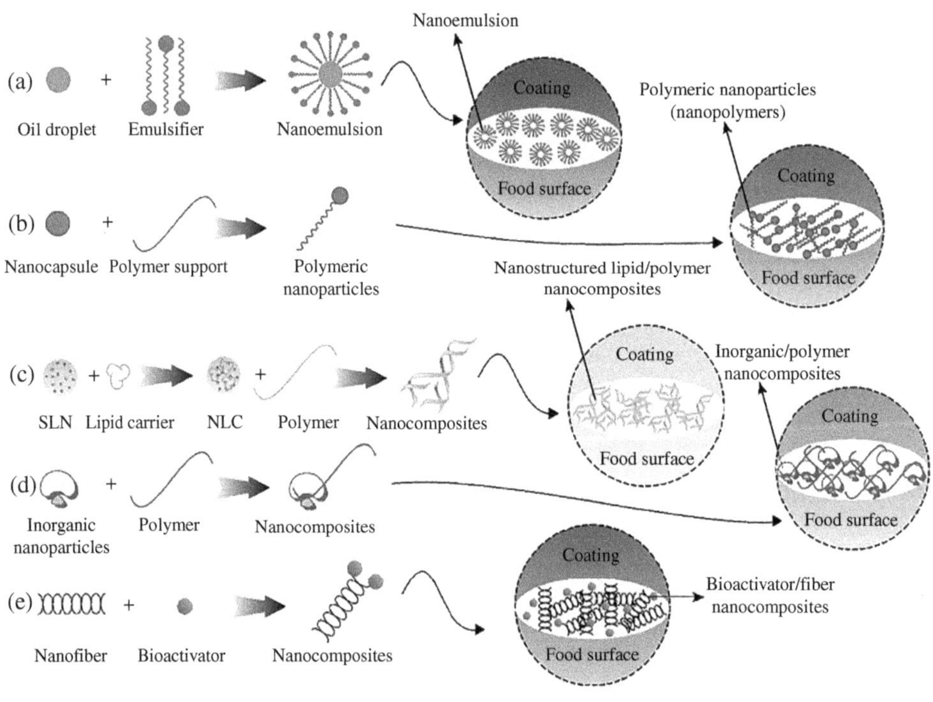

Figure 5.1 Synthesis techniques for various nanocarriers and their mechanisms in food edible coatings; (a) nanoemulsion formed from oil droplets and emulsifier, (b) polymeric nanoparticles fabricated polymeric support, (c) nanocomposites formed from solid lipid nanoparticles/nanostructured lipid carriers in a polymeric matrix, (d) nanocomposites fabricated from inorganic nanoparticles and a polymeric matrix, and (e) nanocomposites formed from nanofibers and a bio activator (*Source:* Jafarzadeh et al. (2021b)/with permission of Elsevier)

step is important for the development of thin coating layers. In the last step, evaporation, solvent, and excess liquid are evaporated from the food surface, and the procedure can take place at room temperature or using drying equipment (Md Nor and Ding, 2020; Suhag et al., 2020).

The dipping process is often used to apply edible coatings to fresh food products such as fruits and vegetables. Adding antimicrobial and antioxidant agents to the coating solution helps to improve microbial stability and regulate the browning of freshly cut or peeled fruit. Despite providing good uniformity to the coatings, this method presents some problems, such as the generation of a thick coating, which in some cases can cause difficulties in the food respiration rates. In addition, other drawbacks of the dipping mechanism are the dilution of the coating and external layer and the degradation of its functionality (Suhag et al., 2020).

Incorporating nanoparticles or nanomaterials in edible films is being widely explored and has shown promising results. Nanoparticles or, nanofibers, or nanomaterials are incorporated into films, providing functionality to them. Therefore, nanocomponents can be introduced into the polymer matrix to produce nanocomposites with improved barrier and mechanical properties, and thermal and dimensional stability (Umaraw et al., 2020). In recent research conducted by Arroyo et al. (2020), nanoparticles of ZnO were incorporated in chitosan and alginate edible coating, which was able to reduce lesions/injuries on fruits and improve the keeping quality of guava. In this study, the authors used the dipping method. Each guava was immersed in the film-forming solution (previously prepared), removed, and placed on a grid to remove the excess coating, and then dried at room temperature. Divya et al. (2018) studied chitosan nanoparticles and their use as an edible coating on vegetables (tomato, chilly, and brinjal). The vegetables were coated by dipping method, and this nanomaterial was able to decrease weight loss and also become an ideal antifungal, antioxidant, and coating agent.

5.3.2 Spraying

Spraying is another common method used in the coating application of food products. This method increases the liquid surface through the formation of droplets and distributes them on the food surface using a set of nozzles. Spraying technology provides a uniform thickness coating and is a great option for multilayer applications. In the food industry, three spraying techniques have been used for coating. First is air spray atomization, in which a high-velocity air stream surrounds the fluid that is at low velocity, and this fluid–air friction induces atomization. Second, air-assisted airless atomization, where partial atomization of the fluid occurs first by air-assisted airless spray guns, and then the atomization is completed with small amounts of compressed air. This technique helps to minimize some of the problems faced by using high-viscosity and high-solid coatings, as well as those related to heating and application of higher fluid pressures when working with viscous materials. Third, pressure atomization, which is also known as air less atomization technique because, in this case, air is not used. Instead, the application of the edible coating on food products is made by pressure (Suhag et al., 2020).

In the spraying method, some critical parameters affect the efficiency of the process, such as the atomization pressure, viscosity, surface temperature and surface tension (related to the coating solution), temperature, air and polymeric-fluid flows, and spray piston. In addition, nozzle shape and design, which are crucial for flow rate, final droplet size, distance, and angle, influence the final coating quality. Furthermore, drying time, drying temperature, and drying method are determinant factors for spraying systems (Suhag et al., 2020).

An emerging new spray system, called electrospray, has several advantages over mechanical atomizers. Regarding the size of electrospray droplets, hundreds of micrometers down to several tens of nanometers can be achieved, which is in line with the new nanotechnologies that have been developed for food edible coatings. In addition, it is possible a size distribution of nearly monodisperse, and droplet generation/size can be controlled by liquid-flow rate and voltage at the capillary nozzle (to a certain extent). Because of these characteristics, electrospray has been used for micro- and nano-thin-film deposition, micro- or nanoparticle production, and micro- or nanocapsule formation. This spraying technology can deliver a single-step, low-energy, and low-cost process that generates products with unique properties, which makes it promising to be used in edible coating systems (Andrade et al., 2012).

5.3.3 Layer-by-Layer

The basis of the layer-by-layer method is an alternating deposition of oppositely charged polyelectrolytes, which leads to greater control over the properties and functionality of the coating. Through this technique, a multilayer coating is produced, such as a first layer composed of a polysaccharide or protein solution and a second layer of lipid. It also allows composite coatings to be formed, as well as producing uniform coatings with good compactness (Md Nor and Ding, 2020). Through the layer-by-layer electrostatic deposition technique, a multilayer edible coating is obtained. Food products are dipped in two solutions with oppositely charged polyelectrolytes; however, between them, there is a drying step to remove excess solution. The principle of this technique is related to the electrostatic force generated by the variation of charge on the coating surface, which drives the self-assembly of each one (Adiletta et al., 2021).

Layer-by-layer has become a novel nanoscale approach, which is mainly based on the electrostatic deposition of charged materials (such as nanoparticles, polymers, proteins, and polyelectrolytes) onto the assembly surface. Different combinations can provide various layer-by-layer assemblies. In recent years, this method has been applied to prepare biodegradable edible coatings, which resulted in notable food quality improvements (Chen et al., 2020).

Multilayer coating allows different materials to be used, and thus, its properties can be improved, such as barrier, mechanical, and active properties (Costa et al., 2018). Since its discovery, the layer-by-layer method has been explored in several areas and then applied in electrical, optical and biomedical devices, biosensors, and drug delivery systems (Arnon-Rips and Poverenov, 2018). In the preparation of edible coatings, the layer-by-layer technique has been used mainly in fresh agricultural products such as fruits and vegetables, whether whole (Chen et al., 2020) or fresh-cut (Basaglia et al., 2021; Treviño-Garza et al., 2017). However, technological advances over the years have also allowed its application in processed foods such as French fries (Daraei Garmakhany et al., 2014).

In addition, as the concept of nanosystems has advanced, research with nano multilayered coatings has been highlighted in recent years, especially as edible coatings. Souza et al. (2015) evaluated the effect of an alginate/chitosan nano-multilayer coating, developed by electrostatic layer-by-layer self-assembling, on the quality, and shelf life of fresh-cut mangoes. Results showed that these nano-multilayered coatings improved the microbiological and physicochemical quality of the fruits, being able to extend their shelf life up to 8 days.

5.3.4 Vacuum Impregnation

The vacuum impregnation method is based on the same principle as the dipping technique but applies an extra advance to the technique. This advance is a vacuum environment during the food dipping stage. The submersion takes place in an airtight vacuum chamber connected to vacuum pumps instead of immersing the food in a normal tank, as in the dipping method. After the vacuum stage, the food proceeds to atmospheric restoration while remaining immersed in the coating solution under atmospheric pressure. Through vacuum impregnation, it is possible to enrich fruit with vitamins and minerals, preserve the natural flavor and heat-sensitive compounds, and prevent discoloration due to oxygen removal and the application of antimicrobial and antioxidant agents. This technique has been applied mainly to minimally processed fruits and vegetables. However, in the case of whole fresh fruits, especially tropical ones, its implementation is still limited (Md Nor and Ding, 2020). Xu et al. (2020) have applied the vacuum impregnation method to develop nanocomposite coatings to improve the properties of several products. It managed to improve the mechanical, hydrophobicity, and thermal properties of Chinese fir wood by vacuum impregnation of nano-silica sol. Nevertheless, in the food sector, the development of nanosystems for edible coating through this method is deficient. Therefore, to date, research on edible nanocoating by vacuum impregnation is limited.

5.3.5 Foaming

Foaming is mainly used in emulsion-type coatings. In this method, the coating is added with a foaming agent, and then compressed air is blown into the air of the applicator tank. In order to obtain uniform distribution, extensive tumbling action is used to break the foam. To achieve a fast drying process, the applicability of this method is limited to emulsions that contain little water. The use of coatings obtained through this technique finds greater applicability in bakery and confectionery industries, unlike fruits, whose foam technique is still limited (Md Nor and Ding, 2020). Furthermore, some common drawbacks of emulsions are instability and lack of transparency. However, recent research about coatings/films is focused on nanoemulsions, which, besides preventing these problems, are generally more effective in extending shelf life due to the greater transport of active components to the food surface (Maan et al., 2021).

Emulsions are generally made of two immiscible liquids, in general oil and water, forming systems with a dispersed and continuous phase. In recent years, nanotechnology has emerged as a new improvement strategy for coatings and films. Usually, coatings based on plant-source nanoemulsions show better water barrier mechanical, optical, and microstructural properties compared to those based on conventional emulsions (macro- or micro-emulsions). Nanocoatings incorporated with antimicrobial and antioxidant compounds provide a gradual and controlled release of bioactive throughout food storage (better than conventional emulsions), thus increasing their bioactivity, extending shelf life, and improving nutritional quality. Therefore, nanotechnology has been applied as a new tool to improve coating properties and functionalities. Nanoemulsions appear as nanoscale droplets (particle radius <100 nm) dispersed in a solution, which changes the physical properties of the coating and leads to, for example, a reduction in moisture migration, gas exchange, oxidative reactions, and microbiological growth. In addition, nanoscale emulsions for edible coatings can be potential vehicles for active components, such as vitamins and nutraceuticals. The droplet size in nanoemulsions is important for the improvement of the bioavailability of added compounds. Studies have mentioned the importance of modulating nanoemulsions' composition and structure to achieve efficient delivery of bioactives. So, the choice of emulsifier is another important criterion in nanoemulsions due to its influence on bioavailability. Lastly, research on this topic has intensified recently, demonstrating the scientific community's increased interest in nanoscale emulsion systems (de Oliveira Filho et al., 2021; Pandey et al., 2022).

The bioavailability of active compounds can be improved by controlling the composition and structure of oil-in-water nanoemulsions, that is, by controlling the factors related to the design of these nanoemulsions (Zhang et al., 2020a), as shown in Figure 5.2.

5.3.6 Fluidized-Bed Processing

The fluid bed method is widely used to apply thin layers of coating material to dry particles of low density and/or small size. In a fluidized coating process, the coating solution and suspension are sprayed onto the surface of the fluidized powder through various nozzles to obtain a shell-like structure. This method comprises three categories: top spray, bottom spray, and rotary fluidized bed. However, in the food industry, the most effective is the traditional top-spraying process. Some of the features related to this method are:

1) a particulate material of a size greater than 100 μm in the fluidized bed,
2) the use of a liquid binder to enhance the stability of the material, and
3) application to both continuous and batch processes.

The fluidized-bed method has been used by the food industry in a wide range of products, such as functional ingredients (yeasts and enzymes), preservatives (acids and salts), vitamins and minerals, flavors, and spices. In addition to food and additive encapsulation and seed coating (for protection against mechanical and biological damage) (Suhag et al., 2020).

With the evolution of nanotechnology over the years, changes have been made to introduce nanosystems to the fluidized-bed method. This is widely used in the food and pharmaceutical industries, and to prevent the formation of agglomerates, new research has focused on the development of techniques for coating fine particles. An innovative alternative is based on the combined application of a fluidized bed and an aerosol atomizer. Thus, the use of very small aerosol droplets allows the coating of fine particles without severe agglomeration in a fluidized bed process (Zhang et al., 2020b).

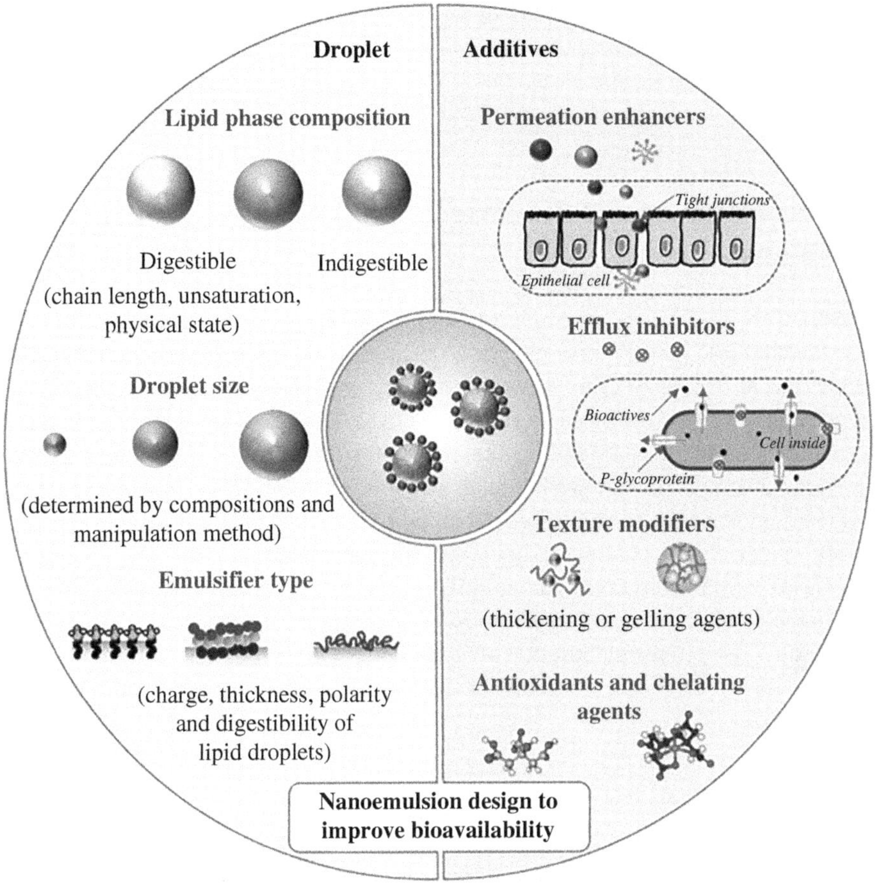

Figure 5.2 The composition or structure of oil-in-water nanoemulsions can be designed to enhance the bioavailability of bioactives (*Source:* Zhang et al. (2020b)/with permission of Elsevier)

5.3.7 Panning

In the panning method, the food to be coated is placed in a large rotating bowl called a pan. From there, a layering solution is sprayed or sprinkled into this bowl and the product is tumbled. In this way, it is possible to distribute the coating solution evenly on the food product surface. Drying of the coating layers takes place by the use of air with force and ambient temperature or higher. There are three types of panning coating applications. The first, hard panning, covers a continuous application of sugar syrup on the surface that dries and crystallizes. The second, soft panning, uses corn syrup and sugar mixture syrup to obtain the coating, which is dried by application of dry sugar to generate thick and soft layers (in smaller numbers). The third is described as chocolate panning and uses a fat-based layer around a core (Suhag et al., 2020).

The panning method makes it possible to apply thin or thick layers to hard and usually circular/oval particles. In the food industry, these panning techniques are mainly applied in confectionery, for example, in the coating of peanuts, nutmeats, and almonds.

Moreover, this method allows the coating of a large number of food products in a single batch (Suhag et al., 2020). However, to date, very few advances have been explored with respect to nanotechnology in this method over the past few years. Thus, research in this direction is limited.

5.4 Nanocomposites and Bio-Based Materials for Edible Coatings

Edible coatings are thin layers of edible materials applied to food products to protect them from mechanical damage and physical, chemical, and microbiological activities. Thus, edible coatings provide several beneficial effects, such as inhibiting the senescence process and weight loss in fruits and vegetable products, as well as reducing mechanical and microbiological damage, extending the commercial validity of the food products (Falguera et al., 2011). Various materials such as plant residues, flours, gums, and agro-industrial residues from the fruit and vegetable industry can be used as suitable sources of polymeric materials for coating formation. Many of these materials, such as agro-industrial waste, contain bioactive compounds that add antioxidant and antimicrobial properties to the edible coatings. Nanocomposites and nanoparticles, with active or reinforcing properties, used in the elaboration of edible coatings constitute an excellent green technology. These materials are nontoxic, biodegradable, and environmentally friendly (Galus et al., 2020).

5.4.1 Polymers and Their Blends

Many polymers have been used in the production of edible coatings. Polysaccharides, proteins, and lipids (waxes) obtained from plants, animals, or microorganisms are commonly used in the production of edible coatings. The cellulose derivatives, starch, alginate, pectin, chitosan, pullulan, carrageenan, and various gums are the polysaccharides widely used in the development of edible coatings, while among the proteins soy protein, wheat gluten, corn zein, sunflower proteins, gelatin, whey, casein, and keratin are commonly used (Falguera et al., 2011). These materials can form a cohesive and continuous matrix.

The use of these polymers in the elaboration of edible coatings is due to their properties such as biodegradable, nontoxic, antibacterial, biocompatible, transparent, antifungal, good mechanical properties, high water vapor permeability, and a barrier to gases. Polysaccharides and proteins have shown excellent mechanical and structural properties, but because they are hydrophilic. They have a poor barrier capacity against moisture transfer. Thus, coatings produced with those polymers generally exhibit very low moisture barrier properties. The ability of proteins to form networks, plasticity, and elasticity are advantageous properties in the preparation of coatings. The incorporation of lipids into edible coatings favors the water vapor barrier property but provides reduced mechanical strength and greater oxygen permeability (Tan et al., 2021).

Edible coatings with beneficial properties can be obtained by combining two or more biopolymers. The hydrophobic–hydrophilic characteristics of the biomolecules must be considered in the preparation of blends to improve the physical, mechanical, and barrier properties of the coatings (Dhumal and Sarkar, 2018; Falguera et al., 2011; Galus et al., 2020). In this context, composite coatings prepared from the combination of two or more film-forming substances generally have physical, mechanical, and barrier properties superior to those of one-component material. Furthermore, in the formulation of coatings, various other substances, such as plasticizers, crosslinking agents, emulsifiers, and reinforcements, can be used to improve or modify the basic functionality of the material (Galus et al., 2020).

5.4.2 Nanocomposites and Nanoparticles

The properties of the biopolymer-based coatings can be improved by using different methods. Nowadays, nanotechnology is applied with great results in the science of polymers and materials. Nanometer-sized particles are dispersed in a polymer matrix, forming nanocomposite coatings as reinforcements to improve mechanical, thermal, optical, and physicochemical properties when compared to pure or conventional polymers. In addition, nanoparticles are also added to the polymer matrix to act as antioxidants and antimicrobials (Perera et al., 2022).

According to Chaudhry et al. (2008), the use of nanocomposite in materials that come into contact with food, such as edible coatings, can be divided into four categories: (i) enhancers of improved packaging properties, such as gas barrier, mechanics, etc.; (ii) as active components that have additional properties such as antimicrobials, antioxidants or oxygen scavengers; (iii) particles that make the packaging "intelligent" due to the incorporation of nano-sensors; and (iv) biopolymers of biodegradable polymer–nanomaterials composites with improved characteristics through nanofills. Among the numerous nanoparticles used to functionalize polymeric materials, metal oxides such as TiO_2, ZnO, and MgO have also been exploited for the preparation of antimicrobial packaging coatings and films due to their strong antimicrobial activity with high stability compared with organic antimicrobial agents.

5.4.3 Agro-Industrial Waste as Coating-Forming Materials

The agro-industrial residues resulting from the processing of fruits and vegetables consist of seeds, bagasse, and peels. This material has a high content of relevant compounds such as fiber, polysaccharides, vitamins, minerals, and bioactive compounds (López-Marcos et al., 2015; Silva et al., 2017). The discarding of these residues directly in the soil can cause a serious environmental problem since it presents a high amount of water and fermentable sugars, providing enormous production of oxygen during its degradation, in addition to providing uncontrolled fermentation (Paraman et al., 2015).

Agro-industrial residues can be used to obtain biopolymers, renewable, and biodegradable, to be used in the production of edible coatings, reducing the environmental impact and increasing the added value. The application of agro-industrial residues in food packaging systems, including coatings and films, brings some advantages, such as antioxidant and antimicrobial action, improved barrier properties, improving quality of food products (Bayram et al., 2021). In this context, the flour obtained from tuber peels, such as potatoes and cassava, is a source of starch from agroindustry residues. Flours obtained from other residues of the processing of fruits and vegetables are also used as material for the preparation of edible films and coatings. Ferreira et al. (2016) report that fruit and vegetable residues, including edible and nonedible parts, were used as raw materials to develop biodegradable films and coatings. The formulated films and coatings exhibited promising characteristics and showed good performance when applied to acerola. Torres-León et al. (2018) developed biodegradable coatings and films by using mango peel and antioxidant extracts of seed kernel. Peaches with this coating showed a reduction in the transfer of gases (ethylene and CO_2) when compared to uncoated peaches, promoting an increase in the shelf life of treated fruits. So, based on the results of these articles, as well as others existing in the literature, the feasibility of using agro-industry residues in the elaboration of edible films and coatings to be applied in foods is evidenced.

5.4.4 Active Edible Coatings and Agro-Industrial Waste as a Source of Bioactive Compounds

The production of active edible coatings has also been a focus of research investigation. These coatings are obtained by incorporating bioactive compounds in the coating matrix to provide antioxidant, antibacterial, antifungal, and nutraceutical action, among others (Arnon-Rips et al., 2021; Galus et al., 2020). Within this context, agro-industrial residues, especially those from fruits and vegetable processing or wine and beer industries, appear as sources of bioactive compounds because they present in their composition significant amounts of phenolic compounds, such as tannins, phenolic acids and flavonoids, and other components. In addition, there is the potential use of the residues to obtain the extracts or isolate specific active compounds for application in food products, especially in edible films and coatings, reducing industrial, economic, and environmental impact (Salgado et al., 2015).

5.5 The Essential Properties of Bio-Nanocomposite Coatings

Edible coatings are applied to food surfaces to prevent the growth of bacteria, yeast, and mold. The effectiveness of these coatings in reducing microbial growth is mainly due to the type, concentration, and properties of the biopolymer-based coatings, as well as the type, size, shape, concentration, chelation, release, and photocatalytic properties of the embedded nanoparticles. Furthermore, the synergistic effect of coating materials and nanoparticles also influences the antimicrobial activity of edible coatings (Xing et al., 2019).

In addition to antimicrobial properties, bio-nanocomposite edible coatings contribute to reducing weight loss and prolonging product shelf life, as they are semipermeable barriers to moisture and gases, such as carbon dioxide and oxygen. The increase in barrier properties of edible coatings is due to the formation in the polymer matrix of a complex diffusion path that reduces the permeability to gas and water, as well as the bond between nanoparticles and biopolymers, reducing their interaction with the water molecules. The high proportion and homogeneous dispersion of the nanoparticles favor the mechanical and thermal resistance of the packaging material as a function of the molecular mobility and relaxation of the polymer. The bio-nanocomposite edible coatings also improve the texture and appearance of the product (Jafarzadeh et al., 2021b; Perera et al., 2022).

These properties are essential in increasing self-life, reducing microbial growth, and increasing the quality and safety of food products during storage and transport. Thus, bio-nanocomposite coatings can be applied to the surfaces of fruits and vegetables but also have the potential to be used in packaging for many other food products (Jafarzadeh et al., 2021a). Therefore, bio-nanocomposites are considered a promising alternative and new generation of food packaging materials, especially because they present essential characteristics such as biocompatibility, antimicrobial activity, antioxidant action, biodegradability, and improved mechanical, optical, and barrier properties. In addition, bio-nanocomposites provide sustainable and ecologically correct food packaging (Perera et al., 2022).

5.6 Coatings as Substance Releasers or Carriers in Food Models

Over the years, we have seen how technology advances in different sectors, and this would be no different for the food industry, which constantly seeks to evolve according to emotions, sensations, tastes, and even some trends marked by consumers. High food demands have also affected the development of new technologies in the food packaging industry. The greatest advances are marked mainly by those resources that strongly seek to replace or at least reduce the use of petroleum-based plastic materials and incorporate improvements in the safety, quality, and storage of products. The extension of shelf life could be performed by the application of appropriate smart active packaging and preservation methods from new technologies (Jafarzadeh et al., 2021a). Given this, nanotechnology has shown a growing potential in improving the properties of packaging materials. The development of nanosystems has advanced through the incorporation of compounds that lead to functional modifications in edible coatings, integrating nanoemulsions, polymeric nanoparticles, nanofibers, nanofillers, polymeric nanocomposites (inorganic–organic), and other nanostructured substances systems (Jafarzadeh et al., 2021a).

According to the EU (EU, 2011a, 2011b), only nanomaterials that are explicitly approved in their nano-form are allowed to be used independently if in the direct food contact layer or behind a functional barrier (Bott et al., 2018). Thus, smart packaging is a technology that includes active and intelligent packaging or a combination of both. Active packaging tech involves the addition of particles to a packaging material that can interact directly with food and protect it from UV, oxygen, ethylene, or microbiological contamination (Nikolic et al., 2021; Rai et al., 2019; Vilela et al., 2018). Intelligent packaging materials can provide information about the nature of the food inside a package, such as its quality, maturity, or contamination status (Cheng et al., 2022).

Active films/coatings are typically manufactured using edible biopolymers such as polysaccharides, proteins, and lipid substances in multiple forms that can be obtained from natural plants or animal sources (simple or complex, single or multilayered) (Zhang et al., 2022). The aim of applying these coatings is to create a thin layer on surfaces with a similar effect to modified atmosphere packaging, reducing the respiratory rate and weight loss, hence preserving the color and firmness of coated foods (Maringgal et al., 2020; Sánchez-González et al., 2011; Teixeira et al., 2022).

Tests have shown that the intrinsic mechanical, chemical, and antibacterial properties of biopolymers alone are not efficient unless they are combined with other additives or components that strengthen these properties (Zhang and Jiang, 2020). Nevertheless, they can constitute an ideal carrier for active agents, with the advantage of being adaptable in terms of a controlled release system and the possibility of combining several polymers through blending or multilayer assembly (Arroyo et al., 2020; Cosme Silva et al., 2017; Romanazzi et al., 2017).

Natural compounds with antimicrobial and antioxidant activity, such as essential oils, enzymes, organic acids, acid salts, fatty acids, metals, chelating agents, nanoparticles, and others, have been incorporated into the biopolymers to extend the shelf-life of the food (Sutil et al., 2022; Vilela et al., 2018).

5.6.1 Application on Fresh/Fresh-Cut Fruits and Vegetables

Fruit quality is not only related to storage but also to management and climatic conditions during the production phase. Hence, the necessity to use other processes to prevent microbiological deterioration and minimize the physiological and biochemical changes responsible for postharvest degradation since, generally in normal conditions, fruits dry quickly with the loss of rigidity (Arroyo et al., 2020; Farcuh et al., 2018; Jafarzadeh et al., 2021a).

Previous research has illustrated that when essential oils (EOs) are directly applied to fruits and vegetables, organoleptic characteristics can be negatively impacted due to their high reactivity and low flavor threshold; thus, the importance of setting a matrix with a controlled but constant release of these volatile EOs under various application conditions (Guo et al., 2021; Pedrós-Garrido et al., 2020). In the literature, the use of coatings applying nanocomposite antimicrobial compounds within packaging systems is effective as they have been used in different fruits such as guavas, papaya, mango, apple, and strawberry, among others. In addition, they have improved properties like mechanical, barrier, thermal, and antimicrobial properties (Arroyo et al., 2020; Ayala-Zavala et al., 2013; Demetrio et al., 2019; Jafarzadeh et al. 2021a; Yin et al., 2019). According to Yuan et al. (2022), surface coating of zein nanoparticles "has been burgeoning in recent years to give full play to the advantages of zein as a delivery material for bioactive compounds" the actual applications of zein nanoparticles-based edible films in the preservation of fruits, vegetables, and meat have been reported (Zhang et al., 2022).

5.6.2 Application on Meat and Its Derivatives

There are different types of intelligent packaging, such as time-temperature indicators, gas leakage indicators, and relative humidity or freshness sensors, among others, being the most common (Müller et al., 2017). The inhibition of microbial contamination can be done by either direct or indirect contact between packaging material and foods in which the antimicrobial packaging releases a volatile antimicrobial agent into the headspace of the packaging (Bahmid et al., 2021; Yildirim et al., 2018). According to Fang et al. (2017), the most important active packaging systems applied to meat and meat products are antimicrobial, antioxidant, and carbon dioxide emitting/generating packaging as one of the main indicators of spoilage in protein-rich food products, such as meat and fish, is represented by an increase in total volatile basic nitrogen concentration (Alamdari et al., 2021).

As in any other food, it is important to consider the organoleptic properties of the food. Therefore, in meat products or their derivatives, the use of natural compounds widely used in fruits or vegetables, such as EOs, may present limitations at the time of their application since they affect odor and flavor due to the migration of their aroma (unless desired). This was one of the conclusions obtained by Jayasena and Jo (2013) in their review, where they evaluated several scientific works on the application of essential oils as potential antimicrobial agents in meat and meat products.

Another research led by Rahnemoon et al. (2021) observed that encapsulated pomegranate peel extract into calcium alginate nanosphere in one of their model proposal improved antimicrobial activity on chicken meat and enhanced its shelf-life, especially during long time storage (two weeks). Although the antimicrobial effect during storage of meat products continues to be a fairly broad field of research, chitosan stands out as basic raw material with various chemical, physical, and biological properties (solubility, film formation, viscosity, ion binding, and antimicrobial properties) (Priyadarshi and Rhim, 2020; Wang et al., 2018).

5.6.3 Application on Dairy

In the dairy industry, cheese has attracted the attention of many researchers because its surface has a higher probability of being contaminated by microorganisms because of their suitable acidity conditions and high-water activity (Jafarzadeh et al., 2021b; Proulx et al., 2017). As in the dairy industry, contamination with *L. monocytogenes* is not only associated with the use of raw milk but with the production of artisanal cheese, especially those with high pH values (Silva et al., 2022).

So, what features should smart cheese packaging have, and what does it depend on to be successful? According to Pantaleao et al. (2007), the optimal packaging for cheese must have a barrier to moisture that controls the water loss rate in cheese, with the moisture permeating out of the package; also, impermeability to oxygen and UV is important that minimizes the activity of bacteria and fungi and simultaneously leads to good textural and sensory properties of cheese. A study conducted by Silva et al. (2018) concluded that the inclusion of lactic acid bacteria (LAB) into edible coatings and films could lead to an inhibition of pathogens through competition for space and nutrients and/or through the production of antimicrobial substances. Furthermore, proteins, polysaccharides, and lipids are the most used materials in the development of edible films and coatings for application on cheese (Costa et al., 2018).

"Nanomaterials," due to their very small sizes, can impart various properties compared to larger systems. In the case of preserving fresh cheeses, for example, proper packages are required to prevent moisture loss, and, in the case of cheeses with some fat content, the presence of light and O_2 impair their quality. To protect them, novel technologies of edible coating have emerged as alternative packages, the bio-nanocomposites. In addition, several attempts have also been made to avoid microbiological contamination, such as applying nanoparticles (ZnO, Ag, CuO, etc.) with antimicrobial activity to these systems, nanofibers, nanocapsules, and nanoemulsions. Due to the complex structure of lipids and proteins in cheese composition, more investigations must be done to deliver nanosystems suitable for dairy products such as cheese (Jafarzadeh et al., 2021b).

5.6.4 Bakery Application

The most important physical and chemical changes in bakery products result in the loss of freshness and the impairment of texture and taste induced by bacteria, yeast, and mold growths (Melini and Melini, 2018). Some more numerical results will be good. EOs such as oregano essential oil, cinnamaldehyde, Carvacrol, star anise essential oil and thymol, garlic extract, and grapefruit seed extract are being used as either a biopolymer-based matrix or sachet (Qian et al., 2021). In their review, Qian et al. (2021) studied the factors affecting bakery products' shelf-life and emphasized some food packaging innovations, such as nanomaterials, biodegradable, and edible coatings. For example, Ferreira Saraiva et al. (2016) developed edible coatings of potato starch, inverted sugar, and sucrose, which were able to increase up to 130% the shelf-life of mini panettones.

Gregirchak et al. (2020) used probiotics such as LAB in wheat bread to protect the bread from mold spoilage and also increase its shelf life. The incorporation of LAB in a coating containing alginate and whey provided the guarantee of their good viability for 120 hours of storage.

5.7 General Safety Aspects Related to the Development of Food Contact Materials

Recently, research about new materials, processes, and components on micro or nanoscales that will come into direct contact with food has been carried out (who they are, provide some references). However, some questions need to be very well clarified before being commercially applied. For instance, questions such as (i) security protocols, (ii) level of toxicity, and (iii) migration. Each country has established protocols, regulations, and legislation that should be observed and complied with, independent of the destination of the food.

In the object of study of this chapter, edible coatings, the reader needs to understand that the search for new materials is not random. Before starting research with new materials, some approaches were elucidated clearly and objectively, such as the broad understanding of what coating means and its function. In addition to application processes, that is, existing methodologies and their modifications in such a way that it can adapt to the type of product to be applied, detailed knowledge of the product, and the actions of these coverages as mentioned above: toxicity, safety, migration, which can be summed up in a single word: Quality.

Coatings can be defined as a film, with or without additives, that coats the food, improving its quality and shelf life (Díaz-Montes and Castro-Muñoz, 2021; Galus and Kadzińska, 2015; Han et al., 2013; Rojas-Graü et al., 2009). These coatings can be developed from natural polymers such as chitosan and alginate coatings applied to guavas (Arroyo et al., 2020), chitosan-caseinate in apples (Volpe et al., 2017), chitosan-gelatin in shrimps (Farajzadeh et al., 2016), pectin-alginate supplemented with essential oils (Guerreiro et al., 2015b), alginate-carrageenan (Tavassoli-Kafrani et al., 2016).

A quick search on specialized websites' publication of articles using the keywords: edible coatings applied to fruits and vegetables in the last three years (2018–2021) shows 15,900 articles. Expanding the search by entering the keywords toxicity, cytotoxicity, safety, and migration, in the same period, we found 17,000 articles. Research with proposals to obtain new materials and methodologies to preserve food has also been a director of commercial application of these processes. It appears that not only the search to improve the shelf life of the products but also the safety of these materials in contact with food through the optimization of methodologies, meeting the norms of each place where the product is intended to be preserved.

5.8 Toxicological Aspects, Safety, and Components Migration Related to Edible Coatings

Applying coatings to food is nothing new. Since the 12th and 13th centuries, waxes were applied to citrus fruits, extending to apples and pears in the 1930s (Baldwin et al., 1995; Debeaufort et al., 1998). In meats and cheeses, there are reports that since the middle ages, lipid coatings were used for coating (Debeaufort et al., 1998). How can a food contact-safe coating or covering be characterized? What are the criteria to be observed?

In the segment of synthetic and natural polymers used as packaging in direct contact with food, edible coatings can be highlighted. The development of food packaging must follow Good Manufacturing Practices and be recognized as GRAS (generally recognized as safe), respecting Food and Drug Administration (FDA) designation (USA). Detailed information can be found on the FDA's official website (Galus et al., 2020).

When referring to toxicity, a parameter that can be decisive is the migration concentration of the compounds present in the coating or that are incorporated. In terms of toxicity, we cannot separate the migration studies of compounds from packaging to product and from product to packaging. Therefore, all material, nano or macro, in contact with food to be approved for its use must undergo migration and toxicity tests. For a long time, the global migration limit was 60 mg/kg. With the updating of food safety regulations, this value is no longer adequate for reality. Control becomes essential to avoid excessive contamination of food (Grob et al., 2010).

Technological advances have made it necessary to rigorously approve food contact material. In search of material improvements, as mentioned in this chapter, some critical points become necessary to consider, such as the toxicity and migration of the components (Grob et al., 2010). Analyzing the evolution of research, it has been found that coatings that were previously produced using vegetable waxes, such as carnauba, underwent improvement processes with the incorporation of antioxidant compounds (Divya et al., 2018; Ponce et al., 2008; Sánchez et al., 2014), essential oils (Azevedo et al., 2014; Guerreiro et al., 2015a, 2015b; Riveros et al., 2016), antimicrobials (Arroyo et al., 2020; Mantilla et al., 2013; Ponce et al., 2008; de Jesus Salas-Méndez et al., 2019; Valencia-Chamorro et al., 2011), nanoparticles (Arroyo et al., 2020), functional ingredients and probiotics (Pandhi et al., 2019), and layer-by-layer configuration (Arnon-Rips and Poverenov, 2018; Bilbao-Sainz et al., 2018; Yan et al., 2019). It is important to emphasize that many of the coatings developed can cause allergies to the consumer, making it mandatory to include information in the labeling about the direct or indirect presence of allergens (Galus and Kadzińska, 2015; Grob et al., 2010).

An important point to consider in the use of new materials for the development of films or coatings for packaging is the scale. If we think about the evolution of packaging technology in recent years, we can see a significant amount of new materials developed at the nanometer scale, aiming to add improvements in mechanical and barrier properties, besides antimicrobial and antioxidant actions (Alfei et al., 2020; de Lima Júnior et al., 2016). These modifications, for example, may come from the application of nanocomposites and the combination of biopolymers developed layer-by-layer (Arnon-Rips and Poverenov, 2016; Mantilla et al., 2013). Because of the exposure, we must consider two important points: (i) environmental contamination by discarding the packaging in the environment and (ii) the toxicological effect of these packaging (films or edible coatings) after consumption on the body. These considerations reinforce the need for careful studies on cytotoxicity, toxicity, and migration of components, thus evaluating their commercial use. Emphasizing that it does not matter whether the nanomaterial or nanocomposite is of natural or synthetic origin. Since the allergenic effects are independent of the type and origin of the molecules, Andrelina Maria Pinheiro Santos et al. (2022) discussed these topics in the chapter that dealt with the biodegradability assessment of biopolymer-based films, present in the book "Nanotechnology-Enhanced Food Packaging." This book deeply addressed the topic of toxicological aspects, safety, and component migration in nanosystems (Parameswaranpillai et al., 2022).

5.9 Conclusion and Prospects

In addition to increasing food shelf life and improving and maintaining quality and safety, edible coatings have been developed to meet the current demand for more sustainable processes, nonchemical preservation methods, healthy lifestyles, reliability, and practicality. Innovative strategies, such as using nanosystems, bio-nanocomposites, active components, and a combination of coating with other conservation treatments, have been applied to develop new edible coating alternatives. Currently, raw materials as by-products and waste have become a sustainable way of providing cheaper and environment-friendly materials. In addition, nanotechnology has emerged as a new trend for coatings. Several updates, such as nanocomposites, nanoemulsions, nanoparticles, nanofibers, lipid nanoparticles, nanostructured carriers, and their integration, have been made to the main coating application methods to include the nanosystems demand. Due to these advances, a variety of edible coatings have been applied to different types of food. However, it is extremely necessary to make a thorough analysis of aspects related to safety, toxicity, component migration, and the legislation and protocols established by each country.

Acknowledgment

The research conducted in this chapter is funded by the "Irish Research Council Government of Ireland Postgraduate Scholarship" under grant number GOIPG/2022/1392.

Conflicts of Interest

The authors declare no conflict of interest.

References

Adiletta, G., Di Matteo, M., & Petriccione, M. (2021). Multifunctional role of chitosan edible coatings on antioxidant systems in fruit crops: a review. *International Journal of Molecular Sciences*, *22*(5), 2633. https://doi.org/10.3390/ijms22052633

Alamdari, N. E., Aksoy, B., Aksoy, M., Beck, B. H., & Jiang, Z. (2021). A novel paper-based and pH-sensitive intelligent detector in meat and seafood packaging. *Talanta*, *224*, 121913. https://doi.org/10.1016/j.talanta.2020.121913

Alfei, S., Marengo, B., & Zuccari, G. (2020). Nanotechnology application in food packaging: a plethora of opportunities versus pending risks assessment and public concerns. *Food Research International*, *137*, 109664. https://doi.org/10.1016/j.foodres.2020.109664

Andrade, R. D., Skurtys, O., & Osorio, F. A. (2012). Atomizing spray systems for application of edible coatings. *Comprehensive Reviews in Food Science and Food Safety*, *11*(3), 323–337. https://doi.org/10.1111/j.1541-4337.2012.00186.x

Arnon-Rips, H., Cohen, Y., Saidi, L., Porat, R., & Poverenov, E. (2021). Covalent linkage of bioactive volatiles to a polysaccharide support as a potential approach for preparing active edible coatings and delivery systems for food products. *Food Chemistry*, *338*, 127822. https://doi.org/10.1016/j.foodchem.2020.127822

Arnon-Rips, H., & Poverenov, E. (2016). Biopolymers-Embedded Nanoemulsions and Other Nanotechnological Approaches for Safety, Quality, and Storability Enhancement of Food Products: Active Edible Coatings and Films. In A. M. Grumezescu (Ed.), *Nanotechnology in the Agri-Food Industry, Emulsions* (pp. 329–363). Academic Press. https://doi.org/10.1016/B978-0-12-804306-6.00010-6

Arnon-Rips, H., & Poverenov, E. (2018). Improving food products' quality and storability by using layer by layer edible coatings. *Trends in Food Science & Technology*, *75*, 81–92. https://doi.org/10.1016/j.tifs.2018.03.003

Arroyo, B. J., Bezerra, A. C., Oliveira, L. L., Arroyo, S. J., de Melo, E. A., & Santos, A. M. P. (2020). Antimicrobial active edible coating of alginate and chitosan add ZnO nanoparticles applied in guavas (*Psidium guajava* L.). *Food Chemistry*, *309*, 125566. https://doi.org/10.1016/j.foodchem.2019.125566

Ayala-Zavala, J. F., Silva-Espinoza, B. A., Cruz-Valenzuela, M. R., Leyva, J. M., Ortega-Ramírez, L. A., Carrazco-Lugo, D. K., Pérez-Carlón, J. J., Melgarejo-Flores, B. G., González-Aguilar, G. A., & Miranda, M. R. A. (2013). Pectin–cinnamon leaf oil coatings add antioxidant and antibacterial properties to fresh-cut peach. *Flavour and Fragrance Journal*, *28*(1), 39–45. https://doi.org/10.1002/ffj.3125

Azevedo, A. N., Buarque, P. R., Cruz, E. M. O., Blank, A. F., Alves, P. B., Nunes, M. L., & de Aquino Santana, L. C. L. (2014). Response surface methodology for optimisation of edible chitosan coating formulations incorporating essential oil against several foodborne pathogenic bacteria. *Food Control*, *43*, 1–9. https://doi.org/10.1016/j.foodcont.2014.02.033

Bahmid, N. A., Dekker, M., Fogliano, V., & Heising, J. (2021). Development of a moisture-activated antimicrobial film containing ground mustard seeds and its application on meat in active packaging system. *Food Packaging and Shelf Life*, *30*, 100753. https://doi.org/10.1016/j.fpsl.2021.100753

Baldwin, E. A., Nisperos-Carriedo, M. O., & Baker, R. A. (1995). Edible coatings for lightly processed fruits and vegetables. *Hort Science*, *30*(1), 35–37.

Basaglia, R. R., Pizato, S., Santiago, N. G., Maciel de Almeida, M. M., Pinedo, R. A., & Cortez-Vega, W. R. (2021). Effect of edible chitosan and cinnamon essential oil coatings on the shelf life of minimally processed pineapple (smooth cayenne). *Food Bioscience*, *41*, 100966. https://doi.org/10.1016/j.fbio.2021.100966

Bayram, B., Ozkan, G., Kostka, T., Capanoglu, E., & Esatbeyoglu, T. (2021). Valorization and application of fruit and vegetable wastes and by-products for food packaging materials. *Molecules*, *26*(13), 4031. https://doi.org/10.3390/molecules26134031

Bilbao-Sainz, C., Chiou, B., Punotai, K., Olson, D., Williams, T., Wood, D., Rodov, V., Poverenov, E., & McHugh, T. (2018). Layer-by-layer alginate and fungal chitosan based edible coatings applied to fruit bars. *Journal of Food Science*, *83*(7), 1880–1887. https://doi.org/10.1111/1750-3841.14186

Bott, J., Störmer, A., & Albers, P. (2018). Investigation into the release of nanomaterials from can coatings into food. *Food Packaging and Shelf Life*, *16*, 112–121. https://doi.org/10.1016/j.fpsl.2018.03.004

Chaudhry, Q., Scotter, M., Blackburn, J., Ross, B., Boxall, A., Castle, L., Aitken, R., & Watkins, R. (2008). Applications and implications of nanotechnologies for the food sector. *Food Additives & Contaminants: Part A*, *25*(3), 241–258. https://doi.org/10.1080/02652030701744538

Chen, F., Zhang, J., Chen, C., Kowaleguet, M. G. G. M., Ban, Z., Fei, L., & Xu, C. (2020). Chitosan-based layer-by-layer assembly: towards application on quality maintenance of lemon fruits. *Advances in Polymer Technology*, *2020*, 1–10. https://doi.org/10.1155/2020/7320137

Cheng, H., Xu, H., Julian McClements, D., Chen, L., Jiao, A., Tian, Y., Miao, M., & Jin, Z. (2022). Recent advances in intelligent food packaging materials: principles, preparation and applications. *Food Chemistry*, *375*, 131738. https://doi.org/10.1016/j.foodchem.2021.131738

Cosme Silva, G. M., Silva, W. B., Medeiros, D. B., Salvador, A. R., Cordeiro, M. H. M., da Silva, N. M., Santana, D. B., & Mizobutsi, G. P. (2017). The chitosan affects severely the carbon metabolism in mango (*Mangifera indica* L. cv. palmer) fruit during storage. *Food Chemistry*, *237*, 372–378. https://doi.org/10.1016/j.foodchem.2017.05.123

Costa, M. J., Maciel, L. C., Teixeira, J. A., Vicente, A. A., & Cerqueira, M. A. (2018). Use of edible films and coatings in cheese preservation: opportunities and challenges. *Food Research International*, *107*, 84–92. https://doi.org/10.1016/j.foodres.2018.02.013

Daraei Garmakhany, A., Mirzaei, H. O., Maghsudlo, Y., Kashaninejad, M., & Jafari, S. M. (2014). Production of low fat french-fries with single and multi-layer hydrocolloid coatings. *Journal of Food Science and Technology*, *51*(7), 1334–1341. https://doi.org/10.1007/s13197-012-0660-9

Debeaufort, F., Quezada-Gallo, J.-A., & Voilley, A. (1998). Edible films and coatings: tomorrow's packagings: a review. *Critical Reviews in Food Science and Nutrition*, *38*(4), 299–313. https://doi.org/10.1080/10408699891274219

Demetrio, A. A., Dos Santos, M., Melo, E. A., & Santos, A. M. P. (2019). A produção do conhecimento nas ciências da saúde 2. *Atena Editora*. https://doi.org/10.22533/at.ed.999193004

Dhumal, C. V., & Sarkar, P. (2018). Composite edible films and coatings from food-grade biopolymers. *Journal of Food Science and Technology*, *55*(11), 4369–4383. https://doi.org/10.1007/s13197-018-3402-9

Díaz-Montes, E., & Castro-Muñoz, R. (2021). Edible films and coatings as food-quality preservers: an overview. *Foods*, *10*(2), 249. https://doi.org/10.3390/foods10020249

Divya, K., Smitha, V., & Jisha, M. S. (2018). Antifungal, antioxidant and cytotoxic activities of chitosan nanoparticles and its use as an edible coating on vegetables. *International Journal of Biological Macromolecules*, *114*, 572–577. https://doi.org/10.1016/j.ijbiomac.2018.03.130

EU (2011a). Commission recommendation of 18 October 2011 on the definition of nanomaterial (2011/696/EU). *Official Journal of the European Union*, *275*, 38–40.

EU (2011b). Commission regulation (EU) no 10/2011 of 14 January 2011 on plastic materials and articles intended to come into contact with food. *Official Journal of the European Union*, *22*(1).

Falguera, V., Quintero, J. P., Jiménez, A., Muñoz, J. A., & Ibarz, A. (2011). Edible films and coatings: structures, active functions and trends in their use. *Trends in Food Science & Technology*, *22*(6), 292–303. https://doi.org/10.1016/j.tifs.2011.02.004

Fang, Z., Zhao, Y., Warner, R. D., & Johnson, S. K. (2017). Active and intelligent packaging in meat industry. *Trends in Food Science & Technology*, *61*, 60–71. https://doi.org/10.1016/j.tifs.2017.01.002

Farajzadeh, F., Motamedzadegan, A., Shahidi, S.-A., & Hamzeh, S. (2016). The effect of chitosan-gelatin coating on the quality of shrimp (*Litopenaeus vannamei*) under refrigerated condition. *Food Control*, *67*, 163–170. https://doi.org/10.1016/j.foodcont.2016.02.040

Farcuh, M., Rivero, R. M., Sadka, A., & Blumwald, E. (2018). Ethylene regulation of sugar metabolism in climacteric and non-climacteric plums. *Postharvest Biology and Technology*, *139*, 20–30. https://doi.org/10.1016/j.postharvbio.2018.01.012

Ferreira, M. S., Fai, A. E. C., Andrade, C. T., Picciani, P. H., Azero, E. G., & Gonçalves, É. C. (2016). Edible films and coatings based on biodegradable residues applied to acerolas (*Malpighia punicifolia* L.). *Journal of the Science of Food and Agriculture*, *96*(5), 1634–1642. https://doi.org/10.1002/jsfa.7265

Ferreira Saraiva, L. E., de Oliveira Melo Naponucena, L., da Silva Santos, V., Silva, R. P. D., de Souza, C. O., Evelyn Gomes Lima Souza, I., de Oliveira Mamede, M. E., & Druzian, J. I. (2016). Development and application of edible film of active potato starch to extend mini panettone shelf life. *LWT*, *73*, 311–319. https://doi.org/10.1016/j.lwt.2016.05.047

Galus, S., Arik Kibar, E. A., Gniewosz, M., & Kraśniewska, K. (2020). Novel materials in the preparation of edible films and coatings—a review. *Coatings*, *10*(7), 674. https://doi.org/10.3390/coatings10070674

Galus, S., & Kadzińska, J. (2015). Food applications of emulsion-based edible films and coatings. *Trends in Food Science & Technology*, *45*(2), 273–283. https://doi.org/10.1016/j.tifs.2015.07.011

Gardesh, A. S. K., Badii, F., Hashemi, M., Ardakani, A. Y., Maftoonazad, N., & Gorji, A. M. (2016). Effect of nanochitosan based coating on climacteric behavior and postharvest shelf-life extension of apple cv. *Golab Kohanz*. *Food Science and Technology*, *70*, 33–40. https://doi.org/10.1016/j.lwt.2016.02.002

Gregirchak, N., Stabnikova, O., & Stabnikov, V. (2020). Application of lactic acid bacteria for coating of wheat bread to protect it from microbial spoilage. *Plant Foods for Human Nutrition*, *75*(2), 223–229. https://doi.org/10.1007/s11130-020-00803-5

Grob, K., Camus, P., Gontard, N., Hoellinger, H., Joly, C., Macherey, A. C., Masset, D., Nesslany, F., Régnier, J. F., Riquet, A. M., Saillard, P., & Ribera, D. (2010). Need for a better safety evaluation of food contact materials produced from resins. *Food Control*, *21*(5), 763–769. https://doi.org/10.1016/j.foodcont.2009.11.006

Guerreiro, A. C., Gago, C. M. L., Faleiro, M. L., Miguel, M. G. C., & Antunes, M. D. C. (2015a). Raspberry fresh fruit quality as affected by pectin- and alginate-based edible coatings enriched with essential oils. *Scientia Horticulturae*, *194*, 138–146. https://doi.org/10.1016/j.scienta.2015.08.004

Guerreiro, A. C., Gago, C. M. L., Faleiro, M. L., Miguel, M. G. C., & Antunes, M. D. C. (2015b). The effect of alginate-based edible coatings enriched with essential oils constituents on *Arbutus unedo* L. fresh fruit storage. *Postharvest Biology and Technology*, *100*, 226–233. https://doi.org/10.1016/j.postharvbio.2014.09.002

Guo, Q., Du, G., Jia, H., Fan, Q., Wang, Z., Gao, Z., Yue, T., & Yuan, Y. (2021). Essential oils encapsulated by biopolymers as antimicrobials in fruits and vegetables: a review. *Food Bioscience*, *44*, 101367. https://doi.org/10.1016/j.fbio.2021.101367

Han, J., Elgowainy, A., Cai, H., & Wang, M. Q. (2013). Life-cycle analysis of bio-based aviation fuels. *Bioresource Technology*, *150*, 447–456. https://doi.org/10.1016/j.biortech.2013.07.153

Jafarzadeh, S., Mohammadi Nafchi, A., Salehabadi, A., Oladzad-abbasabadi, N., & Jafari, S. M. (2021a). Application of bio-nanocomposite films and edible coatings for extending the shelf life of fresh fruits and vegetables. *Advances in Colloid and Interface Science*, *291*, 102405. https://doi.org/10.1016/j.cis.2021.102405

Jafarzadeh, S., Salehabadi, A., Mohammadi Nafchi, A., Oladzadabbasabadi, N., & Jafari, S. M. (2021b). Cheese packaging by edible coatings and biodegradable nanocomposites; improvement in shelf life, physicochemical and sensory properties. *Trends in Food Science & Technology*, *116*, 218–231. https://doi.org/10.1016/j.tifs.2021.07.021

Jayasena, D. D., & Jo, C. (2013). Essential oils as potential antimicrobial agents in meat and meat products: a review. *Trends in Food Science & Technology*, *34*(2), 96–108. https://doi.org/10.1016/j.tifs.2013.09.002

de Jesus Salas-Méndez, E., Vicente, A., Pinheiro, A. C., Ballesteros, L. F., Silva, P., Rodríguez-García, R., Hernández-Castillo, F. D., de Lourdes Virginia Díaz-Jiménez, M., Flores-López, M. L., Villarreal-Quintanilla, J. Á., Peña-Ramos, F. M., Carrillo-Lomelí, D. A., & de Rodríguez, D. J. (2019). Application of edible nanolaminate coatings with antimicrobial extract of flourensia cernua to extend the shelf-life of tomato (*Solanum lycopersicum* L.) fruit. *Postharvest Biology and Technology*, *150*, 19–27. https://doi.org/10.1016/j.postharvbio.2018.12.008

Kumarihami, H. M. P. C., Kim, Y.-H., Kwack, Y.-B., Kim, J., & Kim, J. G. (2022). Application of chitosan as edible coating to enhance storability and fruit quality of kiwifruit: a review. *Scientia Horticulturae*, *292*, 110647. https://doi.org/10.1016/j.scienta.2021.110647

de Lima Júnior, R. G., Vinhas, G. M., Souto-Maior, R. M., Santos, A. M. P., Santos, E. J. P., & de Almeida, Y. M. B. (2016). Desenvolvimento e caracterização de filmes à base de poli (3-hidroxibutirato) aditivado com ZnOnano. *Brazilian Journal of Food Technology*, *19*. https://doi.org/10.1590/1981-6723.8215

López-Marcos, M. C., Bailina, C., Viuda-Martos, M., Pérez-Alvarez, J. A., & Fernández-López, J. (2015). Properties of dietary fibers from agroindustrial coproducts as source for fiber-enriched foods. *Food and Bioprocess Technology*, *8*(12), 2400–2408. https://doi.org/10.1007/s11947-015-1591-z

Maan, A. A., Reiad Ahmed, Z. F., Iqbal Khan, M. K., Riaz, A., & Nazir, A. (2021). Aloe vera gel, an excellent base material for edible films and coatings. *Trends in Food Science & Technology*, *116*, 329–341. https://doi.org/10.1016/j.tifs.2021.07.035

Mantilla, N., Castell-Perez, M. E., Gomes, C., & Moreira, R. G. (2013). Multilayered antimicrobial edible coating and its effect on quality and shelf-life of fresh-cut pineapple (*Ananas comosus*). *LWT-Food Science and Technology*, *51*(1), 37–43. https://doi.org/10.1016/j.lwt.2012.10.010

Maringgal, B., Hashim, N., Tawakkal, I. S. M. A., & Muda Mohamed, M. T. (2020). Recent advance in edible coating and its effect on fresh/fresh-cut fruits quality. *Trends in Food Science & Technology*, *96*, 253–267. https://doi.org/10.1016/j.tifs.2019.12.024

Md Nor, S., & Ding, P. (2020). Trends and advances in edible biopolymer coating for tropical fruit: a review. *Food Research International*, *134*, 109208. https://doi.org/10.1016/j.foodres.2020.109208

Melini, V., & Melini, F. (2018). Strategies to extend bread and GF bread shelf-life: from sourdough to antimicrobial active packaging and nanotechnology. *Fermentation*, *4*(1), 9. https://doi.org/10.3390/fermentation4010009

Müller, K., Bugnicourt, E., Latorre, M., Jorda, M., Echegoyen Sanz, Y., Lagaron, J., Miesbauer, O., Bianchin, A., Hankin, S., Bölz, U., Pérez, G., Jesdinszki, M., Lindner, M., Scheuerer, Z., Castelló, S., & Schmid, M. (2017). Review on the processing and properties of polymer nanocomposites and nanocoatings and their applications in the packaging, automotive and solar energy fields. *Nanomaterials*, *7*(4), 74. https://doi.org/10.3390/nano7040074

Nikolic, M. V., Vasiljevic, Z. Z., Auger, S., & Vidic, J. (2021). Metal oxide nanoparticles for safe active and intelligent food packaging. *Trends in Food Science & Technology*, *116*, 655–668. https://doi.org/10.1016/j.tifs.2021.08.019

de Oliveira Filho, J. G., Miranda, M., Ferreira, M. D., & Plotto, A. (2021). Nanoemulsions as edible coatings: a potential strategy for fresh fruits and vegetables preservation. *Foods*, *10*(10), 2438. https://doi.org/10.3390/foods10102438

Pandey, V. K., Islam, R. U., Shams, R., & Dar, A. H. (2022). A comprehensive review on the application of essential oils as bioactive compounds in nano-emulsion based edible coatings of fruits and vegetables. *Applied Food Research*, *2*(1), 100042. https://doi.org/10.1016/j.afres.2022.100042

Pandhi, S., Kumar, A., & Alam, T. (2019). Probiotic edible films and coatings: concerns, applications and future prospects. *Journal of Packaging Technology and Research*, *3*(3), 261–268. https://doi.org/10.1007/s41783-019-00069-6

Pantaleao, I., Pintado, M., & Pocas, M. (2007). Evaluation of two packaging systems for regional cheese. *Food Chemistry*, *102*(2), 481–487. https://doi.org/10.1016/j.foodchem.2006.05.058

Paraman, I., Sharif, M. K., Supriyadi, S., & Rizvi, S. S. H. (2015). Agro-food industry byproducts into value-added extruded foods. *Food and Bioproducts Processing*, *96*, 78–85. https://doi.org/10.1016/j.fbp.2015.07.003

Parameswaranpillai, J., Krishnankutty, R. E., Jayakumar, A., Rangappa, S. M., Siengchin, S., & (Eds.). (2022). *Nanotechnology-Enhanced Food Packaging*. Wiley. https://doi.org/10.1002/9783527827718

Pedrós-Garrido, S., Clemente, I., Calanche, J. B., Condón-Abanto, S., Beltrán, J. A., Lyng, J. G., Brunton, N., Bolton, D., & Whyte, P. (2020). Antimicrobial activity of natural compounds against listeria spp. and their effects on sensory attributes in salmon (*Salmo salar*) and cod (*Gadus morhua*). *Food Control*, *107*, 106768. https://doi.org/10.1016/j.foodcont.2019.106768

Perera, K. Y., Jaiswal, S., & Jaiswal, A. K. (2022). A review on nanomaterials and nanohybrids based bio-nanocomposites for food packaging. *Food Chemistry*, *376*, 131912. https://doi.org/10.1016/j.foodchem.2021.131912

Ponce, A. G., Roura, S. I., del Valle, C. E., & Moreira, M. R. (2008). Antimicrobial and antioxidant activities of edible coatings enriched with natural plant extracts: in vitro and in vivo studies. *Postharvest Biology and Technology*, *49*(2), 294–300. https://doi.org/10.1016/j.postharvbio.2008.02.013

Priyadarshi, R., & Rhim, J.-W. (2020). Chitosan-based biodegradable functional films for food packaging applications. *Innovative Food Science & Emerging Technologies*, *62*, 102346. https://doi.org/10.1016/j.ifset.2020.102346

Proulx, J., Sullivan, G., Marostegan, L. F., VanWees, S., Hsu, L. C., & Moraru, C. I. (2017). Pulsed light and antimicrobial combination treatments for surface decontamination of cheese: favorable and antagonistic effects. *Journal of Dairy Science*, *100*(3), 1664–1673. https://doi.org/10.3168/jds.2016-11582

Qian, M., Liu, D., Zhang, X., Yin, Z., Ismail, B. B., Ye, X., & Guo, M. (2021). A review of active packaging in bakery products: applications and future trends. *Trends in Food Science & Technology*, *114*, 459–471. https://doi.org/10.1016/j.tifs.2021.06.009

Rahnemoon, P., Sarabi-Jamab, M., Bostan, A., & Mansouri, E. (2021). Nano-encapsulation of pomegranate (*Punica granatum* L.) peel extract and evaluation of its antimicrobial properties on coated chicken meat. *Food Bioscience*, *43*, 101331. https://doi.org/10.1016/J.FBIO.2021.101331

Rai, M., Ingle, A. P., Gupta, I., Pandit, R., Paralikar, P., Gade, A., Chaud, M. V., & dos Santos, C. A. (2019). Smart nanopackaging for the enhancement of food shelf life. *Environmental Chemistry Letters*, *17*(1), 277–290. https://doi.org/10.1007/s10311-018-0794-8

Resende, N. S., Gonçalves, G. A. S., Reis, K. C., Tonoli, G. H. D., & Boas, E. V. B. V. (2018). Chitosan/cellulose nanofibril nanocomposite and its effect on quality of coated strawberries. *Journal of Food Quality*, *2018*, 1–13. https://doi.org/10.1155/2018/1727426

Riveros, C. G., Nepote, V., & Grosso, N. R. (2016). Thyme and basil essential oils included in edible coatings as a natural preserving method of oilseed kernels. *Journal of the Science of Food and Agriculture*, *96*(1), 183–191. https://doi.org/10.1002/jsfa.7080

Rojas-Graü, M. A., Soliva-Fortuny, R., & Martín-Belloso, O. (2009). Edible coatings to incorporate active ingredients to fresh-cut fruits: a review. *Trends in Food Science & Technology*, *20*(10), 438–447. https://doi.org/10.1016/j.tifs.2009.05.002

Romanazzi, G., Feliziani, E., Baños, S. B., & Sivakumar, D. (2017). Shelf life extension of fresh fruit and vegetables by chitosan treatment. *Critical Reviews in Food Science and Nutrition*, *57*(3), 579–601. https://doi.org/10.1080/10408398.2014.900474

Salgado, P. R., Ortiz, C. M., Musso, Y. S., Di Giorgio, L., & Mauri, A. N. (2015). Edible films and coatings containing bioactives. *Current Opinion in Food Science*, *5*, 86–92. https://doi.org/10.1016/j.cofs.2015.09.004

Sánchez, L. C. A., Real, C. P. V., & Perez, Y. B. (2014). Effect of an edible crosslinked coating and two types of packaging on antioxidant capacity of castilla blackberries. *Food Science and Technology (Campinas)*, *34*(2), 281–286. https://doi.org/10.1590/fst.2014.0047

Sánchez-González, L., Vargas, M., González-Martínez, C., Chiralt, A., & Cháfer, M. (2011). Use of essential oils in bioactive edible coatings: a review. *Food Engineering Reviews*, *3*(1), 1–16. https://doi.org/10.1007/s12393-010-9031-3

Santos, A. M. P., Del Carmen, B., Arroyo, J., de Souza Cavalcante Carnaval, L., & de Almeida Melo, E. (2022). Biodegradability Assessment of Biopolymer-Based Films. In J. Parameswaranpillai, R. E. Krishnankutty, A. Jayakumar, S. M. Rangappa, & S. Siengchin (Eds.), *Nanotechnology-Enhanced Food Packaging* (pp. 307–325). Wiley. https://doi.org/10.1002/9783527827718.ch13

Silva, J. G. S., Rebellato, A. P., Greiner, R., & Pallone, J. A. L. (2017). Bioaccessibility of calcium, iron and magnesium in residues of citrus and characterization of macronutrients. *Food Research International*, *97*, 162–169. https://doi.org/10.1016/j.foodres.2017.04.005

Silva, S. P. M., Ribeiro, S. C., Teixeira, J. A., & Silva, C. C. G. (2022). Application of an alginate-based edible coating with bacteriocin-producing lactococcus strains in fresh cheese preservation. *LWT*, *153*, 112486. https://doi.org/10.1016/j.lwt.2021.112486

Silva, C. C. G., Silva, S. P. M., & Ribeiro, S. C. (2018). Application of bacteriocins and protective cultures in dairy food preservation. *Frontiers in Microbiology*, *9*, 594. https://doi.org/10.3389/fmicb.2018.00594

Souza, M. P., Vaz, A. F. M., Cerqueira, M. A., Texeira, J. A., Vicente, A. A., & Carneiro-da-Cunha, M. G. (2015). Effect of an edible nanomultilayer coating by electrostatic self-assembly on the shelf life of fresh-cut mangoes. *Food and Bioprocess Technology*, *8*(3), 647–654. https://doi.org/10.1007/s11947-014-1436-1

Suhag, R., Kumar, N., Petkoska, A. T., & Upadhyay, A. (2020). Film formation and deposition methods of edible coating on food products: a review. *Food Research International*, *136*, 109582. https://doi.org/10.1016/j.foodres.2020.109582

Sutil, G. A., Andrade, K. S., Rebelatto, E. A., & Lanza, M. (2022). Effects of incorporation of pure or multicomponent active agents in biopolymers for food packaging using supercritical CO_2. *Trends in Food Science & Technology*, *120*, 349–362. https://doi.org/10.1016/j.tifs.2022.01.025

Tan, C., Han, F., Zhang, S., Li, P., & Shang, N. (2021). Novel bio-based materials and applications in antimicrobial food packaging: recent advances and future trends. *International Journal of Molecular Sciences*, *22*(18), 9663. https://doi.org/10.3390/ijms22189663

Tavassoli-Kafrani, E., Shekarchizadeh, H., & Masoudpour-Behabadi, M. (2016). Development of edible films and coatings from alginates and carrageenans. *Carbohydrate Polymers*, *137*, 360–374. https://doi.org/10.1016/j.carbpol.2015.10.074

Teixeira, R. F., Balbinot Filho, C. A., & Borges, C. D. (2022). Essential oils as natural antimicrobials for application in edible coatings for minimally processed apple and melon: a review on antimicrobial activity and characteristics of food models. *Food Packaging and Shelf Life*, *31*, 100781. https://doi.org/10.1016/j.fpsl.2021.100781

Torres-León, C., Vicente, A. A., Flores-López, M. L., Rojas, R., Serna-Cock, L., Alvarez-Pérez, O. B., & Aguilar, C. N. (2018). Edible films and coatings based on mango (var. Ataulfo) by-products to improve gas transfer rate of peach. *LWT*, *97*, 624–631. https://doi.org/10.1016/j.lwt.2018.07.057

Treviño-Garza, M. Z., García, S., Heredia, N., Alanís-Guzmán, M. G., & Arévalo-Niño, K. (2017). Layer-by-layer edible coatings based on mucilages, pullulan and chitosan and its effect on quality and preservation of fresh-cut pineapple (*Ananas comosus*). *Postharvest Biology and Technology*, *128*, 63–75. https://doi.org/10.1016/j.postharvbio.2017.01.007

Umaraw, P., Munekata, P. E. S., Verma, A. K., Barba, F. J., Singh, V. P., Kumar, P., & Lorenzo, J. M. (2020). Edible films/coating with tailored properties for active packaging of meat, fish and derived products. *Trends in Food Science & Technology*, *98*, 10–24. https://doi.org/10.1016/j.tifs.2020.01.032

Valencia-Chamorro, S. A., Palou, L., del Río, M. A., & Pérez-Gago, M. B. (2011). Antimicrobial edible films and coatings for fresh and minimally processed fruits and vegetables: a review. *Critical Reviews in Food Science and Nutrition*, *51*(9), 872–900. https://doi.org/10.1080/10408398.2010.485705

Vilela, C., Kurek, M., Hayouka, Z., Röcker, B., Yildirim, S., Antunes, M. D. C., Nilsen-Nygaard, J., Pettersen, M. K., & Freire, C. S. R. (2018). A concise guide to active agents for active food packaging. *Trends in Food Science & Technology*, *80*, 212–222. https://doi.org/10.1016/j.tifs.2018.08.006

Volpe, S., Torrieri, E., & Cavella, S. (2017). Use of chitosan and chitosan-caseinate coating to prolong shelf life of minimally processed apples. *Italian Journal of Food Science/Rivista Italiana Di Scienza Degli Alimenti*, *30*(5), 30–35.

Wang, H., Qian, J., & Ding, F. (2018). Emerging chitosan-based films for food packaging applications. *Journal of Agricultural and Food Chemistry*, *66*(2), 395–413. https://doi.org/10.1021/acs.jafc.7b04528

Xing, Y., Li, W., Wang, Q., Li, X., Xu, Q., Guo, X., Bi, X., Liu, X., Shui, Y., Lin, H., & Yang, H. (2019). Antimicrobial nanoparticles incorporated in edible coatings and films for the preservation of fruits and vegetables. *Molecules*, *24*(9), 1695. https://doi.org/10.3390/molecules24091695

Xu, E., Zhang, Y., & Lin, L. (2020). Improvement of mechanical, hydrophobicity and thermal properties of Chinese fir wood by impregnation of nano silica sol. *Polymers*, *12*(8), 1632. https://doi.org/10.3390/polym12081632

Yan, J., Luo, Z., Ban, Z., Lu, H., Li, D., Yang, D., Aghdam, M. S., & Li, L. (2019). The effect of the layer-by-layer (LBL) edible coating on strawberry quality and metabolites during storage. *Postharvest Biology and Technology*, *147*, 29–38. https://doi.org/10.1016/j.postharvbio.2018.09.002

Yildirim, S., Röcker, B., Pettersen, M. K., Nilsen-Nygaard, J., Ayhan, Z., Rutkaite, R., Radusin, T., Suminska, P., Marcos, B., & Coma, V. (2018). Active packaging applications for food. *Comprehensive Reviews in Food Science and Food Safety*, *17*(1), 165–199. https://doi.org/10.1111/1541-4337.12322

Yin, C., Huang, C., Wang, J., Liu, Y., Lu, P., & Huang, L. (2019). Effect of chitosan- and alginate-based coatings enriched with cinnamon essential oil microcapsules to improve the postharvest quality of mangoes. *Materials*, *12*(13), 2039. https://doi.org/10.3390/ma12132039

Yuan, Y., Ma, M., Xu, Y., & Wang, D. (2022). Surface coating of zein nanoparticles to improve the application of bioactive compounds: a review. *Trends in Food Science & Technology*, *120*, 1–15. https://doi.org/10.1016/j.tifs.2021.12.025

Zhang, R., Hoffmann, T., & Tsotsas, E. (2020a). Novel technique for coating of fine particles using fluidized bed and aerosol atomizer. *Processes*, *8*(12), 1525. https://doi.org/10.3390/pr8121525

Zhang, W., & Jiang, W. (2020). Antioxidant and antibacterial chitosan film with tea polyphenols-mediated green synthesis silver nanoparticle via a novel one-pot method. *International Journal of Biological Macromolecules*, *155*, 1252–1261. https://doi.org/10.1016/j.ijbiomac.2019.11.093

Zhang, W., Jiang, H., Rhim, J.-W., Cao, J., & Jiang, W. (2022). Effective strategies of sustained release and retention enhancement of essential oils in active food packaging films/coatings. *Food Chemistry*, *367*, 130671. https://doi.org/10.1016/j.foodchem.2021.130671

Zhang, R., Zhang, Z., & McClements, D. J. (2020b). Nanoemulsions: an emerging platform for increasing the efficacy of nutraceuticals in foods. *Colloids and Surfaces B: Biointerfaces*, *194*, 111202. https://doi.org/10.1016/j.colsurfb.2020.111202

6

Multilayer Flexible Films for Bio-Based Food Packaging

Maria José Costa[1,2], *Victor Souza*[1], *Lorenzo Pastrana*[1], *and Miguel Ângelo Cerqueira*[1]

[1] *International Iberian Nanotechnology Laboratory, Braga, Portugal*
[2] *Centre of Biological Engineering, University of Minho, Braga, Portugal*

6.1 Introduction

Food packaging is present in our daily lives, and in the last years, packaging become essential due to its functionalities, namely protection, convenience, and communication (Bauer et al., 2021). To fulfill the customers' needs and to be considered a solution for the new customers' lifestyle where convenience, information, and traceability become more valued, new solutions and improvement of the existing packaging need to be performed (Cerqueira et al., 2019). Moreover, today, the demands from the industry are higher and more demanding for food packaging. The cost, production methods, sustainability, materials, and packaging weight are all topics that need to be considered (Cerqueira et al., 2018a). To improve food packaging and assure commercial feasibility new solutions are studied and explored, using different approaches to improve the packaging performance (e.g., mechanical properties and gas properties). Methods using blends, composites, and multilayers appeared as possible solutions and presented significant improvements in packaging properties (Cerqueira et al., 2018a).

Packaging materials can be based on different materials and present different forms, but all of them present advantages and drawbacks. In most cases, the final application dictates the selection of the material used to produce the packaging and the form that it gets. In packaging systems, primary packaging is very important to preserve the quality of the products and guarantee their safety when consumed. One of the most explored forms is flexible packaging, which has been used for multiple applications due to its lightweight and mechanical features. According to Niaounakis (2020), flexible packaging refers to "packaging structures that are capable of being flexed or bent, such that they are pliant and yieldable in response to externally applied forces." This kind of packaging structure can alter its shape according to the packed product and can be in the form of a bag, pouch, liner, or

Nanotechnology for Sustainable Food Packaging, First Edition. Edited by C. Anandharamakrishnan, Jeyan A. Moses, and M. Maria Leena.

overwrap. Flexible packaging systems can exist in different forms, such as monolayer, coated monolayer, or multilayer, where the layer has a specific function (e.g., barrier and seal layers) according to the application.

The use of a barrier layer is related to the poor permeability of the packaging materials based on polymers (both petroleum-based and bio-based) when compared to other materials such as glass or metal. Therefore, researchers and the industry have been working to find new and innovative technological solutions that could help improve those properties. Some ways to improve their barrier properties are incorporating additives (e.g., nanoparticles), blending different materials, and using layers or coatings from different materials (Zabihzadeh Khajavi et al., 2020).

Nowadays, the most common way is to produce multilayer films, where the gas permeability is highly decreased by the combination of different materials in the form of layers. For example, ethylene vinyl alcohol is often used as an inner layer (typically less than 10 μm) of the multilayer film to guarantee the barrier to oxygen, but also metalized films have been used. However, these complex systems are very difficult to recycle (Niaounakis, 2020). Currently, the proposed solution to overcome this issue is to decrease the number of layers and used materials that can be physically or chemically recycled in a conventional process.

The same approach has been presented for bio-based and biodegradable packaging, where the use of layers can help improve the properties of the films. Therefore, it is proposed that several bio-based and biodegradable materials that not being compatible can be used in combination in the form of a multilayer (Jariyasakoolroj et al., 2020; Mujtaba et al., 2022). For example, polyesters such as polylactide, polyhydroxyalkanoates (PHAs), or paper can be coated with polysaccharides, proteins, or waxes aiming to improve their barrier to oxygen or water, respectively.

One of the approaches that emerged in the last 30 years in the packaging industry was the use of nanotechnology (Cerqueira et al., 2018b). The use of different nanosized fillers has been proposed to improve the properties of materials, and as well as for petroleum-based packaging, bio-based packaging materials can also use nanotechnology to overcome some of these challenges. Nanoparticles, nanofibers, and nanocoatings have been shown to be a good solution to reinforce packaging materials (improving mechanical, thermal, and barrier properties), but also a way to develop active (e.g., antimicrobial and scavenging compounds) an intelligent (e.g., indicators and nano barcodes and radio frequency identification active tags) solutions that could add value to the packaging and thus improve quality and safety of food products.

In this chapter, we addressed the most recent developments of multilayer flexible films for food applications using biopolymers. We describe the different production technologies and materials used and the most explored applications focused on the reinforcement of the materials and the development of active and intelligent packaging systems. Additionally, some examples of food applications are presented.

6.2 Bio-Based Materials and Methodologies

The design of multilayer structures should focus on at least three levels of knowledge: the materials used, the production technologies, and the intended final application. The following sections of this chapter are focused on the different bio-based materials used in multilayer flexible packaging and the processing technologies that can be used in their production. Then, some examples related to the reinforcement of the materials and the development of active and intelligent materials will be presented.

6.2.1 Bio-Based Materials

Packaging is one of the main reasons for the large plastic consumption worldwide (Macena et al., 2021). The use of petrol-based materials to produce the plastic used in packaging led to environmental problems related to waste management, water, and carbon footprint. To reduce the environmental impact and overcome the inherent limitations of petrol-based plastics, bio-based and/or biodegradable materials became the subject of study as an eco-friendly solution, able to face the new customer demands and also to provide innovative packaging systems (e.g., active and intelligent packaging) (Cerqueira et al., 2016; Cerqueira et al., 2019). Bio-based materials can be extracted from different sources like biomass, synthesized from bio-derived monomers, and produced by microorganisms; some specific materials are presented in Figure 6.1 (Cerqueira et al., 2019; Chausali et al., 2022; Morinval and Averous, 2022).

Each of these bio-based materials has unique properties, which can be seen as a positive feature (e.g., strength, flexibility); however, they can also limit their use (e.g., low moisture and/or gas barrier). The use of the multilayer method could help to overcome the limitations of each material and improve their properties. Regarding food packaging, the biomaterials most explored and reported in literature come from three different sources (i.e., extracted from natural sources, synthesized, and produced by microorganisms).

The most studied polysaccharides are chitosan, alginate, starch, and cellulose derivatives, while beeswax, carnauba wax, and free fatty acids are the most used lipids, and

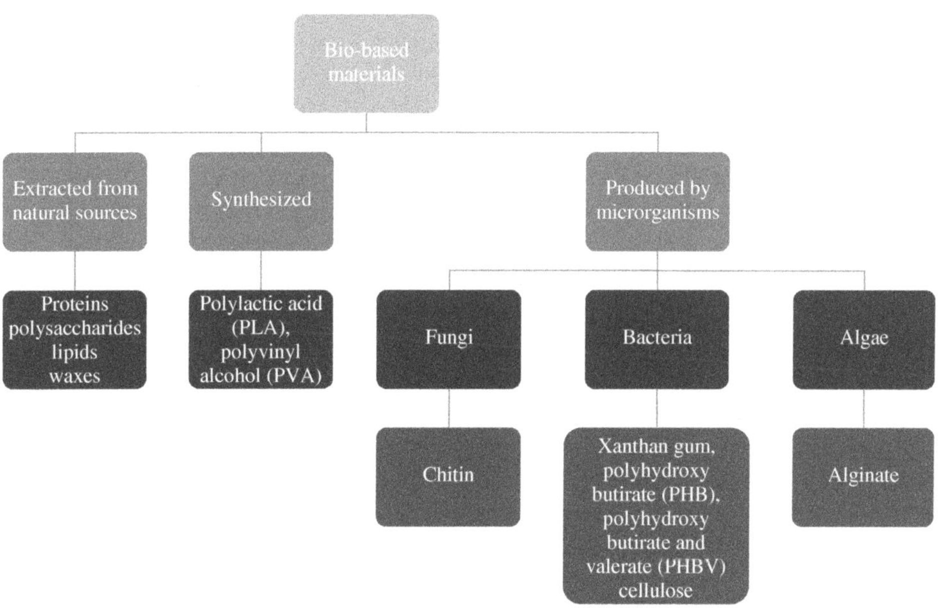

Figure 6.1 Examples of bio-based materials used in food packaging.

regarding proteins, whey, casein, and gluten are widely studied. Finally, polylactic acid (PLA), is the most used synthetic biopolymers, while among the ones produced by microorganisms are the polyhydroxybutyrates (PHB) and bacterial cellulose the most studied (Păuşescu et al., 2022).

The mentioned biomaterials are used to produce films and that can go from some micrometers to the nanoscale in thickness. But they can also be used in different formats, such as particles, rods, and fibers at different scales. They should be considered since they can help improve the barrier and mechanical properties of the materials. In this regard, besides cellulose nanocrystals (CNCs) and nanofibers that are widely used, also inorganic materials can be used.

6.2.2 Production Methodologies

Some of the methodologies used to produce multilayer structures focusing on flexible films for food packaging applications will be discussed in detail in this section.

6.2.2.1 Solution Casting and Layer-by-Layer (LbL) Technique

Solution casting is one of the easiest methods to produce thin films, thus commonly used at laboratory scale, but also suitable to produce multilayer structures. Basically, it consists of dissolving the polymer or biopolymer in the proper solvent; in this step, other ingredients of interest can also be incorporated to improve some functional properties (i.e., plasticizers, active compounds, or nano-fillers); after its complete homogenization, the film forming suspension (FFS) is poured onto an inert surface (e.g., glass mold, Petri dishes, and Teflon plates, to mention a few) and left to dry. After solvent evaporation, the resulting films can be detached and are ready to be used (Pola et al., 2016).

To form bilayers or multilayers structures, the solution casting technique needs to be repeated for each layer to be assembled. Thus, after casting the first layers (outer film) onto the inert surface, the system is let to dry, but not completely, only until a firm but still adhesive surface is obtained. Then, the FFS of the second polymer is cast on the top of the first layer, and both layers are left to dry (Ferreira et al., 2016; Phan The et al., 2008; Souza et al., 2022). The electrostatic interactions and hydrogen bondings formed between the polymers composing the bilayers and multilayer films are responsible for their good assembling. Strategies to enhance the interactions between the layers are to use a crosslinking agent, such as calcium chloride, or by chemically modifying the polymers to obtain stronger hydrogen bonding (Luo and Wang, 2014; Zhang et al., 2021; Zhuang et al., 2018). The solution casting method is mainly used at the laboratory scale; however, it is possible to scale up the process to an industrial scale by coupling with a wet coating station with a roll-to-roll system (Vasile, 2018). Another simple and versatile method to prepare multilayer thin films is the LbL technique, first presented by Decher et al. (1991). LbL is based on the sequential adsorption of oppositely charged molecules onto a substrate guided by electrostatic interaction and hydrogen bonding, resulting in a multilayer structure with layers reaching up to a few nanometers (Cerqueira et al., 2018a; Manna et al., 2009). The final properties and functionalities of the multilayer are affected by the adsorbing material used in each layer, and the condition used to prepare the multilayer, the number of layers incorporated and their sequence (Calinoiu et al., 2019; Cerqueira et al., 2018a). The LbL technique is an interesting approach to producing a multilayer structure incorporated with active substances for their controlled release since it sustains the permeation of small molecules while trapping larger molecules, and this semi-permeable feature can be regulated by the experimental parameters during the assembly (Calinoiu et al., 2019). The conventional assembly methods used for the formation of multilayers based on LBL are the dipping and spray methods; however, recently, these production methods have also been revised to form nanolayer structures (Richardson et al., 2015). The application of these techniques is limited to polymers that can be easily dissolved in organic/inorganic solvents at milder temperatures and are mainly used to cast biodegradable polymers such as chitosan, alginate, cellulose derivates, agar, to mention a few.

6.2.2.2 Thermoplastic "Dry" Process—Extrusion

Extrusion is one of the most important polymer processing techniques commonly used at industrial scale. It consists of the application of high energy input for a short time with controlled pressure and temperature, allowing the polymers to melt (i.e., become formable) and be subsequently formed through the extruder die. In the packaging field, the main processes used are injection molding, film extrusion, and extrusion coating (Müller et al., 2017).

Other additives, such as plasticizers, preservatives, and nanomaterials, can be incorporated during the process, enabling the production of different types of nanocomposites (Azeredo et al., 2011; Wahyuningtiyas and Suryanto, 2018). To achieve an exfoliated nanocomposite using an extrusion process, the extruder and screw configurations are responsible for the good dispersion quality, as the exfoliation is favored at high shear rates and longer residence time results in better dispersion (Müller et al., 2017).

In the development of bio-based biodegradable polymers, these melting processes have a serious limitation due to the intrinsic lack of thermal stability of many compounds and biopolymers used for this purpose, especially those with bioactive properties (natural antimicrobial and antioxidant compounds) that can be lost during the process (i.e., degraded and evaporated) (Vasile, 2018). In these cases, alternative methods are preferred such as coating techniques. The application of layers/multilayers can also be done using extrusion coating. This technique is commonly applied on foils, paper, and petroleum-based polymers, and it consists of extruding a web directly into the tip of a roll-to-roll conveyor where the substrate is passing. To ensure adequate adhesion of the extruded polymer to the substrate, high temperature is necessary (Müller et al., 2017).

Coextrusion technology is another way to obtain multilayer films. Through this process, it is possible to produce films with up to thousands of layers with layer thicknesses in the micro- and nanoscale (Cerqueira et al., 2018a). In coextrusion processes, it is important to use materials with similar rheology to ensure a consistent melt curtain and even distribution of the individual layers; if one polymer is more viscous than the other, there is the risk of this polymer pushing the other out of the edges of the melt curtain (Vlachopoulos and Polychronopoulos, 2019).

6.2.2.3 Coating Technologies

Coatings are defined as a coherent layer formed onto a substrate, while nanocoatings are differentiated by their thickness, as the latter are ultra-thin layers on the nanoscale, that is, <1–100 nm thick (Müller et al., 2017; Vasile, 2018). These coatings can improve the material, for example, by enhancing its gas-barrier properties, hydrophilic/hydrophobic or oleophobic properties, and insulating or conductive properties (Müller et al., 2017). Moreover, active compounds can also be incorporated into the coating material to create additional functions such as antimicrobial and antioxidant properties. Examples of compounds used for this purpose are antimicrobial nanoparticles (metal oxides nanoparticles—silver, zinc oxide, and titanium dioxide, to mention a few) and bioactive organic agents (bioactive extracts from herbs and spices, chitosan, essential oils, peptides, to mention a few) (Bhargava et al., 2020; Enescu et al., 2019; Soltani Firouz et al., 2021). More details on the development of active food packaging are presented in Section 6.3.2.

The formation of multilayer structures through the application of coatings of different polymers is commonly used to produce flexible multilayer food packaging. The substrates can be coated using a variety of techniques, and the simplest one is probably the bench coating using a multicoater (e.g., commercialized by RK Print Coat Instruments Ltd., UK) (Khwaldia et al., 2014). Other more robust techniques can also be used to coat polymeric substrates, such as spray technology, spin coating, dispersing coating, and electrospinning, among others. More details on these technologies are found in the following sections.

Moreover, prior to the coating application, the substrate can also go through a surface modification to increase the interactions between the coating solution and the substrate (Silvestre et al., 2017). Surface modifications can be done by physical, chemical (wet), or biological methods; however, physical methods are preferred over the others as they offer greater precision, easier industrial scalability, simple process control as no liquid reagents are needed, and are environmentally friendly (Goddard and Hotchkiss, 2007; Vasile, 2018). The most used physical methods are the ionized gas treatments with plasma, corona or flame, and UV and gamma-ray irradiation. Surface modifications result in the formation of reactive species (such as oxygenated groups—carbonyl, hydroxyl, carboxyl, ester groups and carboxylic acid, or nitrogen-containing groups) to react with the coating molecules (polymer, bioactive compounds, and nanoparticles) as the radical sites are available for coupling covalently or noncovalently immobilized (Vasile, 2018). As this subject is not the aim of this book chapter, those readers more interested in surface modification can get more information from existing books and reviews (Silvestre et al., 2017; Goddard and Hotchkiss, 2007; Sun and Chmielewski, 2017).

Spray Technology

Spray technologies can be used to functionalize the surface of polymer materials by the application of single or multiple layers of coating material onto the substrate.

The traditional process to spray a solution onto a substrate is by the use of spraying guns, which can be manual or automatic with air compressors. In fact, various configurations are found in the market to adjust to a wide range of applications and processes. One common gun used is the high volume low pressure, which has the advantages of low maintenance cost, high transfer efficiency, versatile applications (suitable for a variety of materials), reduced consumption of coating material, controllable spraying, and it is portable (Vasile, 2018).

A novel spray technology that couples ultrasounds in the atomization process is produced and commercialized by Sono-Tek (USA). This new technology is known as ultrasonic spray coating, and compared to other coating systems, the ultrasonic spray nozzle is more precise, controllable, and environmentally friendly (Sono-Tek Corporation, 2022). Moreover, the ultrasonic nozzles are suitable for depositing precision nanotechnology coatings, are economical, with good reproducibility efficiency, and are capable of producing droplets in the micrometer range (Abbas et al., 2021; Riemer, 2011).

Coating nanoparticles with spray technology is not easy once, when conventional coating systems are used (e.g., pressure spray coating, spin coating, etc.), the particles tend to agglomerate and separate from the liquid. When ultrasonic spray coating technology is used, this does not occur due to the continuous vibrations that are automatically generated, which break the agglomerates, even holding nanoparticles dispersed in the suspension for several hours (Riemer, 2011).

The ultrasonic energy concentrated in the atomizing surface of the nozzle is responsible for breaking the liquid (coating solution/suspension) into a spray (fine mist of the coating solution) of tiny drops homogenously dispersed onto the substrate (Sono-Tek Corporation, 2022). Opposite to the broad drop size distribution generally obtained in the conventional spray coating, a narrower distribution with a smaller medium size is achieved with the ultrasonic nozzle from Sono-Tek (Riemer 2011).

Nanostructured Layers by EHDP

Electrohydrodynamic processing (EHDP) is a technology that comprises two micro and nanofabrication processes, namely electrospinning and electrospraying (Fabra et al., 2013; Vasile, 2018). EHDP is suitable to produce polymer- and biopolymer-based fiber/particle with diameters within the submicron and nano range through the action of an external high voltage electric field forced on a polymer solution, which cannot be achieved by using conventional-generating technologies (Cerqueira et al., 2018a; Müller et al., 2017). The electrospinning process refers to those capable of producing nonwoven mats of continuous polymer ultrathin fibers (electrospun material), while electrospraying produces noncontinuous structures, such as nanoparticles in the form of capsules or beads (electrospray material) (Fabra et al., 2016). Electrospraying is a cost-effective technique that breaks up the coating solution into fine particles using uniform electrohydrodynamic force (Müller et al., 2017). Different nanostructure morphologies (electrospun or electrospray material) are achieved by changing the coating solution properties (viscosity, conductivity, and surface tension) and process parameters (applied voltage, tip-to-collector distance, and flow rate) (Cerqueira et al., 2018a).

In comparison to conventional deposition and coating techniques, EHDP techniques have the advantages of high deposition efficiency (up to 80%) and a reduction of the process steps without requiring a controlled pressure or temperature environment (Müller et al., 2017). Electrospun ultrathin fiber-like morphology presents the intrinsic ability to reinforce the mechanical and barrier properties of the polymer and bio-based polymer matrices when smartly applied (Cerqueira et al., 2018a); moreover, due to its high specific surface area and porosity, it is also capable of encapsulating active substances for efficient and prolonged controlled release of active compounds (Vasile, 2018).

In multiplayer structured packaging materials, once the nanofibers (electrospun) adhere by physical forces onto the substrate, no need for tie layers are needed, thus highlighting the potential and advantage of EHDP to generate materials with improved passive and active/bioactive properties by the direct application of the coating onto the substrate or producing interlayers of electrospun material. By controlling the coating solution properties and operational parameters, it is also possible to control the thickness of the layers applied to meet specific requirements (e.g., mechanical and barrier properties) (Cerqueira et al., 2018a; Fabra et al., 2013).

The production of multilayer structured flexible films using exclusively bio-based materials that keep their biodegradability and/or renewability is suitable to be achieved through EHDP once there is no need for synthetic adhesives or tie layers to assemble the different layers when both coatings and interlayers are prepared by EHDP (Cerqueira et al., 2018a; Gaona-Sánchez et al., 2021). Another big advantage of having layers/interlayers in nanostructured form produced by EHDP is that they do not potentially affect the overall mechanical performance of the structure by not affecting flexibility and toughness with improved crack resistance and no need for compatibilizers (Fabra et al., 2013). This technology can be applied in several fields; however, only recently has it been scaled up, and both techniques (i.e., electrospinning and electrospraying) can be manufactured at any scale via multi-nozzle injectors through companies such as Bioinicia S.L. (www.bioinicia.com).

Post-treatment with a thermal annealing process can also be applied to improve the final properties of multilayer flexible films produced with EHDP, and this step consists of the thermal compression of the composite in temperatures below the melting temperature of the biopolymers aiming to improve the adhesion between layers (Lasprilla-Botero et al., 2018).

6.2.2.4 Other Techniques

Gas phase processes are another class of methods used to coat flexible substrates with polymeric layers with thickness on the nanometer scale. According to the deposition mechanisms, they are classified as physical vapor deposition (PVD) or chemical vapor deposition (CVD) (Müller et al., 2017). The former is based on the transfer of the solid coating material into the vapor phase under vacuum, followed by its condensation on the top of the substrate, while the latter is based on chemical reactions between the substrate and the volatile precursor, which by reacting and/or decomposing on the substrate surface produce the desired deposition/coating (Müller et al., 2017; Vasile, 2018).

The PVD is commonly used to deposit thin metal films on the surface of polymers to achieve good barrier properties to gases/light required for some types of foods; the typical materials used are metals, metalloids and their oxides, nitrides, and carbides (Müller et al., 2017).

Surface modification of the substrates enables the deposition of even inorganic or polymer materials; thus plasma enhanced chemical vapor deposition (PECVD) was created, which combines a pretreatment of the surface with plasma followed by the CVD process (Müller et al., 2017).

The application of silicon oxide (SiO_x) coating on polymer substrates has been investigated recently as an alternative to metalized polymers as this material is transparent, non-toxic, suitable to microwave and with outstanding barrier properties (Song et al., 2018). PECVD is a suitable technique for fabricating this SiO_x coating, and it was successfully used to prepare a layered Poly-L-Lactic Acid (PLLA)/SiO_x film with high gas, water, and UV light barrier properties and enhanced perm selectivity (Müller et al., 2017; Song et al., 2018).

Atomic layer deposition (ADL) is a modified version of the CVD process developed by VTT (Valtion Teknillinen Tutkimuskeskus) (Vasile, 2018). It consists of a surface-controlled layer-by-layer (LbL) deposition process based on repeated self-limited gas reactions, suitable to produce inorganic moisture and gas barrier coatings (e.g., SiO_2 and ZnO) onto both traditional petroleum-based polymers and bio-based ones (Vartiainen et al., 2014). The coatings produced by the ADL method are thin, light, conformal and, as the coating follows the contours of the coated material, they are pinhole-free with excellent gas barrier properties (Vasile, 2018).

6.3 Multilayer Structures

6.3.1 Reinforcing Materials

To be used as packaging, a multilayer structure must present a specific physical resistance (mechanical properties) and water and gas permeability, depending on the purpose of application (i.e., type of food, type of packaging, and environmental conditions). Several studies are focused on how to overcome the inherent and specific limitations of biomaterials, and the results showed that using multilayer structures and nanomaterials could improve mechanical and barrier properties. Some of these studies are described in Table 6.1.

A biopolymer can form a coating and be part of the film but can, in some cases, be used as a nanomaterial. The nanomaterials are added to one layer to help enhance their performance or, in some cases, represent the layer itself. Starch is listed among the most used biopolymers to form mono and multilayer films. Starch is a carbohydrate considered a good biopolymer to be used for packaging applications due to its low cost, availability, and biodegradability, but it presents some limitations due to its water sensitivity, low moisture barrier, and low tensile strength (La Fuente Arias et al., 2021; Trinh et al., 2021). Nevertheless, its use as thermoplastic starch (TPS) has been widely explored. Another widely used biopolymer is PLA. PLA is a biodegradable thermoplastic that, due to its good mechanical properties and biodegradability properties, is appealing to be used in food packaging, although it has a poor oxygen barrier and brittleness that, together with its high cost, represent some limitations to its utilization (Trinh et al., 2021). Recently, Trinh et al. (2021) used different processes (i.e., reactive extrusion, compression molding, and dip coating) to produce multilayer films with the TPS and PLA. TPS was modified using maleic anhydride (MA) and/or incorporating hydrophilic montmorillonite nanoclays. The optimized multilayer film with nano clays filled maleated TPS and PLA (MTPS-3C/PLA) displayed a significant improvement in the moisture barrier of 1,300% compared to TPS films and 3,300% in oxygen barrier compared to neat PLA films. These multilayer films were composed of compostable and bio-based ingredients and showed that the incorporation of MA and nanoclays into TPS and the assembling with PLA significantly reinforced the films improving their moisture and gas barrier properties. These results were achieved because MA decreased the crystallinity and polarity of TPS, improving the interfacial interaction with PLA, improving the adhesive strength between each layer and the nano clays at optimal loading, also enhanced the physico-mechanical and barrier properties of the multilayer assembly (Trinh et al., 2021).

In the study of Gao et al. (2018), multilayer coextrusion was used to organize and align graphite nanoplatelets (GNPs) in a PLA matrix. The authors used a forced-assembled multilayer coextrusion methodology to produce 128 alternating layers of neat PLA and PLA/GNPs (in different ratios). The multilayering process reduced the thickness of the GNPs and increased their alignment, indicating improved dispersion and possible exfoliation, which was confirmed by the improvements in both mechanical and barrier

Table 6.1 Biodegradable multilayers, composition, production methods, and reinforcing properties

Biomaterial/Matrix Composition	Production Method	Main Conclusions	Reference
Thermoplastic corn starch (TPS), polycaprolactone (PCL)	Compression molding	Higher mechanical and barrier properties compared to TPS	Ortega-Toro et al. (2015)
Plasticized wheat starch (PWS) poly(lactic acid) (PLA), polyesteramide (PEA), poly(ε-caprolactone) (PCL), poly(butylene succinate adipate) (PBSA), and poly(hydroxybutyrate-co-valerate) (PHBV)	Coextrusion and compression molding	Slightly improve mechanical properties Satisfactory water resistance properties compared to PWS	Martin et al. (2001)
Gelatin, PLA	Heat compression molding	Higher tensile strength and lower WVP values when compared to PLA films	Martucci and Ruseckaite (2010)
PLA, PBSA	Forced assembly coextrusion	Higher water vapor and gas barrier	Messin et al. (2017)
PLA, wheat gluten proteins (WG)	High-pressure homogenization of WG Corona treatment of PLA Wet casting and spin coating for WG coating thickness and hot pressing to shape multilayers	O_2 barrier improved 2,000% Water vapor barrier improved by 20%	Rocca-Smith et al. (2019)
PLA, PHBV	Coextrusion		Boufarguine et al. (2013)

properties of the multilayer when compared to both neat PLA or monolayer PLA/GNPs. In terms of barrier properties, these PLA/GNP multilayer films presented similar performances as PLA/nanoclay monolayer films but only at 1/10 of the filler loading; thus, the force-assembled multilayer method is a technology suitable to produce lightweight and strong packaging materials for food and industrial applications (Gao et al., 2018).

Another interesting biopolymer is chitosan. It is one of the most abundant polysaccharides obtained from the partial deacetylation of chitin and is biodegradable, biocompatible, and presents antimicrobial activity (La Fuente Arias et al., 2021). Recently, Wang et al. (2022) used carboxymethyl chitosan (CMCC)/carboxymethyl cellulose sodium (CMCS) + PLA and nano zinc oxide nanoparticles (ZnONPS) and produced a multilayer coated paperboard (CMCC/CMCS + PLA/ZnO NPs) and realized that the oxygen transmission rate (OTR) was reduced by 99% and the barrier properties regarding soybean and water vapor were doubled when compared with a single PLA coating. Also, the oil resistance increased 28 times, and the water vapor barrier properties were increased by twofold when compared to a single CMCC/CMCS coating layer. The barrier properties when ZnO NPs were present in the coating were improved (i.e., oxygen transfer rate and water vapor transmission rate (WVTR) were reduced by more than 50%) due to the effect of ZnO_2 NPs, that filled the gap in the PLA layer and blocked the oxygen penetration and due to the increase in tortuosity.

Bilayers formed only by mixing two different polymers without any cross-linker can be considered polyelectrolyte complexes (PEC), as they are formed simultaneously by mixing oppositely charged polyelectrolyte polymers (Luo and Wang, 2014). Chitosan is a good example of a biopolymer suitable to produce these PEC; once in the acidic medium, the amino groups on the backbone are protonated, changing chitosan into a cationic polyelectrolyte capable of forming PEC with negatively charged polyelectrolytes, such as pectin (Souza et al., 2022), fucopol (Ferreira et al., 2016; Ferreira et al., 2018), hyaluronic acid (Manna et al., 2009), carrageenan (Luo and Wang, 2014), and so forth. Abbas et al., (2021) studied the oxygen and UV barrier properties of nano-ZnO-coated polyethylene terephthalate (PET) and poly(3-hydroxybutyrate-co-3-hydroxyhexanoate) (PHBHHx) films produced using ultrasonic spray coating. The authors tested a coating solution composed of 2.5 wt.% ZnO NPs in polyvinyl alcohol (PVA) aqueous solution, which was applied in different numbers of coating passes (layers), namely 3, 5, 10, and 50 times (Abbas et al., 2021). ZnO NPs coating presented distinguish behavior depending on the substrate they were applied onto. Due to the hydrophobic characteristic of PHBHHx, the nanoparticles were not well adhered to its surface, which resulted in noncoated areas and worse results compared to coated PET films. Improvement in the oxygen barrier properties was only depicted for PET films coated with 50 layers of Zn ONPs, with a reduction of 5% in the OTR. Regarding the UV barrier properties, a significant increase was observed in the UV blocking characteristic for all films with the application of the nanocoatings. In order to enhance the adherence of the coating onto the substrate and improve gas barrier properties, the substrates need to pass through a surface modification (i.e., a pretreatment with plasma or corona) to enable chemical interactions between the polymer and the coating solution, surpassing the only physical deposition achieved (Abbas et al., 2021).

A good example of the use of electrospinning to produce multilayer packaging is the three layers based on wheat gluten (WG) and electrospun poly-hydroxybutyrates (PHB) developed by Fabra et al. (2015). In this study, plasticized WG was produced in compression mold and coated with electrospinning of PHB or poly-hydroxybutyrate-co-valerate copolymer with 3% valerate content (PHBV3) on both sides of the WG film, followed by an annealing post-treatment in a hot press at 145 or 160 °C during 1 minute. The addition of PHB and PHBV3 outer layer to plasticized WG film resulted in a significant increase in both barriers to oxygen and water vapor while improving the mechanical properties. However, the transparency was negatively affected, as the composites presented slightly lower transparency than stand-alone WG films. Overall, the films were a bit more yellowish and opaque due to the amount of deposited polymer and the temperature used in the posttreatment process.

In another study designed by the same group, a superhydrophobic bilayer coating based on annealed electrospun ultrathin poly(ε-caprolactone) (PCL) fibers and electrospray nanostructured silica microplates (SiO_2) were developed on corona treated low-density polyethylene (LDPE) films, followed by annealing posttreated, as depicted in Figure 6.2 (Lasprilla-Botero et al., 2018). This is a good example of the application of both electrospinning and electrospraying techniques with thermal annealing posttreatment, which resulted

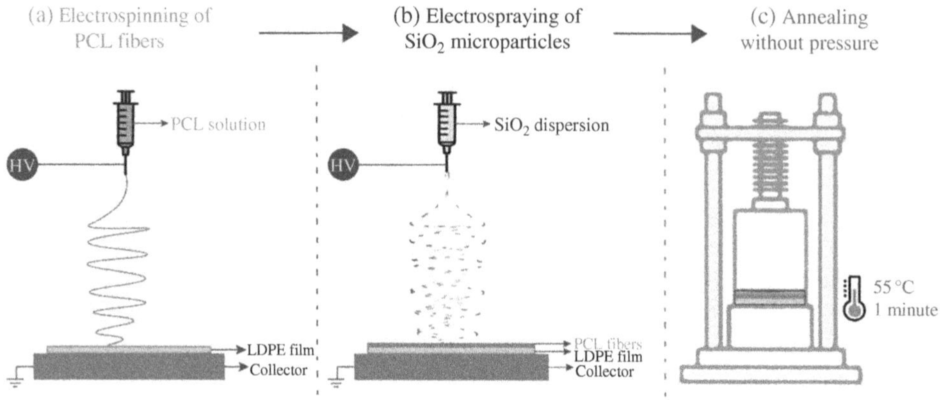

Figure 6.2 Schematic illustration of the multilayer film production: (a) depositing of the electrospun polycaprolactone (PCL) on corona-treated LDPE; (b) deposition of silica microparticles on PCL/LDPE; (c) thermal annealing posttreatment of the multilayer films. HV: high voltage; PCL: poly(ε-caprolactone); LDPE: low-density poly-ethylene (Published under an open access Creative Common CC BY license) (*Source:* Lasprilla-Botero et al. (2018)/MDPI/CC BY 4.0)

in bonded coating layers to themselves and to the LDPE substrate that could not be delaminated by hand. The application of the PCL layer promoted a higher apparent contact angle than the stand-alone LDPE surface, which increased from ≅67° to ≅128°. However, with the thermal treatment, this hydrophobic enhancement was reduced, especially when it was applied at higher temperatures for longer times (i.e., >55° for 5 minutes). This behavior was explained by the reduction of the inter-fiber roughness with the thermal treatment that promotes inter-material adhesion in the ultrathin film. To achieve this superhydrophobic surface, additional topographical features were created by adding the SiO_2 electrospray particles on top of the PCL fiber, resulting in a superhydrophobic surface with a contact angle of up to ≅157°. In terms of gas barrier properties, the multilayer structure presented values of permeance to D-limonene (considered a standard to assess aroma barrier) in the same magnitude, however, with a slight improvement on the aroma barrier for the composite LDPE/PCL and a decreased, to a small extent, when the electrospray SiO_2 was also applied (Lasprilla-Botero et al., 2018). EHDP is, therefore, a suitable methodology to create permanent and thermally stable superhydrophobic films to be applied in food packaging applications with easy-emptying properties for high water-content foods.

Cellulose is the most abundant polymer on Earth. It is biodegradable and biocompatible with low toxicity. It can be extracted from different sources like wood and cotton and presents several properties that make it suitable to use as food packaging material, like its good mechanical properties, high strength, and low cost (Păușescu et al., 2022). Besides its use as the main material for the production of paper and cellulose-based films, it is also possible to produce nanocrystals and nanofibers. Figueroa-Lopez et al. (2020) used PHAs with high-barrier coatings of CNCs to produce active multilayer films. PHBV loaded with Oregano essential oil (OEO) and ZnONPS, produced using electrospinning, was deposited onto the PHA film and a CNC coating was applied using casting using 2% of nanocellulose. These films presented high antimicrobial and antioxidant properties for up to 15 days and showed to be safe in alcoholic and acid food simulants and also nontoxic to Caco-2 cells.

The combination of more than one technique to produce multilayer structured films is also a tendency, as in the work of Vartiainen et al. (2016), in which multilayer films were fabricated by combining three different techniques/materials for the application of intermediate layers: (i) dispersion coating (cellulose nanofibrils, CNF), (ii) ADL (Al_2O_3; aluminum oxide), and (iii) extrusion coating (polyglycolic acid, PGA). The substrate used was bio-HDPE (high density polyethylene), on which all the different layers were deposited, and as the outer/final layer, bio-LDPE was extruded coated. Intermediate layers of CNF, Al_2O_3, and PGA presented 2 μm, 35 nm, and 29 μm, respectively, which illustrate how ADL is suitable to produce layers on a nanometer scale. Multilayer structured with intermediate layers presented reductions in both OTR and WVTR when compared to the bio-HDPE/LPDE bilayer. In fact, multilayer presented OTR at 23 °C/50% RH reduction from 2130 $cm^3/(m^2$ day) for bio-HDPE/LPDE to 8, 650, and 0.9 $cm^3/(m^2$ day) respectively for intermediate layer of CNF, Al_2O_3, and PGA, respectively. The superior value for OTR of the multilayer containing the ADL Al_2O_3 layer is probably related to the extrusion coating of bio-LDPE once the authors reported OTR values around only 10–12 $cm^3/(m^2$ day) for the bio-HDPE/Al_2O_3 before the addition of the bio-LDPE extruded layer. During the extrusion coating, high melt temperatures and low coating speed are used, which could have a detrimental effect on the bio-HDPE surface onto which the thin Al_2O_3 coating was deposited (Vartiainen et al., 2016).

6.3.2 Active Materials

Active packaging can be defined as a system where there is an active modification of the internal package environment through physical, chemical, or biological interactions between the food product, the packaging, and the internal environment, with the purpose of improving food safety, quality, and shelf-life (López-Carballo et al., 2012; Yam and Zhu, 2012). When producing active packaging, the focus can be on oxygen, ethylene and carbon dioxide scavenging, moisture control, microbial growth, and controlled release of antimicrobials are also widely explored when developing active packaging (López-Carballo et al., 2012). In active packaging with controlled release, active compounds (e.g., antimicrobials and antioxidants) and/or flavors are incorporated into the packaging and then released into the food product in a controlled way.

The most common structure is the sandwich structure, presenting three layers. One layer is the barrier layer, which is in contact with the environment and will act as a barrier to moisture, oxygen, and microorganisms. The second can be an active layer that is in the middle and will contain active compounds like antimicrobials and antioxidants, and the inner layer in contact with the food will be the control layer that protects the food and the active compounds and will control the active compounds release rate into the food (Figure 6.3) There are other multilayer structures where the active layer can be in other position, and the control layer is inexistent (Wang et al. 2022). The nature of the active compounds (e.g., polarity, volatility, solubility), the environmental factors (e.g., pH, temperature, relative humidity, and light exposure), and the food properties (e.g., surface hydrophobicity, porosity, and moisture), will have a big impact on the functional performance of the packaging, besides the matrix by itself.

Also, PVA is a biodegradable polymer that is nontoxic, water-soluble and presents high transparency and good chemical and mechanical resistance, adhesive and antistatic

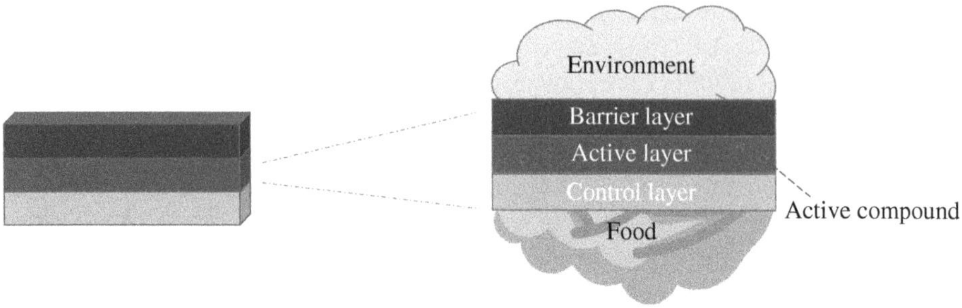

Figure 6.3 Schematic representation of a multilayer sandwich structure.

properties (Păuşescu et al., 2022). A study using electrospinning technique and casting produced nanofibers and coatings, respectively, using PHBV film as matrix and PVA with incorporated bacteriophage Felix O1 (F01) deposited as nanofiber or as coating. The results demonstrated that both delivery systems with F01 exhibit promising properties to inhibit and control *Salmonella* contamination in food and can be considered a potential solution to be used in food packaging (Costa et al., 2021).

Jipa et al. (2012) used bacterial cellulose (BC) and incorporated sorbic acid (SA), an antimicrobial agent, and compared monolayer films with BC and PVA and multilayers coated with BC membrane. The results showed a significant decrease in the SA release rate in multilayer films, showing the potential of this new multi-layered film to be used as active antimicrobial food packaging material. Also using a cellulose matrix, Martins et al. (2018) developed bio-based bilayer films using two different techniques (i.e., solvent casting and electrospinning). Ethylcellulose was deposited on microfibrillated cellulose films, and cinnamaldehyde was used as an antimicrobial compound and incorporated into the ethylcellulose layer. After characterization, results showed that the bilayer films produced using the solvent casting method exerted antimicrobial activity of cinnamaldehyde against *Listeria monocytogenes* and *Salmonella Typhimurium* and displayed a significant decrease in WVTR, improving barrier properties and observed hydrophobic properties in the surface.

Cerqueira et al. (2016) also used cinnamaldehyde and developed a three-multilayer bio-based structure using PHBV and zein in the interlayer through the electrospinning method. The electrospun multilayer structures (PHBV + zein/CNMA + PHBV) displayed antimicrobial activity against *L. monocytogenes* and showed reduced transport properties (water vapor permeability [WVP] and oxygen permeability) related to CNMA hydrophobic character, showing great potential for food applications. Another study already mentioned in a previous section that also uses electrospinning reported that a multilayer coating paperboard (CMCS/CMC + PLA/ZnO NPs) coated with 1.5% ZnO nanoparticles presented antimicrobial activity toward *Escherichia coli* and *Staphylococcus aureus* with a 99% reduction. This multilayer coating shows interesting properties to be used in the food industry as a packaging for fast-food.

González and Alvarez Igarzabal (2013) used isolated soy protein (SPI) and PLA to produce bilayer films loaded with natamycin and thymol. The mechanical properties were improved significantly compared to SPI films; the water barrier was also improved, and the films exhibited a marked growth inhibition, presenting antifungal and antibacterial properties against *Aspergillus sp*, *Saccharomyces cerevisiae, E. coli* ATCC 25922, and *S. aureus* ATCC 25923.

6.3.3 Intelligent Materials

Intelligent packaging can detect, recognize, and record changes in the food product environment. It can use interactive devices to guarantee traceability and provide counterfeit protection through labels, tags, codes, RFID, sensors, and indicators that collect information regarding the environment (e.g., pH, time, temperature, and CO_2) and, depending on the recognized changes, it has a response with a signal; in the specific case of indicators, these provide quantitative information through a color change but are unable to store data of measurement and time (Ma et al., 2022; Sobhan et al., 2021). This type of packaging communicates to improve decision-making regarding the best actions to be performed to achieve the desired results for food quality and safety. This can be performed, for example, by monitoring and improving distribution and storage conditions (Fuertes et al., 2016; Yam, 2012).

In the literature, it can be seen that bio-based materials are mostly reported to produce sensors and indicators for intelligent packaging. Anthocyanins are the most common natural dye used as an indicator for pH changes, being safe and eco-friendly. Anthocyanins exhibit halochromic properties due to the reaction of flavylium cations that are dependent on a pH equilibrium; as a result, the color changes with pH changes (Păuşescu et al., 2022). An interesting approach was reported recently, where a biodegradable structure with a bio pH-sensitive dye was synthesized through acid-catalyzed condensation. Chitosan was the main polymer blended with PVA, starch, and cellulose, and anthocyanidin was used as a dye. The results showed no toxicity. Chitosan-PVA-dye films exhibited the best thermal stability and showed pH-responsive behavior (Păuşescu et al., 2022).

Jamróz et al. (2022) produced a bilayer film based on furcellaran, chitosan, and gelatin hydrolysates incorporated with the nanofillers montmorillonite and silver nanoparticles (AgNPs), for reinforcement and active action purposes, respectively, and with the bioactive compounds curcumin and rosemary essential oil (REO). The first layer was composed of furcellaran enriched with curcumin/REO (tested in three different levels), AgNPs (green synthesized using yerba mate extract), and montmorillonite, while the second layer contained a mixture of chitosan with gelatin hydrolysate (1:1). The bilayer bio-nanocomposites were successfully produced using two-step solution casting method. The presence of curcumin and AgNPs improved the UV light barrier properties of the film, and bilayers incorporated with the intermediate level of curcumin/REO presented the highest antimicrobial property, while the antioxidant activities of the films were correlated to the AgNP and curcumin loads. Moreover, films also presented potential use as indicators for monitoring the freshness of food products, as their color changes with pH value (from orange to red when pH increased from 2 to 10). The composites presented lower stiffness, and at the lowest load of ingredients, the WVTR was also reduced. This is an outstanding example of bio nanocomposites with potential use as smart food packaging (Jamróz et al., 2022), where the use of multilayer structure associated with nanotechnology improved the material's properties and added active properties.

6.4 Food Applications

There are different food applications where multilayer flexible packaging can be applied successfully, namely food packaging films. One of the approaches proposed by Lin et al. (2019), was the use of modified liposomes with chrysanthemum essential oil prepared with chitosan and pectin using LbL electrostatic deposition method directly in the food products. Three layers were used, where chitosan was used as the second layer and pectin as the third layer. A concentration of 0.050 mg/mL of chitosan and pectin was used, and the results showed that compared to mono and double layers, this system was more stable. The application of the multilayer liposome system was performed by spraying the chicken sample, and high antibacterial activity against *Campylobacter jejuni* was observed after 14 days of storage and in a temperature range from 4 °C to 37 °C, without impact on the chicken (i.e., sensory and color surface properties).

A recent study used a three-layer system with carvacrol or ferulic acid incorporated into PVA (internal layer) and with PLA as the two external layers obtained by thermocompression. The multilayer showed extended stretchability compared to PLA films, and the antimicrobial capacity was evaluated using packaged beef meat samples for 17 days at 5 °C; the total viable bacteria remained below 6 log (acceptable limit), and *E. coli* was below the detection limit while in the control the counts were between 1.5 and 1.8 log (Andrade et al., 2022).

Nowzari et al., (2013) compared the effect of a chitosan (1%, w/v)—gelatin (3%, w/v) coating vs film on the rancidity in rainbow trout at 4 °C for 16 days. Composite and bilayer (using a two-step casting technique) methods were used, and the results showed an extension of rainbow trout shelf-life after 16 days under refrigerated conditions. The CFU counts of total viable counts and Gram-negative psychrophilic bacteria were below the 7 log, while in the control samples, this limit was achieved after eight days of storage. Also, an antioxidant effect was observed, although the coating performed better than the film regarding lipid oxidation. These results showed a possible packaging solution to be used for refrigerated fish food products.

Multilayer structures were also used as indicators by Sobhan et al. (2021), who developed an exco-friendly color indicator produced by casting based on nanocellulose/chitosan-based film with methyl red, coated with PLA to monitor beef spoilage over time under refrigerated conditions (4 °C) (Figure 6.4). An apparent color change was observed

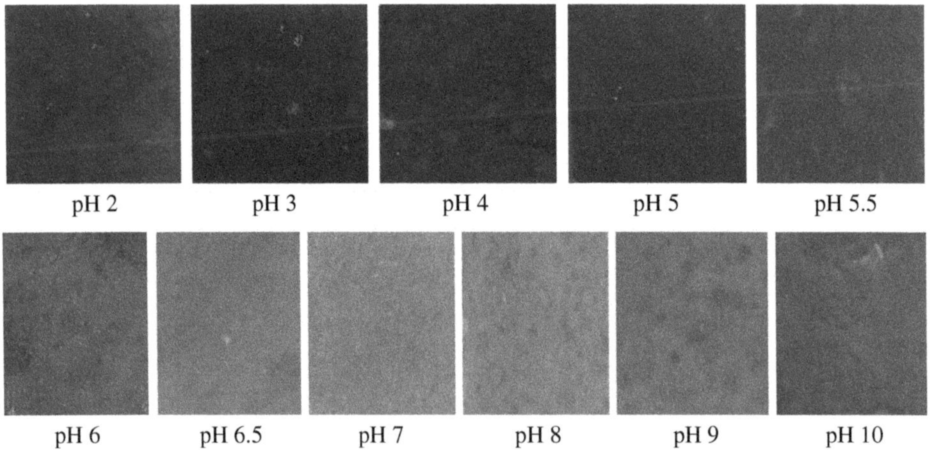

Figure 6.4 Color response of the PLA/Nanocellulose Chitosan Mixture (NCM) film with respect to different pH buffer solutions (pH 2–10) (*Source:* Sobhan et al., (2021) / with permission of Royal Society of Chemistry.)

after five days, indicating a threshold for the beef spoilage linked with the pH increase correlated with a total viable count increase. These results show the potential of using this system as a color indicator for smart food packaging and real-time monitoring.

6.5 Conclusion

The use of packaging in food products is one of the most effective ways to decrease food waste and is, without any doubt, one of the paths to act against food insecurity. However, recent sustainability issues related to petroleum-based plastics and unclear information regarding paper-based packaging and bioplastics have brought to discussion among stakeholders on how the sustainability assessment should be performed. It is clear that bio-based materials are one of the solutions to avoid using fossil sources. Still, in some cases and according to the applications, their final properties need to be highly improved for commercial viability. In the case of flexible packaging, the use of multilayer structures using nanotechnology clearly shows to be a promising strategy to overcome some of the issues related to the use of bio-based materials for food packaging. In this chapter, we showed that the final properties of films can be improved in terms of barrier properties but can also be used to develop active and intelligent packaging systems. In the case of intelligent packaging, although several works have been done using bio-based materials the use of multilayer systems is unexplored and should be addressed in the following years. Other of the lacks in the field are the studies related to migration in food simulants to confirm compliance with regulation and real food applications.

Acknowledgments

The authors acknowledge the financial support of the project "Agenda VIIAFOOD: Platform for the Agri-Food valorisation, industrialization and commercial innovation", with the reference n.º CC644929456-00000040, co-funded by Component C5 – Capitalisation and Business Innovation under the Portuguese Resilience and Recovery Plan, through the NextGenerationEU Fund.

References

Abbas, M., Buntinx, M., Deferme, W., Reddy, N., & Peeters, R. (2021). Oxygen gas and UV barrier properties of nano-ZnO-coated pet and phbhhx materials fabricated by ultrasonic spray-coating technique. *Nanomaterials*, *11*(2), 1–14. https://doi.org/10.3390/nano11020449

Andrade, J., González-Martínez, C., & Chiralt, A. (2022). Antimicrobial PLA-PVA multilayer films containing phenolic compounds. *Food Chemistry*, *375*, 131861. https://doi.org/10.1016/J.FOODCHEM.2021.131861

Azeredo, H. M. C., Mattoso, L. H. C., & McHugh, T. H. (2011). Nanocomposites in Food Packaging—A Review. In B. Reddy (Ed.), *Advances in Diverse Industrial Applications of Nanocomposites* (p. 550). InTech. DOI: 10.5772/1931

Bauer, A. S., Tacker, M., Uysal-Unalan, I., Cruz, R. M. S., Varzakas, T., & Krauter, V. (2021). Recyclability and redesign challenges in multilayer flexible food packaging—a review. *Foods*, *10*(11). https://doi.org/10.3390/foods10112702

Bhargava, N., Sharanagat, V. S., Mor, R. S., & Kumar, K. (2020). Active and intelligent biodegradable packaging films using food and food waste-derived bioactive compounds: a review. *Trends in Food Science and Technology*, *105*(July), 385–401. https://doi.org/10.1016/j.tifs.2020.09.015

Boufarguine, M., Guinault, A., Miquelard-Garnier, G., & Sollogoub, C. (2013). PLA/PHBV films with improved mechanical and gas barrier properties. *Macromolecular Materials and Engineering*, *298*(10), 1065–1073. https://doi.org/10.1002/mame.201200285

Calinoiu, L. F., Ştefanescu, B. E., Pop, I. D., Muntean, L., & Vodnar, D. C. (2019). Chitosan coating applications in probiotic microencapsulation. *Coatings*, *9*(3), 1–21. https://doi.org/10.3390/COATINGS9030194

Cerqueira, M. A., Fabra, M. J., Castro-Mayorga, J. L., Bourbon, A. I., Pastrana, L. M., Vicente, A. A., & Lagaron, J. M. (2016). Use of electrospinning to develop antimicrobial biodegradable multilayer systems: encapsulation of Cinnamaldehyde and their physicochemical characterization. *Food and Bioprocess Technology*, *9*(11), 1874–1884. https://doi.org/10.1007/S11947-016-1772-4/TABLES/3

Cerqueira, M. A., Torres-Giner, S., & Lagaron, J. M. (2018a). Nanostructured Multilayer Films. In M. A. Cerqueira, J. M. Lagarón, L. M. Pastrana, & A. A. Vicente (Eds.), *Nanomaterials for Food Packaging* (pp. 147–171). Taylor & Francis Group. https://doi.org/10.1016/B978-0-323-51271-8.00006-1

Cerqueira, M. A., Vicente, A. A., & Pastrana, L. M. (2018b). Chapter 1—Nanotechnology in Food Packaging: Opportunities and Challenges. In M. Â. P. R. Cerqueira, J. M. Lagaron, L. M. Pastrana Castro, & A. A. M. de Oliveira Soares Vicente (Eds.), *Micro and Nano Technologies* (pp. 1–11). Elsevier. https://doi.org/10.1016/B978-0-323-51271-8.00001-2

Cerqueira, M. A., Fuciños, P., & Pastrana, L. M. (2019). Nanotechnology as a Way for Bio-based and Biodegradable Food Packaging with Enhanced Properties. In Ó. L. Ramos, R. N. Pereira, M. A. Cerqueira, J. A. Teixeira, & A. A. Vicente (Eds.), *Advances in Processing Technologies for Bio-based Nanosystems in Food* (pp. 233–258). Taylor and Francis. https://doi.org/10.1201/9781315177328-11/NANOTECHNOLOGY-WAY-BIO-BASED-BIODEGRADABLE-FOOD-PACKAGING-ENHANCED-PROPERTIES-MIGUEL-CERQUEIRA-PABLO-FUCI

Chausali, N., Saxena, J., & Prasad, R. (2022). Recent trends in nanotechnology applications of bio-based packaging. *Journal of Agriculture and Food Research*, *7*, 100257. https://doi.org/10.1016/J.JAFR.2021.100257

Costa, M. J., Pastrana, L. M., Teixeira, J. A., Sillankorva, S. M., & Cerqueira, M. A. (2021). Characterization of PHBV films loaded with FO1 bacteriophage using polyvinyl alcohol-based nanofibers and coatings: a comparative study. *Innovative Food Science & Emerging Technologies*, *69*, 102646. https://doi.org/10.1016/j.ifset.2021.102646

Decher, G., Hong, J., & Gutenberg-universitt, J. (1991). *Gero Decher* and Jong-Dal Hong Institut fur Physikalische Chemie; Johannes Gutenberg-Universitt welder weg 11; D-6500 maim. Federal Republic of Germany AbstraG*, *327*, 321–327.

Enescu, D., Cerqueira, M. A., Fucinos, P., & Pastrana, L. M. (2019). Recent advances and challenges on applications of nanotechnology in food packaging. A literature review. *Food and Chemical Toxicology*, *134*(May), 110814. https://doi.org/10.1016/j.fct.2019.110814

Fabra, M. J., Busolo, M. A., López-Rubio, A., & Lagarón, J. M. (2013). Nanostructured biolayers in food packaging. *Trends in Food Science & Technology*, *31*(1), 79–87. https://doi.org/10.1016/j.tifs.2013.01.004

Fabra, M. J., López-Rubio, A., & Lagaron, J. M. (2015). Three-layer films based on wheat gluten and electrospun PHA. *Food and Bioprocess Technology*, *8*(11), 2330–2340. https://doi.org/10.1007/s11947-015-1590-0

Fabra, M. J., López-Rubio, A., & Lagaron, J. M. (2016). Use of the electrohydrodynamic process to develop active/bioactive bilayer films for food packaging applications. *Food Hydrocolloids*, *55*, 11–18. https://doi.org/10.1016/j.foodhyd.2015.10.026

Ferreira, A. R. V., Torres, C. A. V., Freitas, F., Sevrin, C., Grandfils, C., Reis, M. A. M., Alves, V. D., & Coelhoso, I. M. (2016). Development and characterization of bilayer films of FucoPol and chitosan. *Carbohydrate Polymers*, *147*, 8–15. https://doi.org/10.1016/j.carbpol.2016.03.089

Ferreira, A. R. V., Bandarra, N. M., Moldão-Martins, M., Coelhoso, I. M., & Alves, V. D. (2018). FucoPol and chitosan bilayer films for walnut kernels and oil preservation. *LWT—Food Science and Technology*, *91*(December 2017), 34–39. https://doi.org/10.1016/j.lwt.2018.01.020

Figueroa-Lopez, K. J., Torres-Giner, S., Angulo, I., Pardo-Figuerez, M., Escuin, J. M., Bourbon, A. I., Cabedo, L., Nevo, Y., Cerqueira, M. A., & Lagaron, J. M. (2020). Development of active barrier multilayer films based on electrospun antimicrobial hot-tack food waste derived poly(3-hydroxybutyrate-co-3-hydroxyvalerate) and cellulose nanocrystal interlayers. *Nanomaterials*, *10*(12), 2356. DOI: 10.3390/NANO10122356

Fuertes, G., Soto, I., Carrasco, R., Vargas, M., Sabattin, J., & Lagos, C. (2016). Intelligent packaging systems: sensors and nanosensors to monitor food quality and safety. *Journal of Sensors*, *2016*. https://doi.org/10.1155/2016/4046061

Gao, Y., Picot, O. T., Tu, W., Bilotti, E., & Peijs, T. (2018). Multilayer coextrusion of graphene polymer nanocomposites with enhanced structural organization and properties. *Journal of Applied Polymer Science*, *135*(13), 1–11. https://doi.org/10.1002/app.46041

Gaona-Sánchez, V. A., Calderón-Domínguez, G., Morales-Sánchez, E., Moreno-Ruiz, L. A., Terrés-Rojas, E., de la Salgado-Cruz, M., P., Escamilla-García, M., & Barrios-Francisco, R. (2021). Physicochemical and superficial characterization of a bilayer film of zein and pectin obtained by electrospraying. *Journal of Applied Polymer Science*, *138*(12), 1–15. https://doi.org/10.1002/app.50045

Goddard, J. M., & Hotchkiss, J. H. (2007). Polymer surface modification for the attachment of bioactive compounds. *Progress in Polymer Science*, *32*(7), 698–725. https://doi.org/10.1016/j.progpolymsci.2007.04.002

González, A., & Alvarez Igarzabal, C. I. (2013). Soy protein—poly (lactic acid) bilayer films as biodegradable material for active food packaging. *Food Hydrocolloids*, *33*(2), 289–296. https://doi.org/10.1016/J.FOODHYD.2013.03.010

Jamróz, E., Janik, M., Marangoni, L., Vieira, R. P., Tkaczewska, J., Kawecka, A., Szuwarzyński, M., Mazur, T., Jasińska, J. M., Krzyściak, P., & Juszczak, L. (2022). Double-layered films based on furcellaran, chitosan, and gelatin hydrolysates enriched with AgNPs in yerba mate extract, montmorillonite, and curcumin with rosemary essential oil. *Polymers*, *14*(20), 4283. https://doi.org/10.3390/polym14204283

Jariyasakoolroj, P., Leelaphiwat, P., & Harnkarnsujarit, N. (2020). Advances in research and development of bioplastic for food packaging. *Journal of the Science of Food and Agriculture*, *100*(14), 5032–5045. https://doi.org/10.1002/jsfa.9497

Jipa, I. M., Stoica-Guzun, A., & Stroescu, M. (2012). Controlled release of sorbic acid from bacterial cellulose based mono and multilayer antimicrobial films. *LWT*, *47*(2), 400–406. https://doi.org/10.1016/J.LWT.2012.01.039

Khwaldia, K., Basta, A. H., Aloui, H., & El-Saied, H. (2014). Chitosan-caseinate bilayer coatings for paper packaging materials. *Carbohydrate Polymers*, *99*(2014), 508–516. https://doi.org/10.1016/j.carbpol.2013.08.086

La Fuente Arias, C. I., Kubo, M. T. K., Tadini, C. C., & Augusto, P. E. D. (2021). Bio-based multilayer films: a review of the principal methods of production and challenges. *Critical Reviews in Food Science and Nutrition*, *63*(14), 1–17. https://doi.org/10.1080/10408398.2021.1973955

Lasprilla-Botero, J., Torres-Giner, S., Pardo-Figuerez, M., & álvarez-Láinez, M., & Lagaron, J. M. (2018). Superhydrophobic bilayer coating based on annealed electrospun ultrathin poly(ε-caprolactone) fibers and electrosprayed nanostructured silica microparticles for easy emptying packaging applications. *Coatings*, *8*(5). https://doi.org/10.3390/coatings8050173

Lin, L., Gu, Y., Sun, Y., & Cui, H. (2019). Characterization of chrysanthemum essential oil triple-layer liposomes and its application against campylobacter jejuni on chicken. *LWT*, *107*, 16–24. https://doi.org/10.1016/J.LWT.2019.02.079

López-Carballo, G., Gómez-Estaca, J., Catalá, R., Hernández-Muñoz, P., & Gavara, R. (2012). Active antimicrobial food and beverage packaging. *Emerging Food Packaging Technologies*, 27–54. https://doi.org/10.1533/9780857095664.1.27

Luo, Y., & Wang, Q. (2014). Recent development of chitosan-based polyelectrolyte complexes with natural polysaccharides for drug delivery. *International Journal of Biological Macromolecules*, *64*, 353–367. https://doi.org/10.1016/j.ijbiomac.2013.12.017

Ma, Y., Yang, W., Xia, Y., Xue, W., Wu, H., Li, Z., Zhang, F., Qiu, B., & Fu, C. (2022). Properties and applications of intelligent packaging indicators for food spoilage. *Membranes*, *12*(5), 1–17. https://doi.org/10.3390/membranes12050477

Macena, M. W., Carvalho, R., Cruz-Lopes, L. P., & Guiné, R. P. F. (2021). Plastic food packaging: perceptions and attitudes of Portuguese consumers about environmental impact and recycling. *Sustainability*, *13*(17), 9953. https://doi.org/10.3390/SU13179953

Manna, U., Bharani, S., & Patil, S. (2009). Layer-by-layer self-assembly of modified hyaluronic acid/chitosan based on hydrogen bonding. *Biomacromolecules*, *10*(9), 2632–2639. https://doi.org/10.1021/bm9005535

Martin, O., Schwach, E., Avérous, L., & Couturier, Y. (2001). Properties of biodegradable multilayer films based on plasticized wheat starch. *Starch/Staerke*, *53*(8), 372–380. https://doi.org/10.1002/1521-379X(200108)53:8<372::AID-STAR372>3.0.CO;2-F

Martins, V. D. F., Cerqueira, M. A., Fuciños, P., Garrido-Maestu, A., Curto, J. M. R., & Pastrana, L. M. (2018). Active bi-layer cellulose-based films: development and characterization. *Cellulose*, *25*(11), 6361–6375. https://doi.org/10.1007/s10570-018-2021-y

Martucci, J. F., & Ruseckaite, R. A. (2010). Three-layer sheets based on gelatin and poly(lactic acid), part 1: preparation and properties. *Journal of Applied Polymer Science*, *118*(5), 3102–3110. https://doi.org/10.1002/app.32751

Messin, T., Follain, N., Guinault, A., Sollogoub, C., Gaucher, V., Delpouve, N., & Marais, S. (2017). Structure and barrier properties of multinanolayered biodegradable PLA/PBSA films: confinement effect via forced assembly Coextrusion. *ACS Applied Materials and Interfaces*, *9*(34), 29101–29112. https://doi.org/10.1021/ACSAMI.7B08404/ASSET/IMAGES/LARGE/AM-2017-08404V_0011.JPEG

Morinval, A., & Averous, L. (2022). Systems based on biobased thermoplastics: from bioresources to biodegradable packaging applications. *Polymer Reviews*, *62*(4), 653–721. https://doi.org/10.1080/15583724.2021.2012802

Mujtaba, M., Lipponen, J., Ojanen, M., Puttonen, S., & Vaittinen, H. (2022). Trends and challenges in the development of bio-based barrier coating materials for paper/cardboard food packaging; a review. *Science of The Total Environment*, *851*, 158328. https://doi.org/10.1016/j.scitotenv.2022.158328

Müller, K., Bugnicourt, E., Latorre, M., Jorda, M., Echegoyen Sanz, Y., Lagaron, J., Miesbauer, O., Bianchin, A., Hankin, S., Bölz, U., Pérez, G., Jesdinszki, M., Lindner, M., Scheuerer, Z., Castelló, S., & Schmid, M. (2017). Review on the processing and properties of polymer nanocomposites and Nanocoatings and their applications in the packaging, automotive and solar energy fields. *Nanomaterials*, *7*(4), 74. https://doi.org/10.3390/nano7040074

Niaounakis, M. (2020). 1—flexible plastic packaging and recycling. In M. Niaounakis (Ed.), *Plastics Design Library* (pp. 1–20). William Andrew Publishing. https://doi.org/10.1016/B978-0-12-816335-1.00001-3

Nowzari, F., Shábanpour, B., & Ojagh, S. M. (2013). Comparison of chitosan–gelatin composite and bilayer coating and film effect on the quality of refrigerated rainbow trout. *Food Chemistry*, *141*(3), 1667–1672. https://doi.org/10.1016/J.FOODCHEM.2013.03.022

Ortega-Toro, R., Morey, I., Talens, P., & Chiralt, A. (2015). Active bilayer films of thermoplastic starch and polycaprolactone obtained by compression molding. *Carbohydrate Polymers*, *127*, 282–290. https://doi.org/10.1016/J.CARBPOL.2015.03.080

Păușescu, I., Dreavă, D. M., Bîtcan, I., Argetoianu, R., Dăescu, D., & Medeleanu, M. (2022). Bio-based pH indicator films for intelligent food packaging applications. *Polymers*, *14*(17). https://doi.org/10.3390/POLYM14173622

Phan The, D., Debeaufort, F., Luu, D., & Voilley, A. (2008). Moisture barrier, wetting and mechanical properties of shellac/agar or shellac/cassava starch bilayer bio-membrane for food applications. *Journal of Membrane Science*, *325*(1), 277–283. https://doi.org/10.1016/j.memsci.2008.07.052

Pola, C. C., Medeiros, E. A. A., Pereira, O. L., Souza, V. G. L., Otoni, C. G., Camilloto, G. P., & Soares, N. F. F. (2016). Cellulose acetate active films incorporated with oregano (*Origanum vulgare*) essential oil and organophilic montmorillonite clay control the growth of phytopathogenic fungi. *Food Packaging and Shelf Life*, *9*, 69–78. https://doi.org/10.1016/j.fpsl.2016.07.001

Richardson, J. J., Björnmalm, M., & Caruso, F. (2015). Technology-driven layer-by-layer assembly of nanofilms. *Science*, *348*(6233). https://doi.org/10.1126/science.aaa2491

Riemer, J. (2011) Ultrasonic spray coating of nanoparticles. *Global Solar Technology, February*, 26–28. https://www.yumpu.com/en/document/view/10787768/ultrasonic-spray-coating-of-nanoparticles-sono-tekRocca-Smith, J. R., Pasquarelli, R., Lagorce-Tachon, A., Rousseau, J., Fontaine, S. S., Aguié-Béghin, V., Debeaufort, F., & Karbowiak, T. (2019). Toward Sustainable PLA-Based Multilayer Complexes with Improved Barrier Properties. *ACS Sustainable Chemistry & Engineering*, *7*(4). DOI: 10.1021/acssuschemeng.8b04064

Silvestre, C., Cimmino, S., Stolcru, E., & Vasile, C. (2017). Application of Radiation Technology to Food Packaging. In Y. Sun, & A. G. Chmielewski (Eds.), *Applications of Ionizing radiation in materials processing* (pp. 461–484). Institute of Nuclear Chemistry and Technology.

Sobhan, A., Muthukumarappan, K., Wei, L., Zhou, R., & Tummala, H. (2021). Development of a polylactic acid-coated nanocellulose/chitosan-based film indicator for real-time monitoring of beef spoilage. *Analytical Methods, 13*(23), 2612–2623. https://doi.org/10.1039/d1ay00365h

Soltani Firouz, M., Mohi-Alden, K., & Omid, M. (2021). A critical review on intelligent and active packaging in the food industry: research and development. *Food Research International, 141*(July 2020), 110113. https://doi.org/10.1016/j.foodres.2021.110113

Song, S., Wang, Y., Liang, M., Qi, X., Yang, J., Jin, Y., & Dong, T. (2018). Mechanical and gas barrier properties of poly(L-lactic acid) by plasma-enhanced chemical vapor deposition of SiO x. *Polymer-Plastics Technology and Engineering, 57*(6), 581–590. https://doi.org/10.1080/03602559.2017.1329439

Sono-Tek Corporation (2022) Ultrasonic coating technology. Available at: https://www.sono-tek.com/ultrasonic-coating/ (accessed 01 October 2022).

Souza, V. G. L., Mello, I. P., Khalid, O., Pires, J. R. A., Rodrigues, C., Alves, M. M., Santos, C., Fernando, A. L., & Coelhoso, I. (2022). Strategies to improve the barrier and mechanical properties of pectin films for food packaging: comparing nanocomposites with bilayers. *Coatings, 12*(2), 108. https://doi.org/10.3390/coatings12020108

Trinh, B. M., Chang, C. C., & Mekonnen, T. H. (2021). Facile fabrication of thermoplastic starch/poly (lactic acid) multilayer films with superior gas and moisture barrier properties. *Polymer, 223*. https://doi.org/10.1016/J.POLYMER.2021.123679

Vartiainen, J., Vähä-Nissi, M., & Harlin, A. (2014). Biopolymer films and coatings in packaging applications—a review of recent developments. *Materials Sciences and Applications, 05*(10), 708–718. https://doi.org/10.4236/msa.2014.510072

Vartiainen, J., Shen, Y., Kaljunen, T., Malm, T., Vähä-Nissi, M., Putkonen, M., & Harlin, A. (2016). Bio-based multilayer barrier films by extrusion, dispersion coating and atomic layer deposition. *Journal of Applied Polymer Science, 133*(2), 1–6. https://doi.org/10.1002/app.42260

Vasile, C. (2018). Polymeric nanocomposites and nanocoatings for food packaging: a review. *Materials, 11*(10), 1834. https://doi.org/10.3390/ma11101834

Vlachopoulos, J., & Polychronopoulos(2019). Viscosity and Unidirectional Melt Flows. In J. Vlachopoulos & N. D. Polychronopoulos (Eds.), *Rheology and Technology of Polymer Extrusion* (1, pp. 2–1 to 2–45). Polydynamics Inc.

Wahyuningtiyas, N. E., & Suryanto, H. (2018). Properties of cassava starch based bioplastic reinforced by nanoclay. *Journal of Mechanical Engineering Science and Technology, 2*(1), 20–26. https://doi.org/10.17977/um016v2i12018p020

Wang, F. J., Wang, L. Q., Zhang, X. C., Ma, S. F., & Zhao, Z. C. (2022). Study on the barrier properties and antibacterial properties of cellulose-based multilayer coated paperboard used for fast food packaging. *Food Bioscience, 46*, 101398. https://doi.org/10.1016/J.FBIO.2021.101398

Wang, Q., Chen, W., Zhu, W., McClements, D. J., Liu, X., & Liu, F. (2022). A review of multilayer and composite films and coatings for active biodegradable packaging. *Npj Science of Food, 6*(1). https://doi.org/10.1038/s41538-022-00132-8

Yam, K. L. (2012). Intelligent packaging to enhance food safety and quality. *Emerging Food Packaging Technologies*, 137–152. https://doi.org/10.1533/9780857095664.2.137

Yam, K. L., & Zhu, X. (2012). Controlled release food and beverage packaging. *Emerging Food Packaging Technologies*, 13–26. https://doi.org/10.1533/9780857095664.1.13

Zabihzadeh Khajavi, M., Ebrahimi, A., Yousefi, M., Ahmadi, S., Farhoodi, M., Mirza Alizadeh, A., & Taslikh, M. (2020). Strategies for producing improved oxygen barrier materials appropriate for the food packaging sector. *Food Engineering Reviews*, *12*(3), 346–363. https://doi.org/10.1007/s12393-020-09235-y

Zhang, B., Liu, Y., Wang, H., Liu, W., Cheong, K., & leong, & Teng, B. (2021). Characterization of seaweed polysaccharide-based bilayer films containing essential oils with antibacterial activity. *Lwt*, *150*(December 2020), 111961. https://doi.org/10.1016/j.lwt.2021.111961

Zhuang, C., Jiang, Y., Zhong, Y., Zhao, Y., Deng, Y., Yue, J., Wang, D., Jiao, S., Gao, H., Chen, H., & Mu, H. (2018). Development and characterization of nano-bilayer films composed of polyvinyl alcohol, chitosan and alginate. *Food Control*, *86*, 191–199. https://doi.org/10.1016/j.foodcont.2017.11.024

7

Nanoencapsulation and Nanodelivery Through Food Packaging

Vimala S. K. Bharathi[1,2] *and Digvir S. Jayas*[1,2]

[1] *Department of Biosystems Engineering, University of Manitoba, Winnipeg, Manitoba, Canada*
[2] *President's Office, University of Lethbridge, Lethbridge, Alberta, Canada*

7.1 Introduction

The emerging need for safe, wholesome, high-quality, and nutritious food with minimum use of synthetic compounds has driven the research community to focus on the application of bioactive components from natural sources in food packaging. To overcome the practical challenges associated with the bioactive components (such as volatility, hydrophobicity, poor solubility, oxidative instability, adverse effects on food matrix and sensory attributes, and rapid degradation as a result of environmental stresses), the encapsulation of bioactive components is performed. Compared to microencapsulation, nanoencapsulated particles are more efficient due to their high surface-to-volume ratio. Nanoencapsulation is the process of enclosing a solid, liquid, or gaseous core material inside a wall material to form nanosized encapsulated materials. In food-packaging applications, other than protecting core materials, nanoencapsulation also facilitates improved storage stability and solubility, controlled and prolonged release, targeted delivery, prevention of nontarget interaction, and control of unpleasant flavors and odors. Moreover, nanoencapsulation of components with antimicrobial and antioxidant activities in food packaging can help tackle the problems of microbial contamination and lipid oxidation in food materials (Chelladurai and Jayas, 2019).

The nanoencapsulation process requires the identification of appropriate wall materials and suitable techniques, depending on the core material and end-use. For instance, cinnamon essential oil has been successfully encapsulated in chitosan nanoparticles using ionic gelification method (Hu et al., 2015), polyvinyl alcohol/β-cyclodextrin using electrospinning method (Wen et al., 2016), β-cyclodextrin using coprecipitation method (Petrović et al., 2010) and alginate using double-stage millifluidic method (Farahmand et al., 2022). The wall materials used in food-packaging applications generally include polysaccharides, proteins, lipids, and other organic and inorganic materials (Shishir et al., 2018).

Nanotechnology for Sustainable Food Packaging, First Edition. Edited by C. Anandharamakrishnan, Jeyan A. Moses, and M. Maria Leena.

The nanoencapsulation and nano delivery systems employ the application of various nanocarriers such as lipid-based (nanoemulsions, nanostructured phospholipids, and nano lipids), nature-inspired (cyclodextrins and caseins), special equipment-based (electrospraying, electrospinning, nanofluidics, and nano-spray drying), biopolymer-based (single biopolymeric nanoparticles, complexed, and conjugated biopolymers), and other miscellaneous (nanocrystals, chemical polymer nanoparticles, nanostructured surfactants, and inorganic) (Assadpour and Jafari, 2019). The preparation of nanoencapsulated particles involves either disintegrating large materials into nanosized materials by size reduction (top-down approach) or self-assembly of smaller-sized molecules or particles to form a large particle of a few nanometers (bottom-up approach). The top-down approach requires sophisticated tools and has less control over the structure and size of the nanomaterial, while the bottom-up approach consumes less energy and has better control over the size and structure of the nanomaterial. Emulsification, emulsification-solvent evaporation, and extrusion are some of the top-down approaches, while spray drying, supercritical fluid, coacervation, nanoprecipitation, inclusion complexation, and electrospinning are the bottom-up approaches (Augustin and Sanguansri, 2009; Shishir et al., 2018).

In order to analyze the morphology and particle size of the developed nanoencapsulated products, techniques such as electron microscopy, scanning probe microscopy, and light scattering are adopted (Jafari and Esfanjani, 2017). To evaluate their thermal properties (thermal stability, freezing, melting, heat energy storage, glass transition temperature, and decomposition), techniques such as differential scanning calorimetry, differential thermal analysis, thermo-gravimetric analysis, and evolved gas detection are used (Bharathi et al., 2018; Vishwakarma et al., 2016). To investigate their physicochemical characteristics, nuclear magnetic resonance, Fourier transform infrared spectroscopy, Raman spectroscopy, and X-ray diffraction techniques are commonly used. Ultraviolet (UV) spectroscopy, high-performance liquid chromatography, and gas chromatography–mass spectrometry are generally used to evaluate the encapsulation efficiency (percentage of core material encapsulated in regard to the total core material) and loading capacity (mass ratio of core material in regard to the total mass of the encapsulated product). Surface charge measured as zeta potential aids in understanding the electrostatic interaction, potential stability as well as mobility of the colloids or suspension (Anandharamakrishnan, 2014). The characteristics of core and wall materials, the encapsulation technique adopted, as well as the instrumental, solution, and ambient parameters, are crucial in determining the characteristics of the nanoencapsulated materials. For instance, Volić et al. (2022) encapsulated thyme essential oil with different combinations of wall materials (denatured whey protein concentrate and native whey protein concentrate, solely or in combination with alginate) and reported that the antimicrobial efficacy (against *Salmonella* serotype Enteritidis, *Staphylococcus aureus*) of encapsulated materials containing alginate were lower than those with whey protein concentrate only. They reported that the formation of alginate–protein complexes in the encapsulated materials containing alginate led to lower oil retention, thus lowering the efficacy. Yang et al. (2021) reported that the pH, ratio of core to wall material, wall material concentration, and the type of emulsifier agent used impacted the encapsulation efficiency of the cinnamon essential oil nanocapsules prepared using chitosan and whey protein isolate as wall materials through nanoprecipitation method.

The hunt to ameliorate the functionalities of the encapsulated products has led to the development of new formulations (coencapsulation of core materials or combined application of different wall materials), advancements in techniques, and encapsulating systems. For instance, Niu et al. (2022) coencapsulated an antimicrobial (cinnamaldehyde essential oil), and an antioxidant (chlorogenic acid) material in chitosan nanoparticles, which can have a potential active application in food packaging to prevent oxidation and microbial infection. Zein is one of the widely used wall materials because of its film-forming abilities, biocompatibility, and biodegradability; however, electrospun nanofibers have poor water resistance. To overcome this drawback, Niu et al. (2020) added ethyl cellulose into zein and prepared zein-ethyl cellulose hybrid nanofibers at various concentrations of $0:1, 1:2, 1:1, 1:2$, and $1:0$ to encapsulate cinnamon essential oil. The researchers reported that the addition of ethyl cellulose reduced the free hydrophilic group in the fibers by forming hydrogen bonds between the hydroxyl groups of ethyl cellulose and the amino groups of zein, thus improving the water resistance of fibers.

As of August 31, 2022, the number of publications with the words (in the title, abstract, and keywords) "nano encapsulation" or "nano delivery" and "food packaging" in the Scopus database showed a gradually increasing trend from 2016 (Figure 7.1), and the number of publications in 2020 was about 13 times more than those in 2010. The rapidly growing literature on nanoencapsulation in food-packaging applications clearly depicts the growing interest in the area. Among the total 235 publications from 2006 to August 2022, 82 are original articles, excluding reviews, books, book chapters, conference papers, editorials, notes, and short surveys. The original articles mainly focused on the preparation (optimization), characterization, and functionality of nanocapsules, incorporation into synthetic or biopolymer-based packaging materials, and their interaction with selected food materials. This chapter highlights the recent applications of nanoencapsulation in the packaging of perishable and nonperishable foods.

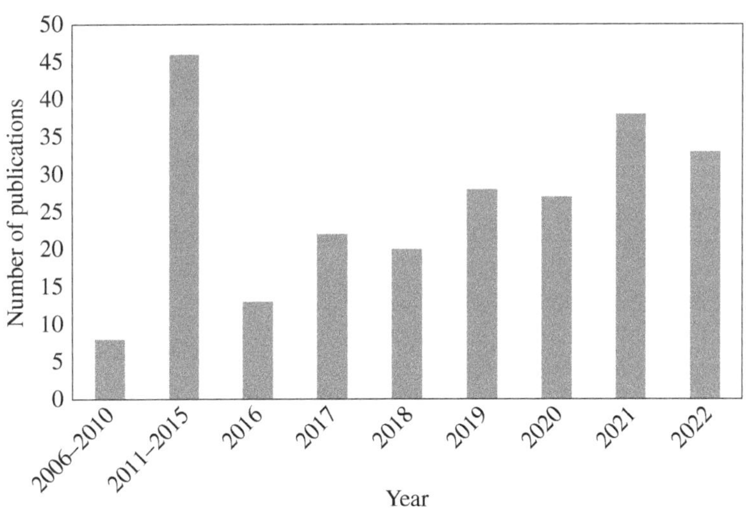

Figure 7.1 Number of publications in the Scopus database with the keywords "nanoencapsulation," "nano delivery," and food packaging

7.2 Food-Packaging Applications

7.2.1 Active Packaging

Active packaging alters the environment of the packaging to prolong the shelf-life, maintain the quality of the food products, or both. The desired results are obtained by the incorporation of various components that exhibit antimicrobial, antioxidant, oxygen scavenging, ethylene scavenging, moisture scavenging, carbon dioxide emitting, ethanol generating, or flavor release/absorption capabilities into the polymer structure (Yildirim et al., 2018). Depending on the method of incorporation of components, the active packaging design can be direct incorporation into films, coating, or adsorption onto the surface, immobilization, addition in the form of sachets or pads, application of bioactive polymers as films (Almasi et al., 2021).

Nanoencapsulation of the bioactive components has proven to improve bioavailability and stability, which is a major concern in developing active packaging materials. In addition to facilitating the controlled release of active components into the packaging environment, nanoencapsulation also allows the incorporation of water-insoluble compounds. The addition of hydrophobic bioactive compounds to hydrophilic polymer matrices and vice-versa often results in the loss of structural integrity of the film due to the low affinity. Thus, encapsulation of those components before addition to the polymer matrix provides additional mechanical strength to the developed films. For example, Assis et al. (2018) found that the addition of nanoencapsulated β-carotene in a cassava starch film improved elongation at break capability of the developed film compared with films developed with free β-carotene and provided stability to β-carotene against light and temperature. Moreover, cassava starch film incorporated with 5% nanoencapsulated β-carotene effectively prevented the formation of peroxides in sunflower oil even after 30 days of storage. Besides, nanoencapsulation regulates the migration of essential oil from films, which prolongs the exerted activity. For example, Zhang et al. (2019) compared the microbial stability, lipid oxidation, pH, and sensory properties of the ready-to-cook pork chops coated with chitosan containing un-encapsulated *Paulownia Tomentosa* essential oil with those containing nanoencapsulated essential oil, for 16 days at 4 °C and reported that the encapsulation resulted in the sustained release of the essential oil, thus, enhancing the antimicrobial, antioxidant, and sensory properties.

Generally, essential oil and phytochemicals alter the cell wall and cytoplasmic membrane structural integrity, which leads to diffusion and coagulation of cytoplasm in bacterial cells (Man et al., 2019). Incorporation of different combinations of antimicrobial compounds into the packaging material can enhance the antimicrobial activity against an extensive range of organisms. Homayonpour et al. (2021) dip-coated sardine fillets with an edible coating containing nano-chitosan and cumin essential oil-loaded nanoliposomes. The combined antimicrobial effect of nano-chitosan and cumin essential oil maintained an acceptable total viable count and decelerated the growth of pseudomonads and lactic acid bacteria during a 16-day storage period.

7.2.2 Intelligent Packaging

An intelligent packaging system serves a very specific purpose of communicating the condition of the food product or the environment, facilitating decision-making to achieve quality and safe products. In general, intelligent packaging can be categorized into indicators (temperature, pH, and gas concentration), sensors (chemical, gas, and biosensors), and data carriers (one- and two-dimensional barcodes and radiofrequency identification tags) (Ghaani et al., 2016). Bioactive components such as curcumin and anthocyanin are sensitive to changes in pH; thus, they are often used in developing intelligent packaging systems to ensure the quality of food products (Etxabide et al., 2021; Lin et al., 2022). However, these bioactive compounds are unstable in their free form and react with oxygen and light, leading to degradation. Encapsulation provides an additional barrier to the highly active indicator compounds, which extends the applicability of the developed packaging (Lin et al., 2022).

Pounds et al. (2021) used a glycerol-based core-shell film to develop intelligent packaging that shows the pH of the food products in real-time. Precisely, the researchers encapsulated a color-indicative dye using a pH-responsive dendrimer comonomer prepared with glycerol and diamine, which responded to changes in the pH of the food material. The use of stimuli-responsive dendrimers as the nanocarrier and delivery system provides flexibility, high reactivity, and biocompatibility. Similarly, Ibrahim et al. (2021) prepared a packaging film with nano polymer encapsulated methyl red as a pH indicator label for fish fillets. The color of the indicator label shifted from yellow (acceptable) to red (not acceptable) over six color shades, according to the pH, indicating the microbial growth in fish fillets.

Electrospun nanofiber mats have gained attention recently in developing sensors and indicators due to their unique structural benefits compared to conventional counterparts such as coating and casting. The porous nature of the nanofibers allows effective encapsulation of indicator dyes and provides a higher surface area for the adsorption of reactive molecules (Pakolpakcil et al., 2018). Moreira et al. (2018) developed a film based on phycocyanin encapsulated in electrospun polylactic acid and polyethylene oxide fibers that showed immense potential in indicating pH in the range of 1–10. Moreover, the color pattern varied with the thickness of the fiber used. The pH sensitivity of alizarin was explored to monitor the freshness of trout fish by encapsulating alizarin in cellulose acetate (Aghaei et al., 2018) and zein (Aghaei et al., 2020) nanofibers synthesized using electrospinning.

Time–temperature indicators (TTIs) indirectly reveal the quality of the food materials through the history of temperature during storage, transportation, and retail. The temperature-dependent quality of the food product is indicated by a visible color change in the indicator rather than an expiry date on the packaging (Galagan and Su, 2008). Azra et al. (2015) developed a thermo-responsive fiber mat as a TTI that has the potential to be incorporated into packaging material. The thermoplastic polyurethane nanofiber mat prepared using electrospinning was employed for encapsulation of indicator solution containing nopal, glycerol, and water. At a temperature above 60 °C, the nanofiber mat contracts due to the heat-induced relaxation of a nonequilibrium stretched state, resulting in the release of indicator solution. The release of indicator solution from pores causes a change in the color of the fiber mat from white to red, indicating the time–temperature history of the product. Among various types of TTIs available such as microbiological, chemical reaction, enzymatic, and diffusion systems, enzyme-based showed high sensitivity to changes in the temperature of the environment. Immobilization of enzymes in nanofibers through electrospinning improves the stability and sensitivity of the enzymes and provides a broader working temperature range than the free enzymes (Forghani et al., 2021).

With an appropriate selection of wall and core materials, encapsulation facilitates maintaining as well as monitoring the quality of the food materials, thus developing active-intelligent food packaging. For instance, naringin-polylactic acid coaxially electrospun fiber coating on salmon fillets reduced the rate of total volatile basic nitrogen (TVB-N) assay and extended the shelf life of the fillets by four days. Meanwhile, the fibers' color gradually shifted from yellow to lavender and then to purple, with an increase in TVB-N content (Li et al., 2021). Similarly, Qi et al. (2022) incorporated nisin (active–antibacterial) and anthocyanins of pomegranate and *Clitoria ternatea* (active–antioxidant; intelligent) into ethylene vinyl alcohol. The developed packaging film extended the shelf life of the shrimp by 100% at 4 °C storage, along with monitoring the freshness of the shrimp through color change at the pH range of 2–12.

7.3 Perishable Food Applications

7.3.1 Fruits and Vegetables

The increasing consumer demand for fresh and nutritious food has skyrocketed the fruits and vegetables market. However, the high moisture content and continued metabolic activity make fruits and vegetables a highly perishable commodity with a significantly short storage life. Moreover, enzymatic browning, microbial growth, tissue softening, higher respiration rate, mechanical damage, higher ethylene concentration, and extrinsic factors (temperature, gas composition, and relative humidity) are the prominent hurdles involved in marketing minimally processed products (Ghidelli and Pérez-gago, 2018). Encapsulation of active components can aid in enhancing the quality and safety of minimally processed fruits and vegetables. The effects of the nanoencapsulation-based packaging materials on some of the fruits and vegetables are given in Table 7.1.

Bio-based packaging and coating containing essential oil have shown promising results in maintaining the physicochemical and nutritional qualities of fruits and vegetables (Gago et al., 2022; Roshandel-hesari et al., 2022). The cinnamon essential oil loaded with polyvinyl alcohol and β-cyclodextrin nanofibrous film enhanced the shelf-life of strawberries. Precisely, the strawberries packed with the nanofibrous film were compared with the positive (packed with fresh-keeping film) and negative (unpacked strawberries) controls at 21 °C for six days, and the results revealed that the unpacked strawberries and those packed with fresh-keeping film spoiled on day 4 and 6, respectively, whereas those packed with nanofibrous film did not spoil even after six days (Wen et al., 2016).

In recent years, several published works have explored the possibility of enhancing the quality and safety of fruits and vegetables. However, the emerging need for convenient and healthy foods and the higher susceptibility of fruits and vegetables to deterioration demands a feasible alternative to enhance their shelf life on a commercial scale. Nanoencapsulated edible coatings and biopolymer-based packaging materials, in combination with techniques like modified atmosphere packaging, can provide maximal quality and safe products with minimal nutrient loss.

7.3.2 Meat

Meat and meat products can benefit from an active packaging material containing antimicrobials and antioxidants as the lipid, moisture, and protein content of the meat and meat products promote microbial growth and lipid oxidation upon storage. Microbial growth is

Table 7.1 Nanoencapsulation-based packaging materials for perishable food applications

Products	Core Material	Wall Material	Packaging	Results	Reference
Fruits and Vegetables					
Mandarin	Carvacrol: oregano: cinnamon essential oil mixture (70 : 10 : 20)	β-cyclodextrin—inclusion complex	Sprayed on internal surface of the cardboard box	• Extended shelf life by one wk • Controlled microbial growth and reduction in weight loss by 1.6-fold	López-Gómez et al. (2020a)
Fresh tomato				• Reduced decay incidence from 15% (Control) to 2% after the storage period (six d at 8 °C, followed by 12 d at 25 °C)	Buendía–Moreno et al. (2020a)
Cherry tomato				• Extended shelf life by four d at 8 °C • Reduced decay incidences and maintained firmness up to 24 d	Buendía–Moreno et al. (2019)
Green, red, and yellow bell pepper				• Reduced enterobacteria by 2 logs after 18 d • Lowered mold counts by 1 log unit • Better firmness after 18 d	Buendía–Moreno et al. (2020b)
Grapes and nectarines				• Reduced dehydration and product weight loss after 30 d • Maintained the physicochemical quality and reduced microbial growth by 0.5–1.4 log units than control • Acceptable sensory quality for more than 30 and 25 d at 2 °C, for grapes and nectarines, respectively	López-Gómez et al. (2020b)
Flat Peaches			Sprayed on recycled Kraft paper and the sheet was placed in a glass jar	• Reduced ethylene production by up to 70% after 30 d (26 d at 2 °C and four d at 22 °C) • Better physicochemical qualities • Increased total antioxidant capacity	López-Gómez et al. (2020c)

(*Continued*)

Table 7.1 (Continued)

Products	Core Material	Wall Material	Packaging	Results	Reference
Strawberries	Ho wood or cinnamon essential oil	chitosan nanocapsules	Thermoplastic starch films	• Extended shelf life without fungal infection and reduced weight loss (stored at 4 °C for 15 d)	Ferreira et al. (2021)
Cucumber	Clove essential oil	Chitosan nanoparticles	Gelatin nanofiber mat coated on kraft	• Reduction in *Escherichia coli* biofilm population. • Maintained color and flavor of cucumber for >four d at 12 °C	Cui et al. (2018)
Meat and Meat Products					
Vacuum-packed sausages	Garlic essential oil (GEO)	Soybean phosphatidylcholine-based nanoliposomes	Chitosan or whey protein films	• Retarded the growth of main spoilage bacteria and lipid oxidation • Reduced aerobic plate count to 3.69 log CFU/g on day 50 (control = 6.21 CFU/g)	Esmaeili et al. (2020)
Ground beef	*Cinnamodendron dinisii* schwanke essential oil	Zein	Chitosan films	• Lowered production of malondialdehyde and oxidative degradation • Stabilized free radicals and the formation of reactive oxygen species in a 12-d study at 6 ± 2 °C	Xavier et al. (2021)
Chicken meat	Pomegranate peel extract	Alginate nanospheres	Coating	• Improved microbial inhibitory effects during the storage for two wk at 4 °C	Rahnemoon et al. (2021)
Chilled pork	Cinnamon essential oil	Chitosan nanoparticles	Sprayed on low-density polyethylene	• Reduced microbial growth, pH, peroxide, and 2-thiobarbituric acid values after storage period (4 °C for 15 d)	Hu et al. (2015)
Fresh beef and sucuk (Turkish-type fermented sausage)	Thymol, eugenol, or carvacrol	Montmorillonite or halloysite nanoclays	Linear low-density polyethylene	• Maintained fresh beef color up to four d at 4 °C • Thymol-incorporated films exhibited bactericidal effect on *E. coli* during beef storage • Vacuum packaging with active films, suppressed native microflora of the sucuk	Tornuk et al. (2015)
Lamb meat	*Satureja khuzestanica* essential oil	Soy-lecithin based nanoliposomes	Chitosan coating	• Retarded microbial growth • Reduced chemical spoilage and 2-thiobarbituric acid values • Deferred red color fading in lamb meat	Pabast et al. (2018)

Chicken breast fillet	Ginger essential oil	Sodium caseinate-based nanoemulsion	Coating	• Decreased of total aerobic psychrophilic bacterial count during 12 d storage at 4 °C • Lowest color difference and cooking loss	Noori et al. (2018)
Chicken	Thyme essential oil	β-cyclodextrin ε-polylysine nanoparticles	Gelatin nanofibers	• Antimicrobial activity against *Campylobacter jejuni* • Less oxidation in chicken • Lower total volatile basic nitrogen (TVB-N) and pH values	Lin et al. (2018)
Pork	Pomegranate peel extract	Chitosan nanoparticles	Zein film	• Restricted growth of *Listeria monocytogenes* at 4 °C	Cui et al. (2020)
Beef fillet	Red onion extract	*Alyssum homolocarpum* and *Lepidium sativum* seed gum	Coating	• Extended the shelf life from 4 to 16 d, reduced pH, peroxide, and thiobarbituric value	Sarvinehbaghi et al. (2021)
Chicken fillet	GEO	Phospholipid (liposomes)	Chitosan film	• Improved mechanical properties and water resistance of the film • Reduced *Staphylococcus aureus*, psychrotrophic bacteria, coliforms, and total viable count	Kamkar et al. (2021)
Dairy Products					
Kashar cheese	Pomegranate seed oil	Polyvinyl alcohol	Coated on aluminum foil	• Reduced total mesophilic aerobic bacteria load by 1.3 log units • Increased the oxidative stability of food samples	Kutlu et al. (2022)
Karish cheese	Thyme essential oil	Chitosan liposomes	Coating	• Shelf life extended from two to four wk	Al-Moghazy et al. (2021)
Kerry gold cheddar cheese	Nisin	Poly-γ-glutamic acid/chitosan nanoparticles	Polyethylene oxide nanofibers	• Antimicrobial activity against *L. monocytogenes* without affecting the sensory attributes	Cui et al. (2017)
Cheshire cheddar cheese	Moringa oil	Chitosan nanoparticles	Gelatin nanofibers	• Antimicrobial activity against *S. aureus* and *L. monocytogenes* on cheese at 4 and 25 °C for 10 d, without affecting the sensory attributes	Lin et al. (2019)

the main cause of spoilage of meat and meat products, resulting in textural defects, off-flavors, and off-odors. In addition, oxidation of myoglobin under refrigeration conditions is another indicator of meat spoilage. *Escherichia coli* and *Salmonella* are the most common causes of food-borne outbursts related to meat-based products; *S. aureus, Clostridium perfringens, Clostridium botulinum, Bacillus cereus,* and *Listeria monocytogenes* are also reported to cause meat-associated outbreaks (Omer et al., 2018). Nitrite and nitrate are the most used food additives in cured meat products to inhibit microbial growth and to develop the distinctive color and flavor of the meat. However, the N-nitroso compounds formed in the process are carcinogenic and detrimental to human health (Ferysiuk and Wójciak, 2020). Essential oils possess excellent antimicrobial and antioxidant characteristics and can be used as a substitute for nitrites. However, the interaction of the essential oils with the high fat and protein content of the meat can result in a reduction in their properties and can also lead to a negative effect on the organoleptic attributes of the meat (Ji et al., 2021). Thus, the encapsulation of essential oils and application into packaging material facilitates the shelf-life extension of meat, poultry, and fish. For instance, thyme essential oil encapsulated in chitosan-gelatin nanofibers showed a bactericidal effect against *C. perfringens* and had color and sensory properties on the sausage, similar to those treated with 120 ppm nitrite (Vafania et al., 2019). Selected examples of the application of nanoencapsulation-based packaging on the shelf-life extension of meat and meat products are summarized in Table 7.1.

Temperature fluctuation during transportation and storage can escalate the chemical reactions and microbial growth of the meat products. To address this problem, phase change materials can be encapsulated and used in the packaging materials. Karim et al. (2022) developed zein nanofibers containing tetradecane (phase change material) and cinnamaldehyde (antioxidant and antimicrobial agent) and showed that the developed nanofiber-based packaging has the potential to maintain the sausage temperature for about 8 hours when the refrigerator was switched off, at the same time inhibiting the growth of *E. coli* and *S. aureus* and preserving the color, texture, peroxide, and thiobarbituric acid values.

7.3.3 Dairy Products

Dairy products are prone to contamination by undesirable pathogens during processing and storage. Major pathogens of concern in dairy products include *E. coli, S. aureus, L. monocytogens, Salmonella* spp., and *Campylobacter* spp. (Buyser et al., 2001). Electrospun nanofiber mats loaded with various drugs and essential oils are incorporated into packaging materials in the form of coatings or active layers on the surface of various cheese products to enhance their microbiological safety (Table 7.1). For instance, thymol-loaded nanofibers coating was shown to have potential application as antimold surface material on Kashar cheese (Tatlisu et al., 2019). Nisin incorporated amaranth protein isolate: pullulan nanofibers inactivated the microorganisms, including *Salmonella Typhimurium, L. monocytogenes*, and *Leuconostoc mesenteroids* after 142, 120, and 170 hours, respectively, on fresh cheese (Soto et al., 2019). However, only limited studies have been performed on the application of nano encapsulated products on the packaging of dairy products.

7.4 Stored Grains

The loss of stored grain because of pests is significant globally. To reduce the occurrence of food loss as a result of insects and microorganisms, synthetic agents are used. However, excessive use of chemical agents has drawbacks, including but not limited to the evolution of resistant insects, toxicity of nontarget species, residual toxicity, and negative effects on the environment and humans. To overcome these drawbacks, the use of biopesticides is being explored. Biopesticides can be plant-, biochemical- or microbial-based (Kumar, 2012). Plant-based essential oils have been explored for their insecticidal and insect-repellent characteristics. In addition, essential oils have lower mammalian toxicity than synthetic insecticides and are environmentally friendly (Mossa, 2016). However, their properties, such as high volatility, rapid vaporization in response to light, heat, pressure or airflow, and reduced water solubility, restrict their wide application. To address these drawbacks, encapsulation of essential oils has been performed, and several published works have shown promising results against stored grain pests. Because of their smaller size, nanoencapsulation enhances the efficacy of the essential oils. Table 7.2 lists some of those published works conducted with nanoencapsulated essential oils and their results for pest control.

Encapsulation of essential oils also leads to improved efficacy over an extended period. For instance, Yang et al. (2009) showed that as a result of slow and persistent release of encapsulated active components, the control efficiency of garlic essential oil (GEO) encapsulated with polyethylene glycol against *Tribolium castaneum* was over 80% at the end of five months, while the control efficiency of free GEO at the same concentration was only 11%. Similarly, *Artemisia annua* essential oil encapsulated with silica MCM-41 showed prolonged stability and fumigant toxicity by 23 days against *Glyphodes pyloalis*, as compared to the crude essential oil (Oftadeh et al., 2021). Moreover, *Zataria multiflora* essential oil encapsulated with gum-maltodextrin (nanoemulsions) was 4.32 times more toxic than pure oil against the moth *Ephestia kuehniella* (Emamjomeh et al., 2017). Encapsulation of *Mentha piperita* essential oils in Chitosan-cinnamic acid nanogel inhibited *Aspergillus flavus* at 800 ppm, whereas nonencapsulated oil failed to suppress the growth of mycelium even at high concentrations (3,000 ppm) (Beyki et al., 2014). Coencapsulation of two or more insecticides can provide a synergistic effect against the insects. For example, Campos et al. (2018) reported that combined encapsulation of carvacrol and linalool in chitosan nanoparticles functionalized with β-cyclodextrin provided higher efficacy against *Helicoverpa armigera* and *Tetranychus urticae*, than the encapsulated individual components. Despite the application of essential oils in insect control, they can also be employed to prevent insect infestation. For example, Sabbour (2019) compared the effects of various nanoencapsulated essential oils such as rosemary, catnip, garlic, and citronella on the prevention of *Callosbruchus maculatus* and *Callosobruchus chinensis* and observed more than 85% repellency for both the species, in the presence of encapsulated rosemary oil. The packaging films incorporated with encapsulated essential oils can facilitate the repelling of insects during the distribution, transportation, and storage of food materials.

The encapsulated essential oils were also explored for their antifungal characteristics. *Pimenta dioica* essential oil encapsulated in chitosan nanoemulsion showed antifungal and

Table 7.2 Effect of encapsulated essential oils on stored grain pests

Essential Oil	Wall Material	Stored Grain Pest	Treated Stage	Concentration	Effects	Reference
Cardamom	Polyethylene glycol	*Oryzaephilus surinamensis*	Adults	12.5 μL/cm^2	After 96 h of treatment, about 85% and 81% mortality were observed in contact and fumigant toxicity, respectively	Hemalatha and Kumar (2021)
Cinnamon	Mesoporous silica nanoparticles	*Corcyra cephalonica*	6th instar larvae	66.57 mg	Only 45% pupation; No emergence of adults	Attia et al. (2020)
Rosemary	Urea and formaldehyde	*Callosobruchus maculatus*	3rd instar larvae	2 g foam (sprayed with 0.5% oils) added to 100 g cowpea	Accumulated mortality after seven d was 82.2%; Number of eggs laid per female reduced to 6.4 ± 9.89 in treated samples as compared to 299.9 ± 9.89 in control	Sabbour (2019)
		Callosobruchus chinensis			Accumulated mortality after seven d was 89.7%; Number of eggs laid per female reduced to 8.3 ± 1.0 in treated samples as compared to 298.9 ± 7.89 in control	
Peppermint	Chitosan nanoparticles	*Tribolium castaneum* (Herbst)	Adult	34.79 μL/L air	*In vivo* Acetylcholinesterase enzyme inhibition of 52.4%	Rajkumar et al. (2020)
		Sitophilus oryzae (L.)		57.47 μL/L air	*In vivo* Acetylcholinesterase enzyme inhibition of 37.8%	
Melissa officinalis L.	Chitosan nanoparticles	*T. castaneum*	Adult	LD_{50} = 0.048 μL/mL air with 24 h exposure		Upadhyay et al. (2019)
Mentha longifolia	Glyceryl monostearate	*Ephestia kuehniella*	5th instar larva	20,000 ppm	79.5% mortality in 24 h	Louni et al. (2018)
Cuminum cyminum L.	Poly urea-formaldehyde	*T. castaneum*	Adult	LC_{50} = 16.25 ppm with seven d exposure		Negahban et al. (2012a)

Rosmarinus officinalis	Poly-Ɛ-(caprolactone)	*T. castaneum*	Adult	27.76 μL/L air	96.6% mortality in 24 h (fumigant toxicity)	Khoobdel et al. (2017)
Mentha spicata L.	Sodium alginate	*T. castaneum*	Adult	LC_{50} = 9.279 μL/mL with 72 h exposure		Heydarzadeh et al. (2019)
Mentha pulegium L.				LC_{50} = 6.793 μL/mL with 72 h exposure		
Citronella	Urea-formaldehyde	*Rhyzopertha dominica* (F.)	–	Accumulated adult mortality of 98.7% with seven d exposure (0.5% in control) Number of eggs laid per female reduced to 13.9 ± 7.1, as compared to 295.7 ± 9.19 in control. Percentage of adult emergence was 0.1% (100% in control)		Sabbour (2020)
Moringa				Accumulated adult mortality of 83.8% with seven d exposure (0.5% in control) Number of eggs laid per female reduced to 28.9 ± 5.2, as compared to 295.7 ± 9.19 in control. Percentage of adult emergence was 2% (100% in control).		
Goatweed				Accumulated adult mortality of 90.9% with seven d exposure (0.5% in control) Number of eggs laid per female reduced to 20.4 ± 5.4, as compared to 295.7 ± 9.19 in control. Percentage of adult emergence was 1% (100% in control).		
Artemisia sieberi Besser	Urea-formaldehyde	*T. castaneum*	Adult	LC_{50} = 11.24 ppm with seven d exposure		Negahban et al. (2012b)

antiaflatoxigenic activity against *A. flavus* present in maize, in addition to inhibiting the lipid peroxidation in maize without altering germination and sensory properties. Moreover, the encapsulated oil showed better efficacy at lower doses (1.6 and 1.0 μL/mL for antifungal and antiaflatoxigenic activity, respectively) than the pure oil (2.5 and 1.5 μL/mL for antifungal and antiaflatoxigenic activity, respectively) (Chaudhari et al., 2022). Similarly, chitosan nanoemulsion-based *Pogostemon cablin* essential oil was active against mold-induced biodeterioration and aflatoxin B_1 biosynthesis in maize seeds for 30 days, in addition to inhibiting biosynthesis of ergosterol and *A. flavus* cellular integrities (Roshan et al., 2022a). Several other published papers have reported the use of selected encapsulated essential oil against a broad spectrum of common storage fungi (Kujur et al., 2021; Tiwari et al., 2022). However, the antifungal activity against varying species of fungi is dose dependent. For instance, the minimum inhibitory concentration of chitosan-based *Toddalia asiatica* essential oil was reported to be 1 μg/mL against *Alternaria alternata, Cladosporium cladosporioides, Fusarium equiseti, Fusarium graminearum,* and *Fusarium poae*; was 1.5 μg/mL against *Aspergillus fumigatus, Aspergillus ochraceus, Fusarium oxysporum,* and *Fusarium verticillioides*; was 2 μg/mL against *A. flavus, Aspergillus niger,* and *Penicillum chrysogenum* (Roshan et al., 2022b). A similar report was provided by Das et al. (2019) on encapsulated *Coriandrum sativum* essential oil against 15 common storage molds.

The type of formulation, wall material, encapsulation technique, and process parameters determine the level of active components encapsulated and their release, thus determining the efficacy of the encapsulated product. Costa et al. (2014) prepared different formulations of neem oil, including nanoformulated powder, soluble powder, and emulsifiable concentrate against *Zabrotes subfasciatus* and reported that emulsifiable concentrate of neem oil exhibited the highest mortality of *Z. subfasciatus*, whereas, nanoformulated powder exhibited better UV stability.

Encapsulated thyme essential oil, effective against *Sitophilus zeamais*, was tested for 200 days on the quality of the corn and Barros et al. (2022) reported that the essential oil did not affect the pH, conductivity, leached solids, water content, and staining of corn grains, but increase in concentration increased the acidity and soluble solid contents. The encapsulated essential oils have the potential to be used as pesticides against stored grain pests; however, knowledge about their effect on the stored grain, their effect on long-term storage, possibility and cost associated with commercial applications is required.

Other plant-based insecticides have been explored for application in the prevention and control of stored grain pests. Cysteine protease, extracted from the seed coat of *Albizia procera*, nanoencapsulated with graphene quantum dots, showed a better decrease in life cycle attributes of *Sitotroga cerealella* (Olivier) (egg, larvae, pupae, and adult eclosion) as compared to the control (without treatment) and nonencapsulated protein (Batool et al., 2021).

To disrupt the mating process of the pests, synthetic sex pheromones are being explored. Few studies have been reported on the encapsulation of sex pheromones to facilitate the controlled, sustained, and environmentally safe release of pheromones (Bhagat et al., 2013; Danaye-Tous et al., 2022; Gu et al., 2010; Kong et al., 2009). However, the published reports on stored grain pests are limited. Additional studies are required to explore the possibility of nanoencapsulating natural and synthetic pheromones of stored grain pests.

Although several encapsulated biopesticides have been explored for their oviposition deterrence, ovicidal, larvicidal, pupicidal, adult emergence inhibition, repellency, antifeedant, contact, and fumigant toxicity against a variety of stored grain pests, their commercial application has been limited. According to Isman and Grieneisen (2014), although the proportion of published papers on botanical insecticides has increased from 1.4% (1980) to 21.4% (2012), about one-third of those publications (published in 2011) provided information on chemical data or characterization; while, positive controls were included in only a quarter publication. This implies that the surprisingly larger number of published works limits reproducibility and application in advancement toward commercial application. Thus, future studies on the effect of encapsulated components against stored grain pests must consider the use of a commercially used pesticide as a positive control. In addition, further studies must consider testing the effects of encapsulated compounds on a commercial scale for a long period. This aids in identifying the practical difficulties associated with the application of encapsulated products in stored grain ecosystems, in addition to determining the quantity required per tonne of grain (Demeter et al., 2021). Overall, appropriate application of nanoencapsulation in pest control and prevention facilitates less exposure to the ecosystem and paves the way for advancements in integrated pest management.

7.5 Health and Environmental Impacts of Nanoencapsulated Materials

The nanoencapsulated materials in the food-packaging systems offer several advantages including improved barrier and mechanical properties, providing active and intelligent functionalities, and enhancing the biodegradability of the packaging materials. However, adequate data are not available on nanoencapsulated materials in food packages related to toxicity and ecotoxicity. Over the last decade, researchers have reported the nanomaterials' migration from packaging to food. The migration of silver nanoparticles from packaging materials has been extensively studied regarding their detrimental effect on membrane permeability, mitochondrial function, and reactive oxygen species generation in cells (Akter et al., 2018; Chelladurai and Jayas, 2019). Jokar et al. (2017) showed that a conclusive decision cannot be made about the migration of nanoparticles from polymer matrix based on the existing experimental studies. They concluded that the small size of nanoparticles and the lack of highly sensitive detection techniques for trace amounts may result in false negative results for the migration of engineered nanomaterials from food contact materials. The mathematical model development considers the migration of nanoparticles as a diffusion process and predicts that the nanoparticle migration is negligible except for very small nanoparticles. However, apart from diffusion, many other physical phenomena are yet to be identified as possible mechanisms for the migration of nanoparticles from the packaging matrix. Meanwhile, evidence of migration of nanomaterials from polymer matrix is widely available including silver nanoparticles migration from the pectin nanocomposite coating (Vishnuvarthanan and Rajeswari, 2019), silver nanoparticles migration from polylactic acid migration (Fan et al., 2019), SiO_2 nanoparticles from soluble soybean polysaccharide film (Salarbashi et al., 2021), and ZnO nanoparticles from low-density polyethylene (Bumbudsanpharoke et al., 2019).

Migration studies involving nanomaterials from polymer matrix are challenging due to the lack of highly sensitive analytical methods capable of detecting nanomaterials in addition to the small size and concentration of nanomaterials. Thus, the development of elemental techniques capable of detecting inorganic and organic nanomaterials is necessary to ensure the safety and quality of food materials.

The migrated nanoparticles in the food materials from the packaging material end up in the human body through the gastrointestinal tract. The mechanism of adsorption and the fate of nanomaterial in the gastrointestinal tract is undiscovered due to the complexity of the mechanism. The toxicity of nanomaterials depends on the physicochemical characteristics of the materials. The reduction of the size of a particular material may increase the inherent toxicity of the material. The heightened toxicity is mainly associated with the large surface area, creating a large number of surface atoms available for chemical reactions. In contrast, the accumulation of certain inactive and nontoxic nanomaterial in the cells over time can lead to the development of toxicity in cells. An *in vitro* study on the effect of nanoparticles on intestinal barrier function showed that chronic exposure leads to an increase in reactive oxygen species, intestinal alkaline phosphatase activity, and proinflammatory signaling, which resulted in a decrease in iron, zinc, and fatty acid transport (Guo et al., 2017). Nanoparticle ingestion causes disruption of transport mechanisms in cells by decreasing the absorptive microvilli in the intestinal epithelial cells (Sommi et al., 2021). The cytotoxicity effect of carbon nanomaterial has been identified to be DNA damage, reactive oxygen species generation, and lysosomal damage leading to cell death through apoptosis (Rozhina et al., 2021). Further, carbon nanomaterial inadvertently creates immunological effects such as pulmonary macrophage activation and inflammation in cells (Yuan et al., 2019).

The absorption and toxicity of the nanomaterial are closely associated with the dispersion, transformation, modification, and transportation in the environment. The release of nanomaterials into the environment takes place during the manufacturing process and discarding of products incorporated with nanomaterials. As the product matrix to which the nanomaterial is incorporated breaks down in the environment over time, free nanoparticles are released into the environment. To fully understand the effect of nanomaterials on the environment, detailed studies on the mobility, persistence, degradation, and bioavailability of specific nanomaterials are required. Some nanomaterials are known to react with the macromolecules in the environment and release toxic chemicals that are harmful to plants, animal cells, and algae alike (Yang et al., 2023). The high affinity of heavy metal to TiO_2 nanoparticles resulted in large adsorption of heavy metal on the surface of nanoparticles in sediments, which eventually led to a decrease in metal concentration in the water phase. Moreover, the heavy metal content in *Corbicula fluminea* increased with the increasing TiO_2 nanoparticles in the sediment (Fan et al., 2018). On the other hand, studies have shown that high concentrations of metal–organic framework nanoparticles inhibited the growth of pea seedlings, and an increase in germination was observed at low concentrations (Guan et al., 2021). It should be noted that adequate data are not yet available on the risk assessment of nanoparticles to understand the effect of nanomaterial on the environment. However, the existing environmental toxicity studies and the growing concerns regarding the potential health implications should encourage a proactive approach to ensure that nanomaterial-based products are environmentally sustainable and devoid of health risks.

7.6 Summary and Prospects

Recent research has shown that nanoparticle-based delivery systems such as nanoemulsion, nanofibers, nanoliposomes, and nanogels incorporated into packaging material can be used for controlled delivery of active ingredients into food products. The incorporated packaging system showed improved physiochemical and functional attributes, including mechanical, optical, and biodegradable properties, in addition to enhancing the quality and safety of the products. The addition of nanoencapsulated products into the packaging material is a promising way to enhance the properties of the food without the use of synthetic and carcinogenic components.

However, several issues need to be addressed before nanomaterial-based food-packaging materials are made commercially available and widely accepted. Further studies should be conducted on the toxicity and ecotoxicity of nanoencapsulated particles in packaging materials regarding their migration, accumulation, and effect on cell activity. The lack of research on the surface chemistry of nanoparticles limits the extent of the understanding of the toxic effect of bio-based nanoparticles on health and the environment. Thus, the development of precision analytical protocols that can measure the trace amount of nanomaterial in cells and mathematical models that consider all possible mechanisms of nanoparticle migration have paramount importance in guaranteeing the safety of the packaging material and the food products. Moreover, a general challenge in the commercialization of nano-based food-packaging systems is the lack of validated reference materials to compare the migration studies to assure the accuracy of the results.

References

Aghaei, Z., Emadzadeh, B., Ghorani, B., & Kadkhodaee, R. (2018). Cellulose acetate Nanofibres containing alizarin as a halochromic sensor for the qualitative assessment of rainbow trout fish spoilage. *Food and Bioprocess Technology*, *11*(5), 1087–1095. https://doi.org/10.1007/S11947-017-2046-5/FIGURES/3

Aghaei, Z., Ghorani, B., Emadzadeh, B., Kadkhodaee, R., & Tucker, N. (2020). Protein-based halochromic electrospun nanosensor for monitoring trout fish freshness. *Food Control*, *111*, 107065. https://doi.org/10.1016/J.FOODCONT.2019.107065

Akter, M., Sikder, M. T., Rahman, M. M., Ullah, A. K. M. A., Hossain, K. F. B., Banik, S., Hosokawa, T., Saito, T., & Kurasaki, M. (2018). A systematic review on silver nanoparticles-induced cytotoxicity: physicochemical properties and perspectives. *Journal of Advanced Research*, *9*, 1–16. https://doi.org/10.1016/j.jare.2017.10.008

Almasi, H., Oskouie, M. J., & Saleh, A. (2021). A review on techniques utilized for design of controlled release food active packaging. *Critical Reviews in Food Science and Nutrition*, *61*(15), 2601–2621. https://doi.org/10.1080/10408398.2020.1783199

Al-Moghazy, M., El-sayed, H. S., Salama, H. H., & Nada, A. A. (2021). Edible packaging coating of encapsulated thyme essential oil in liposomal chitosan emulsions to improve the shelf life of Karish cheese. *Food Bioscience*, *43*, 101230. https://doi.org/10.1016/j.fbio.2021.101230

Anandharamakrishnan, C. (2014). *Techniques for Nanoencapsulation of Food Ingredients*. Springer. https://doi.org/10.1007/978-1-4614-9387-7

Assadpour, E., & Jafari, S. M. (2019). An Overview of Biopolymer Nanostructures for Encapsulation of Food Ingredients. In S. M. Jafari (Ed.), *Biopolymer Nanostructures for Food Encapsulation Purposes* (pp. 1–35). Academic Press.

Assis, R. Q., Pagno, C. H., Costa, T. M. H., Flôres, S. H., & de Oliveria Rios, A. (2018). Synthesis of biodegradable films based on cassava starch containing free and nanoencapsulated β-carotene. *Packaging Technology and Science*, *31*(3), 157–166. https://doi.org/10.1002/PTS.2364

Attia, R. G., Rizk, S. A., Hussein, M. A., Abdel Fattah, H. M., Khalil, M. M. H., & Ma'moun, S. A. M. (2020). Effect of cinnamon oil encapsulated with silica nanoparticles on some biological and biochemical aspects of the rice moth, *Corcyra cephalonica* (Staint.) (Lepidoptera: Pyralidae). *Annals of Agricultural Sciences*, *65*(1), 1–5. https://doi.org/10.1016/j.aoas.2020.05.003

Augustin, M. A., & Sanguansri, P. (2009). Nanostructured materials in the food industry. *Advances in Food and Nutrition Research*, *58*, 183–213. https://doi.org/10.1016/S1043-4526(09)58005-9

Azra, C., Alhazov, D., & Zussman, E. (2015). Effect of polymer nanofibers thermoelasticity on deformable fluid-saturated porous membrane. *Polymer*, *58*, 162–169. https://doi.org/10.1016/J.POLYMER.2014.12.062

Barros, F. A. P., Radünz, M., Scariot, M. A., Camargo, T. M., Nunes, C. F. P., de Souza, R. R., Gilson, I. K., Hackbart, H. C. S., Radünz, L. L., Oliveira, J. V., Tramontin, M. A., Radünz, A. L., & Magro, J. D. (2022). Efficacy of encapsulated and non-encapsulated thyme essential oil (*Thymus vulgaris* L.) in the control of *Sitophilus zeamais* and its effects on the quality of corn grains throughout storage. *Crop Protection*, *153*, 105885. https://doi.org/10.1016/j.cropro.2021.105885

Batool, M., Hussain, D., Akrem, A., Haq, M. N., & ul, Saeed, S., & Saeed, Q. (2021). Nanoencaspsulation of cysteine protease for the management of stored grain pest, *Sitotroga cerealella* (Olivier). *Journal of King Saud University—Science*, *33*(4), 101404. https://doi.org/10.1016/j.jksus.2021.101404

Beyki, M., Zhaveh, S., Khalili, S. T., Rahmani-Cherati, T., Abollahi, A., Bayat, M., Tabatabaei, M., & Mohsenifar, A. (2014). Encapsulation of *Mentha piperita* essential oils in chitosan–cinnamic acid nanogel with enhanced antimicrobial activity against *Aspergillus flavus*. *Industrial Crops and Products*, *54*, 310–319. https://doi.org/10.1016/j.lwt.2014.07.054

Bhagat, D., Samanta, S. K., & Bhattacharya, S. (2013). Efficient management of fruit pests by pheromone nanogels. *Scientific Reports*, *3*, 1–8. https://doi.org/10.1038/srep01294

Bharathi, S. K. V., Moses, J. A., & Anandharamakrishnan, C. (2018). Nano and Microencapsulation Using Food Grade Polymers. In T. Gutiérrez (Ed.), *Polymers for Food Applications* (pp. 357–400). Springer. https://doi.org/10.1201/b17388

Buendía-Moreno, L., Soto-Jover, S., Ros-Chumillas, M., Antolinos, V., Navarro-Segura, L., Sánchez-Martínez, M. J., Martínez-Hernández, G. B., & López-Gómez, A. (2019). Innovative cardboard active packaging with a coating including encapsulated essential oils to extend cherry tomato shelf life. *LWT—Food Science and Technology*, *116*, 108584. https://doi.org/10.1016/j.lwt.2019.108584

Buendía-Moreno, L., Sánchez-Martínez, M. J., Antolinos, V., Ros-Chumillas, M., Navarro-Segura, L., Soto-Jover, S., Martínez-Hernández, G. B., & López-Gómez, A. (2020a). Active cardboard box with a coating including essential oils entrapped within cyclodextrins and/or halloysite nanotubes. A case study for fresh tomato storage. *Food Control*, *107*, 106763. https://doi.org/10.1016/J.FOODCONT.2019.106763

Buendía-Moreno, L., Soto-Jover, S., Ros-Chumillas, M., Antolinos-López, V., Navarro-Segura, L., Sánchez-Martínez, M. J., Martínez-Hernández, G. B., & López-Gómez, A. (2020b). An innovative active cardboard box for bulk packaging of fresh bell pepper. *Postharvest Biology and Technology*, *164*, 111171. https://doi.org/10.1016/j.postharvbio.2020.111171

Bumbudsanpharoke, N., Choi, J., Park, H. J., & Ko, S. (2019). Zinc migration and its effect on the functionality of a low density polyethylene-ZnO nanocomposite film. *Food Packaging and Shelf Life*, *20*, 100301. https://doi.org/10.1016/j.fpsl.2019.100301

Campos, E. V. R., Proença, P. L. F., Oliveira, J. L., Pereira, A. E. S., De Morais Ribeiro, L. N., Fernandes, F. O., Gonçalves, K. C., Polanczyk, R. A., Pasquoto-Stigliani, T., Lima, R., Melville, C. C., Della Vechia, J. F., Andrade, D. J., & Fraceto, L. F. (2018). Carvacrol and linalool co-loaded in β-cyclodextrin-grafted chitosan nanoparticles as sustainable biopesticide aiming pest control. *Scientific Reports*, *8*(1), 1–14. https://doi.org/10.1038/s41598-018-26043-x

Chaudhari, A. K., Singh, V. K., Das, S., Deepika, & Dubey, N. K. (2022). Fabrication, characterization, and bioactivity assessment of chitosan nanoemulsion containing allspice essential oil to mitigate *Aspergillus flavus* contamination and aflatoxin B1 production in maize. *Food Chemistry*, *372*, 131221. https://doi.org/10.1016/j.foodchem.2021.131221

Chelladurai, V., & Jayas, D. S. (2019). *Nanoscience and nanotechnology in foods and beverages* (P. 344). CRC Press, Boca Rotan. https://doi.org/10.1201/9781315153162

da Costa, J. T., Forim, M. R., Costa, E. S., De Souza, J. R., Mondego, J. M., & Boiça Junior, A. L. (2014). Effects of different formulations of neem oil-based products on control *Zabrotes subfasciatus* (Boheman, 1833) (Coleoptera: Bruchidae) on beans. *Journal of Stored Products Research*, *56*, 49–53. https://doi.org/10.1016/j.jspr.2013.10.004

Cui, H., Bai, M., Rashed, M. M. A., & Lin, L. (2018). The antibacterial activity of clove oil/chitosan nanoparticles embedded gelatin nanofibers against *Escherichia coli* O157 : H7 biofilms on cucumber. *International Journal of Food Microbiology*, *266*, 69–78. https://doi.org/10.1007/s11947-017-2046-5

Cui, H., Surendhiran, D., Li, C., & Lin, L. (2020). Biodegradable zein active film containing chitosan nanoparticle encapsulated with pomegranate peel extract for food packaging. *Food Packaging and Shelf Life*, *24*, 100511. https://doi.org/10.1016/j.fpsl.2020.100511

Cui, H., Wu, J., Li, C., & Lin, L. (2017). Improving anti-listeria activity of cheese packaging via nanofiber containing nisin-loaded nanoparticles. *LWT—Food Science and Technology*, *81*, 233–242. https://doi.org/10.1016/j.lwt.2017.04.003

Danaye-Tous, A. H., Jafari, S., Heidary-Alizadeh, B., & Farazmand, H. (2022). Efficacy of nanocapsules loaded with *Ectomyelois ceratoniae* (Zeller) (Lepidoptera: Pyralidae) sex pheromone as evaluated in wind tunnel and field trapping experiments. *Journal of Plant Diseases and Protection*, *129*(4), 853–860. https://doi.org/10.1007/s41348-022-00632-5

Das, S., Singh, V. K., Dwivedy, A. K., Chaudhari, A. K., Upadhyay, N., Singh, P., Sharma, S., & Dubey, N. K. (2019). Encapsulation in chitosan-based nanomatrix as an efficient green technology to boost the antimicrobial, antioxidant and in situ efficacy of *Coriandrum sativum* essential oil. *International Journal of Biological Macromolecules*, *133*, 294–305. https://doi.org/10.1016/j.ijbiomac.2019.04.070

De Buyser, M.-L., Dufour, B., Maire, M., & Lafarge, V. (2001). Implication of milk and milk products in food-borne diseases in France and in different industrialised countries. *International Journal of Food Microbiology*, *67*, 1–17. https://doi.org/10.1016/s0168-1605(01)00443-3

Demeter, S., Lebbe, O., Hecq, F., Nicolis, S. C., Kenne Kemene, T., Martin, H., Fauconnier, M. L., & Hance, T. (2021). Insecticidal activity of 25 essential oils on the stored product pest, sitophilus granarius. *Foods*, *10*(2), 1–13. https://doi.org/10.3390/foods10020200

Emamjomeh, L., Imani, S., Talebi, K. H., Moharramipour, S., & Larijani, K. (2017). Preparation of nanoemulsion formulation of essential oil of *Zataria multiflora* and comparison of contact toxicity with pure essential oil on *Ephestia kuehniella*. *Entomology and Phytopathology*, *85*, 181–190. https://doi.org/10.22092/jaep.2017.109159.1151

Esmaeili, H., Cheraghi, N., Khanjari, A., Rezaeigolestani, M., Basti, A. A., Kamkar, A., & Aghaee, E. M. (2020). Incorporation of nanoencapsulated garlic essential oil into edible films: a novel approach for extending shelf life of vacuum-packed sausages. *Meat Science*, *166*, 108135. https://doi.org/10.1016/j.meatsci.2020.108135

Etxabide, A., Maté, J. I., & Kilmartin, P. A. (2021). Effect of curcumin, betanin and anthocyanin containing colourants addition on gelatin films properties for intelligent films development. *Food Hydrocolloids*, *115*, 106593. https://doi.org/10.1016/J.FOODHYD.2021.106593

Fan, C., Cui, R., Lu, W., Chen, H., Yuan, M., & Qin, Y. (2019). Effect of high pressure treatment on properties and nano–ag migration of PLA-based food packaging film. *Polymer Testing*, *76*, 73–81. https://doi.org/10.1016/j.polymertesting.2019.03.005

Fan, X., Wang, C., Wang, P., Hu, B., & Wang, X. (2018). TiO_2 nanoparticles in sediments: effect on the bioavailability of heavy metals in the freshwater bivalve *Corbicula fluminea*. *Journal of Hazardous Materials*, *342*, 41–50. https://doi.org/10.1016/j.jhazmat.2017.07.041

Farahmand, A., Emadzadeh, B., Ghorani, B., & Poncelet, D. (2022). Droplet-based millifluidic technique for encapsulation of cinnamon essential oil: optimization of the process and physicochemical characterization. *Food Hydrocolloids*, *129*, 107609. https://doi.org/10.1016/j.foodhyd.2022.107609

Ferreira, R. R., Souza, A. G., Quispe, Y. M., & Rosa, D. S. (2021). Essential oils loaded-chitosan nanocapsules incorporation in biodegradable starch films: A strategy to improve fruits shelf life. *International Journal of Biological Macromolecules*, *188*, 628–638. https://doi.org/10.1016/J.IJBIOMAC.2021.08.046

Ferysiuk, K., & Wójciak, K. M. (2020). Reduction of nitrite in meat products through the application of various plant-based ingredients. *Antioxidants*, *9*(8), 711. https://doi.org/10.3390/antiox9080711

Forghani, S., Almasi, H., & Moradi, M. (2021). Electrospun nanofibers as food freshness and time-temperature indicators: a new approach in food intelligent packaging. *Innovative Food Science & Emerging Technologies*, *73*, 102804. https://doi.org/10.1016/J.IFSET.2021.102804

Gago, C., Guerreiro, A., Cruz, S., Martins, N., Cabrita, M. J., Miguel, M. G., Faleiro, M. L., & Antunes, M. D. (2022). 1-Methylcyclopropene and lemongrass essential oil nanocoatings effect on the preservation of cold stored "Rocha" pear. *Postharvest Biology and Technology*, *192*, 111992. https://doi.org/10.1016/J.POSTHARVBIO.2022.111992

Galagan, Y., & Su, W. F. (2008). Fadable ink for time–temperature control of food freshness: novel new time–temperature indicator. *Food Research International*, *41*(6), 653–657. https://doi.org/10.1016/J.FOODRES.2008.04.012

Ghaani, M., Cozzolino, C. A., Castelli, G., & Farris, S. (2016). An overview of the intelligent packaging technologies in the food sector. *Trends in Food Science & Technology*, *51*, 1–11. https://doi.org/10.1016/J.TIFS.2016.02.008

Ghidelli, C., & Pérez-gago, M. B. (2018). Recent advances in modified atmosphere packaging and edible coatings to maintain quality of fresh-cut fruits and vegetables. *Critical Reviews in Food Science and Nutrition*, *58*(4), 662–679. https://doi.org/10.1080/10408398.2016.1211087

Gu, X. L., Zhu, X., Kong, X. Z., & Tan, Y. (2010). Comparisons of simple and complex coacervations for preparation of sprayable insect sex pheromone microcapsules and release control of the encapsulated pheromone molecule. *Journal of Microencapsulation*, *27*(4), 355–364. https://doi.org/10.3109/02652040903221532

Guan, X., Li, Q., Maimaiti, T., Lan, S., Ouyang, P., Ouyang, B., Wu, X., & Yang, S.-T. (2021). Toxicity and photosynthetic inhibition of metal-organic framework MOF-199 to pea seedlings. *Journal of Hazardous Materials*, *409*, 124521. https://doi.org/10.1016/j.jhazmat.2020.124521

Guo, Z., Martucci, N. J., Moreno-Olivas, F., Tako, E., & Mahler, G. J. (2017). Titanium dioxide nanoparticle ingestion alters nutrient absorption in an in vitro model of the small intestine. *NanoImpact*, *5*, 70–82. https://doi.org/10.1016/j.impact.2017.01.002

Hemalatha, N., & Kumar, K. R. (2021). Insecticidal and repellent activity of PEG—cardamom essential oil nanoparticles against *Phoenix dactylifera* infesting insect pest *Oryzaephilus surinamensis*. *European Journal of Molecular & Clinical Medicine*, *8*(3), 2896–2912.

Heydarzadeh, A., Valizadegan, O., Negahban, M., & Mehrkhou, F. (2019). Efficacy of *Mentha spicata* and *Mentha pulegium* essential oil Nanoformulation on mortality and physiology of *Tribolium Castaneum* (Col.: Tenebrionidae). *Journal of Crop Protection*, *8*(4), 501–520.

Homayonpour, P., Jalali, H., Shariatifar, N., & Amanlou, M. (2021). Effects of nano-chitosan coatings incorporating with free /nano-encapsulated cumin (*Cuminum cyminum* L.) essential oil on quality characteristics of sardine fillet. *International Journal of Food Microbiology*, *341*, 109047. https://doi.org/10.1016/J.IJFOODMICRO.2021.109047

Hu, J., Wang, X., Xiao, Z., & Bi, W. (2015). Effect of chitosan nanoparticles loaded with cinnamon essential oil on the quality of chilled pork. *LWT - Food Science and Technology*, *63*(1), 519–526. https://doi.org/10.1016/j.lwt.2015.03.049

Ibrahim, S., Fahmy, H., & Salah, S. (2021). Application of interactive and intelligent packaging for fresh fish shelf-life monitoring. *Frontiers in Nutrition*, *8*, 333. https://doi.org/10.3389/FNUT.2021.677884/BIBTEX

Isman, M. B., & Grieneisen, M. L. (2014). Botanical insecticide research: many publications, limited useful data. *Trends in Plant Science*, *19*(3), 140–145. https://doi.org/10.1016/j.tplants.2013.11.005

Jafari, S. M., & Esfanjani, A. F. (2017). Instrumental Analysis and Characterization of Nanocapsules. In S. M. Jafari (Ed.), *Nanoencapsulation Technologies for the Food and Nutraceutical Industries* (pp. 524–544). Academic Press.

Ji, J., Shankar, S., Royon, F., Salmieri, S., & Lacroix, M. (2021). Essential oils as natural antimicrobials applied in meat and meat products—a review. *Critical Reviews in Food Science and Nutrition*, 1–17. https://doi.org/10.1080/10408398.2021.1957766

Jokar, M., Pedersen, G. A., & Loeschner, K. (2017). Six open questions about the migration of engineered nano-objects from polymer-based food-contact materials: a review. *Food Additives & Contaminants: Part A*, *34*(3), 434–450. https://doi.org/10.1080/19440049.2016.1271462

Kamkar, A., Molaee-aghaee, E., Khanjari, A., Akhondzadeh-basti, A., Noudoost, B., Shariatifar, N., Alizadeh Sani, M., & Soleimani, M. (2021). Nanocomposite active packaging based on chitosan biopolymer loaded with nano-liposomal essential oil: its characterizations and effects on microbial, and chemical properties of refrigerated chicken breast fillet. *International Journal of Food Microbiology*, *342*, 109071. https://doi.org/10.1016/J.IJFOODMICRO.2021.109071

Karim, M., Fathi, M., Soleimanian-Zad, S., & Spigno, G. (2022). Development of sausage packaging with zein nanofibers containing tetradecane produced via needle-less electrospinning method. *Food Packaging and Shelf Life*, *33*, 100911. https://doi.org/10.1016/j.fpsl.2022.100911

Khoobdel, M., Ahsaei, S. M., & Farzaneh, M. (2017). Insecticidal activity of polycaprolactone nanocapsules loaded with *Rosmarinus officinalis* essential oil in *Tribolium castaneum* (Herbst). *Entomological Research*, *47*(3), 175–184. https://doi.org/10.1111/1748-5967.12212

Kong, X. Z., Gu, X., Zhu, X., & Zhang, Z. (2009). Spreadable dispersion of insect sex pheromone capsules, preparation via complex coacervation and release control of the encapsulated pheromone component molecule. *Biomedical Microdevices*, *11*(1), 275–285. https://doi.org/10.1007/s10544-008-9234-z

Kujur, A., Kumar, A., & Prakash, B. (2021). Elucidation of antifungal and aflatoxin B1 inhibitory mode of action of *Eugenia caryophyllata* L. essential oil loaded chitosan nanomatrix against *Aspergillus flavus*. *Pesticide Biochemistry and Physiology*, *172*, 104755. https://doi.org/10.1016/j.pestbp.2020.104755

Kumar, S. (2012). Biopesticides: a need for food and environmental safety. *Biofertilizers & Biopesticides*, *3*(4), 1000e107. https://doi.org/10.4172/2155-6202.1000e107

Kutlu, N., Meral, R., Ekin, M. M., Erim Kose, Y., & Ceylan, Z. (2022). A new application for the valorisation of pomegranate seed oil: nanoencapsulation of pomegranate seed oil into electrospun nanomats for food preservation. *International Journal of Food Science & Technology*, *57*(2), 1074–1082. https://doi.org/10.1111/IJFS.15472

Li, T., Wang, D., Ren, L., Mei, J., Xu, Y., & Li, J. (2021). Preparation of pH-sensitive polylactic acid-naringin coaxial electrospun fiber membranes for maintaining and monitoring salmon freshness. *International Journal of Biological Macromolecules*, *188*, 708–718. https://doi.org/10.1016/j.ijbiomac.2021.08.087

Lin, L., Gu, Y., & Cui, H. (2019). Moringa oil/chitosan nanoparticles embedded gelatin nanofibers for food packaging against *Listeria monocytogenes* and *Staphylococcus aureus* on cheese. *Food packaging and shelf Life*, *19*(August 2018), 86–93. https://doi.org/10.1016/j.fpsl.2018.12.005

Lin, X., Li, N., Xiao, Q., Guo, Y., Wei, J., Jiao, T., Chen, Q., Chen, Q., & Chen, X. (2022). Polyvinyl alcohol/starch-based film incorporated with grape skin anthocyanins and metal-organic framework crystals for colorimetric monitoring of pork freshness. *Food Chemistry*, *395*, 133613. https://doi.org/10.1016/J.FOODCHEM.2022.133613

Lin, L., Zhu, Y., & Cui, H. (2018). Electrospun thyme essential oil/gelatin nanofibers for active packaging against *Campylobacter jejuni* in chicken. *LWT*, *97*, 711–718. https://doi.org/10.1016/j.lwt.2018.08.015

López-Gómez, A., Ros-Chumillas, M., Buendía-Moreno, L., & Martínez-Hernández, G. B. (2020a). Active cardboard packaging with encapsulated essential oils for enhancing the shelf life of fruit and vegetables. *Frontiers in Nutrition*, *7*. https://doi.org/10.3389/fnut.2020.559978

López-Gómez, A., Navarro-Martínez, A., & Martínez-Hernández, G. B. (2020b). Active paper sheets including nanoencapsulated essential oils: a green packaging technique to control ethylene production and maintain quality in fresh horticultural products—a case study on flat peaches. *Foods*, *9*(12), 1904. https://doi.org/10.3390/FOODS9121904

López-Gómez, A., Ros-Chumillas, M., Buendía-Moreno, L., Navarro-Segura, L., & Martínez-Hernández, G. B. (2020c). Active cardboard box with smart internal lining based on encapsulated essential oils for enhancing the shelf life of fresh mandarins. *Foods*, *9*(5). https://doi.org/10.3390/foods9050590

Louni, M., Shakarami, J., & Negahban, M. (2018). Insecticidal efficacy of nanoemulsion containing *Mentha longifolia* essential oil against *Ephestia kuehniella* (Lepidoptera: Pyralidae). *Journal of Crop Protection*, *7*(2), 171–182.

Man, A., Santacroce, L., Jacob, R., Mare, A., & Man, L. (2019). Antimicrobial activity of six essential oils against a group of human pathogens: a comparative study. *Pathogens*, *8*(1). https://doi.org/10.3390/PATHOGENS8010015

Moreira, J. B., Terra, A. L. M., Costa, J. A. V., & de. Morais, M. G. (2018). Development of pH indicator from PLA/PEO ultrafine fibers containing pigment of microalgae origin. *International Journal of Biological Macromolecules*, *118*, 1855–1862. https://doi.org/10.1016/J.IJBIOMAC.2018.07.028

Mossa, A. T. H. (2016). Green pesticides: essential oils as biopesticides in insect-pest management. *Journal of Environmental Science and Technology*, *9*(5), 354–378. https://doi.org/10.3923/jest.2016.354.378

Negahban, M., Moharramipour, S., Zandi, M., & Ali Hashemi, S. (2012a). Fumigant Properties of Nano-Encapsulated Essential Oil from *Artemisia sieberi* on *Tribolium castaneum*. In S. Navarro, H. Banks, D. Jayas, C. Bell, R. Noyes, A. Ferizli, M. Emekci, A. Isikber, & K. Alagusundaram (Eds.), *Proceedings 9th. International Conference on Controlled Atmosphere and Fumigation in Stored Products* (Issue October (pp. 101–105). ARBER Professional Congress Services, Turkey.

Negahban, M., Moharramipour, S., Zandi, M., Hashemi, S. A., & Ziayee, F. (2012b). Nano-Insecticidal Activity of Essential Oil from *Cuminum cyminum* on *Tribolium castaneum*. *Proceedings 9th. International Conference on Controlled Atmosphere and Fumigation in Stored Products, Antalya, Turkey*, October, 63–68.

Niu, B., Chen, H., Wu, W., Fang, X., Mu, H., Han, Y., & Gao, H. (2022). Co-encapsulation of chlorogenic acid and cinnamaldehyde essential oil in pickering emulsion stablized by chitosan nanoparticles. *Food Chemistry: X*, *14*(February), 100312. https://doi.org/10.1016/j.fochx.2022.100312

Niu, B., Zhan, L., Shao, P., Xiang, N., Sun, P., Chen, H., & Gao, H. (2020). Electrospinning of zein-ethyl cellulose hybrid nanofibers with improved water resistance for food preservation. *International Journal of Biological Macromolecules*, *142*, 592–599. https://doi.org/10.1016/j.ijbiomac.2019.09.134

Noori, S., Zeynali, F., & Almasi, H. (2018). Antimicrobial and antioxidant efficiency of nanoemulsion-based edible coating containing ginger (*Zingiber officinale*) essential oil and its effect on safety and quality attributes of chicken breast fillets. *Food Control*, *84*, 312–320. https://doi.org/10.1016/j.foodcont.2017.08.015

Oftadeh, M., Jalali Sendi, J., & Ebadollahi, A. (2021). Biologically active toxin identified from *Artemisia annua* against lesser mulberry pyralid, *Glyphodes pyloalis*. *Toxin Reviews*, *40*(4), 953–961. https://doi.org/10.1080/15569543.2020.1811345

Omer, M. K., Álvarez-Ordoñez, A., Prieto, M., Skjerve, E., Asehun, T., & Alvseike, O. A. (2018). A systematic review of bacterial foodborne outbreaks related to red meat and meat products. *Foodborne Pathogens and Disease*, *15*(10), 598–611. https://doi.org/10.1089/fpd.2017.2393

Pabast, M., Shariatifar, N., Beikzadeh, S., & Jahed, G. (2018). Effects of chitosan coatings incorporating with free or nano-encapsulated satureja plant essential oil on quality characteristics of lamb meat. *Food Control*, *91*, 185–192. https://doi.org/10.1016/j.foodcont.2018.03.047

Pakolpakcil, A., Karaca, E., & Becerir, B. (2018). Investigation of a natural pH-indicator dye for nanofibrous wound dressings. *IOP Conference Series: Materials Science and Engineering*, *460*(1), 12020. https://doi.org/10.1088/1757-899X/460/1/012020

Petrović, G. M., Stojanović, G. S., & Radulović, N. S. (2010). Encapsulation of cinnamon oil in β-cyclodextrin. *Journal of Medicinal Plants Research*, *4*(14), 1382–1390. https://doi.org/10.5897/JMPR10.146

Pounds, K., Jairam, S., Bao, H., Meng, S., Zhang, L., Godinez, S., Savin, D. A., Pelletier, W., Correll, M. J., & Tong, Z. (2021). Glycerol-based dendrimer nanocomposite film as a tunable pH-sensor for food packaging. *ACS Applied Materials and Interfaces*, *13*(19), 23268–23281. https://doi.org/10.1021/ACSAMI.1C05145/ASSET/IMAGES/LARGE/AM1C05145_0010.JPEG

Qi, D., Xiao, Y., Xia, L., Li, L., Jiang, S., Jiang, S., & Wang, H. (2022). Colorimetric films incorporated with nisin and anthocyanins of pomegranate/*Clitoria ternatea* for shrimp freshness monitoring and retaining. *Food Packaging and Shelf Life*, *33*, 100898. https://doi.org/10.1016/j.fpsl.2022.100898

Rahnemoon, P., Sarabi-Jamab, M., Bostan, A., & Mansouri, E. (2021). Nano-encapsulation of pomegranate (*Punica granatum* L.) peel extract and evaluation of its antimicrobial properties on coated chicken meat. *Food Bioscience*, *43*, 101331. https://doi.org/10.1016/J.FBIO.2021.101331

Rajkumar, V., Gunasekaran, C., Paul, C. A., & Dharmaraj, J. (2020). Development of encapsulated peppermint essential oil in chitosan nanoparticles: characterization and biological efficacy against stored-grain pest control. *Pesticide Biochemistry and Physiology*, *170*(June), 104679. https://doi.org/10.1016/j.pestbp.2020.104679

Roshan, A. B., Dubey, N. K., & Mohana, D. C. (2022a). Chitosan nanoencapsulation of *Pogostemon cablin* (Blanco) Benth. Essential oil and its novel preservative effect for enhanced shelf life of stored maize kernels during storage: evaluation of its enhanced antifungal, antimycotoxin, antioxidant activities and possible mode of action. *International Journal of Food Science and Technology*, *57*(4), 2195–2202. https://doi.org/10.1111/ijfs.15289

Roshan, A. B., Venkatesh, H. N., Dubey, N. K., & Mohana, D. C. (2022b). Chitosan-based nanoencapsulation of *Toddalia asiatica* (L.) lam. Essential oil to enhance antifungal and aflatoxin B1 inhibitory activities for safe storage of maize. *International Journal of Biological Macromolecules*, *204*, 476–484. https://doi.org/10.1016/j.ijbiomac.2022.02.026

Roshandel-hesari, N., Mokaber-Esfahani, M., Taleghani, A., & Akbari, R. (2022). Investigation of physicochemical properties, antimicrobial and antioxidant activity of edible films based on chitosan/casein containing *Origanum vulgare* L. essential oil and its effect on quality maintenance of cherry tomato. *Food Chemistry*, *396*, 133650. https://doi.org/10.1016/J.FOODCHEM.2022.133650

Rozhina, E., Batasheva, S., Miftakhova, R., Yan, X., Vikulina, A., Volodkin, D., & Fakhrullin, R. (2021). Comparative cytotoxicity of kaolinite, halloysite, multiwalled carbon nanotubes and graphene oxide. *Applied Clay Science, 205*, 106041. https://doi.org/10.1016/j.clay.2021.106041

Sabbour, M. M. A. (2019). Efficacy of natural oils against the biological activity on *Callosobruchus maculatus* and *Callosobruchus chinensis* (Coleoptera: Tenebrionidae). *Bulletin of the National Research Centre, 43*(1), 1–8. https://doi.org/10.1186/s42269-019-0252-1

Sabbour, M. M. (2020). Effect of some essential oils on *Rhyzopertha dominica* (Coleoptera: Bostrichidae). *International Research Journal of Biological Sciences, 02*(01), 32–37.

Salarbashi, D., Tafaghodi, M., Bazzaz, B. S. F., Mohammad Aboutorabzade, S., & Fathi, M. (2021). pH-sensitive soluble soybean polysaccharide/SiO_2 incorporated with curcumin for intelligent packaging applications. *Food Science & Nutrition, 9*(4), 2169–2179. https://doi.org/10.1002/fsn3.2187

Sarvinehbaghi, M. B., Ahmadi, M., Shiran, M., & Azizkhani, M. (2021). Antioxidant and antimicrobial activity of red onion (*Allium cepa*, L.) extract nanoencapsulated in native seed gums coating and its effect on shelf-life extension of beef fillet. *Journal of Food Measurement and Characterization, 15*(5), 4771–4780. https://doi.org/10.1007/s11694-021-00985-9

Shishir, M. R. I., Xie, L., Sun, C., Zheng, X., & Chen, W. (2018). Advances in micro and nano-encapsulation of bioactive compounds using biopolymer and lipid-based transporters. *Trends in Food Science and Technology, 78*, 34–60. https://doi.org/10.1016/j.tifs.2018.05.018

Sommi, P., Vitali, A., Coniglio, S., Callegari, D., Barbieri, S., Casu, A., Falqui, A., Vigano', L., Vigani, B., Ferrari, F., & Anselmi-Tamburini, U. (2021). Microvilli adhesion: an alternative route for nanoparticle cell internalization. *ACS Nano, 15*(10), 15803–15814. https://doi.org/10.1021/acsnano.1c03151

Soto, K. M., Hernández-Iturriaga, M., Loarca-Piña, G., Luna-Bárcenas, G., & Mendoza, S. (2019). Antimicrobial effect of nisin electrospun amaranth: pullulan nanofibers in apple juice and fresh cheese. *International Journal of Food Microbiology, 295*, 25–32. https://doi.org/10.1016/j.ijbiomac.2021.08.087

Tatlisu, N. B., Yilmaz, M. T., & Arici, M. (2019). Fabrication and characterization of thymol-loaded nanofiber mats as a novel antimould surface material for coating cheese surface. *Food Packaging and Shelf Life, 21*, 100347. https://doi.org/10.1016/j.fpsl.2019.100347

Tiwari, S., Upadhyay, N., Singh, B. K., Singh, V. K., & Dubey, N. K. (2022). Chemically characterized nanoencapsulated *Homalomena aromatica* Schott. Essential oil as green preservative against fungal and aflatoxin B1 contamination of stored spices based on in vitro and in situ efficacy and favorable safety profile on mice. *Environmental Science and Pollution Research, 29*, 3091–3106. https://doi.org/10.1007/s11356-021-15794-2

Tornuk, F., Hancer, M., Sagdic, O., & Yetim, H. (2015). LLDPE based food packaging incorporated with nanoclays grafted with bioactive compounds to extend shelf life of some meat products. *LWT—Food Science and Technology, 64*(2), 540–546. https://doi.org/10.1016/j.lwt.2015.06.030

Upadhyay, N., Singh, V. K., Dwivedy, A. K., Das, S., Chaudhari, A. K., & Dubey, N. K. (2019). Assessment of *Melissa officinalis* L. essential oil as an eco-friendly approach against biodeterioration of wheat flour caused by *Tribolium castaneum* Herbst. *Environmental Science and Pollution Research, 1996*, 14036–14049. https://doi.org/10.1007/s11356-019-04688-z

Vafania, B., Fathi, M., & Soleimanian-Zad, S. (2019). Nanoencapsulation of thyme essential oil in chitosan-gelatin nanofibers by nozzle-less electrospinning and their application to reduce nitrite in sausages. *Food and Bioproducts Processing*, *116*, 240–248. https://doi.org/10.1016/j.fbp.2019.06.001

Vishnuvarthanan, M., & Rajeswari, N. (2019). Food packaging: pectin–laponite–ag nanoparticle bionanocomposite coated on polypropylene shows low O_2 transmission, low ag migration and high antimicrobial activity. *Environmental Chemistry Letters*, *17*(1), 439–445. https://doi.org/10.1007/s10311-018-0770-3

Vishwakarma, G. S., Gautam, N., Babu, J. N., Mittal, S., & Jaitak, V. (2016). Polymeric encapsulates of essential oils and their constituents: a review of preparation techniques, characterization, and sustainable release mechanisms. *Polymer Reviews*, *56*(4), 668–701. https://doi.org/10.1080/15583724.2015.1123725

Volić, M., Pećinar, I., Micić, D., Đorđević, V., Radojica, P., Nedović, V., & Obradović, N. (2022). Design and characterization of whey protein nanocarriers for thyme essential oil encapsulation obtained by freeze-drying. *Food Chemistry*, *386*, 132749. https://doi.org/10.1016/j.foodchem.2022.132749

Wen, P., Zhu, D. H., Wu, H., Zong, M. H., Jing, Y. R., & Han, S. Y. (2016). Encapsulation of cinnamon essential oil in electrospun nanofibrous film for active food packaging. *Food Control*, *59*, 366–376. https://doi.org/10.1016/j.foodcont.2015.06.005

Xavier, L. O., Sganzerla, W. G., Rosa, G. B., da Rosa, C. G., Agostinetto, L., de Veeck, A. P., L., Bretanha, L. C., Micke, G. A., Dalla Costa, M., Bertodi, F. C., Barreto, P. L. M., & Nunes, M. R. (2021). Chitosan packaging functionalized with *Cinnamodendron dinisii* essential oil loaded zein: a proposal for meat conservation. *International Journal of Biological Macromolecules*, *169*, 183–193. https://doi.org/10.1016/J.IJBIOMAC.2020.12.093

Yang, F. L., Li, X. G., Zhu, F., & Lei, C. L. (2009). Structural characterization of nanoparticles loaded with garlic essential oil and their insecticidal activity against *Tribolium castaneum* (Herbst) (Coleoptera: Tenebrionidae). *Journal of Agricultural and Food Chemistry*, *57*(21), 10156–10162. https://doi.org/10.1021/jf9023118

Yang, K., Liu, A., Hu, A., Li, J., Zen, Z., Liu, Y., Tang, S., & Li, C. (2021). Preparation and characterization of cinnamon essential oil nanocapsules and comparison of volatile components and antibacterial ability of cinnamon essential oil before and after encapsulation. *Food Control*, *123*, 107783. https://doi.org/10.1016/j.foodcont.2020.107783

Yang, C., Wen, J., Xue, Z., Yin, X., Li, Y., & Yuan, L. (2023). The accumulation and toxicity of ZIF-8 nanoparticles in *Corbicula fluminea*. *Journal of Environmental Sciences*, *127*, 91–101. https://doi.org/10.1016/j.jes.2022.03.020

Yildirim, S., Röcker, B., Pettersen, M. K., Nilsen-Nygaard, J., Ayhan, Z., Rutkaite, R., Radusin, T., Suminska, P., Marcos, B., & Coma, V. (2018). Active packaging applications for food. *Comprehensive Reviews in Food Science and Food Safety*, *17*(1), 165–199. https://doi.org/10.1111/1541-4337.12322

Yuan, X., Zhang, X., Sun, L., Wei, Y., & Wei, X. (2019). Cellular toxicity and immunological effects of carbon-based nanomaterials. *Particle and Fibre Toxicology*, *16*(1), 18. https://doi.org/10.1186/s12989-019-0299-z

Zhang, H., Li, X., & Kang, H. (2019). Chitosan coatings incorporated with free or nano-encapsulated paulownia tomentosa essential oil to improve shelf-life of ready-to-cook pork chops. *LWT—Food Science and Technology*, *116*, 108580. https://doi.org/10.1016/j.lwt.2019.108580

8

Active Packaging: Concept, Applications, and Regulatory Aspects

A. Vanmathi Mugasundari and Jeyan A. Moses

Computational Modeling and Nanoscale Processing Unit, Department of Food Process Engineering, National Institute of Food Technology, Entrepreneurship and Management, Thanjavur, Government of India, Thanjavur, Tamil Nadu - 613005, India

8.1 Introduction

Traditional food packaging serves multiple purposes, including protection, communication, convenience, and containment (Robertson, 2014). It plays a crucial role in shielding products from adverse environmental conditions such as heat, light, moisture levels, pressure, microorganisms, gaseous emissions, and more. Furthermore, it enhances consumer usability and time efficiency, accommodating products of various sizes and shapes, as highlighted by (Marsh and Bugusu, 2007). In the case of traditional packaging materials that come into contact with food, the primary safety objective is to minimize reactivity. Conversely, innovative packaging systems, such as active and intelligent packaging concepts, rely on the dynamic interaction between the packaging environment and the food itself to actively protect the food. Furthermore, over the past decade, there has been an increased emphasis on consumer preferences for higher quality, fresher, and more convenient food products. Consequently, a range of active packaging (AP) technologies has emerged to deliver improved product quality, maintain freshness, ensure food safety, and address environmental concerns associated with packaging waste and disposal (Ozdemir and Floros, 2004).

The inherent properties of food, encompassing elements like water activity (aw), pH, nutritional composition, presence of antimicrobial substances, and respiration rate, along with external influences like relative humidity (RH), temperature, and the atmosphere's configuration, collectively influence the longevity of food items. Various AP techniques have been employed and these are specifically designed to absorb or release substances into or from the food or its environment, aiming to prolong shelf life and preserve or enhance packaged foods (Mugasundari and Anandakumar, 2022). Significant instances of AP encompass oxygen scavengers, carbon dioxide emitters/absorbers, moisture absorbers, ethylene absorbers, and ethanol emitters, systems for releasing or absorbing flavors,

Nanotechnology for Sustainable Food Packaging, First Edition. Edited by C. Anandharamakrishnan, Jeyan A. Moses, and M. Maria Leena.

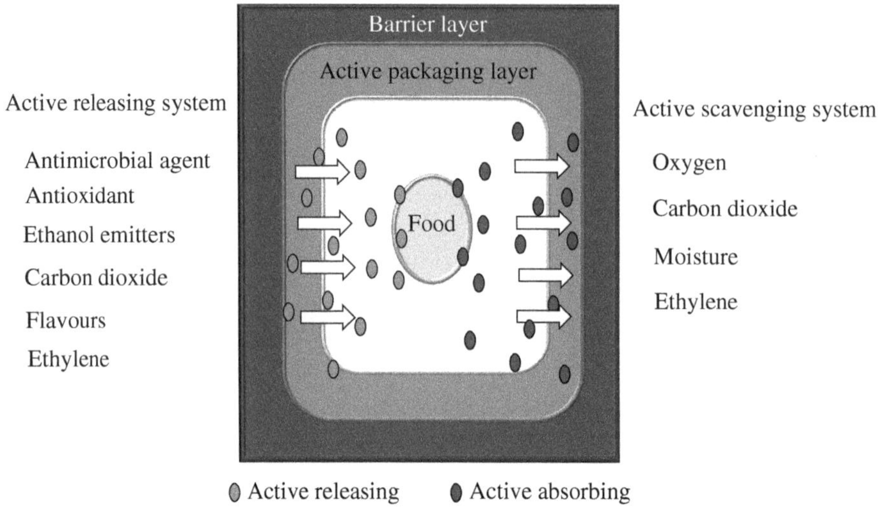

Figure 8.1 Active packaging system

time-temperature indicators, and films containing antimicrobial agents. AP system is shown in Figure 8.1.

This chapter delivers an overview of nanotechnology's role in AP, detailing various nanomaterials contributing active functionalities to food packaging. The focus is on nanomaterials, like metals and metal oxides, which are not inherently "nano" but nonetheless are employed at the nanosized to impart or enhance active functionalities. A particular emphasis is placed on inorganic NPs with antimicrobial properties. The utilization of NPs consists of bioactive compounds for antimicrobial films. Additionally, the chapter explores oxygen and ethylene scavengers on active nanomaterials, highlighting studies testing AP on food systems.

8.2 Types and Mechanisms of Nano Material-Based Active Packaging

8.2.1 Nanoparticles-Based Antimicrobial Packaging Systems

Antimicrobial packaging systems represent groundbreaking innovations developed to suppress the proliferation of spoilage and pathogenic microorganisms within the packaging of food products. Within the realm of AP, there has been a noteworthy surge in enthusiasm for antimicrobial packaging systems over the past few decades. This heightened interest stems from their potential as a promising, supplementary approach to conventional preservation methods aimed at managing unwanted microorganisms in food. Antimicrobial packaging systems can be classified into three primary categories: (i) Systems that release antimicrobial agents, such as organic compounds, directly from the packaging material onto the food surface, enabling their diffusion and necessitating direct interaction with the food. (ii) Emit volatile compounds from the packaging material into the headspace of the packaging without the need for direct contact, and they are absorbed by food. (iii) Incorporated within the polymer structure are nonmigratory antimicrobial polymers containing antimicrobial agents. In such systems, direct contact with food is vital, as the antimicrobial agent is intentionally engineered to remain within the polymer without migration. This group includes surface-modified films that inherently possess antimicrobial characteristics (Yildirim, 2011).

In the past, antimicrobial compounds were commonly integrated directly into food. Nonetheless, the direct introduction of antimicrobial substances to food might compromise their effectiveness, as interactions with the food's composition could lead to reduced efficacy. Additionally, both in fresh and processed foods, microbial deterioration predominantly manifests at the surface. Consequently, a more efficient approach could involve the controlled release of antimicrobial agents through AP onto the product's surface rather than distributing them throughout the entire volume of the food. This approach has the potential to reduce the quantity of such substances needed to achieve the desired antimicrobial effect. Nanotechnology, with its enabling capabilities, presents fresh opportunities for the creation of novel antimicrobial packaging systems and the augmentation of antimicrobial properties in active substances. Nanosized antimicrobial agents possess a substantial surface-to-volume ratio and heightened surface activity, potentially enabling them to combat unwanted microorganisms more efficiently compared to their larger-scale counterparts, whether micro or macro in size (Morones et al., 2005; Padmavathy and Vijayaraghavan, 2008; Radusin et al., 2016).

Antimicrobial properties have been detected in numerous NPs, particularly metals like silver (Ag) (Abolghasemi et al., 2016; Fernández et al., 2009; Nobile et al., 2001), gold (Au) (Lima et al., 2013; Pagno et al., 2015), and copper (Cu) (Conte et al., 2013), as well as metal oxides such as titanium dioxide (TiO_2) (Bodaghi et al., 2013; Chawengkijwanich and Hayata, 2008), zinc oxide (ZnO) (Jin and Gurtler, 2011), and magnesium oxide (MgO) (Jin and He, 2011) Table 8.1 provides NPs -based antimicrobial food-packaging systems. Additionally, beyond inorganic NPs, natural compounds like chitin/chitosan have demonstrated antimicrobial efficacy when employed at the nano size (Han et al., 2010; Youssef et al., 2016).

8.2.1.1 Popular Metal Nanoparticles

Silver Silver (Ag) has served to preserve food and beverages for centuries, and the potent antimicrobial properties of silver ions and salts are widely recognized (Duncan, 2011). The incorporation of Ag-NPs) into polymer matrices presents a mostly appealing option, as they exhibit chemical stability, allowing for controlled release over extended storage durations. Abolghasemi et al. (2016) conducted a study to explore the antimicrobial effects of polyvinyl chloride-poly ethylene (PVC-PE) laminate trays coated with Ag-NPs using inkjet printing were significant. The silver-coated packaging was found to markedly decrease microbial proliferation, including bacterial counts, *Staphylococcus aureus,* and *Escherichia coli*, in packaged beef. This resulted in an extension of the product's shelf life from day 2 to 7 when stored in refrigerator condition. Nobile et al. (2001) researchers utilized an active polyethylenc (PE) film incorporating nanoscale silver (Ag) deposited via plasma in a 90-nm thickness, achieved through a silver-polyethyleneoxide (PEO) coating. They evaluated the antimicrobial efficacy of this Ag-PEO film against

Table 8.1 Examples of nanoparticles-based antimicrobial packaging systems

Active Substances	Packaging Materials	Target Microorganisms	Food Application	Reference
Silver	Polyvinyl chloride-poly ethylene (PVC-PE) laminate trays	*Staphylococcus aureus*, and *Escherichia coli*	Minced beef	Abolghasemi et al. (2016)
	Polyethylene (PE) film	*Alicyclobacillus acidoterrestris*	Apple juice	Nobile et al. (2001)
	Polyvinyl chloride (PVC)	*S. aureus, E. coli, P. fluorescens, and Bacillus cereus*	Raw chicken, raw beef, and cooked ham	Azlin-Hasim et al. (2016)
Gold	Faujasite zeolite	*E. coli* and *Salmonella typhi*	*In vitro*	Lima et al. (2013)
	Starch-based biofilms			Pagno et al. (2015)
Copper	Chitosan films	*Salmonella typhimurium*, and *S. aureus*	*In vitro*	Cárdenas et al. (2009)
	Polylactic acid (PLA)	*Pseudomonas* spp.	Dairy products	Conte et al. (2013)
	Agar-based bionanocomposite films	*Listeria monocytogenes* and *E. coli*	*In vitro*	Shankar et al. (2014)
	Cellulose fibers	*Saccharomyces cerevisiae*	Pineapple and melon juice	Llorens et al. (2012)
Metal Oxide Nanoparticles				
TiO_2	Polyethylene (PE)-	*S. aureus*, and *E. coli*	*In vitro*	Yang et al. (2010)
	Low-density polyethylene (LDPE)	*Pseudomonas* spp. and *Rhodotorula mucilaginosa*	Fresh pears	Bodaghi et al. (2013)
	LDPE	*Botrytis cinerea*	Kiwi fruit	Hu et al. (2011)
MgO + nisin + ZnO	Glass bottles	*E. coli* and *Salmonella stanley*	Fruit juices	Jin and He (2011)
ZnO	Calcium alginate films	*S. aureus* and *S. typhimurium*	Ready-to-eat poultry products	Akbar and Anal (2014)
	Chitosan film	*Mold* and *yeast*	Soft cheese	Youssef et al. (2016)
ZnO + basil leaf essential oil (BEO)	Fish skin gelatin/fish protein isolate films	*Enterobacteriaceae, Pseudomonas*	Sea bass slices	Arfat et al. (2015)
Bioactive Compounds				
Carvacrol and thymol essential oils	LDPE	*Botrytis cinerea*	Strawberries	Campos-Requena et al. (2015)

Alicyclobacillus acidoterrestris, a robust food spoilage bacterium resistant to thermal and acidic conditions, in both apple juice and malt extract broth (MEB). Notably, the film exhibited superior antimicrobial activity in the MEB culture medium compared to apple juice.

The antimicrobial characteristics of Ag-NPs in a solution were evaluated against diverse bacterial strains, encompassing *S. aureus*, *E. coli*, *Pesudomonas fluorescens*, *microflora*, and *Bacillus cereus,* derived from raw chicken, beef, and ham. Additionally, the microflora derived from meat exhibited a higher resistance to Ag-NPs when compared to the pure culture bacteria. Notably, the most effectiveness of Ag-NPs was observed in the case of *P. fluorescens*. Furthermore, reduced antimicrobial efficacy may arise when Ag-NPs are incorporated directly into polymer matrices because most of the NPs do not come into direct interact with the food. Moreover, an uneven distribution of NPs within the biopolymer can further diminish their antimicrobial impact (Azlin-Hasim et al., 2016). In their study, Fernández et al. (2009) examined the antibacterial hybrid materials centered on silver (Ag) through an in-situ process of silver nitrate adsorbed onto cellulose fibers. Their study disclosed that both the fiber structure and the method employed for Ag reduction had a significant impact on the resulting Ag NPs ' morphology, subsequently influencing their antimicrobial capabilities. *E. coli* and *S. aureus* were conducted *in vitro*, and a notable decrease of around 4 log CFU/g was observed. The utilization of cellulose-Ag-NP absorbent pads proved to be effective against aerobic and lactic acid bacteria, *Pseudomonas* spp., and *Enterobacteriaceae* in beef meat extrudates (Fernández et al., 2010b). Moreover, besides the microbial reduction, this approach also delayed the ripening of the fruit, resulting in an enhancement of its visual appearance.

Gold Gold nanoparticles (Au-NPs) are well known for their antibacterial and antifungal properties, as highlighted by Lima et al. (2013), who reported exceptional biocidal efficacy when gold-NPs were supported on zeolites, particularly faujasite zeolite. After a 90-minute exposure to culture media, only 5–10% of *E. coli* and *Salmonella typhi* colonies remained viable. Gomashe and Dharmik (2012) demonstrated the potent antibacterial properties of gold-NPs integrated into quinoa starch-based biofilms, achieving inhibition rates of 99% against *E. coli* and 98% against *S. aureus* while also enhancing mechanical, optical, and morphological characteristics. Additionally, Thirumurugan et al. (2013) observed an antimicrobial synergistic effect when combining Au-NPs with peptides from bacteriocins produced by Lactobacilli, along with nisin, significantly enhancing the antimicrobial activity against various food spoilage microorganisms compared to individual use of bacteriocins or nisin with gold-NPs. The findings suggested that the synergistic use of Ag-NPs significantly boosted the antimicrobial effectiveness of bacteriocins produced by Lactobacilli against *Micrococcus luteus*, *B. cereus*, *Klebsiella pneumoniae*, *Proteus mirabilis*, *E. coli*, *S. aureus* surpassing the individual efficacy of these bacteriocins. Moreover, this cooperative impact was more pronounced than when combining nisin with gold-NPs.

Copper Copper (Cu) is commonly found in several foods in the form of ions or salts (Costa et al., 2016). When compared to silver, bulk copper exhibits impressive antimicrobial characteristics (Eide, 2000; Shankar and Rhim, 2014). Numerous research studies have provided evidence of the antimicrobial properties of Cu-NPs against pathogens. Shankar et al. (2014) illustrated the antimicrobial potential of agar-based bio nanocomposite biopolymer infused with various copper salts. These films displayed robust antimicrobial properties against *E. coli* and *Listeria monocytogenes*. Llorens et al. (2012) successfully demonstrated the efficacy of cellulose/copper composites in inhibiting *Saccharomyces cerevisiae* in laboratory conditions. Furthermore, these nanocomposite permeable materials exhibited remarkable antibacterial and antifungal properties when tested in pineapple and watermelon juice, resulting in a reduction in yeasts and molds by approximately 4 log. Conte et al. (2013) synergized the bioactive properties of Cu-NPs with polylactic acid (PLA), creating packaging solutions for dairy products. The PLA-Cu films exhibited a slowdown in the growth of *Pseudomonas* spp. (*in vitro*) and the primary spoilage microorganisms found in fior di latte samples. Importantly, the researchers proved that the presence of Cu-NPs did not impact the typical flora in dairy products, thereby preserving the sensory characteristics of the products.

8.2.1.2 Metal Oxide Nanoparticles

Titanium Dioxide Titanium dioxide (TiO_2) is recognized for its inert nature, nontoxicity, cost-effectiveness, eco-friendliness, and its demonstrated ability to deactivate a diverse range of microorganisms in numerous applications (Chawengkijwanich and Hayata, 2008; Yemmireddy et al., 2015). In their study, Yang et al. (2008) explored the impact of TiO_2-NPs on the antimicrobial characteristics within polyethylene (PE)-based films as the substrate. Their findings demonstrated a noteworthy enhancement in antibacterial efficacy through UV irradiation. This enhancement resulted in an inhibition rate of 95% (*S. aureus*), and 89% (*E. coli*), respectively, in comparison to the TiO_2-PE film without UV exposure. Bodaghi et al. (2013) illustrated the photo degradation-driven antimicrobial impact achieved by integrating TiO_2-NPs into low-density polyethylene (LDPE) films. This innovative approach was tested against *Rhodotorula mucilaginosa* and *Pseudomonas* spp., which are prominent microorganisms associated with fruits and vegetables. Furthermore, the efficacy of these photocatalytic films was verified in the context of fruit packaging, leading to a notable reduction in mesophilic bacteria and yeast cell counts within packaging containing fresh pears.

In a study by Hu et al. (2011), Ag-NPs, TiO_2-NPs, and montmorillonite NPs were incorporated into the LDPE film, and their impact on the kiwi throughout cold storage was assessed. The antimicrobial properties of these nanocomposite-based packaging films were evaluated using a conidial germination test involving *Botrytis cinerea*, a common fungus found on kiwifruit. The results clearly demonstrated the effectiveness of the applied AP in inhibiting the germination of *B. cinerea*. Utilizing HDPE films infused with nano-Ag/TiO_2 was proven effective in extending the shelf life of bread by up to six days, primarily due to its efficacy against yeast, molds, and bacteria.

Magnesium Oxide Magnesium oxide (MgO)-NPs have been documented for their antimicrobial properties against specific vegetative bacteria (Jin and He, 2011), as well as bacterial spores (Koper et al., 2002). Jin and He (2011) showcased the remarkable efficacy of MgO-NPs against pathogens was demonstrated by achieving reductions of above 7 logs for *E. coli O157 : H7* and *Salmonella stanley*. Additionally, the researchers examined the impact of MgO-NPs in combination with other antimicrobial agents such as nisin and ZnO-NPs. The antimicrobial potential of ZnO, MgO, and metallic oxides was explored against various spoilage microorganisms commonly found in fruit juices.

Zinc Oxide Zinc oxide (ZnO) aids in employing zinc addition and is additionally employed in the lining of food cans, where it plays a vital role in averting spoilage and maintaining the vibrant color of the contents. Extensive research has been conducted to explore the antibacterial properties of ZnO-NP against pathogens, with numerous studies delving into this aspect (Emamifar et al., 2010). Numerous studies have explored the use of ZnO-NPs in food packaging, with one notable investigation led by Akbar and Anal (2014). They incorporated ZnO-NPs (approximately 50 nm in size) into calcium alginate films and assessed their antibacterial efficacy in ready-to-eat poultry products. The results were promising, as the film containing 3 mg/mL of ZnO-NPs proved highly effective for AP. When applied to poultry meat and subjected to challenge tests at $8 \pm 1\,°C$, it achieved a significant 2-log reduction in bacterial counts (*S. aureus* and *Salmonella typhimurium* at 10^6–10^7 CFU/mL) within 24 hours. *S. typhimurium* and *S. aureus* cells were identified in the wrapped meat, while the control group without the active film showed no such reduction in bacterial populations.

Formulated a casting bio nanocomposite packaging film by incorporating chitosan (CH), carboxymethyl cellulose (CMC), and ZnO-NPs. This CH/CMC/ZnO film was assessed for its antimicrobial properties as packaging for Egyptian soft white cheese. The results revealed that the films exhibited significantly enhanced antimicrobial effectiveness against both Gram-positive and Gram-negative bacteria, yeast, and mold in comparison to the control film (CH/CMC). As a result, this innovation played a pivotal role in prolonging the shelf life of soft cheese (Youssef et al., 2016).

ZnO-NPs have been subjected to testing in combination with essential oils, renowned for their antimicrobial properties. For instance, Arfat et al. (2015) employed gelatin/fish protein films infused with ZnO-NPs (3%) and basil leaf essential oil (BEO) were utilized to wrap sea bass slices. Throughout a 12-day storage period at 4 °C, the fish enveloped in this AP displayed the slowest growth of *Enterobacteriaceae,* and *Pseudomonas*. This was in contrast to the films containing only BEO, ZnO-NPs alone, and the control film (fish protein isolate/fish skin gelatin). Moreover, sensory evaluation extended the sea bass's shelf life to 12 days for the ZnO-NP/BEO films, contrasting with the control's 6-day shelf life. The inclusion of BEO also contributed to antioxidant properties, mitigating lipid oxidation.

8.2.2 Bioactive Compounds

Certain nanomaterials, even without inherent antimicrobial attributes, can serve as carriers for antimicrobials, particularly essential oils. They play a crucial role in safeguarding these antimicrobial agents against adverse environmental or processing factors while facilitating a controlled and gradual release of these substances over time (Blanco-Padilla et al., 2014; Zhang et al., 2010). Cellulose-based films containing carbon nanotubes with allyl isothiocyanate (AITC), an essential oil found in mustard, displayed antimicrobial properties. These films effectively combated *Salmonella choleraesuis* and mitigated microbial deterioration in packed chicken meat, as demonstrated in the study (Dias et al., 2013).

Thymol and carvacrol essential oils have been incorporated into LDPE films that contain organically modified montmorillonite nanocomposites. These EO-infused clay/polymer nanocomposite films were used to pack strawberries contaminated with *B. cinerea*, commonly known as gray mold. Remarkably, these films demonstrated an impressive capability to effectively suppress the development of gray mold in the strawberries (Campos-Requena et al., 2015). Han et al. (2010) conducted a study wherein montmorillonite was produced through an ion exchange process involving chitosan and Na+−MMT. It was observed that these chitosan-Na+montmorillonite exhibited greater effects on bacteria compared to chitosan or Na+montmorillonite alone. Furthermore, these nanocomposites demonstrated enhanced thermal stability, suggesting their possible utilization in creating natural, biodegradable packaging materials endowed with antimicrobial characteristics. In a study conducted by Imran et al. (2012), nano-liposomes encapsulating nisin peptides, inherently small in size, were incorporated into hydroxypropyl methylcellulose biodegradable film. *In vitro* assessments have shown their effective antimicrobial effects against the food-borne pathogen *Listeria monocytogenes*. However, it is crucial to acknowledge that these systems are still in the developmental stage, and additional research is necessary for further progress and comprehension.

8.2.3 Nanoscaled Oxygen Scavengers

Oxygen sensitivity is a significant concern for various food products, as it has the potential to deteriorate product quality and reduce shelf life through oxidation (Choe and Min, 2006). Additionally, it can create favorable conditions for the proliferation of aerobic microorganisms, leading to alterations in color (Larsen et al., 2006; Nannerup et al., 2004), changes in sensory attributes (Jacobsen, 1999), and nutritional degradation (Van Bree ct al., 2012). Within the food industry, oxygen-sensitive food items are typically packaged through methods like modified atmosphere packaging (MAP) or gas flushing. Despite these efforts, various factors can contribute to the presence of residual oxygen in the package, including oxygen originally dissolved within the food, which may later escape into the headspace, oxygen permeating through the packaging film, inadequate removal during the process, or improper sealing (de Abreu et al., 2012; Gibis and Rieblinger, 2011).

Hence, the utilization of oxygen scavengers (OS) has emerged as a crucial technology in AP. These scavengers serve the dual purpose of eliminating oxygen from food packages and enhancing barrier properties, effectively functioning as active barriers (Sängerlaub et al., 2013). Mu et al. (2013) have made a nano-iron oxygen scavenger. They combined Fe-NPs with $CaCl_2$, NaCl, and activated carbon, then encapsulated them in sachets. This nanoscale scavenger, with an average particle size of 110 nm, exhibited an oxygen scavenging capacity nearly 1.4 times higher than its larger, microsized counterpart (approximately 20 μm). The researchers implemented a nanosized oxygen scavenger system for packaging roasted sunflower seeds and walnuts, both highly prone to lipid oxidation. Throughout a 120-day storage period, their experiments demonstrated that samples consisting of nanoscaled oxygen scavengers exhibited the lower value of *p*-anisidine and peroxide among the treatment conditions. Busolo and Lagaron (2012) assessed iron-infused kaolinite was evaluated as an oxygen-scavenging preservative in packaging material, specifically High-density polyethylene and Linear low-density polyethylene (HDPE and LLDPE) films. These films played a dual role in combating oxygen by actively reacting with molecular oxygen while also passively hindering its diffusion. Consequently, the authors suggest that this innovative oxygen scavenger shows significant potential for enhancing food-packaging applications.

Titanium dioxide (TiO_2) has been identified as another viable option for oxygen scavenging (Costa et al., 2016). Xiao-e et al. (2004) studied incorporating TiO_2-NPs into various polymer matrices. They illustrated the oxygen-scavenging capabilities of nanocrystalline TiO2 films under UV light exposure. Mills et al. (2006) illustrated the oxygen-scavenging capabilities of nanocrystalline TiO_2 films under UV light exposure. The nanocrystalline titania particles were integrated into an ethyl cellulose (EC) packaging film. In their study, Yu et al. (2004) utilized an *in situ* fusion technique to integrate minor volumes of Pd or Pt NPs into different commercially available thermoplastic polymers. Their findings indicated that even modest concentrations (approximately 2%) of hydrogen in the purging gas were effective in significantly decreasing oxygen permeation through the polymer films infused with metal NPs. This catalytic approach demonstrated high efficacy and was not contingent on the chemical composition of the polymer matrix.

In their study, Hutter et al. (2016) showcased the ability of a palladium-based oxygen scavenger (OS) film to preserve the red color of cooked cured ham for a remarkable 21-day storage period, even when the packages were exposed to continuous light for 24 hours a day. In stark contrast, samples that were packaged without the OS film lost their red color within just two hours after being packaged. The utilization of the Pd-based oxygen scavenger (OS) in bakery products yielded notable benefits, prolonging the shelf life by three to four times. This extension was attributed to the significant retardation of mold growth, which was effectively suppressed for a duration of 8–10 days. In contrast, bread samples lacking the OS film displayed mold growth within just two days of storage. The incorporation of nano-TiO_2 or nano-aluminum oxide (Al_2O_3) significantly prolonged the oxygen-scavenging capacity, even with reduced levels of calcium ascorbate (from 35% to 15%). Additionally, it was observed the inclusion of Al_2O_3 or TiO2 NPs, along with edible oil in the formulations based on laccase or ascorbate, enhances oxygen removal from the film (Rüegg et al., 2022). Figure 8.2 represents the experimental setup of a palladium-based oxygen scavenger. Table 8.2 outlines packaging systems with nanoscale oxygen scavenging.

8.2.4 Nanoscaled Ethylene Scavengers

The primary metabolic process that takes place in numerous fruits and vegetables post-harvesting (Fernández et al., 2010a). In this process, ethylene, a natural plant hormone, plays a pivotal role in accelerating respiration, which in turn leads to the ripening and

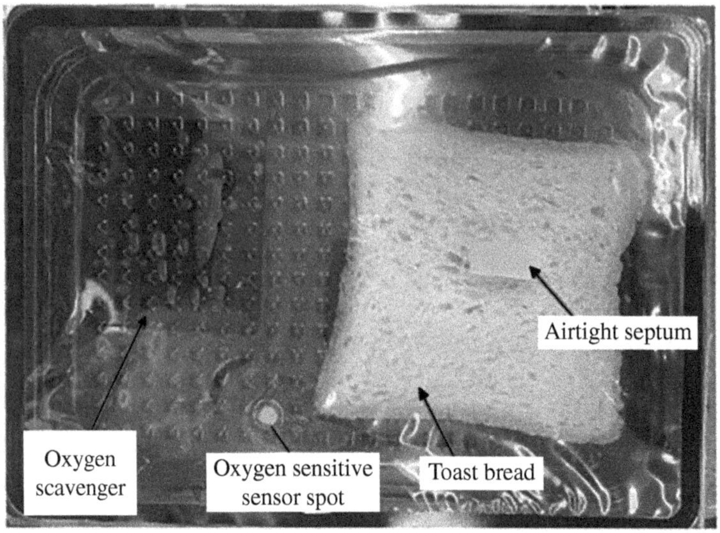

Figure 8.2 Experimental setup to assess the impact of a palladium-based oxygen scavenger on the shelf life of bakery products. The setup includes a tray containing an oxygen scavenger attached to the lidding film, a slice of toast bread, an airtight septum, and an oxygen-sensitive sensor spot. (*Source:* Rüegg et al., (2022) / ELSEVIER / CC BY 4.0)

Table 8.2 Packaging systems with nanoscale oxygen scavenging

S. No.	Active Substances	Packaging Materials	Potential Applications	Reference
1	Iron	Kaolinite + HDPE + LDPE	Substantial potential for improving food-packaging applications	Busolo and Lagaron (2012)
2	Palladium	PET	Bakery products	
3	TiO_2 or Al_2O_3	TiO_2 or Al_2O_3 + edible oil to the CaAsc/enzyme-coated PET film	Enhancing the rate of oxygen removal from the packages	Rüegg et al. (2022)
4	Pd-platinum	LLDPE, PP, PET, nylon	Regulate the transport of oxygen using active membrane	Yu et al. (2004)

softening of the product tissues and an acceleration of senescence. Conversely, excessive ethylene accumulation can result in the yellowing of green vegetables and trigger undesirable reactions. Therefore, the exclusion of ethylene gas from the packaging becomes essential to preserve acceptable visual and sensory quality, ultimately extending the shelf life (López-Rubio et al., 2004; Ozdemir and Floros, 2004). The most important agent for eliminating ethylene is potassium permanganate ($KMnO_4$) (Smith et al., 2009), which oxidizes ethylene into ethanol and acetate (Vermeiren et al., 1999). Alternatively, ethylene can be eliminated by physical absorbance on carbon or zeolite (Smith et al., 2009), which can be incorporated into packaging films.

Metals like silver (Ag) or palladium (Pd) and metal oxides such as titanium oxide (TiO_2) or zinc oxide (ZnO) have garnered attention because of their appealing physicochemical characteristics. Johnson Matthey scientists made a significant discovery in the form of a palladium (Pd)—enhanced material (Smith et al., 2009). Its ethylene adsorption capacity, measured by 4,162 μL/g under conditions of 20 °C and 100% RH, was found to be considerably larger than potassium permanganate scavengers in small quantities and high RH (Terry et al., 2007).

The researchers showcased the efficacy of the Pd-promoted material in scavenging ethylene, whether externally applied or internally generated by bananas or avocados, resulting in the suppression of ethylene-induced ripening in climacteric fruits. In another study, different authors coated PVC packaging films (control) with nano-ZnO NPs and assessed their impact on preserving the quality of freshly cut "Fuji" apples (Li et al., 2011). Following the cutting of apples, ethylene levels increased due to wound-induced respiration. However, associated to the control film, the nano-ZnO film effectively slowed down ethylene production. Additionally, packaging the apples in the NPs -coated film significantly minimizes fruit deterioration, attributed to decreased accretion of malondialdehyde (MDA), and inhibition of enzyme activities.

Other NPs known to impact ethylene production include TiO_2 NPs and Ag-NPs. TiO_2 NPs can react with ethylene in water (H_2O) and carbon dioxide (CO_2) (Costa et al., 2016), while Ag-NPs have remained as ethylene inhibitors in various studies (Beyer, 1976; Fernández et al., 2010a; Veen, 1979). For instance, Zi-sheng et al. (2013) examined the influence of LDPE-TiO_2 on strawberry quality. They were able to illustrate a significant inhibition of ethylene production in strawberries packaged with LDPE films containing nano-TiO_2. Additionally, this packaging approach resulted in reduced fruit deterioration, weight loss, softening, and titratable acidity. Fernández et al. (2010a) engineered cellulose-based Ag-NP to manage spoilage microflora in absorption pads. Similar outcomes were achieved when employing the nano-packaging material with strawberry (Yang et al. 2010) and bayberry (Wang et al. 2010). Packaging systems with nanoscale ethylene scavenging are presented in Table 8.3.

Table 8.3 Packaging systems with nanoscale ethylene scavenging

S. No.	Active Substances	Packaging Materials	Food Applications	Reference
1	ZnO	PVC	Fuji apples	Li et al. (2011)
2	TiO_2	LDPE	Strawberries	Zi-sheng et al. (2013)
3	TiO_2 + Ag + Montmorillonite	LDPE	Kiwi	Hu et al. (2011)
4	Palladium	Zeolite	Bananas and avocados	Smith et al. (2009)

8.3 Nanomaterial-Based Active Packaging Enhancements in MAP

Nanocomposite films made from LDPE and containing varying concentrations of silver (Ag) NPs (0.5% and 1% of concentration) were produced through extrusion. The study evaluated the microbial effect of chicken breast fillets enveloped in Ag/LDPE nanocomposite films, and MAP (gas mixture of 60% N_2, and 40% CO_2). It was observed that the tensile strength of the Ag/LDPE nanocomposite films was notably lower in comparison to control (uncoated Ag-NPs). Regardless, the concentration of Ag-NPs effectively extended the shelf life of the chicken breast fillets and significantly improved oxidative stability compared to the control films ($p < 0.05$). These findings suggest that LDPE nanocomposite films containing Ag NPs have potential applications as antimicrobial packaging materials for food (Azlin-Hasim et al., 2015). Various concentrations of Ag-NPs (0.25, 0.50, and 1.00 mg/mL) were dispersed within an 8% sodium alginate solution before applying the coating to the cheese. The MAP consisted of a gas mixture (65% N_2, 30% CO_2, and 5% O_2). The Ag-nanocomposite coating and MAP effectively extended the shelf life of Fior di latte cheese. The effects between antimicrobial NPs and the initial headspace within packaging could facilitate the distribution of dairy products (Gammariello et al., 2011). Silver-based antibacterial hybrid materials were created by reducing silver nitrate (1%) adsorbed onto cellulose fibers through thermal and UV treatments. Microscopic analysis confirmed uniform dispersion of Ag-NPs. These materials released silver ions (60 ppm) into beef meat exudates. When used as absorbent pads in MAP, they consistently reduced microbial contamination, lowering major microbial groups by an average of 1 log CFU/g throughout storage. Silver significantly reduced total aerobic bacteria and Pseudomonas spp., while lactic acid bacteria were less affected. Enterobacteriaceae levels remained undetectable with silver. Importantly, these materials did not alter meat color or significantly impact microbial loads (Fernández et al., 2010b).

8.4 Emerging Concepts and Future Trends

8.4.1 Smart and Intelligent Packaging

In the realm of smart and intelligent technologies, nanomaterials are applied to sense alterations in the biochemical or microbial composition of food. This includes the detection of specific pathogens as they develop within the food or the identification of particular gases emanating from food spoilage (Kuswandi et al., 2011). Within the domain of intelligent packaging, NPs can be utilized as responsive particles within packaging materials to provide real-time information regarding the condition of the packaged product. These nanosensors, also known as nanoscale sensors, have the capability to react to external stimuli and alterations in order to relay, convey, and recognize product information, all with the objective of ensuring its quality and safety.

8.4.2 Nanosensors

Nanosensor-equipped packaging is engineered to monitor the environmental conditions both inside and outside of food products, pellets, and containers across the entire supply chain. For instance, this specialized packaging can continuously track parameters like temperature and humidity and communicate this data by means of, for instance, altering its color. In plastic packaging containing nanosensors, the detection of gases released by deteriorating food prompts a color change in the packaging, serving as a warning signal. These so-called nanosensors possess the ability to react to alterations in the surroundings, such as temperature or humidity within storage facilities, levels of oxygen exposure, as well as the presence of degradation products or microbial contaminants. At present, nanosensors are employed for identifying toxins, food-borne pathogens, chemicals, contaminants and pesticides within various food products (Shafiq et al., 2020).

8.4.2.1 Nanosensors for Detection of Toxins

In contemporary applications, advanced nanomaterials, like carbon nanotubes and NPs (including Ag, Au, Cu, and ZnO), are extensively utilized in electrochemical sensors and biosensors for the detection of various toxins in foods (Ra et al., 2012; Xiang et al., 2011). A specific group of hazardous substances known as aflatoxins can contaminate food, originating from *Aspergillus flavus* and *Aspergillus parasiticus*. Au NPs, modified with antibodies tailored for aflatoxin B1, have demonstrated efficacy in identifying this toxin. Using superparamagnetic beads incorporating antibodies against aflatoxin M1 in milk, along with gold nanoprobes (Radoi et al., 2008).

8.4.2.2 Nanosensors for Food Pathogen Detection

Detection of food-borne pathogens in food substances mainly involves identifying the bacterial genetic material or complete bacterial cells. While traditional microbiological techniques for pathogen control are highly dependable, they are simultaneously intricate (Kumar et al., 2018). Nanotechnology enables the integration of nanosensors, facilitating the detection of miscellaneous pathogenic microorganisms found in different food items (Ranjan et al., 2014). The nano bioluminescent spray, known for its high efficiency as a nano biosensor, emits a visible glow to facilitate the straightforward identification of pathogen strains in various food products. Comprising diverse magnetic NPs, the spray engages with pathogens within food substances, generating a distinct visual color that is easily detectable for prompt analysis (Arshak et al., 2007). Magnetic iron oxide NPs were employed for the extraction of DNA from the pathogenic bacterium *L. monocytogenes* in milk (Yang et al., 2010).

8.4.2.3 Nanosensors for Chemical and Pesticide Residue Detection

Ag-NP-conjugated nanosensors employing colorimetric and fluorometric techniques have been utilized to detect organophosphorus and carbamate pesticides (Liu et al., 2012). Similarly, nanosensors utilizing voltammetry, including nanocomposite films (gold/zirconium dioxide) and Au NPs-based fluorescent nanosensors have been established for the identification of pesticides (Varshney et al., 2007; Wang and Irudayaraj, 2008). In the context of food safety, when food dyes and preservatives exceed acceptable limits, they can pose toxicity risks.

8.5 Regulatory Considerations and Safety

The utilization of various nanomaterials has numerous advantages; nevertheless, concurrently, they present significant threats to human health and cytotoxic effects (Higashisaka et al., 2015; Saleem and Zaidi, 2020). Toxic elements are present in nanomaterials composition. For instance, certain nanomaterials have the capacity to enter the skin, causing health issues in both humans and animals. NPs possess the capability to trigger changes in the genetic and protein makeup, even within plants, thus influencing their rate of growth (Rizwan et al., 2017).

Zebrafish have been utilized as a model organism to assess the toxicity of nanomaterials, overcoming obstacles associated with establishing and sustaining cell cultures for both animal and human cells. This vertebrate fish species exhibits fast reproductive abilities. Authorization to evaluate nanomaterial toxicity using zebrafish has been granted by both the US Food and Drug Administration (FDA) and the European Union (EU) (Jeevanandam et al., 2019). A comprehensive examination of various endpoints of toxicity explicitly states that NPs did not induce mortality or defects in the zebrafish behavior. The recent assessment of cytotoxicity for zinc oxide NPs, synthesized through green hydrothermal methods, has been conducted both *in vitro* and *in vivo*. The findings distinctly reveal that these green-synthesized NPs exhibit higher biocompatibility and lower toxicity compared to their chemically synthesized counterparts (Shubha et al., 2019).

The accumulation of NPs originating from nanopackaging or nanoprocessing has been established in both food and humans (Abo-Elseoud et al., 2018). Consequently, it is imperative to strictly adhere to risk assessment procedures during the processing of food items (Fortunati et al., 2019; Ghosh et al., 2019). While nanotechnology has made significant strides, the obstacles to creating sustainable solutions. With the increasing prevalence of nanotechnology in the food industry, there is a pressing need to educate the public about potential threats linked with nanomaterials to health and the environment. Many countries have established regulations to address the NPs, ensuring the safety of nano products in the food and agriculture sector (Amenta et al., 2015). Moreover, regulatory authorities like the FDA and the Environmental Protection Agency (EPA) have implemented standards that oversee commercializing food products, placing emphasis on health, quality, and safety (Kookana et al., 2014).

8.6 Conclusion

Various NPs have demonstrated their ability to impart multiple beneficial properties to food-packaging materials. These include scavenging oxygen and ethylene, along with providing antimicrobial activity. This presents a promising opportunity to uphold or enhance the quality of food products. While numerous nanomaterials, especially those integrated into antimicrobial packaging systems, have been examined against microorganisms *in vitro*, it is crucial to conduct food-specific tests. This is significant because the effectiveness of active compounds may be influenced by the components present in the food. For the successful integration of active nanomaterials into food packaging, several factors must be considered. Further, the cost should align with the benefits conferred to the particular food product. Additionally, addressing legislative and regulatory issues is essential, and achieving widespread consumer acceptance is a key requirement.

References

Abo-Elscoud, W. S., Hassan, M. L., Sabaa, M. W., Basha, M., Hassan, E. A., & Fadel, S. M. (2018). Chitosan nanoparticles/cellulose nanocrystals nanocomposites as a carrier system for the controlled release of repaglinide. *International Journal of Biological Macromolecules*, *111*(2017), 604–613. https://doi.org/10.1016/j.ijbiomac.2018.01.044

Abolghasemi, M. M., Sobhi, M., & Piryaei, M. (2016). Preparation of a novel green optical pH sensor based on immobilization of red grape extract on bioorganic agarose membrane. *Sensors and Actuators, B: Chemical.* https://doi.org/10.1016/j.snb.2015.10.038

Akbar, A., & Anal, A. K. (2014). Zinc oxide nanoparticles loaded active packaging, a challenge study against *Salmonella typhimurium* and *Staphylococcus aureus* in ready-to-eat poultry meat. *Food Control*, *38*(1), 88–95. https://doi.org/10.1016/j.foodcont.2013.09.065

Amenta, V., Aschberger, K., Arena, M., Bouwmeester, H., Botelho Moniz, F., Brandhoff, P., Gottardo, S., Marvin, H. J. P., Mech, A., Quiros Pesudo, L., Rauscher, H., Schoonjans, R., Vettori, M. V., Weigel, S., & Peters, R. J. (2015). Regulatory aspects of nanotechnology in the agri/feed/food sector in EU and non-EU countries. *Regulatory Toxicology and Pharmacology*, *73*(1), 463–476. https://doi.org/10.1016/j.yrtph.2015.06.016

Arfat, Y. A., Benjakul, S., Vongkamjan, K., Sumpavapol, P., & Yarnpakdee, S. (2015). Shelf-life extension of refrigerated sea bass slices wrapped with fish protein isolate/fish skin gelatin-ZnO nanocomposite film incorporated with basil leaf essential oil. *Journal of Food Science and Technology*, *52*(10), 6182–6193. https://doi.org/10.1007/s13197-014-1706-y

Arshak, K., Adley, C., Moore, E., Cunniffe, C., Campion, M., & Harris, J. (2007). Characterisation of polymer nanocomposite sensors for quantification of bacterial cultures. *Sensors and Actuators, B: Chemical*, *126*(1), 226–231. https://doi.org/10.1016/j.snb.2006.12.006

Azlin-Hasim, S., Cruz-Romero, M. C., Morris, M. A., Cummins, E., & Kerry, J. P. (2015). Effects of a combination of antimicrobial silver low density polyethylene nanocomposite films and modified atmosphere packaging on the shelf life of chicken breast fillets. *Food Packaging and Shelf Life*, *4*, 26–35. https://doi.org/10.1016/j.fpsl.2015.03.003

Azlin-Hasim, S., Cruz-Romero, M. C., Morris, M. A., Padmanabhan, S. C., Cummins, E., & Kerry, J. P. (2016). The potential application of antimicrobial silver polyvinyl chloride nanocomposite films to extend the shelf-life of chicken breast fillets. *Food and Bioprocess Technology*, *9*(10), 1661–1673. https://doi.org/10.1007/s11947-016-1745-7

Beyer, E. M. (1976). A potent inhibitor of ethylene action in plants. *Plant Physiology*, *58*(3), 268–271. https://doi.org/10.1104/pp.58.3.268

Blanco-Padilla, A., Soto, K. M., Hernández Iturriaga, M., & Mendoza, S. (2014). Food antimicrobials nanocarriers. *Scientific World Journal*, *2014*(1), 837215. https://doi.org/10.1155/2014/837215

Bodaghi, H., Mostofi, Y., Oromiehie, A., Zamani, Z., Ghanbarzadeh, B., Costa, C., Conte, A., & Del Nobile, M. A. (2013). Evaluation of the photocatalytic antimicrobial effects of a TiO_2 nanocomposite food packaging film by in vitro and in vivo tests. *Lwt*, *50*(2), 702–706. https://doi.org/10.1016/j.lwt.2012.07.027

Busolo, M. A., & Lagaron, J. M. (2012). Oxygen scavenging polyolefin nanocomposite films containing an iron modified kaolinite of interest in active food packaging applications. *Innovative Food Science and Emerging Technologies*, *16*, 211–217. https://doi.org/10.1016/j.ifset.2012.06.008

Campos-Requena, V. H., Rivas, B. L., Pérez, M. A., Figueroa, C. R., & Sanfuentes, E. A. (2015). The synergistic antimicrobial effect of carvacrol and thymol in clay/polymer nanocomposite films over strawberry gray mold. *Lwt*, *64*(1), 390–396. https://doi.org/10.1016/j.lwt.2015.06.006

Cárdenas, G., Díaz, V. J., Meléndrez, M. F., Cruzat, C. C., & García Cançino, A. (2009). Colloidal cu nanoparticles/chitosan composite film obtained by microwave heating for food package applications. *Polymer Bulletin*, *62*(4), 511–524. https://doi.org/10.1007/s00289-008-0031-x

Chawengkijwanich, C., & Hayata, Y. (2008). Development of TiO_2 powder-coated food packaging film and its ability to inactivate *Escherichia coli* in vitro and in actual tests. *International Journal of Food Microbiology*, *123*(3), 288–292. https://doi.org/10.1016/j.ijfoodmicro.2007.12.017

Choe, E., & Min, D. B. (2006). Chemistry and reactions of reactive oxygen species in foods. *Critical Reviews in Food Science and Nutrition*, *46*(1), 1–22. https://doi.org/10.1080/10408390500455474

Conte, A., Longano, D., Costa, C., Ditaranto, N., Ancona, A., Cioffi, N., Scrocco, C., Sabbatini, L., Contò, F., & Del Nobile, M. A. (2013). A novel preservation technique applied to fiordilatte cheese. *Innovative Food Science and Emerging Technologies*, *19*, 158–165. https://doi.org/10.1016/j.ifset.2013.04.010

Costa, C., Conte, A., Alessandro, M., & Nobile, D. (2016). Use of Metal Nanoparticles for Active Packaging Applications. In J. Barros-Velazquez (Ed.), *Antimicrobial Food Packaging*. Elsevier Inc. https://doi.org/10.1016/B978-0-12-800723-5.00031-0

de Abreu, D. A. P., Cruz, J. M., & Losada, P. P. (2012). Active and intelligent packaging for the food industry. *Food Reviews International*, *28*(2), 146–187. https://doi.org/10.1080/87559129.2011.595022

Dias, M. V., De Fátima, F., Soares, N., Borges, S. V., De Sousa, M. M., Nunes, C. A., De Oliveira, I. R. N., & Medeiros, E. A. A. (2013). Use of allyl isothiocyanate and carbon nanotubes in an antimicrobial film to package shredded, cooked chicken meat. *Food Chemistry*, *141*(3), 3160–3166. https://doi.org/10.1016/j.foodchem.2013.05.148

Duncan, T. V. (2011). Applications of nanotechnology in food packaging and food safety: barrier materials, antimicrobials and sensors. *Journal of Colloid and Interface Science*, *363*(1), 1–24. https://doi.org/10.1016/j.jcis.2011.07.017

Eide, D. J. (2000). Metal ion transport in eukaryotic microorganisms: insights from *Saccharomyces cerevisiae*. *Advances in Microbial Physiology*, *43*, 1–38. https://doi.org/10.1016/s0065-2911(00)43001-8

Emamifar, A., Kadivar, M., Shahedi, M., & Soleimanian-Zad, S. (2010). Evaluation of nanocomposite packaging containing Ag and ZnO on shelf life of fresh orange juice. *Innovative Food Science and Emerging Technologies*, *11*(4), 742–748. https://doi.org/10.1016/j.ifset.2010.06.003

Fernández, A., Picouet, P., & Lloret, E. (2010a). Cellulose-silver nanoparticle hybrid materials to control spoilage-related microflora in absorbent pads located in trays of fresh-cut melon. *International Journal of Food Microbiology*, *142*(1–2), 222–228. https://doi.org/10.1016/j.ijfoodmicro.2010.07.001

Fernández, A., Picouet, P., & Lloret, E. (2010b). Reduction of the spoilage-related microflora in absorbent pads by silver nanotechnology during modified atmosphere packaging of beef meat. *Journal of Food Protection*, *73*(12), 2263–2269. https://doi.org/10.4315/0362-028X-73.12.2263

Fernández, A., Soriano, E., López-Carballo, G., Picouet, P., Lloret, E., Gavara, R., & Hernández-Muñoz, P. (2009). Preservation of aseptic conditions in absorbent pads by using silver nanotechnology. *Food Research International*, *42*(8), 1105–1112. https://doi.org/10.1016/j.foodres.2009.05.009

Fortunati, E., Mazzaglia, A., & Balestra, G. M. (2019). Sustainable control strategies for plant protection and food packaging sectors by natural substances and novel nanotechnological approaches. *Journal of the Science of Food and Agriculture*, *99*(3), 986–1000. https://doi.org/10.1002/jsfa.9341

Gammariello, D., Conte, A., Buonocore, G. G., & Del Nobile, M. A. (2011). Bio-based nanocomposite coating to preserve quality of Fior di latte cheese. *Journal of Dairy Science*, *94*(11), 5298–5304. https://doi.org/10.3168/jds.2011-4161

Ghosh, C., Bera, D., & Roy, L. (2019). Role of Nanomaterials in Food Preservation. In R. Prasad (Ed.),. https://doi.org/10.1007/978-3-030-16534-5_10

Gibis, D., & Rieblinger, K. (2011). Oxygen scavenging films for food application. *Procedia Food Science*, *1*(Icef 11), 229–234. https://doi.org/10.1016/j.profoo.2011.09.036

Gomashe, A. V., & Dharmik, P. G. (2012). Synergistic effect of gold nanoparticles and Bacteriocin against food blemishing microbes: a novel approach for food packaging material preparation. *Global Journal For Research Analysis*, *3*(5), 1–3. https://doi.org/10.15373/22778160/may2014/55

Han, Y. S., Lee, S. H., Choi, K. H., & Park, I. (2010). Preparation and characterization of chitosan-clay nanocomposites with antimicrobial activity. *Journal of Physics and Chemistry of Solids*, *71*(4), 464–467. https://doi.org/10.1016/j.jpcs.2009.12.012

Higashisaka, K., Yoshioka, Y., & Tsutsumi, Y. (2015). Applications and safety of nanomaterials used in the food industry. *Food Safety*, *3*(2), 39–47. https://doi.org/10.14252/foodsafetyfscj.2015005

Hu, Q., Fang, Y., Yang, Y., Ma, N., & Zhao, L. (2011). Effect of nanocomposite-based packaging on postharvest quality of ethylene-treated kiwifruit (*Actinidia deliciosa*) during cold storage. *Food Research International*, *44*(6), 1589–1596. https://doi.org/10.1016/j.foodres.2011.04.018

Hutter, S., Rüegg, N., & Yildirim, S. (2016). Use of palladium based oxygen scavenger to prevent discoloration of ham. *Food Packaging and Shelf Life*, *8*, 56–62. https://doi.org/10.1016/j.fpsl.2016.02.004

Imran, M., Revol-Junelles, A. M., René, N., Jamshidian, M., Akhtar, M. J., Arab-Tehrany, E., Jacquot, M., & Desobry, S. (2012). Microstructure and physico-chemical evaluation of nano-emulsion-based antimicrobial peptides embedded in bioactive packaging films. *Food Hydrocolloids*, *29*(2), 407–419. https://doi.org/10.1016/j.foodhyd.2012.04.010

Jacobsen, C. (1999). Sensory impact of lipid oxidation in complex food systems. *Lipid—Fett*, *101*(12), 484–492. https://doi.org/10.1002/(sici)1521-4133(199912)101:12<484::aid-lipi484>3.0.co;2-h

Jeevanandam, J., Chan, Y. S., & Danquah, M. K. (2019). Zebrafish as a model organism to study nanomaterial toxicity. *Emerging Science Journal*, *3*(3), 195–208. https://doi.org/10.28991/esj-2019-01182

Jin, T., & Gurtler, J. B. (2011). Inactivation of Salmonella in liquid egg albumen by antimicrobial bottle coatings infused with allyl isothiocyanate, nisin and zinc oxide nanoparticles. *Journal of Applied Microbiology*, *110*(3), 704–712. https://doi.org/10.1111/j.1365-2672.2011.04938.x

Jin, T., & He, Y. (2011). Antibacterial activities of magnesium oxide (MgO) nanoparticles against foodborne pathogens. *Journal of Nanoparticle Research*, *13*(12), 6877–6885. https://doi.org/10.1007/s11051-011-0595-5

Kookana, R. S., Boxall, A. B. A., Reeves, P. T., Ashauer, R., Beulke, S., Chaudhry, Q., Cornelis, G., Fernandes, T. F., Gan, J., Kah, M., Lynch, I., Ranville, J., Sinclair, C., Spurgeon, D., Tiede, K., & Van Den Brink, P. J. (2014). Nanopesticides: guiding principles for regulatory evaluation of environmental risks. *Journal of Agricultural and Food Chemistry*, *62*(19), 4227–4240. https://doi.org/10.1021/jf500232f

Koper, O. B., Klabunde, J. S., Marchin, G. L., Klabunde, K. J., Stoimenov, P., & Bohra, L. (2002). Nanoscale powders and formulations with biocidal activity toward spores and vegetative cells of Bacillus species, viruses, and toxins. *Current Microbiology*, *44*(1), 49–55. https://doi.org/10.1007/s00284-001-0073-x

Kumar, S., Shukla, A., Baul, P. P., Mitra, A., & Halder, D. (2018). Biodegradable hybrid nanocomposites of chitosan/gelatin and silver nanoparticles for active food packaging applications. *Food Packaging and Shelf Life*, *16*(November 2017), 178–184. https://doi.org/10.1016/j.fpsl.2018.03.008

Kuswandi, B., Wicaksono, Y., Jayus, Abdullah, A., Heng, L. Y., & Ahmad, M. (2011). Smart packaging: sensors for monitoring of food quality and safety. *Sensing and Instrumentation for Food Quality and Safety*, *5*(3–4), 137–146. https://doi.org/10.1007/s11694-011-9120-x

Larsen, H., Westad, F., Sørheim, O., & Nilsen, L. H. (2006). Determination of critical oxygen level in packages for cooked sliced ham to prevent color fading during illuminated retail display. *Journal of Food Science*, *71*(5). https://doi.org/10.1111/j.1750-3841.2006.00048.x

Li, X., Li, W., Jiang, Y., Ding, Y., Yun, J., Tang, Y., & Zhang, P. (2011). Effect of nano-ZnO-coated active packaging on quality of fresh-cut "Fuji" apple. *International Journal of Food Science and Technology*, *46*(9), 1947–1955. https://doi.org/10.1111/j.1365-2621.2011.02706.x

Lima, E., Guerra, R., Lara, V., & Guzmán, A. (2013). Gold nanoparticles as efficient antimicrobial agents for *Escherichia coli* and *Salmonella typhi*. *Chemistry Central Journal*, *7*(1), 1–7. https://doi.org/10.1186/1752-153X-7-11

Liu, D., Chen, W., Wei, J., Li, X., Wang, Z., & Jiang, X. (2012). A highly sensitive, dual-readout assay based on gold nanoparticles for organophosphorus and carbamate pesticides. *Analytical Chemistry*, *84*(9), 4185–4191. https://doi.org/10.1021/ac300545p

Llorens, A., Lloret, E., Picouet, P. A., Trbojevich, R., & Fernandez, A. (2012). Metallic-based micro and nanocomposites in food contact materials and active food packaging. *Trends in Food Science and Technology*, *24*(1), 19–29. https://doi.org/10.1016/j.tifs.2011.10.001

López-Rubio, A., Almenar, E., Hernandez-Muñoz, P., Lagarón, J. M., Catalá, R., & Gavara, R. (2004). Overview of active polymer-based packaging technologies for food applications. *Food Reviews International*, *20*(4), 357–387. https://doi.org/10.1081/FRI-200033462

Zi-sheng, L., Qing-yang, Y., & Dongdong, L. (2013). Influence of nano-TiO2 modified LDPE film packaging on quality of strawberry. *Modern Food Science And Technology*, *29*(10), 2340–2345.

Del Nobile, M. A., Cannarsi, M., Altieri, C., Sinigaglia, M., Favia, P., Iacoviello, G., & D'Agostino, R. (2001). Effect of Ag-containing nano-composite active packaging system on survival of *Alicyclobacillus acidoterrestris*. *Journal of Food Science*, *69*(8), E379–E383.

Marsh, K., & Bugusu, B. (2007). Food packaging—roles, materials, and environmental issues: scientific status summary. *Journal of Food Science*. https://doi.org/10.1111/j.1750-3841.2007.00301.x

Mills, A., Doyle, G., Peiro, A. M., & Durrant, J. (2006). Demonstration of a novel, flexible, photocatalytic oxygen-scavenging polymer film. *Journal of Photochemistry and Photobiology A: Chemistry*, *177*(2–3), 328–331. https://doi.org/10.1016/j.jphotochem.2005.06.001

Morones, J. R., Elechiguerra, J. L., Camacho, A., Holt, K., Kouri, J. B., Ramírez, J. T., & Yacaman, M. J. (2005). The bactericidal effect of silver nanoparticles. *Nanotechnology*, *16*(10), 2346–2353. https://doi.org/10.1088/0957-4484/16/10/059

Mu, H., Gao, H., Chen, H., Tao, F., Fang, X., & Ge, L. (2013). A nanosised oxygen scavenger: preparation and antioxidant application to roasted sunflower seeds and walnuts. *Food Chemistry*, *136*(1), 245–250. https://doi.org/10.1016/j.foodchem.2012.07.121

Mugasundari, A. V., & Anandakumar, S. (2022). Shelf life extension of bread using ethanol emitters with different packaging materials. *Journal of Food Processing and Preservation*, *46*(12), 1–13. https://doi.org/10.1111/jfpp.17143

Nannerup, L. D., Jakobsen, M., Van Den Berg, F., Jensen, J. S., Møller, J. K. S., & Bertelsen, G. (2004). Optimizing colour quality of modified atmosphere packed sliced meat products by control of critical packaging parameters. *Meat Science*, *68*(4), 577–585. https://doi.org/10.1016/j.meatsci.2004.05.009

Ozdemir, M., & Floros, J. D. (2004). Active food packaging technologies. *Critical Reviews in Food Science and Nutrition*, *44*(3), 185–193. https://doi.org/10.1080/10408690490441578

Padmavathy, N., & Vijayaraghavan, R. (2008). Enhanced bioactivity of ZnO nanoparticles—an antimicrobial study. *Science and Technology of Advanced Materials*, *9*(3). https://doi.org/10.1088/1468-6996/9/3/035004

Pagno, C. H., Costa, T. M. H., De Menezes, E. W., Benvenutti, E. V., Hertz, P. F., Matte, C. R., Tosati, J. V., Monteiro, A. R., Rios, A. O., & Flôres, S. H. (2015). Development of active biofilms of quinoa (*Chenopodium quinoa* W.) starch containing gold nanoparticles and evaluation of antimicrobial activity. *Food Chemistry*, *173*, 755–762. https://doi.org/10.1016/j.foodchem.2014.10.068

Ra, M., Gade, A., Gaikwad, S., Marcato, P. D., & Durán, N. (2012). Biomedical applications of nanobiosensors: the state-of-the-art. *Journal of the Brazilian Chemical Society*, *23*(1), 14–24. https://doi.org/10.1590/s0103-50532012000100004

Radoi, A., Targa, M., Prieto-Simon, B., & Marty, J. L. (2008). Enzyme-linked immunosorbent assay (ELISA) based on superparamagnetic nanoparticles for aflatoxin M1 detection. *Talanta*, *77*(1), 138–143. https://doi.org/10.1016/j.talanta.2008.05.048

Radusin, T., Ristic, I., Pilic, B., & Novakovic, A. (2016). Antimicrobial nanomaterials for food packaging applications. *Food and Feed Research*, *43*(2), 119–126. https://doi.org/10.5937/ffr1602119r

Ranjan, S., Dasgupta, N., Chakraborty, A. R., Melvin Samuel, S., Ramalingam, C., Shanker, R., & Kumar, A. (2014). Nanoscience and nanotechnologies in food industries: opportunities and research trends. *Journal of Nanoparticle Research*, *16*(6), 2464. https://doi.org/10.1007/s11051-014-2464-5

Rizwan, M., Ali, S., Qayyum, M. F., Ok, Y. S., Adrees, M., Ibrahim, M., Zia-ur-Rehman, M., Farid, M., & Abbas, F. (2017). Effect of metal and metal oxide nanoparticles on growth and physiology of globally important food crops: a critical review. *Journal of Hazardous Materials*, *322*, 2–16. https://doi.org/10.1016/j.jhazmat.2016.05.061

Robertson, G. L. (2014). Food Packaging. In N. K. Van Alfen (Ed.), *Encyclopedia of Agriculture and Food Systems* (pp. 232–249). Elsevier. https://doi.org/10.1016/B978-0-444-52512-3.00063-2

Rüegg, N., Röcker, B., & Yildirim, S. (2022). Application of palladium-based oxygen scavenger to extend the mould free shelf life of bakery products. *Food Packaging and Shelf Life*, *31*(November 2021), 100771. https://doi.org/10.1016/j.fpsl.2021.100771

Saleem, H., & Zaidi, S. J. (2020). Developments in the application of nanomaterials for water treatment and their impact on the environment. *Nanomaterials*, *10*(9), 1764. https://doi.org/10.3390/nano10091764

Sängerlaub, S., Gibis, D., Kirchhoff, E., Tittjung, M., Schmid, M., & Müller, K. (2013). Compensation of pinhole defects in food packages by application of iron-based oxygen scavenging multilayer films. *Packaging Technology and Science*, *26*(1), 17–30. https://doi.org/10.1002/pts.1962

Shafiq, M., Anjum, S., Hano, C., Anjum, I., & Abbasi, B. H. (2020). An overview of the applications of nanomaterials and nanodevices in the food industry. *Foods*, *9*(2), 1–27. https://doi.org/10.3390/foods9020148

Shankar, S., & Rhim, J. W. (2014). Effect of copper salts and reducing agents on characteristics and antimicrobial activity of copper nanoparticles. *Materials Letters*, *132*, 307–311. https://doi.org/10.1016/j.matlet.2014.06.014

Shankar, S., Teng, X., & Rhim, J. W. (2014). Properties and characterization of agar/CuNP bionanocomposite films prepared with different copper salts and reducing agents. *Carbohydrate Polymers*, *114*, 484–492. https://doi.org/10.1016/j.carbpol.2014.08.036

Shubha, P., Gowda, M. L., Namratha, K., Manjunatha, H. B., & Byrappa, K. (2019). In vitro and in vivo evaluation of green-hydrothermal synthesized ZnO nanoparticles. *Journal of Drug Delivery Science and Technology*, *49*, 692–699. https://doi.org/10.1016/j.jddst.2018.12.017

Smith, A. W. J., Poulston, S., Rowsell, L., Terry, L. A., & Anderson, J. A. (2009). A new palladium-based ethylene scavenger to control ethylene-induced ripening of climacteric fruit. *Platinum Metals Review*, *53*(3), 112–122. https://doi.org/10.1595/147106709X462742

Terry, L. A., Ilkenhans, T., Poulston, S., Rowsell, L., & Smith, A. W. J. (2007). Development of new palladium-promoted ethylene scavenger. *Postharvest Biology and Technology*, *45*(2), 214–220. https://doi.org/10.1016/j.postharvbio.2006.11.020

Thirumurugan, A., Ramachandran, S., & Shiamala Gowri, A. (2013). Combined effect of bacteriocin with gold nanoparticles against food spoiling bacteria—an approach for food packaging material preparation. *International Food Research Journal*, *20*(4), 1909–1912.

Van Bree, I., Baetens, J. M., Samapundo, S., Devlieghere, F., Laleman, R., Vandekinderen, I., Noseda, B., Xhaferi, R., De Baets, B., & De Meulenaer, B. (2012). Modelling the degradation kinetics of vitamin C in fruit juice in relation to the initial headspace oxygen concentration. *Food Chemistry*, *134*(1), 207–214. https://doi.org/10.1016/j.foodchem.2012.02.096

Varshney, M., Li, Y., Srinivasan, B., & Tung, S. (2007). A label-free, microfluidics and interdigitated array microelectrode-based impedance biosensor in combination with nanoparticles immunoseparation for detection of Escherichia coli O157 : H7 in food samples. *Sensors and Actuators, B: Chemical*, *128*(1), 99–107. https://doi.org/10.1016/j.snb.2007.03.045

Veen, H. (1979). Effects of silver on ethylene synthesis and action in cut carnations. *Planta*, *145*(5), 467–470. https://doi.org/10.1007/BF00380101

Vermeiren, L., Devlieghere, F., Van Beest, M., De Kruijf, N., & Debevere, J. (1999). Developments in the active packaging of foods. *Trends in Food Science and Technology*, *10*(3), 77–86. https://doi.org/10.1016/S0924-2244(99)00032-1

Wang, C., & Irudayaraj, J. (2008). Gold nanorod probes for the detection of multiple pathogens. *Small*, *4*(12), 2204–2208. https://doi.org/10.1002/smll.200800309

Wang, K., Jin, P., Shang, H., Li, H., Xu, F., Hu, Q., & Zheng, Y. (2010). A combination of hot air treatment and nano-packing reduces fruit decay and maintains quality in postharvest Chinese bayberries. *Journal of the Science of Food and Agriculture*, *90*(14), 2427–2432. https://doi.org/10.1002/jsfa.4102

Xiang, L., Zhao, C., & Wang, J. (2011). Nanomaterials-based electrochemical sensors and biosensors for pesticide detection. *Sensor Letters*, *9*(3), 1184–1189. https://doi.org/10.1166/sl.2011.1369

Xiao-e, L., Green, A. N. M., Haque, S. A., Mills, A., & Durrant, J. R. (2004). Light-driven oxygen scavenging by titania/polymer nanocomposite films. *Journal of Photochemistry and Photobiology A: Chemistry*, *162*(2–3), 253–259. https://doi.org/10.1016/j.nainr.2003.08.010

Yang, B. X., Shi, J. H., Pramoda, K. P., & Goh, S. H. (2008). Enhancement of the mechanical properties of polypropylene using polypropylene-grafted multiwalled carbon nanotubes. *Composites Science and Technology*. https://doi.org/10.1016/j.compscitech.2008.05.001

Yang, D., Zhu, L., & Jiang, X. (2010). Electrochemical reaction mechanism and determination of Sudan I at a multi wall carbon nanotubes modified glassy carbon electrode. *Journal of Electroanalytical Chemistry*, *640*(1–2), 17–22. https://doi.org/10.1016/j.jelechem.2009.12.022

Yang, F. M., Li, H. M., Li, F., Xin, Z. H., Zhao, L. Y., Zheng, Y. H., & Hu, Q. H. (2010). Effect of nano-packing on preservation quality of fresh strawberry (*Fragaria ananassa* duch. cv fengxiang) during storage at 4 °c. *Journal of Food Science*, *75*(3), C236–C240. https://doi.org/10.1111/j.1750-3841.2010.01520.x

Yemmireddy, V. K., Farrell, G. D., & Hung, Y. C. (2015). Development of titanium dioxide (TiO_2) Nanocoatings on food contact surfaces and method to evaluate their durability and photocatalytic bactericidal property. *Journal of Food Science*, *80*(8), N1903–N1911. https://doi.org/10.1111/1750-3841.12962

Yildirim, S. (2011). Active Packaging for Food Biopreservation. In C. Lacroix (Ed.), *In Protective Cultures, Antimicrobial Metabolites and Bacteriophages for Food and Beverage Biopreservation*. Woodhead Publishing Limited. https://doi.org/10.1533/9780857090522.3.460

Youssef, A. M., El-Sayed, S. M., El-Sayed, H. S., Salama, H. H., & Dufresne, A. (2016). Enhancement of Egyptian soft white cheese shelf life using a novel chitosan/carboxymethyl cellulose/zinc oxide bionanocomposite film. *Carbohydrate Polymers*, *151*, 9–19. https://doi.org/10.1016/j.carbpol.2016.05.023

Yu, J., Liu, R. Y. F., Poon, B., Nazarenko, S., Koloski, T., Vargo, T., Hiltner, A., & Baer, E. (2004). Polymers with palladium nanoparticles as active membrane materials. *Journal of Applied Polymer Science*, *92*(2), 749–756. https://doi.org/10.1002/app.20013

Zhang, L., Jiang, Y., Ding, Y., Daskalakis, N., Jeuken, L., Povey, M., O'Neill, A. J., & York, D. W. (2010). Mechanistic investigation into antibacterial behaviour of suspensions of ZnO nanoparticles against *E. Coli*. *Journal of Nanoparticle Research*, *12*(5), 1625–1636. https://doi.org/10.1007/s11051-009-9711-1

9

Intelligent Packaging: Concept, Applications, and Regulatory Aspects

Kalpani Y. Perera[1,2,*], *Luana S. C. Carnaval*[1,2,*], *Amit K. Jaiswal*[1,2], *and Swarna Jaiswal*[1,2]

[1] *Centre for Sustainable Packaging and Bioproducts (CSPB), School of Food Science and Environmental Health, Technological University Dublin, Dublin, Ireland*
[2] *Sustainability and Health Research Institute (SHRI), Technological University Dublin, Dublin, Ireland*

9.1 Introduction

According to the World Health Organization (WHO), worldwide, an estimated 600 million—almost 1 in 10 people—fall ill due to contaminated food each year, and 420,000 deaths occur every year by foodborne contamination. In addition, US$110 billion is lost each year in productivity and medical expenses because of unsafe food (World Health Organization, 2022). Thus, food quality is a crucial factor to control the effect of the food on consumers' health and so requires great attention from the food industry (Yousefi et al., 2019). Considering the scenario of the modern food industry, food packaging proved to be an essential tool to ensure the integrity, safety, and quality of the food product during storage and transport to avoid the action of unfavorable conditions such as moisture, light, physical impacts, and microorganisms. In addition to extending food shelf life, packaging has become one of the most relevant means of advertising (Caon et al., 2017; Hu and Miao, 2022).

Conventional food-packaging systems have certain limitations in terms of extending shelf-life and maintaining food safety. Thus, new advances are emerging, and with them, innovative technologies have been developed in the area of food packaging. Among them are intelligent packaging systems (Caon et al., 2017; Hu and Miao, 2022). This technology is capable of performing important functions such as tracking, detection, sensing, recording, and communication. This provides consumers with the basic tools to deal with food and alerts about possible problems related to them. Another point is that despite being two different terms, smart and active packaging are not mutually exclusive, as in some approaches, both are present, and thus the two packaging systems can work synergistically (Figure 9.1).

* Authors equally contributed.

Nanotechnology for Sustainable Food Packaging, First Edition. Edited by C. Anandharamakrishnan, Jeyan A. Moses, and M. Maria Leena.

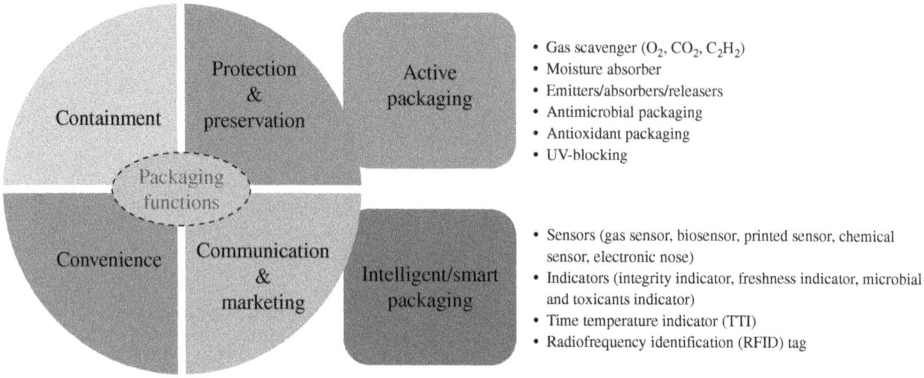

Figure 9.1 Types of active and smart packaging and key characteristics of packaging functions (*Source:* Alizadeh-Sani et al. (2020)/with permission of Elsevier)

An active system (e.g., containing antioxidant and antimicrobial agents) seeks to slow the decline in food quality and prolong its shelf life, while an intelligent system aims to monitor and report the quality status of a food product. In this way, efforts have been made to develop packages that contain both functions, thus meeting the demand of consumers and the industry for increasingly safer products (Hu and Miao, 2022).

As a response to the significant development of today's society and the growing concerns of consumers regarding the quality and safety of food products, new packaging trends have been developed over the last few years. A quick search on ScienceDirect (a website specialized in publishing articles), using the keywords intelligent/smart food packaging, showed more than 10,000 articles from 2019 to now. Intelligent technology provides consumers and manufacturers with product tracing through the critical points in a food supply chain, aiming not only to monitor food integrity but also to reduce waste and save resources/energy. Furthermore, smart packaging based on natural colorants and biodegradable materials (biopolymeric film matrices) has emerged as a promising alternative for use in food packaging due to their low or nontoxicity, eco-friendly, easy preparation, biodegradability, low cost, availability, renewable, and pollution-free features (Alizadeh-Sani et al., 2020). Considering all the new trends related to intelligent packaging systems, different types of indicators/sensors have been developed seeking, for instance, to monitor the product's freshness, microbial growth, and any chemical variations (pH changes, presence/concentration of gases, and volatile compounds), pathogens or their metabolites/toxins, time–temperature changes, and humidity (Alizadeh-Sani et al., 2020; Choi et al., 2017; Fang et al., 2017; Prietto et al., 2017; Xiao-wei et al., 2018).

The design, production, and application of structures and systems, through the control of their shape and size at the nanoscale, has emerged as a strong tool for smart and active food-packaging systems that interact with the food product. Nanotechnology has been applied in many fields, including the food industry. Due to the emerging interest in nanosystems applied to food packaging, about 400 ~ 500 nanotechnology-based food-packaging products are estimated, and about 20% of all food packaging will be used within the next decade. Approaches of nanoscale manufacturing cover, for example, nanomaterial-based (e.g., nanocomposites and nanometals) and nanofabrication-based platforms (e.g., nano electrospinning and nano coating). In this scenario, nanoscale manufacturing has shown great potential to overcome the limitations of current/traditional food-packaging systems and as an innovative tool for future smart food packaging (Park et al., 2020). This chapter will give an overview of the concepts and types of intelligent packaging, its applications, commercial point of view, regulatory aspects, and future perspectives, including a nanotechnological approach to these topics.

9.2 Concepts of Intelligent Packaging

The concept of smart packaging, capable of monitoring in real time the condition of food along the supply chain, has been a topic of growing interest in the food industry due to the increasing demand from consumers for safety, reliability, and convenience (Kalpana et al., 2019). The European Food Safety Authority (EFSA) defines intelligent packaging materials as "materials and articles which monitor the condition of packaged food or the environment surrounding the food" (European Food Safety Authority (EFSA), 2009).

Therefore, intelligent packaging can provide consumers with information about food deterioration caused by a variety of factors, such as temperature fluctuation, variation in gas concentration (e.g., fruit and vegetables), and growth of microorganisms in the packaging environment. These factors lead to modifications in the food quality status, which can be monitored by consumers through colorimetric, electrical, or other signals generated by these changes (Caon et al., 2017; Hu and Miao, 2022). Sani et al. (2021) developed carbohydrate-based films (chitin nanofiber and methylcellulose) containing pH-sensitive red barberry anthocyanins aiming for application as colorimetric indicators to detect food spoilage due to their sensitivity to pH changes and ammonia gas production. The authors identify that these biodegradable films showed both active (antioxidant and antimicrobial) and intelligent (color-changing tool) functions, being able to protect and monitor the freshness of products like meat and fish.

With the development of nanotechnology, nanomaterials have been increasingly applied in the food industry sector, including for intelligent packaging (Figure 9.2). In this context, nanotechnology-based indicators/sensors are developed and incorporated into smart food packaging (Hu and Miao, 2022). Due to the unique optical properties and high surface reactivity of nanomaterials, their performance has been shown to be superior to the traditional indicators/sensors (Primožič et al., 2021).

Lastly, we can summarize that smart food packaging provides the consumer with extra attributes that can help to ensure food safety and quality, monitoring and exchanging information. However, it can also track/trace the products through blockchain technology, ensure quality standards throughout the supply chain, improve recall procedures, and reduce waste (Sobhan et al., 2021). Therefore, it is not a surprise that the market for intelligent food packaging is growing rapidly and is expected to reach about $62.54 billion by 2023 (Sobhan et al., 2020).

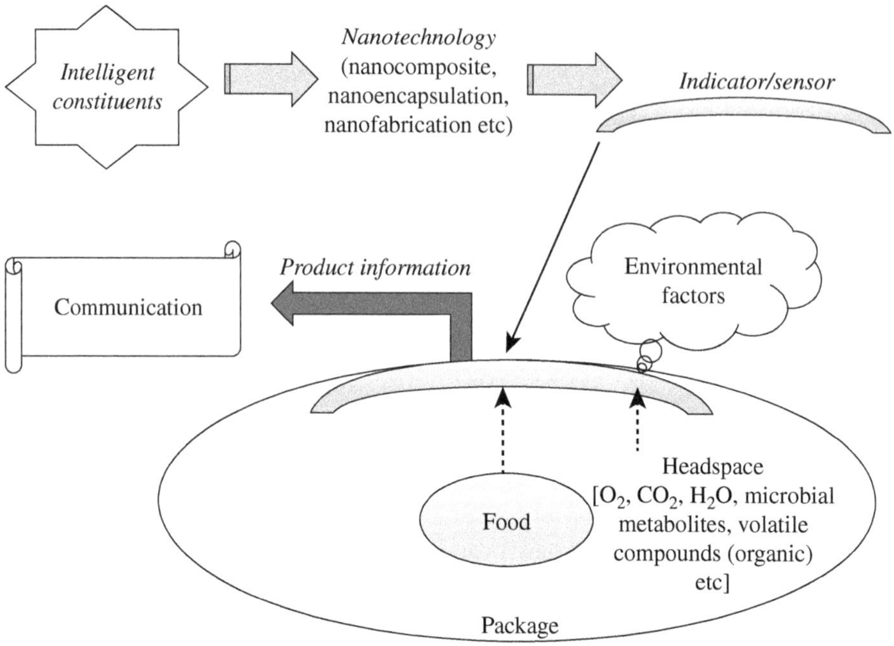

Figure 9.2 Demonstration of intelligent packaging concept and its association with nanotechnology (*Source:* Sharma et al. (2017)/Frontiers Media/CC BY 4.0)

9.3 Types of Intelligent Packaging

Intelligent packaging has the ability to communicate; in this way, it can monitor in real-time the conditions of the packaged product and transmit information to consumers. Three main technologies are used for smart packaging systems: (i) data carriers, which are specifically intended to help track the path taken by food products along the food supply chain; (ii) indicators, which inform the consumers about any property related to food quality status, and most of them are based on colorimetric dyes; and (iii) sensors, capable of rapid detection of specific analytes by transforming information into a quantifiable output signal proportional to that measurement, which is then processed by appropriate software. In addition, a subdivision based on another approach is possible:

1) Environmental conditions: monitoring of conditions that can lead to changes in food quality, e.g., time–temperature indicators, gas leak indicators, and humidity sensors;
2) Quality characteristics or quality indicator compounds: direct monitoring of the food quality attributes, for example, biosensors and freshness sensors/indicators;
3) Data carriers: systems used only to store and transfer data, for example, barcode labels and radio frequency identification (RFID) tags (Azeredo and Correa, 2021; Müller and Schmid, 2019). The different types of smart packaging will be discussed in detail below, and Table 9.1 shows some examples of intelligent systems.

Table 9.1 Types of intelligent packaging

Type of Intelligent Packaging	Packaging Material	Properties of Packaging Material	Reference
pH indicator	pH-sensitive films based on corn starch and anthocyanins extracted from black bean seed coat and red cabbage	Film color changes from pink to purple and blue, as a function of the pH.	Prietto et al. (2017)
Colorimetric indicator sensitive to changes in pH and ammonia gas production	Films based on immobilizing red barberry anthocyanins within composite chitin nanofiber and methylcellulose matrices	The indicator changes from reddish/crimson → pink → yellow with increasing pH, and from pink → yellow with increasing ammonia vapor concentration; Good antioxidant and antimicrobial activity.	Alizadeh-Sani et al. (2020)
Freshness indicator sensitive to hydrogen sulfide (H_2S, released during meat spoilage)	Material based on gellan gum-capped silver nanoparticles and colorimetric strategy related to the ultrastrong binding ability of Ag with H_2S to form Ag_2S.	The device enables the analysis of H_2S with a limit of detection of 0.81 μM; also it exhibited excellent selectivity and visible color change from yellow to colorless.	Zhai et al. (2019)
Aldehyde-sensitive colorimetric sensor	Sensor label constructed using printable inks on paper medium; On-packaging colorimetric sensor labels can detect aldehyde emission of apples based on Methyl Red.	Under exposure to ripening apples, the label changes color from yellow to orange, and then to red.	Kim et al. (2018)
Microfluidic biosensor for detection of *Salmonella typhimurium*	Sensor-based on immunomagnetic separation, fluorescence labeling, and smartphone video processing	Biosensor able to quantitatively detect *S. typhimurium* ranging from 1.4×10^2 to 1.4×10^6 CFU/mL, with a lower detection limit of 58 CFU/mL	Wang et al. (2019)
Chemical sensor	Hybrid chemiresistor sensor based on inorganic nanofibers of SiO_2 : ZnO and single-walled carbon nanotubes functionalized with 3,5 dinitrophenyls.	Ability to convert radio-frequency identification tags into wireless sensors to be potentially employed as intelligent packaging to monitor biogenic amines released from chicken, beef, and fish.	Andre et al. (2021)

9.3.1 Indicators

Properties related to food quality can be reported to consumers through the smart packaging system with an indicator mechanism. Indicators bring information about the conditions to which the food has been exposed and/or its actual quality status. The presence/absence or concentration of a target substance or the extent of a reaction (as a result of its interaction with food components) are related to the information that these indicators provide. Most of them are based on colorimetric and fluorescent dyes to detect pH changes, temperature, and freshness (Azeredo and Correa, 2021). Therefore, indicators are a simple tool to ensure food safety and to reduce losses and their associated costs, for example, product replacement or repair. According to the specificity of a given product and the factor to be controlled, the type of information involved in the indicators can be of a qualitative or semi-quantitative nature. Furthermore, they can be classified as direct, for example, indicators of time, temperature, humidity, damage and freshness, biosensors, or indirect, for example, tracking or traceability indicators (Balbinot-Alfaro et al., 2019). The characteristics of the main types of indicators applied in the food sector will be discussed in the following text.

9.3.1.1 Time Temperature Indicators

Most deteriorative changes in foods are closely related to an alteration in temperature, which deeply influences the shelf life. Thus, temperature control and monitoring are essential for food stability, especially for refrigerated and frozen foods. Excess, deviations, and fluctuations in temperature along the distribution chain can negatively interfere with the food texture, as well as allow microbial growth (Azeredo and Correa, 2021).

Two types of temperature indicators have been developed: the simple temperature indicators and the time-temperature integrators (TTIs). The first, known as critical temperature indicators, can indicate whether food products have remained above or below a critical temperature along storage and thus warn consumers of possible deterioration. The second type, the TTIs, can provide information on the full temperature history of a particular product, recording thermal profiles throughout the food supply chain and indicating residual shelf life. As the kinetics of food deterioration are influenced by both temperature and time, TTIs have attracted greater attention than simple indicators. Usually, the cumulative effects of temperature and time generate an irreversible color change in these time–temperature indicators, which makes it easier for the consumer to evaluate the food quality. Thus, data is recorded in terms of irreversible changes (mechanical deformation, color movement, or color development) that characterize the safety/quality status of a food product to consumers (Figure 9.3(a) and (b)). The principle of TTIs is based on the detection of time and temperature-dependent mechanical, chemical, electrochemical, enzymatic, or microbiological changes in a given food product, whose measurement is usually expressed as a visible response (Azeredo and Correa, 2021; Hu and Miao, 2022; Kalpana et al., 2019; Müller and Schmid, 2019).

Nowadays, with the advance of nanotechnology in several sectors, thermochromic nanomaterials have been applied as temperature indicators in smart food-packaging systems, e.g., gold nanoparticle-based TTIs (generation of different colorimetric signals), silver nanoparticle-based TTI (color signal connected to nanoparticles surface morphology),

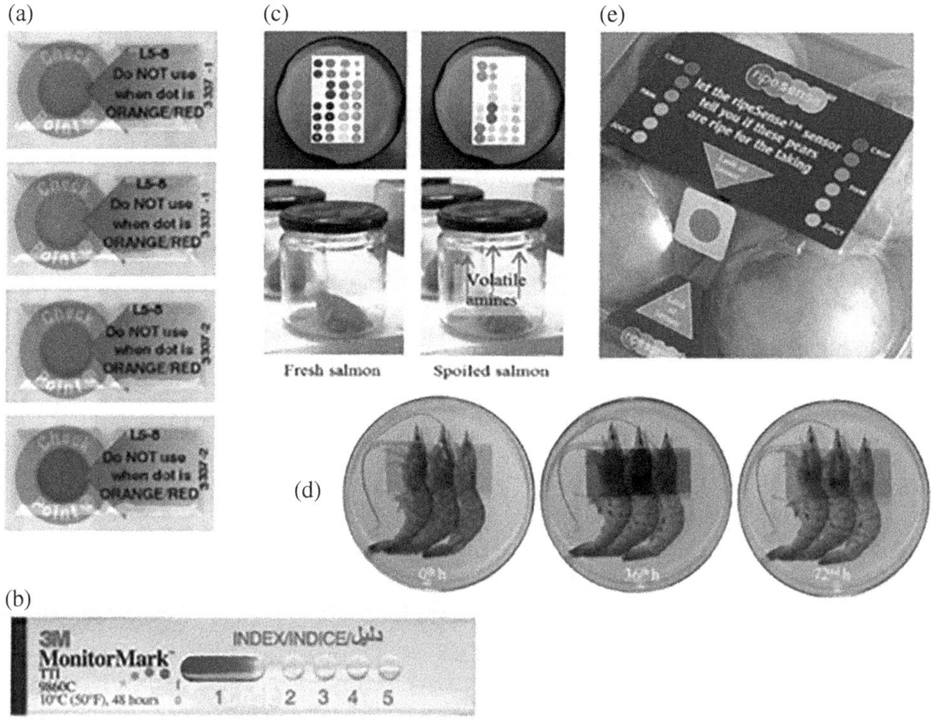

Figure 9.3 Examples of indicators are (a) Vitsab™ L5–8 Smart TTI Seafood Label (Vitsab International AB, Sweden) (*Source:* Vitsab International AB/https://vitsab.com/); (b) TTI Monitor Mark™ (3 M, United States); (c) fish freshness indicator based on volatile amines; (d) pH indicator for shrimp, based on anthocyanins; (e) fruit ripeness indicator ripeSense™, based on aroma volatiles released by the fruit (*Source:* Azeredo and Correa, (2021)/ with permission of Elsevier)

polydiacetylene/silica nanocomposite-based TTI (color change from blue to red according to temperature) (Hu and Miao, 2022).

9.3.1.2 Leak Indicators

Factors such as respiration of fresh foods, leaks, and gas influx can change the headspace composition in a package. Once this happens, the quality and safety of the food inside the package are compromised. Thereby, the long-established modified atmosphere technique has been increasingly applied to maintain the quality of fresh foods. Many perishable foods require the exclusion of oxygen and high concentrations of carbon dioxide to maintain and improve their stability. However, modified atmosphere packaging has limitations that reduce its effectiveness. Once its integrity is compromised, the protective atmosphere is lost, and the shelf life of the packaged product is reduced (Taoukis and Tsironi, 2016).

Modified atmosphere technology is based on changing the atmosphere (i.e., gas concentration levels) inside the package to extend food shelf-life. However, leaks, defects, or deformation in the package integrity can lead to variations in the optimal concentration of gases, consequent food spoilage, microbial growth, and oxidative degradation, and such variations drastically affect food quality. Consequently, this will result in negative effects on taste, nutrition, texture, and overall product acceptability. Therefore, intelligent systems aiming to maintain and monitor the integrity of packages, the so-called leak indicators, have been widely applied (Kalpana et al., 2019). The two most common leak indicators usually applied for intelligent packaging systems are CO_2 and O_2 indicators (Taoukis and Tsironi, 2016). For instance, Jang and Won (2014) developed a new pressure-activated compartmented oxygen indicator for intelligent food packaging, which showed high stability and proved to be an efficient colorimetric oxygen indicator (color change in proportion with the concentration of oxygen). The authors concluded that this device was able to work as an intelligent tool to detect leaks in modified atmosphere packaging.

9.3.1.3 pH Indicators

Intelligent packaging with a pH indicator is one of the new trends in the food industry, which has been standing out in the development of innovative packaging systems due to its sensitivity, safety, quick answer, low cost, compact size, and noninvasive. These pH indicators can be embedded in packaging materials or attached to the inside or outside of a package. Usually, pH indicators are made up of two parts: a solid base and a dye that is sensitive to pH changes. These dyes can be extracted from different sources of fruits and vegetables, the so-called natural pigments, or can be obtained from synthetic pigments, such as bromocresol purple, bromocresol green, and chlorophenol red. When the food deterioration process begins, there is a pH shift that leads to a color change in the packaging (Figure 9.3(d)). This phenomenon works as an indicator of product quality and guarantees its safety to the consumer (Balbinot-Alfaro et al., 2019). Therefore, once food spoilage occurs, many microbial metabolites, such as organic acids and CO_2 gas, are produced. These metabolites affect the pH of the food and its surroundings inside the package, so quality monitoring through pH change proves to be an effective method to ensure food safety for us consumers (Yousefi et al., 2019).

Synthetic pigments are more difficult to apply due to their possible toxicity and threat to consumer health. Because of that, natural dyes have been preferred because of their low toxicity and ease of preparation and obtaining from renewable and nonpolluting sources. Among plant-sourced dyes, anthocyanins, a water-soluble flavonoid pigment that is isolated mainly from flowers and fruits, have become very popular as indicators in smart packaging due to their remarkable color change at different pHs (Balbinot-Alfaro et al., 2019). Other natural colorants that have revealed excellent potential as pH indicators are curcumin, betacyanin, carminic acid, carotenoids, and chlorophylls (Alizadeh-Sani et al., 2020). In addition, the use of these natural compounds as pH indicators also shows multifunctional applications because they provide extra benefits to packaging systems as antioxidant and antimicrobial properties (Kalpana et al., 2019).

Prietto et al. (2017) developed pH-sensitive films with great stability based on corn starch and anthocyanins extracted from black bean seed coats and red cabbage. Alizadeh-Sani et al. (2020) prepared films based on immobilizing red barberry anthocyanins within composite chitin nanofiber and methylcellulose matrices, and these films were able to detect food spoilage through a pH-sensitive response, besides showing good antioxidant and antimicrobial activities. Lastly, according to Hu and Miao (2022), nanomaterial-based pH indicators have also been fabricated and disseminated in recent innovations in the food-packaging sector.

9.3.1.4 Gas Indicators

Maintaining the quality and safety of foods using modified atmosphere technology directly depends on the integrity of the packaging, an essential requirement in these cases. Gas indicators provide a noninvasive and nondestructive alternative to obtain qualitative or semi-quantitative information on changes in the concentration of certain gases (oxygen, carbon dioxide, water vapor, ethanol, hydrogen sulfide, and other gases) through visual responses (Vanderroost et al., 2014), usually devices based on colorimetric substances, redox dyes, reducing compounds, and alkaline components. Gas indicators provide information about the quality status of the food depending on the indoor atmosphere. Changes in the atmosphere inside the packaging react with/are detected by a sensor, while the indicator displays the quality condition. Changes in the food atmosphere can occur on the basis of enzymatic or chemical reactions, fat oxidation, browning reactions, gas generation from microorganism's metabolism, or gas transference through the packaging (Müller and Schmid, 2019).

The presence of carbon dioxide produces an anaerobic environment that inhibits the growth of most bacteria and fungi; in addition, a high concentration of CO_2 and a low concentration of O_2 is used in a modified atmosphere to prolong the shelf life of fruits and vegetables. On the other hand, CO_2 is also a spoilage metabolite. Therefore, the status of food quality can be reflected through CO_2 indicators, being able to indicate leaking in modified atmosphere systems and freshness of some food, for example. Generally, CO_2 indicators are based on the color change of a pH indicator dye (colorimetric) or the CO_2-induced fluorescence change of a luminescent dye. In relation to O_2, an essential gas for certain biological activities, its presence may negatively impact food quality and accelerate spoilage. These activities cover respiration, fat oxidation, browning reactions, pigment oxidation, growth of aerobic microorganisms, etc. Thus, the limitation of O_2 concentration is preferred in food-packaging systems (modified atmosphere packaging). In this context, O_2 indicators are essential to detect leaks which may occur anytime during transportation and storage, and to monitor the content and overall exposure to oxygen that may impair food quality and safety. Lastly, nanotechnology has also been incorporated into the gas indicator approach, and one of the most widely used nanoparticles-based devices for this type of application is a UV-based colorimetric O_2 indicator (Hu and Miao, 2022).

9.3.1.5 Freshness Indicators

Freshness indicators are classified as direct food quality indicators because they deliver information on microbiological and/or chemical changes caused by food spoilage. For instance, considering meat products, their freshness indicators may be based on amines or hydrogen sulfide. Regarding seafood, volatile amines are often the basis for assessing microbial degradation (Azeredo and Correa, 2021; Kalpana et al., 2019). In general, we can mention some of the most common quality-indicating metabolites: glucose, organic acids, ethanol, volatile nitrogen compounds, biogenic amines, carbon dioxide, ATP degradation products, and sulfuric compounds (Müller and Schmid, 2019).

Monitoring microbial or chemical deterioration may also be indirectly done by evaluating the resulting pH changes (derived from the production of organic acids, volatiles, or CO_2 dissolution). Because of that, most freshness indicators are based on pH-sensitive dyes, synthetic or natural (Figure 9.3(c)). In this context, natural food-grade dyes such as curcumin have received great attention due to their biodegradability and consumer perception of healthy and safe products (Azeredo and Correa, 2021; Kalpana et al., 2019).

When referring to fruit ripeness, freshness indicators are frequently applied to obtain information about the degree of ripeness, which is based on the release of aroma compounds or ethylene by the fruit on ripening (Azeredo and Correa, 2021). For example, the ripeSense® sensor changes color (Figure 9.3(e)), which is initially red and gradually turns to orange and finally yellow, by reacting to the aroma released by fruit as it ripens. This device was the world's first intelligent sensor label able to change its color according to fruit ripening (Ripesense, 2022).

9.3.2 Data Carriers

9.3.2.1 Barcode

A barcode can be described as a machine-readable pattern of parallel bars and spaces, so-called one-dimensional or 1-D (Figure 9.4(a)), or an array of dots and spaces, so-called two-dimensional or 2-D (Figure 9.4(b)) positioned in order to represent hidden, encoded data. From there, the information is decoded by an optical scanner that transmits it to a system where it is stored and processed (Azeredo and Correa, 2021; Kalpana et al., 2019). As an intelligent tool, barcodes can be applied to monitor inventory, control product movement, and thus trace product history (Kalpana et al., 2019). Moreover, these devices bring

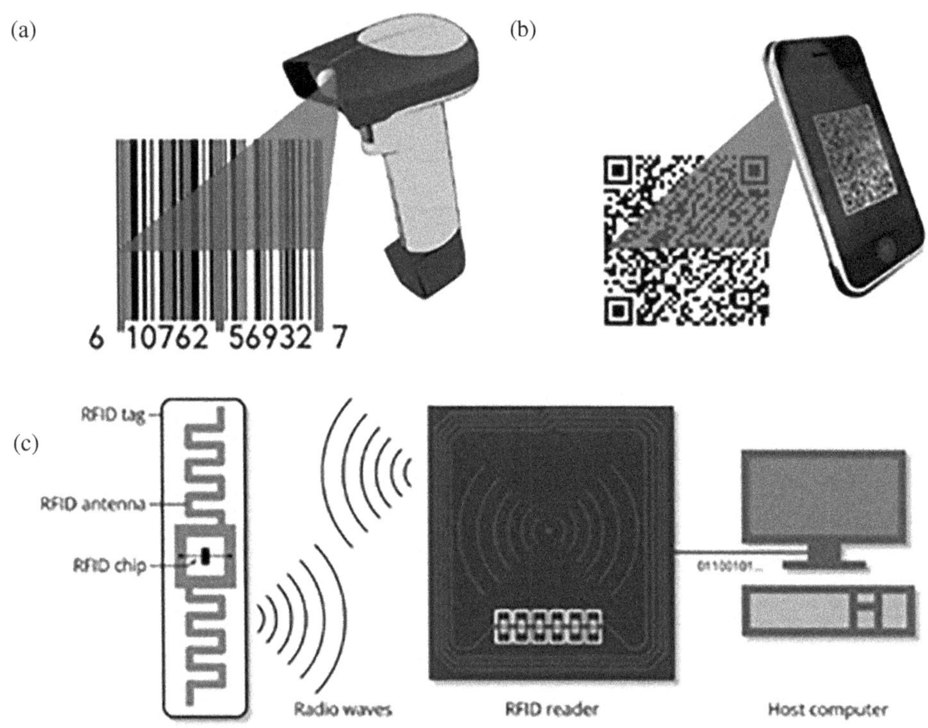

Figure 9.4 Examples of data carriers are (a) 1-D barcode, (b) 2-D barcode, and (c) RFID tag and its components (*Source:* Azeredo and Correa, 2021/with permission of Elsevier)

some benefits such as low cost, ease of use, as well as wide and easy use in stock control, recording, and checkout (Müller and Schmid, 2019).

Depending on the type, barcodes present different storage capacities: one-dimensional/1-D and two-dimensional/2-D, as mentioned previously. In view of this, it is important to emphasize that the two-dimensional barcodes, for example, the QR (quick response) Codes, provide more memory capacity, and this offers greater convenience for retailers, consumers, and the market in general (Müller and Schmid, 2019). The information encoded in the barcodes covers food packing date, batch/lot number, package weight, nutritional information, cooking instructions, and the website address of the food manufacturer, for example. In addition, they are suitable devices to identify the origin of food products (Fang et al., 2017).

9.3.2.2 Radio-Frequency Identification

An RFID system (Figure 9.4(c)), RFID, is considered more advanced and convenient, and its working principle is based on wireless communication between a tag attached to the product and an interrogator. The component of the system includes an RFID antenna for communication connected to a chip for identification and data storage, a reader (so-called interrogator) for wave emission and reception upon backscattering from the tag, and, finally, a device such as a computer or a cellphone to save and process data (Azeredo and Correa, 2021; Caon et al., 2017; Kalpana et al., 2019). In addition to collecting information through automated identification for tracking food products, RFID, with the help of sensors, can detect certain compounds such as water, ammonia, ethanol, toluene, and biogenic amines associated with food spoilage (Yousefi et al., 2019). For instance, smart radio-frequency labels with sensors able to measure temperature, humidity, and the presence of volatile amine compounds can be used to quantify the freshness of fish (Smits et al., 2012).

Usually, RFID tags are classified into two major types based on the power supply: active tags, which present a longer communication range, greater processing power, and a high operating frequency but are bulky and more expensive due to the embedded battery, and passive tags, which do not require any internal power source (no battery), have reduced size, low cost and unlimited life, however, present low levels of signal strength. Another classification characterizes these devices as low-, intermediate-, and high-frequency RFIDs (Caon et al., 2017). In general, for food-packaging applications, both low-frequency (125–134 kHz) and high-frequency (13.56 MHz) ranges are used. Applications of RFID tags include horticultural products, meat and fishery products, dairy and bakery products, beverages, and other foods (Kalpana et al., 2019). In comparison to barcodes, RFID tags are more expensive and require a more powerful electronic information network. However, the information can be loaded electronically on these tags for future changes. In addition to improvement in traceability, inventory management, and quality/safety processes, which helps the entire food supply chain (Müller and Schmid, 2019).

9.3.3 Sensors

Sensors are devices that detect and convert one form of signal to another form using a transducer. Among sensors, the chemical and the biosensors have been highlighted as intelligent packaging tools to detect pH, humidity, color, and biological species (Kalpana et al., 2019), and they will be discussed in detail in the following subtopics. Most sensors are composed of two components: the receptor that detects the presence, activity, composition, or concentration of chemical or physical analytes and converts this information into an energy form that can be measured by the second component, the so-called transducer. The latter is also used to transform the measured signal into a useful analytic signal (electrical, chemical, optical, or thermal) (Caon et al., 2017; Müller and Schmid, 2019).

A concept that has been growing in recent years is that of nanosensors. Nanosensors are defined as nanotechnology-enabled sensors that present one of the following features: either the sensor size or its sensitivity is on the nanoscale, or the interaction distance between sensor and object is in nanometers (Caon et al., 2017). Furthermore, bionanosensors, a new trend in smart systems, are the result of combining biosensors (devices that involve a biological sensing element) and nanotechnology (Primožič et al., 2021). So, why do nanoscale technologies represent the next generation of biosensors? We can indicate three main reasons: (i) improving and amplification of the sensitivity, specificity, and detection limits of the assay; (ii) increase in sample throughputs; and (iii) reducing assay complexity and cost. Generally, nanosensors require less amount of analyte to record a response, mainly because of their surface area (Caon et al., 2017).

9.3.3.1 Chemical Sensors

Chemical sensors are able to quickly detect specific analytes by converting chemical information into a quantifiable output signal (proportional to that measurement), which is treated by electronics/software. The receptor, where chemistry takes place, interacts chemically with the analyte, and the transducer converts this chemical data into a measurable signal. Moreover, the classification of chemical sensors is based on transduction principles, which can be electrical, electrochemical, optical, and gravimetric (Azeredo and Correa, 2021). In other words, sensors experience chemical interactions between substrate and analyte molecules, whose results provide information for detection/sensing (Kalpana et al., 2019).

Chemical sensors have been suggested to quantify gases generated by food deterioration. Senapati and Sahu (2020) designed and developed a capacitive sensor to quantify volatiles such as NH_3, TMA, ethanol, and H_2S emitted from chicken meat spoilage. The authors mentioned that this sensor demonstrated potential as a portable, inexpensive, rapid, and accurate quality device. Electrochemical sensors have also been proposed as smart technology. Ghaani et al. (2018) developed a nanosensor able to quantify toxic additives or monomers migrated from packaging into food (or their toxic products generated upon food contact). Authors emphasize the potential of this nanocomposite-sensor as a new analytical tool for the determination of primary aromatic amines (PAAs), toxic compounds that are generated from unreacted monomers during the process of obtaining polyurethane adhesives. Furthermore, Karthika et al. (2021) reported the detection of bisphenol-A (BPA) using an electrochemical sensor based on a molecularly imprinted polymer. According to the authors, BPA can migrate from polycarbonate bottles and epoxy resins and presents the potential to cause endocrine disorders in humans, especially infants and children.

9.3.3.2 Biosensors

Biosensors perform their function based on a receptor combined with a biological material (e.g., enzymes, antigens, antibodies, and nucleic acids) specific to the target analyte and a transducer that transforms the biological signals into a quantifiable electrical response

(Azeredo and Correa, 2021; Kalpana et al., 2019). Biosensors are especially applicable for the detection of certain microbial species in food, particularly pathogenic microorganisms. Another relevant point is that many of the recent innovations involve nanomaterials as the active layer of sensors due to their unique optical/electrical properties and large surface area that can amplify the detection signal (Azeredo and Correa, 2021).

The most common biosensors considered for intelligent food packaging are fluorescent and microfluidics sensors, electrochemical biosensors, and gas biosensors. The fluorescent-based biosensor involves a fluorescent or phosphorescent dye immobilized with a solid polymeric matrix. This type of biosensor can produce a diversity of colors in contact with food pathogens. The microfluidic biosensor has been reported for efficient real-time detection of pathogens, besides its high sensitivity. These biosensors present a system with a miniature structure, which allows them to detect tiny compounds. The electrochemical-based biosensor appears as a promising technology for monitoring food quality, as well as the gas biosensor for the detection of gas leakage and the presence of gas from the spoilage process (Sobhan et al., 2021). Costa et al. (2017) developed an electrochemical biosensor for the detection of the food carcinogen aflatoxin B1. These intelligent devices have been structured with surfaces functionalized with antibodies that respond to aflatoxins (microbial metabolites). Moreover, biosensors that respond to antigens on the surface of pathogenic microorganisms, such as *Escherichia coli*, via surface acoustic wave (SAW) flexible biosensors were reported and developed by Lamanna et al. (2020).

9.4 Nanotechnological Applications in Intelligent Packaging

Nanomaterials greatly increase the packaging material's qualities and boost its properties, resulting in higher performance and more durable features (Perera et al., 2022). Pathogenic bacteria, toxins that contaminate food, additives, vitamins, dyes, fertilizers, pesticides, taste, and aroma can be found employing nanosensors. In the future, the invisible, intelligent tags made of nanoparticles will be very helpful in determining the authenticity of food products (Kumar et al., 2017). The installation of nanosensors on food packaging allows for the exact identification of various food contaminants as well as the monitoring of internal and external product conditions. The nanosensors interact with the consumers through nano communication, which refers to the exchange of information between nanomaterials, which can be done through nanomechanical, acoustic, electromagnetic, and chemical or molecular mass communication (Fuertes et al., 2016).

Due to their difficulty in manipulation, invisible barcodes based on nanoparticles are also currently growing increasingly common. Recently, novel nanoparticle-based encoding systems and nanodisk codes have been reported. A series of surface-enhanced Raman scattering-producing disk pairings known as nanodisk code can be scanned using a Raman microscope. There are many arrays of nanodisks made of silver, gold, and silver-gold heterodimers (Kumar et al., 2017).

Nanoparticles are used in intelligent food-packaging materials combined with other active materials (Table 9.2) as pH sensors or humidity sensors. For instance, Wu et al. (2020) created an active and intelligent film by immobilizing red cabbage anthocyanins in

Table 9.2 Nanotechnological applications in intelligent packaging

Packaging Material	Type of Intelligent Packaging	Properties of Packaging Material	Reference
Konjac glucomannan—oxidized chitin—red cabbage anthocyanins	pH indicator	• Excellent UV-barrier antioxidant and antibacterial properties • Controlled release of anthocyanin • pH-sensitive and showed color changes	Wu et al. (2020)
Bacterial nanocellulose—poly (sulfobetaine methacrylate)	Humidity sensors	• Enhanced UV-barrier properties and moisture-scavenging ability • High antimicrobial activity • Proton motion is relevant for protonic-conduction humidity sensors that monitor food humidity levels	Vilela et al. (2019)
Starch-carbon nano dots—anthocyanin	pH indicator	• Distinctive colors from red to purple, blue-gray, and then yellow in the pH range (pH 1–14) • Good linearity between pH 1–6, and pH 8–12	Kuswandi et al. (2020)
Ethylene vinyl acetate—anthocyanin—rosemary extract—ZnO/Fe-MMT nanoparticles	pH indicator	• Improved thermal stability and mechanical properties • Improved antioxidant activity and antimicrobial properties • pH-sensitive and showed color changes	Eskandarabadi et al. (2019)
Nanostructure cellulose-polypyrrole-ZnO film	Electrical resistance	• Decrease the growth of microbial load in chicken thigh • Control the pH increasing • Increase the shelf life and stabilize rheological properties of the chicken thigh by increasing antioxidant and antimicrobial activity • It is possible to evaluate the time and temperature of the chicken thigh meat holding by scanning and checking changes in the electrical resistance of the packaging film	Pirsa and Shamusi (2019)

Konjac glucomannan—chitosan—nano-ZnO—mulberry anthocyanin	pH indicator	• Enhanced UV–Vis light barrier properties • Tensile strength (TS) of films increased from 42.83 MPa to 52.00 MPa • Strong antioxidant activity and good thermal stability • Good antibacterial activity against *Escherichia coli* and *Staphylococcus aureus*. • Color differences from red to blue in different buffer solutions (pH 2–12)	Sun et al. (2020)
Nanocellulose-bromothymol blue-methyl red	pH indicator	• Quick response to chicken spoilage • The optical color changed from green to bright red when the packed chicken was stored at 4 °C for three days	Lu et al. (2020)
Chitosan-PVA-black carrot anthocyanin—bentonite nanoclays	pH indicator	• Higher thermal stability • Highest bacterial growth inhibition against *E. coli* (75.20%), *Pseudomonas aeruginosa* (69.95%), and *Staphylococcus aureus* (44.38%)	Koosha and Hamedi (2019)
Poly(vinyl alcohol)—chlorogenic Acid Carbon nanodots	pH and Al^{3+} ions indicator	• Strong antioxidant capacity • Sense basic substances(e.g., amines) during spoilage • Good resistance to photobleaching • The fluorescence of the film is sensitive to both the pH and Al^{3+} ions	Zhang et al. (2020)
Soluble soybean polysaccharide-SiO_2—curcumin	pH indicator	• Film thickness, water solubility, and water vapor permeability decreased • Mechanical performance improved • A small amount of SiO_2 nanoparticles were released in food simulant • Good physical properties and antibacterial activity	Salarbashi et al. (2021)

a matrix of konjac glucomannan and oxidized chitin nanocrystals. The integration of chitin nanocrystals is responsible for the produced films' exceptional mechanical and UV-barrier capabilities. These films exhibit outstanding antioxidant and antibacterial characteristics in addition to their pH sensitivity, which is represented by the color change from pink (pH = 2) to green (pH = 12). Further, studies on pH-sensitive packaging materials have been performed by Koosha and Hamedi (2019) on intelligent packaging film with chitosan-PVA films -black carrot anthocyanins (natural pH indicator) and different concentrations of bentonite nano-filler. The films showed high thermal stability, antimicrobial activity, and reduced water vapor permeability with the addition of both bentonite and anthocyanin. The film color changed from pink (pH = 3) to blue (pH = 13).

Through the polymerization of sulfobetaine methacrylate into a nanofibrous network of bacterial nanocellulose, zwitterion films based on poly (sulfobetaine methacrylate) and bacterial nanocellulose were produced. The films have good mechanical properties (Young's modulus 3.1 GPa), are optically transparent, have thermal stability up to 265 °C, and have a 450–559% water absorption capacity. Additionally, the films exhibit UV radiation barrier qualities and antimicrobial properties. The hygroscopic monomer poly (sulfobetaine methacrylate) can take up humidity from the environment. The ability of the films to absorb moisture has a profound impact on foods that are sensitive to humidity. So thus, this packaging material determines the humidity levels of the packaging material. The sulfonic acid groups in the structure of poly (sulfobetaine methacrylate) nanocomposite films are protogenic moieties recognized for facilitating proton migration. From the perspective of intelligent food packaging, this characteristic is crucial.

9.5 Applications of Intelligent Food Packaging for Meat/Fish, Fruits, Vegetables, and Dairy Products

Distinct intelligent packaging materials are used depending on the product type to align with different characteristics particular to the intended food matrix. Fresh meat products are highly perishable food products with a short lifetime. The spoilage of meat and poultry mainly takes place due to microbial degradation and lipid oxidation, which is a result of its high-water activity and fat content (Luo et al., 2022). Thus, many studies on meat product-based intelligent packaging materials have focused on pH indicators and gas indicators to reveal quality deterioration. Thus, total volatile basic amines (TVB-N), which include ammonia, trimethylamines, and dimethylamines, are frequently used as indicators of meat deterioration. The studies of Zhai et al. (2020) developed an intelligent packaging material utilizing low-density polyethylene and curcumin by the melting extrusion method. The low-density polyethylene (LDPE)-curcumin composite film was sensitive to ammonia (NH3), while the limit of detection for NH_3 at 90% relative humidity was 0.18 M. The increasing TVB-N content of the meat samples led to light yellow to light brown color changes in the film, which was utilized to monitor the deterioration of beef at 4 °C. Researchers have also conducted pH-responsive research on packed meat products, using pH-sensitive color-changing indicators to detect certain acidic/alkaline compounds linked to the development of spoilage microorganisms. Shavisi and Shahbazi (2022) developed an intelligent packaging material with the combination of chitosan-gum Arabic nanofiber and 5% Rosa damascena (commonly known as the Damask rose) extract using the electrospinning method. When exposed to the pH buffer solutions, the film displayed visible color responses: pink (pH 1), red (pH 2–3), pale red (pH 4–5), dark khaki (pH 6–8), green (pH 9–10), and brown (pH 11–12). As the chicken fillets spoiled at 4 °C, the white color of the clever film turned khaki. Presently available intelligent packaging for meat products includes, for instance, the dynamic freshness label for meat products created by To-Genkyo, Japan. Special ink on the label alters color according to the quantity of ammonia the meat emits. As the meat ages, the bottom half of the label fills up, resembling an hourglass. The ink prevents the barcode at the bottom from being scanned at the checkout when the meat is no longer fit for consumption (Media, 2011).

Fish products are particularly perishable and vulnerable to microbiological and chemical deterioration while being stored. There is, therefore, a lot of interest in the creation of intelligent packaging materials to monitor their quality, freshness, and safety. Volatile substances such as ammonia, methylamine, dimethylamine, trimethylamine, and total volatile basic nitrogen (TVB-N) are formed during the microbiological decomposition of fish items. As a result of the development of these substances, pH changes are observed. In order to identify changes in seafood quality during storage, numerous researchers are developing intelligent packaging materials that are sensitive to pH changes and/or the production of TVB-N (Cheng et al., 2022). Eze et al. (2022) developed a colorimetric indicator film based on chitosan and broken riceberry phenolic extract for the visual detection of freshness in seafood products. The indicator film was initially orange-red but turned yellow after 24 hours of being enclosed with the shrimp due to an increase in pH brought on by spoilage. In addition, studies on intelligent packaging materials based on TVB-N levels of fish were performed by Zhai et al. (2017). They developed a colorimetric film utilizing starch-polyvinyl alcohol and different concentrations of roselle anthocyanins for real-time monitoring of silver carp freshness. The increase in TVB-N caused apparent color changes in the colorimetric films over time, as illustrated in Figure 9.5(a). This change in color from pink to purple indicates the freshness of the fish in the present, as visible in Figure 9.5(b)

There have also been developed intelligent food-packaging materials for fruits and vegetables. These packing materials can detect when fruits and vegetables are ripening and when they are becoming spoiled. Intelligent packaging materials with pH sensors have been created, as shown later. Through the incorporation of anthocyanin-loaded ovalbumin-carboxymethyl cellulose nano complexes into a corn starch/polyvinyl alcohol matrix, an intelligent colorimetric film was created by Liu et al. (2022) to sense the freshness of mushrooms. The film clearly displayed a colorimetric reaction to pH buffer and CO_2 gas. By visually changing from purple to pink, the intelligent film could keep track of the mushrooms' freshness while they were being stored. Further, Niponsak et al. (2020) developed a ripeness indicator for fresh-cut durian that detects volatile sulfur levels. This indicator was developed using starch, chitosan, and methyl red mixed bromothymol blue. Based on significant volatile changes during storage (e.g., ethanethiol, diethyl disulfide, diethyl trisulfide, 3, 5-dimethyl-1, and 2, 4-trithiolane), the maximum storage period for fresh-cut durian at 30 °C was five days. Fresh-cut durian packaging incorporates this ripeness signal as a response to the scents generated by fruit as it ripens. The indication exposed to the volatile compounds of durian commenced with red, changed to orange after three days,

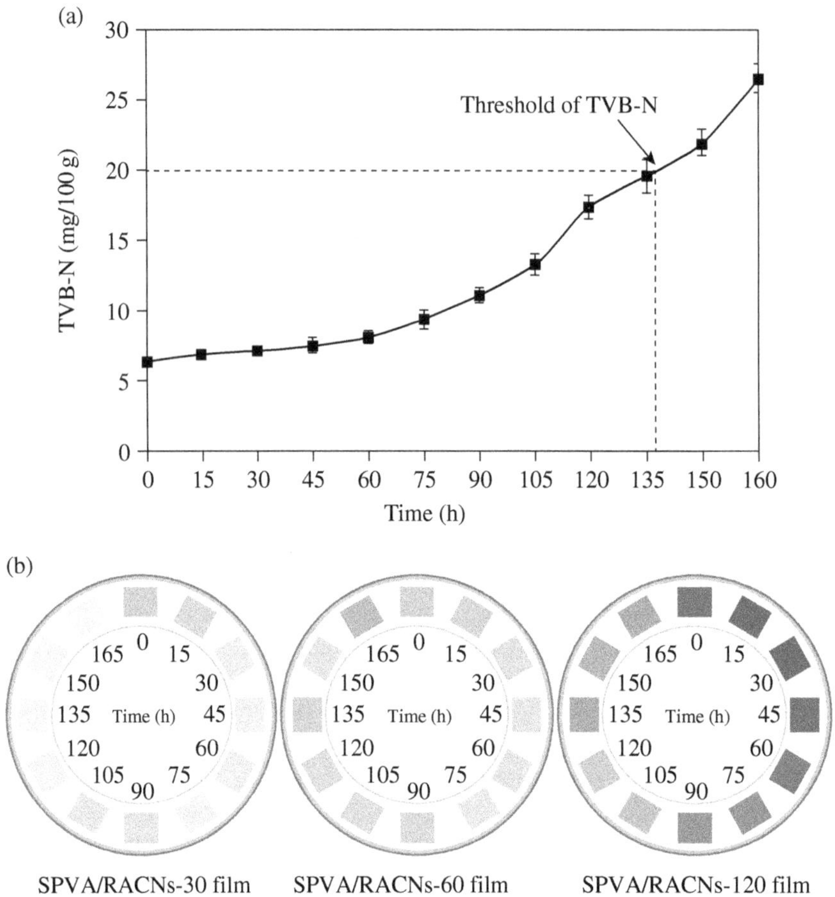

Figure 9.5 (a) The change of TVB-N level of stored silver carp within 165 hours at 4 °C and (b) the corresponding color changes of the colorimetric films starch-polyvinyl alcohol 30 mg RACNs (SPVA/RACNs-30 film), starch-polyvinyl alcohol 60 mg RACNs (SPVA/RACNs-60 film), and starch-polyvinyl alcohol 120 mg RACNs (SPVA/RACNs-120 film) (*Source:* Zhai et al. (2017)/with permission of Elsevier)

and eventually to yellow after five days, indicating the beginning, ripeness, and over-ripeness of fresh-cut durian, correspondingly.

During storage, milk is susceptible to deterioration. A design to implement quality control in the supply chain for milk is provided by intelligent food packaging. pH, microbiological content, and gas concentration changes will all operate in combination with milk spoilage. Chemical and pH sensors are used in intelligent packaging to help consumers detect the quality and safety of packaged milk. During the shelf life of cow milk, Romero et al. (2021) created two distinct prototypes of intelligent packaging based on volatile chemicals, pH, color change, and sensory qualities to determine the shelf life of cow milk. High-density polyethylene was used to make the prototype bottles, and bromothymol blue was used as a pH indicator (the color of the bottle was designed to change when the pH changed). As the pH of the cow milk varied over time, Treatment 1's color shifted from green to yellow. Treatment 2 underwent a color change, going from blue to green and then to yellow. The intelligent packaging detects the deterioration of refrigerated cow milk by changing color during storage conditions.

The studies of Liu et al. (2019) developed a pH-sensitive κ-carrageenan film with different amounts of mulberry polyphenolic extract (MPE). When the pH value increased from 2 to 13, MPE demonstrated a pH-sensitive property, with the color changing from red to purple and then to gray. Carrageenan-MPE films were used to monitor the freshness of milk, and as the milk degraded after being stored at 40 °C for eight hours, the films changed color, as shown in Table 9.3. Further, applications of intelligent food packaging for the different segments of meat/fish, fruits, and vegetables are depicted in Table 9.3.

9.5.1 Commercially Available Intelligent Packaging

The market for worldwide intelligent packaging technology for the food and beverage sector is being driven by the rising demand for fresh and high-quality packaged food, customer convenience, and manufacturers' concern for extended shelf lives of the food products. At US$6.87 billion in 2018, the market for intelligent packaging is anticipated to increase at a CAGR of 8.3% by 2027. The worldwide intelligent packaging market currently includes significant companies, including 3 M Company, Temp Time Corporation, PakSense, American

Table 9.3 The change of acidity, pH values in the milk, and the color response of κ-carrageenan-mulberry polyphenolic extract (MPE) films at different storage times

			Color response of κ-Carrageenan-MPE films		
Time (h)	Acidity (°T)	pH Value	κ-Carrageenan-MPE I film	κ-Carrageenan-MPE II film	κ-Carrageenan-MPE III film
0	17.06 ± 0.12^{e}	6.87 ± 0.02^{a}			
2	17.32 ± 0.04^{e}	6.73 ± 0.02^{a}			
4	17.38 ± 0.01^{e}	6.71 ± 0.01^{a}			
6	17.44 ± 0.02^{e}	6.68 ± 0.01^{ab}			
8	18.45 ± 0.64^{e}	6.50 ± 0.08^{b}			
10	51.65 ± 4.45^{d}	5.13 ± 0.16^{c}			
12	69.40 ± 4.38^{c}	4.64 ± 0.15^{d}			
16	87.15 ± 4.03^{b}	4.13 ± 0.09^{e}			
20	96.35 ± 0.64^{a}	3.96 ± 0.03^{e}			

Values are given as mean ± SD ($n = 3$). Different superscript letters in the same column indicate significantly different ($p < 0.05$).

Source: Liu et al. (2019)/with permission of Elsevier

Thermal Instruments, Avery Dennison, R.R. Donnelly Sons & Company, BASF SE, International Paper, Stora Enso, Huhtamaki Group, Smartrac N.V. (Salgado et al., 2021).

For instance, the Vitsab time/temperature integrator tag, created by COX Technologies, United States, can track the exposure of perishable foods, particularly seafood, to potentially harmful temperatures. The device helps forecast microbial growth in perishables at the grocery store and can also be used in food processing operations when a known limit has been reached for cumulative exposure to "out-of-safe-range" temperatures, the temperature-indicating dots on the tag change from green to yellow (Packaging http://Network.com, 1998). As per Figure 9.6, ripeSense is an intelligent ripeness indicator label ever created for pears by Jenkins Group, New Zealand. The ripeSense sensor responds to the aroma that fruit releases as it ripens by changing color. The sensor starts red and then transitions to orange and then yellow. As a result, customers can select fruits according to their desired level of maturity (Nešić et al., 2019).

To measure vegetable freshness, the Vanprob firm in India created Food fresh™ indicators. They are a specific type of timer that can be programmed to expire after a specified consumption within the period, which can be anything from a few days to several weeks or months. They can be placed on products like jars, vacuum packaging, or cartons using a label applicator and are a practical answer to the issue of storing food after opening (Mohebi and Marquez, 2015).

Further, intelligent packaging is also used in other industries, such as laboratory supplements or medical device transportation companies. 3 M™ MonitorMark™ and 3 M Freeze Watch™ indicators are time–temperature indicators produced at 3 M Company, United States. Monitoring the exposure of products to freezing temperatures can be done

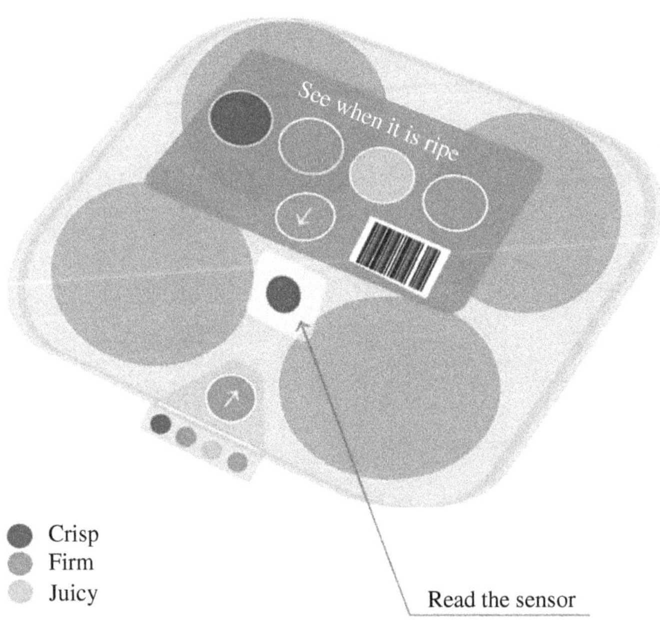

Figure 9.6 RipeSense®, an intelligent ripeness indicator label (*Source:* Anandharamakrishan / John Wiley & Sons / CC BY 4.0)

with 3 M Freeze Watch indicators. The activation temperature aids in the safe shipping of commodities that are freeze-sensitive. At the same time, the 3 M MonitorMark time temperature indicator is a tool for tracking a product's 48-hour exposure to temperatures above the stated limit. Both are still readily accessible from 3 M, Fisher Scientific, and other places (3M, 2022).

While the use of intelligent packaging is commercially available through the abovementioned products, however, the use of it is not widespread. Therefore, further marketing and improvement of intelligent packaging are necessary with the use of environment-friendly intelligent packaging materials.

9.6 Regulatory Aspects

Numerous intelligent packaging systems have been created lately, and it is anticipated that innovation will soon be available commercially. Innovative food-packaging solutions must adhere to regulations to succeed, though. Therefore, the regulations and the legislation associated with intelligent packaging are extremely important. The Commission regulations EC No. 450/2009 deal with the requirements and the endorsement of active and intelligent materials and articles intended to come into contact with food. As per this regulation, the materials and articles that are utilized or combined to create intelligent packaging materials should be evaluated according to the regulations. The intelligent packaging materials, which are positioned on the outer surface of the food without contact with it may be used without concern if they migrate less than 0.01 mg/kg into the food product. The intelligent packaging materials that are in contact with the food product should have minimum migrational levels, and these materials should fall into the community list of authorized substances. In addition, the nonedible parts of these packaging materials should be labeled "DO NOT EAT" and also in symbol (Regulation E.C. No, 2009).

Further, according to the Commission regulations, EC No. 1935/2004, intelligent food-packaging materials should not alter the nutritional content or organoleptic characteristics of food. The consumers shall not be misled by the labeling, advertising, or presentation of any material or information regarding the condition of the product. Additionally, intelligent materials and items must have proper labels that clearly state their intelligence (Regulation No, 2004).

In the United States, intelligent packaging is subject to the same regulations as all food contact items and is governed by the Food and Drug Administration's (FDA) regulatory framework. If considered "food additives," materials used in food-contact applications must acquire premarket regulatory clearance in the United States. Thus, the main regulations globally of intelligent packaging materials revolve around the migrational aspects and the chemical composition of the packaging materials.

9.7 Conclusion and Future Perspective

Intelligent food-packaging materials are developed for the constant monitoring of the food quality during all the steps of food production, that is, manufacturing, storage, and transport.

Innovative, intelligent packaging materials improve food quality, reduce food safety problems, and decrease food waste, which ultimately leads to a healthier, safer, and more sustainable food supply chain. Intelligent packaging materials can take many different forms, including bar codes, sensors, labels, and even films, as was covered in great length in the chapter. The intelligent packaging materials work on different concepts such as pH sensitivity, humidity, gas levels, or microbial contamination of the substances. A limited number of intelligent packaging materials are currently found in the market due to their importance in food safety and communication. However, the widespread application of intelligent packaging is limited to date and is mainly found at the lab scale. This is mainly due to the high production cost and the inability to develop environmentally friendly products. Further, consumer knowledge related to intelligent food packaging is very limited, which makes them reluctant to use the products. Thus, consumer confidence in the safety of the systems also needs to be strengthened. Manufacturers should also realize the importance of intelligent packaging materials for their food products so that they will increase the use of such technologies. Although research on intelligent food packaging is still in its early stages, it is projected that improvements in this field will occur during the next decades due to its significance for consumer communication and food safety. This could result in a brand-new generation of packaging materials that are cost-effective, pragmatic, and sustainable and also have a market edge.

Acknowledgment

The author Luana S. C. Carnaval acknowledges the funding from the "Irish Research Council Government of Ireland Postgraduate Scholarship" under grant number GOIPG/2022/1392, and Kalpani Y. Perera acknowledges the funding from Technological University Dublin - City Campus under the Researcher Award 2021, Dublin, Ireland.

References

3M. (2022). 3M Thermal Exposure Indicators. https://www.3m.com/3M/en_US/p/c/lab-supplies-testing/tests-indicators/thermal-exposure-indicators

Alizadeh-Sani, M., Mohammadian, E., Rhim, J.-W., & Jafari, S. M. (2020). pH-sensitive (halochromic) smart packaging films based on natural food colorants for the monitoring of food quality and safety. *Trends in Food Science & Technology*, *105*, 93–144. https://doi.org/10.1016/j.tifs.2020.08.014

Andre, R. S., Ngo, Q. P., Fugikawa-Santos, L., Correa, D. S., & Swager, T. M. (2021). Wireless tags with hybrid nanomaterials for volatile amine detection. *ACS Sensors*, *6*(6), 2457–2464. https://doi.org/10.1021/acssensors.1c00812

Azeredo, H. M. C., & Correa, D. S. (2021). Smart choices: mechanisms of intelligent food packaging. *Current Research in Food Science*, *4*, 932–936. https://doi.org/10.1016/j.crfs.2021.11.016

Balbinot-Alfaro, E., Craveiro, D. V., Lima, K. O., Costa, H. L. G., Lopes, D. R., & Prentice, C. (2019). Intelligent packaging with pH indicator potential. *Food Engineering Reviews*, *11*(4), 235–244. https://doi.org/10.1007/s12393-019-09198-9

Caon, T., Martelli, S. M., & Fakhouri, F. M. (2017). New Trends in the Food Industry: Application of Nanosensors in Food Packaging. In A. M. Grumezescu (Ed.), *Nanobiosensors* (pp. 773–804). Academic Press. https://doi.org/10.1016/B978-0-12-804301-1.00018-7

Cheng, H., Xu, H., Julian McClements, D., Chen, L., Jiao, A., Tian, Y., Miao, M., & Jin, Z. (2022). Recent advances in intelligent food packaging materials: principles, preparation and applications. *Food Chemistry*, *375*, 131738. https://doi.org/10.1016/j.foodchem.2021.131738

Choi, I., Lee, J. Y., Lacroix, M., & Han, J. (2017). Intelligent pH indicator film composed of agar/potato starch and anthocyanin extracts from purple sweet potato. *Food Chemistry*, *218*, 122–128. https://doi.org/10.1016/j.foodchem.2016.09.050

Costa, M. P., Frías, I. A. M., Andrade, C. A. S., & Oliveira, M. D. L. (2017). Impedimetric immunoassay for aflatoxin B1 using a cysteine modified gold electrode with covalently immobilized carbon nanotubes. *Microchimica Acta*, *184*(9), 3205–3213. https://doi.org/10.1007/s00604-017-2308-y

Eskandarabadi, S. M., Mahmoudian, M., Farah, K. R., Abdali, A., Nozad, E., & Enayati, M. (2019). Active intelligent packaging film based on ethylene vinyl acetate nanocomposite containing extracted anthocyanin, rosemary extract and ZnO/Fe-MMT nanoparticles. *Food Packaging and Shelf Life*, *22*, 100389. https://doi.org/10.1016/j.fpsl.2019.100389

European Food Safety Authority (EFSA) (2009). Opinion of the panel on food contact materials, enzymes, flavourings and processing aids (CEF) guidelines on submission of a dossier for safety evaluation by the EFSA of active or intelligent substances present in active and intelligent materials and arti. *The EFSA Journal*, *1208*(March), 1–11.

Eze, F. N., Jayeoye, T. J., & Singh, S. (2022). Fabrication of intelligent pH-sensing films with antioxidant potential for monitoring shrimp freshness via the fortification of chitosan matrix with broken Riceberry phenolic extract. *Food Chemistry*, *366*, 130574. https://doi.org/10.1016/j.foodchem.2021.130574

Fang, Z., Zhao, Y., Warner, R. D., & Johnson, S. K. (2017). Active and intelligent packaging in meat industry. *Trends in Food Science & Technology*, *61*, 60–71. https://doi.org/10.1016/j.tifs.2017.01.002

Fuertes, G., Soto, I., Vargas, M., Valencia, A., Sabattin, J., & Carrasco, R. (2016). Nanosensors for a monitoring system in intelligent and active packaging. *Journal of Sensors*, *2016*, 1–8. https://doi.org/10.1155/2016/7980476

Ghaani, M., Pucillo, F., Olsson, R. T., Scampicchio, M., & Farris, S. (2018). A bionanocomposite-modified glassy carbon electrode for the determination of 4,4′-methylene diphenyl diamine. *Analytical Methods*, *10*(34), 4122–4128. https://doi.org/10.1039/C8AY01376D

Hu, X., & Miao, M. (2022). Fabrication of Intelligent Packaging Systems Using Nano-Indicators and Sensors. In J. Parameswaranpillai, R. E. Krishnankutty, A. Jayakumar, S. M. Rangappa, & S. Siengchin (Eds.), *Nanotechnology-Enhanced Food Packaging* (pp. 183–211). Wiley. https://doi.org/10.1002/9783527827718.ch8

Jang, N. Y., & Won, K. (2014). New pressure-activated compartmented oxygen indicator for intelligent food packaging. *International Journal of Food Science & Technology*, *49*(2), 650–654. https://doi.org/10.1111/ijfs.12310

Kalpana, S., Priyadarshini, S. R., Maria Leena, M., Moses, J. A., & Anandharamakrishnan, C. (2019). Intelligent packaging: trends and applications in food systems. *Trends in Food Science & Technology*, *93*, 145–157. https://doi.org/10.1016/j.tifs.2019.09.008

Karthika, P., Shanmuganathan, S., Viswanathan, S., & Delerue-Matos, C. (2021). Molecularly imprinted polymer-based electrochemical sensor for the determination of endocrine disruptor bisphenol-A in bovine milk. *Food Chemistry*, *363*, 130287. https://doi.org/10.1016/j.foodchem.2021.130287

Kim, Y. H., Yang, Y. J., Kim, J. S., Choi, D. S., Park, S. H., Jin, S. Y., & Park, J. S. (2018). Non-destructive monitoring of apple ripeness using an aldehyde sensitive colorimetric sensor. *Food Chemistry*, *267*, 149–156. https://doi.org/10.1016/j.foodchem.2018.02.110

Koosha, M., & Hamedi, S. (2019). Intelligent chitosan/PVA nanocomposite films containing black carrot anthocyanin and bentonite nanoclays with improved mechanical, thermal and antibacterial properties. *Progress in Organic Coatings*, *127*, 338–347. https://doi.org/10.1016/j.porgcoat.2018.11.028

Kumar, V., Guleria, P., & Mehta, S. K. (2017). Nanosensors for food quality and safety assessment. *Environmental Chemistry Letters*, *15*(2), 165–177. https://doi.org/10.1007/s10311-017-0616-4

Kuswandi, B., Asih, N. P. N., Pratoko, D. K., Kristiningrum, N., & Moradi, M. (2020). Edible pH sensor based on immobilized red cabbage anthocyanins into bacterial cellulose membrane for intelligent food packaging. *Packaging Technology and Science*, *33*(8), 321–332. https://doi.org/10.1002/pts.2507

Lamanna, L., Rizzi, F., Bhethanabotla, V. R., & De Vittorio, M. (2020). Conformable surface acoustic wave biosensor for E-coli fabricated on PEN plastic film. *Biosensors and Bioelectronics*, *163*, 112164. https://doi.org/10.1016/j.bios.2020.112164

Liu, Y., Qin, Y., Bai, R., Zhang, X., Yuan, L., & Liu, J. (2019). Preparation of pH-sensitive and antioxidant packaging films based on κ-carrageenan and mulberry polyphenolic extract. *International Journal of Biological Macromolecules*, *134*, 993–1001. https://doi.org/10.1016/j.ijbiomac.2019.05.175

Liu, L., Wu, W., Zheng, L., Yu, J., Sun, P., & Shao, P. (2022). Intelligent packaging films incorporated with anthocyanins-loaded ovalbumin-carboxymethyl cellulose nanocomplexes for food freshness monitoring. *Food Chemistry*, *387*, 132908. https://doi.org/10.1016/j.foodchem.2022.132908

Lu, P., Yang, Y., Liu, R., Liu, X., Ma, J., Wu, M., & Wang, S. (2020). Preparation of sugarcane bagasse nanocellulose hydrogel as a colourimetric freshness indicator for intelligent food packaging. *Carbohydrate Polymers*, *249*, 116831. https://doi.org/10.1016/j.carbpol.2020.116831

Luo, X., Zaitoon, A., & Lim, L. T. (2022). A review on colorimetric indicators for monitoring product freshness in intelligent food packaging: indicator dyes, preparation methods, and applications. *Comprehensive Reviews in Food Science and Food Safety*, *21*(3), 2489–2519. September 2021. https://doi.org/10.1111/1541-4337.12942

Media, I. (2011). Intelligent Packaging: with Built-in Freshness Detector. *Interempresas Media, S.L.U.—Grupo Nova Àgora*. Available at: https://www.interempresas.net/Envase/Articulos/47806-Envases-inteligentes-con-un-detector-de-frescor-incorporado.html

Mohebi, E., & Marquez, L. (2015). Intelligent packaging in meat industry: an overview of existing solutions. *Journal of Food Science and Technology*, *52*(7), 3947–3964. https://doi.org/10.1007/s13197-014-1588-z

Müller, P., & Schmid, M. (2019). Intelligent packaging in the food sector: a brief overview. *Foods*, *8*(1), 16. https://doi.org/10.3390/foods8010016

Nešić, A., Cabrera-Barjas, G., Dimitrijević-Branković, S., Davidović, S., Radovanović, N., & Delattre, C. (2019). Prospect of polysaccharide-based materials as advanced food packaging. *Molecules*, *25*(1), 135. https://doi.org/10.3390/molecules25010135

Niponsak, A., Laohakunjit, N., Kerdchoechuen, O., Wongsawadee, P., & Uthairatanakij, A. (2020). Novel ripeness label based on starch/chitosan incorporated with pH dye for indicating eating quality of fresh–cut durian. *Food Control*, *107*, 106785. https://doi.org/10.1016/j.foodcont.2019.106785

Packaging http://Network.com. (1998). COX Technologies' Food Temperature Monitor Offers New Tool for Food Safety. Available at: https://www.packagingnetwork.com/doc/cox-technologies-food-temperature-monitor-off-0001

Park, S., Jeon, Y., Han, T., Kim, S., Gwon, Y., & Kim, J. (2020). Nanoscale manufacturing as an enabling strategy for the design of smart food packaging systems. *Food Packaging and Shelf Life*, *26*, 100570. https://doi.org/10.1016/j.fpsl.2020.100570

Perera, K. Y., Jaiswal, S., & Jaiswal, A. K. (2022). A review on nanomaterials and nanohybrids based bio-nanocomposites for food packaging. *Food Chemistry*, *376*, 131912. https://doi.org/10.1016/j.foodchem.2021.131912

Pirsa, S., & Shamusi, T. (2019). Intelligent and active packaging of chicken thigh meat by conducting nano structure cellulose-polypyrrole-ZnO film. *Materials Science and Engineering: C*, *102*, 798–809. https://doi.org/10.1016/j.msec.2019.02.021

Prietto, L., Mirapalhete, T. C., Pinto, V. Z., Hoffmann, J. F., Vanier, N. L., Lim, L.-T., Guerra Dias, A. R., & da Rosa Zavareze, E. (2017). pH-sensitive films containing anthocyanins extracted from black bean seed coat and red cabbage. *LWT*, *80*, 492–500. https://doi.org/10.1016/j.lwt.2017.03.006

Primožič, M., Knez, Ž., & Leitgeb, M. (2021). (Bio)nanotechnology in food science—food packaging. *Nanomaterials*, *11*(2), 292. https://doi.org/10.3390/nano11020292

Regulation E. C. No (2009). Regulation (EC) No 450/2009 of 29 May 2009 on active and intelligent materials and articles intended to come into contact with food. *Official Journal of European Union*, *135*, 3–11.

Regulation No. (2004). European Parliament and of the Council of 27 October 2004 on Materials and Articles Intended to Come into Contact with Food and Repealing Directives, *Pub. L. No. 1935/2004, No. (2004).*

RIPESENSE. (2022). How Ripe Do You Like It?

Romero, A., Sharp, J. L., Dawson, P. L., Darby, D., & Cooksey, K. (2021). Evaluation of two intelligent packaging prototypes with a pH indicator to determine spoilage of cow milk. *Food Packaging and Shelf Life*, *30*, 100720. https://doi.org/10.1016/j.fpsl.2021.100720

Salarbashi, D., Tafaghodi, M., Bazzaz, B. S. F., Mohammad Aboutorabzade, S., & Fathi, M. (2021). pH-sensitive soluble soybean polysaccharide/SiO_2 incorporated with curcumin for intelligent packaging applications. *Food Science & Nutrition*, *9*(4), 2169–2179. https://doi.org/10.1002/fsn3.2187

Salgado, P. R., Di Giorgio, L., Musso, Y. S., & Mauri, A. N. (2021). Recent developments in smart food packaging focused on biobased and biodegradable polymers. *Frontiers in Sustainable Food Systems*, *5*, 630393. https://doi.org/10.3389/fsufs.2021.630393

Sani, M. A., Tavassoli, M., Hamishehkar, H., & McClements, D. J. (2021). Carbohydrate-based films containing pH-sensitive red barberry anthocyanins: application as biodegradable smart food packaging materials. *Carbohydrate Polymers*, *255*, 117488. https://doi.org/10.1016/j.carbpol.2020.117488

Senapati, M., & Sahu, P. P. (2020). Meat quality assessment using au patch electrode ag-SnO_2/SiO_2/Si MIS capacitive gas sensor at room temperature. *Food Chemistry*, *324*, 126893. https://doi.org/10.1016/j.foodchem.2020.126893

Sharma, C., Dhiman, R., Rokana, N., & Panwar, H. (2017). Nanotechnology: an untapped resource for food packaging. *Frontiers in Microbiology*, *8*, 1735. https://doi.org/10.3389/fmicb.2017.01735

Shavisi, N., & Shahbazi, Y. (2022). Chitosan-gum Arabic nanofiber mats encapsulated with pH-sensitive Rosa damascena anthocyanins for freshness monitoring of chicken fillets. *Food Packaging and Shelf Life*, *32*, 100827. https://doi.org/10.1016/j.fpsl.2022.100827

Smits, E., Schram, J., Nagelkerke, M., Kusters, R., van Heck, G., van Acht, V., Koetse, M., van den Brand, J., & Gerlinck, G. (2012). 4.5.2 development of printed RFID sensor tags for smart food packaging. *Proceedings IMCS*, *2012*, 403–406. https://doi.org/10.5162/IMCS2012/4.5.2

Sobhan, A., Muthukumarappan, K., & Wei, L. (2021). Biosensors and biopolymer-based nanocomposites for smart food packaging: challenges and opportunities. *Food Packaging and Shelf Life*, *30*, 100745. https://doi.org/10.1016/j.fpsl.2021.100745

Sobhan, A., Muthukumarappan, K., Wei, L., Van Den Top, T., & Zhou, R. (2020). Development of an activated carbon-based nanocomposite film with antibacterial property for smart food packaging. *Materials Today Communications*, *23*, 101124. https://doi.org/10.1016/j.mtcomm.2020.101124

Sun, J., Jiang, H., Wu, H., Tong, C., Pang, J., & Wu, C. (2020). Multifunctional bionanocomposite films based on konjac glucomannan/chitosan with nano-ZnO and mulberry anthocyanin extract for active food packaging. *Food Hydrocolloids*, *107*, 105942. https://doi.org/10.1016/j.foodhyd.2020.105942

Taoukis, P., & Tsironi, T. (2016). Smart Packaging for Monitoring and Managing Food and Beverage Shelf Life. In P. Subramaniam (Ed.), *The Stability and Shelf Life of Food* (pp. 141–168). Elsevier. https://doi.org/10.1016/B978-0-08-100435-7.00005-8

Vanderroost, M., Ragaert, P., Devlieghere, F., & De Meulenaer, B. (2014). Intelligent food packaging: the next generation. *Trends in Food Science & Technology*, *39*(1), 47–62. https://doi.org/10.1016/j.tifs.2014.06.009

Vilela, C., Moreirinha, C., Domingues, E. M., Figueiredo, F. M. L., Almeida, A., & Freire, C. S. R. (2019). Antimicrobial and conductive nanocellulose-based films for active and intelligent food packaging. *Nanomaterials*, *9*(7), 980. https://doi.org/10.3390/nano9070980

Wang, S., Zheng, L., Cai, G., Liu, N., Liao, M., Li, Y., Zhang, X., & Lin, J. (2019). A microfluidic biosensor for online and sensitive detection of *Salmonella typhimurium* using fluorescence labeling and smartphone video processing. *Biosensors and Bioelectronics*, *140*, 111333. https://doi.org/10.1016/j.bios.2019.111333

World Health Organization. (2022). Food Safety. Retrieved June 16, 2022, from https://www.who.int/health-topics/food-safety

Wu, C., Li, Y., Sun, J., Lu, Y., Tong, C., Wang, L., Yan, Z., & Pang, J. (2020). Novel konjac glucomannan films with oxidized chitin nanocrystals immobilized red cabbage anthocyanins for intelligent food packaging. *Food Hydrocolloids*, *98*, 105245. https://doi.org/10.1016/j.foodhyd.2019.105245

Xiao-wei, H., Xiao-bo, Z., Ji-yong, S., Zhi-hua, L., & Jie-wen, Z. (2018). Colorimetric sensor arrays based on chemo-responsive dyes for food odor visualization. *Trends in Food Science & Technology*, *81*, 90–107. https://doi.org/10.1016/j.tifs.2018.09.001

Yousefi, H., Su, H.-M., Imani, S. M., Alkhaldi, K., Filipe, M., C. D., & Didar, T. F. (2019). Intelligent food packaging: a review of smart sensing technologies for monitoring food quality. *ACS Sensors*, *4*(4), 808–821. https://doi.org/10.1021/acssensors.9b00440

Zhai, X., Li, Z., Shi, J., Huang, X., Sun, Z., Zhang, D., Zou, X., Sun, Y., Zhang, J., Holmes, M., Gong, Y., Povey, M., & Wang, S. (2019). A colorimetric hydrogen sulfide sensor based on gellan gum-silver nanoparticles bionanocomposite for monitoring of meat spoilage in intelligent packaging. *Food Chemistry*, *290*, 135–143. https://doi.org/10.1016/j.foodchem.2019.03.138

Zhai, X., Shi, J., Zou, X., Wang, S., Jiang, C., Zhang, J., Huang, X., Zhang, W., & Holmes, M. (2017). Novel colorimetric films based on starch/polyvinyl alcohol incorporated with roselle anthocyanins for fish freshness monitoring. *Food Hydrocolloids*, *69*, 308–317. https://doi.org/10.1016/j.foodhyd.2017.02.014

Zhai, X., Wang, X., Zhang, J., Yang, Z., Sun, Y., Li, Z., Huang, X., Holmes, M., Gong, Y., Povey, M., Shi, J., & Zou, X. (2020). Extruded low density polyethylene-curcumin film: a hydrophobic ammonia sensor for intelligent food packaging. *Food Packaging and Shelf Life*, *26*, 100595. https://doi.org/10.1016/j.fpsl.2020.100595

Zhang, X., Wang, H., Niu, N., Chen, Z., Li, S., Liu, S.-X., & Li, J. (2020). Fluorescent poly(vinyl alcohol) films containing chlorogenic acid carbon nanodots for food monitoring. *ACS Applied Nano Materials*, *3*(8), 7611–7620. https://doi.org/10.1021/acsanm.0c01229

10

Biosensors and Nanosensors for Quality Evaluation in Food Packaging

Devika Yadav, Vijayakumar Raja, and C. Anandharamakrishnan

CSIR–National Institute for Interdisciplinary Science and Technology (NIIST), Ministry of Science and Technology–Government of India, Thiruvananthapuram, Kerala - 695019, India

10.1 Introduction

Recently, there have been several cases of infected food products worldwide. In 2012–2015, food-borne illness epidemics caused by *Escherichia coli* O157 : H7, Norovirus, Listeria, Salmonella, and other pathogens in the United States in tuna, chicken, salmon, ground beef, and dairy products created a huge impact (Kim et al., 2018). The quality of food is usually determined by the environment in which the food products are packed and delivered to the consumers. The quality of packed food products is frequently altered by contamination or product degradation. Some fundamental indicators, such as color, odor, and texture, can be used by consumers to determine the quality of products; however, the existence of packaging materials often hinders these decisions (Joo et al., 2020). Integrating food sensors or indicators in the packaging systems is an emerging approach to visualize the shelf life of perishable foods at various stages of the supply chain (Kuswandi et al., 2022). Sensor refers to a device used to sense, trace, or quantify energy or matter by detecting or measuring a physical or chemical property to which the device responds and delivers a continuous output of a signal (Kress-Rogers and Brimelow, 2001). The indicator is much simpler in design and communicates information through a direct, visually apparent change. Sensors or indicators are integrated into conventional packaging materials for the development of smart packaging materials (Mahalik and Nambiar, 2010). Conventional approaches to study quality deterioration, such as measuring the change in weight, appearance, color, and volume, offer various limitations, especially in terms of their accuracy and sensitivity.

Biosensing using bio-recognition elements is a novel approach that results in improving the limit of detection, sensitivity, and robustness. In biosensors, receptors can be antigens, enzymes, hormones, microbes, and nucleic acids, which can be used individually or in combination (Abedi-Firoozjah et al., 2023; Sekhon, 2014). On the other hand, nanosensors are used in food safety and quality, as these are the two most important aspects and have now become the attention of consumer health. Nanosensors use nanomaterials at the

Nanotechnology for Sustainable Food Packaging, First Edition. Edited by C. Anandharamakrishnan, Jeyan A. Moses, and M. Maria Leena.

slightest one dimension lesser or smaller than 100 nm and can be classified as (i) nanotubes and wires (<100 nm in two dimensions), (ii) nanofilms and coatings (<100 in one dimension), (iii) nanoparticles (<100 nm in three dimensions) (Hochella Jr., 2002). Nanosensors have the convenience for the detection of microbial contaminants on food packaging during process, distribution and storage. Integrating the biosensing approach and using nanomaterials (termed bio-nanosensors) can improve the quantification accuracy of microbial contamination, adulteration detection, and the quantification of biochemical reactions taking place inside the packed food systems at trace levels. Therefore, efforts have been made to develop bio-nanosensors for food packaging and quality evaluation, for example, smart packaging materials that can serve as visual support to indicate the quality of food products (Aydınoğlu, 2020).

Integrating the sensors/indicators with packaging systems enables rapid and real-time monitoring of the quality of packed foods. There are different food packaging materials, including indicators or sensors, including CO_2 sensors (Yaparatne et al., 2024), freshness indicators (Kuswandi et al., 2013), time–temperature integrators (TTI) (Wanihsuksombat et al., 2010), pH indicators (Ebrahimi Tirtashi et al., 2019), and so on. Integrating the bio and nanosensors in commercial food packaging systems has challenges, especially in terms of stability. This chapter discusses the applications of biosensors, nanosensors, and bio-nanosensors for quality monitoring in packed foods, challenges associated with integrating bio-nanosensors in food packaging systems and futuristic research requirements.

10.2 General Working Principles of Biosensors, Nanosensors and Bio-Nanosensors

Biosensors are devices that detect and measure biological and chemical constituents (Takhistov, 2009). Nanosensors are sensors with nanoscale dimensions, allowing for highly sensitive detection (Bueno et al., 2017). Bio nanosensors combine the principles of biosensors and nanosensors to enable precise and efficient detection of specific biological molecules or processes at the nanoscale (Perdomo et al., 2021). They have applications in healthcare, environmental monitoring, food industries, food packaging, and many more (Shawon et al., 2020). When it comes to food packaging, bio nanosensors play important roles in ensuring food safety and quality. A biosensor incorporates a biological component, such as enzymes or antibodies, along with a transducer to detect and measure specific substances or biological reactions (Thakur and Ragavan, 2013). Biosensors in food packaging detects specific substances causing spoilage, contamination, or other quality issues (Park et al., 2015b). These sensors are integrated into the packaging material and can provide real-time information about the condition of the food inside. Biosensor consists of a transducer (Figure 10.1), which measures any physical change associated with the reaction including (i) heat output (or absorbed) (calorimetric biosensors), (ii) electrical or electric output (electrochemical biosensors), (iii) redox reaction (amperometric biosensors), (iv) light output or light absorbance between the reactants and products (optical biosensors), or (v) the mass of the reactants or products (Piezoelectric biosensors) (Thakur and Ragavan, 2013). To improve the signal-to-noise ratio in biosensors, a reference baseline signal is often used. This baseline signal is obtained from a similar transducer without any

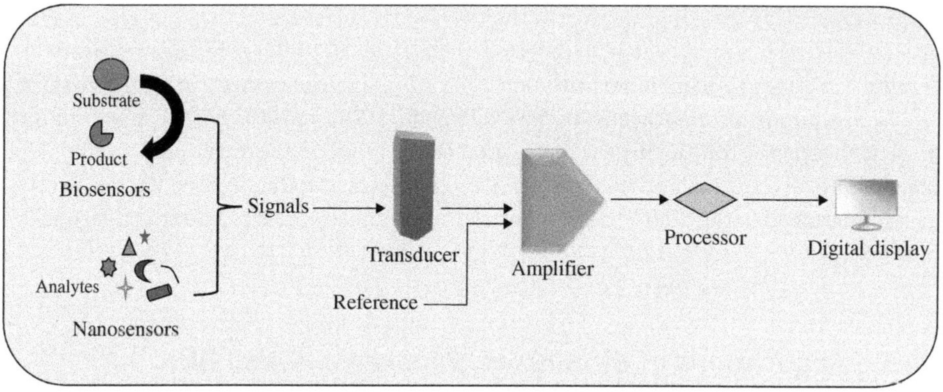

Figure 10.1 Schematic representation of bio/nanosensors component

bio-catalytic membrane, serving as a comparison to the sample signal, and by subtracting the reference signal from the sample signal, the unwanted noise can be minimized, and the difference between the signals becomes more pronounced. This amplified and cleaner output signal provides a more accurate measurement of the analyte concentration. After amplification, the analog signal is typically converted into a digital signal and sent to a microprocessor for further processing. The microprocessor analyses the data, converts it into concentration units, and then comunicates the results to a display device or data storage. This process helps remove noise and convert the raw signal into meaningful concentration values, making it easier to interpret and utilize the biosensor data (Thakur and Ragavan, 2013) (Figure 10.1).

Nanosensors take this a step further by utilizing nanoscale materials or structures to enhance the sensitivity and accuracy of detection. They can detect trace amounts of contaminants or spoilage indicators, helping to ensure that the food is safe for consumption (Patel et al., 2020). The general components of nanosensors are analyte, sensor, transducer, and detector. The working principle of nanosensors involves three main steps: sensing, signal generation, detection and analysis. (i) Sensing: nanosensors are equipped with special materials or molecules that can interact with specific substances present in the food. When these substances come in contact with the nanosensor, they cause a detectable change in their properties, such as electrical conductivity or optical properties (Huang et al., 2021). (ii) Signal generation: the interaction between the substances and the nanosensor generates a signal. This signal can be in the form of an electrical current, a color change, or any other measurable output that indicates the presence or concentration of the targeted substance. (iii) Detection and analysis: the generated signal is then detected and analyzed by a device or system connected to the nanosensor. This device can interpret the signal and provide information about the quality, freshness, or safety of the food. It can also send alerts if any potential issues are detected (Huang et al., 2021).

Generally, nanosensors track the electrical changes in the sensor, and the analyte distributes from the solution to the sensor surface and reacts precisely and efficiently. This alters the physiochemical properties of the transducer surface, which conducts a change in the electrical or optical properties of the transducer surface, and the change converted into an electrical signal which is detected (Denizli et al., 2021). Bio nanosensors are designed to target specific biological molecules or pathogens present in food, providing rapid and accurate information about the food's quality (Sharma et al., 2023). Incorporating these technologies into food packaging, enables to monitor the freshness, quality, and safety of the food throughout its shelf life. This also helps in reducing food waste, preventing food-borne illnesses, and ensuring that consumers have access to safe and high-quality food products.

10.3 Applications of Biosensors, Nanosensors, and Bio-Nanosensors in Food Packaging

10.3.1 Biosensors

Biosensors essentially convert a biological response into a measurable signal, it is used in various fields. A major application is quality control. It can help monitor and assess the quality of food products by detecting indicators like enzyme activity, pH levels, or gas emissions. This real-time information about freshness, spoilage, or contamination helps to ensure that consumers receive high-quality and safe food. Similarly, the second major application is food safety, by detecting harmful substances such as allergens, toxins, or pathogens. They can help to prevent the consumption of contaminated and unsafe products, protecting consumer health. Additionally, biosensors aid in predicting the shelf life of food, allowing for better inventory management and reduced food waste. They can also be used for traceability and authentication, mainly to verify the authenticity and origin of food products. The gas-sensing biosensors are designed to detect the gases emitted by spoilage microorganisms or other contaminants that may be present inside the packed food. For example, if there is a concern about the presence of a gas-producing bacteria in packaged food, CO_2, O_2, ethylene gas sensing biosensors can be employed, and the configuration should be optimized (in terms of detection and quantification limits) to ensure the freshness and quality of the packaged food. The CO_2 sensing biosensor is typically integrated into the food packaging material or placed inside the package, and it consists of a pH-sensitive indicator or an enzyme (carbonic anhydrase) [Meng et al., 2014] or a pH-sensitive dye, (Table 10.1) that reacts with CO_2. As the food undergoes natural respiration or degradation, it produces CO_2, which interacts with the biosensor, and then the biosensor detects the CO_2 concentration and changes color or provides a visual indication on the packaging to show the freshness level (Puligundla et al., 2012). Similarly, O_2 and ethylene sensing biosensors are also integrated into the food packaging material or placed inside the package. It utilizes an oxygen-sensitive indicator (Achatz et al., 2011) or an enzyme (oxygen oxidoreductase or a fluorescent dye) that reacts with oxygen and then biosensor monitors the oxygen levels within the package, which can affect the food's shelf life/quality and it provides a visual indication or changes color to show the remaining oxygen level, ensuring that the package is properly sealed and maintaining an oxygen-controlled environment (Meng et al., 2014). Ethylene sensing biosensor consists of a receptor protein (Janssen et al., 2014) or a conductive polymer (Gal-Oz et al., 2023) that specifically interacts with

Table 10.1 Sensor and biosensors based on metabolites

Metabolites	Food Products	Recognition Element	Indicators	Sensors	References
Glucose and lactic acid	Fermented food, meat	Glucose oxidase Lactate oxidase (LOD) and lactate dehydrogenase (LDH)	Colorimeter based on pH	Electrochemical sensors by redox reaction	Park et al. (2015b), Rathee et al. (2016)
Carbon dioxide	Meat, fermented food	Enzyme carbonic anhydrase	Colorimeter based on pH	Electrochemical sensors by silicon-based polymers	Cammaroto et al. (1998), Abraham (2022)
Oxygen	Meat, vegetables	Enzymes	Optical sensor by fluorescence, colorimeter based on pH-indicator dye	Electrochemical sensor by Zirconia, laser	Nguyen et al. (2019), Park et al. (2015a), Prabha et al. (2021)
Biogenic amines	Fish, meat	Histamine, dopamine, tyramine, tryptamine, spermine, and ethanol-amine	Color-changing pH-sensitive dyes	Electrochemical sensor by enzyme redox reaction	Nami et al. (2024), Olaniyi (2015), Givanoudi et al. (2023)

ethylene gas. Ethylene is a natural plant hormone that can accelerate the ripening and spoilage of fruits and vegetables. The biosensor detects the presence of ethylene gas and provides a visual indication or changes color to alert consumers or food handlers about the ripening process. These biosensors help monitor and maintain optimal gas levels within food packaging, extending the shelf life, preserving freshness, and ensuring the quality of the packaged food (Gal-Oz et al., 2023).

The gas-sensitive elements of biosensors, such as a polymer or a semiconductor, reacts to specific gases. When the food starts to spoil or if there is a contamination issue, the gas-producing bacteria release certain gases as byproducts. Some of the commonly used gas-sensitive elements are (metal oxide materials like tin oxide (SnO_2), titanium dioxide (TiO_2), and zinc oxide (ZnO) due to their high sensitivity to various gases (Fazio et al., 2021; Govardhan and Grace, 2016) including, carbon monoxide (CO), nitrogen dioxide (NO_2), ozone (O_3), methane (CH_4), and volatile organic compounds (VOCs) like benzene and toluene (Olifant, 2023; Rath et al., 2022). These versatilsensors can be tailored to detect specific gases by optimizing the metal oxide layer and sensor design. Metal oxide gas sensors play a vital role in detecting the presence of different gases. Conducting polymers like polyaniline (PANI), polypyrrole (PPy), and polythiophene (PTh) (Zamani et al., 2019) have been used in gas sensors for their ability to change electrical conductivity when exposed to specific gases. On the other hand, carbon nanotubes (CNTs) possess excellent gas-sensing properties due to their high surface area and electrical conductivity. They can detect various gases, including VOCs and toxic gases (Cova et al., 2022; Khatib and Haick, 2022) and metal–organic frameworks (Zhang et al., 2021) are porous materials made up of metal ions or clusters connected by organic ligands. They exhibit high selectivity and sensitivity to specific gases, making them suitable for gas sensing applications causing a measurable change in their properties, such as electrical conductivity or optical properties. This early detection can help prevent the consumption of spoiled or contaminated food, ensuring consumer safety and reducing food waste (Molina-Besch and Olsson, 2022; Tichoniuk et al., 2021).

Enzyme-based biosensors used enzymes such as glucose oxidase, lactate oxidase, urease, lipase, protease, catalase, and amylase. Few of these enzyme sensing systems are also useful for oxygen scavenging in food packaging systems. Oxygen is an important factor in food spoilage; hence, it is practical significance to eradicate oxygen in the package. Use of oxygen scavengers aids in removing the residual oxygen inside the package thus helps to extend their shelf life (Bankar et al., 2009; Dalgaard, 1995; Gómez et al., 1995). Various deoxidizing agents such as iron oxide (iron), oxidative enzymes, ascorbic acid, and light-sensitive dye were also explored in studies (Dai et al., 2001); among all the glucose oxidase is one of the potential oxygen scavengers for enzyme assisted deoxidization. In few studies researchers have directly integrated the deoxidizing material into plastic packaging. Glucose oxidase and alcohol oxidase can be immobilized inside the packaging film to attain deoxidization (Kumar et al., 2019). Glucose oxidase, as an effective oxygen scavenger, catalyzes the reaction of oxygen and glucose, which produce glucuronic acid and has been significantly used to remove residual oxygen in beverages and foods. To improve the stability of the enzyme, several immobilization techniques have been explored. For instance, Ge et al. (2012) developed a novel food packaging material by immobilizing glucose oxidase in Polyvinyl alcohol (PVA)/chitosan/tea extract composite electrospun nanofibrous membrane with an optimum activity retention of 68%. The prepared membrane achieved 73% deoxidization efficiency. In another study, Hanušová et al. (2013) used Ugi reaction for lysozyme and glucose oxidase immobilization with cyclohexyl isocyanide and glutaraldehyde on ionomer and polyamide films partially hydrolyzed by HCl and the results validate that the application of the Ugi reaction is appropriate to immobilize glucose oxidase on commonly used and relatively inert polymer films creating enzymatically active materials.

Several types of deteriorative chemical and biochemical reactions also take place due to fat oxidation, pigment oxidation, and browning reactions. Besides, most of the common food spoilage fungi and bacteria need oxygen for growth (Sandhya, 2010). To overcome this, food industries use either vacuum packaging or in a modified gaseous atmospheric packaging (MAP) to maintain the quality of the food and as well as increase the shelf life by controlling these deteriorative reactions. Quantifying the residual oxygen level in MAP or vacuum-packed foods during storage is an important factor in understanding the degradative processes in food products (Valdés et al., 2009). However, many approaches used for quantification of oxygen are destructive in nature. Furthermore, during processing, packaged foods are tested arbitrarily during a production run, usually one or two in every 300–400 packages (Mills, 2005). The processes is time consuming and costly, and does not ensure that unsampled packages meet safety and quality standards (Smolander et al., 1997). So, the ability to consistently and conveniently monitor the gas composition of a package headspace would also offer a way to evaluate the safety and quality of the enclosed food after departing from the production facility. This suggests that non-invasive leak detection and gas content analysis methods would be highly valuable.(Duncan, 2011).

Food spoilage is caused when gases are produced from microorganism metabolism. These can be detected by numerous sensors, which respond reversibly and quantitatively to the presence of a gaseous analyte by altering specific physical parameters of the sensors and monitored using an external device (Kerry et al., 2006). There are many systems currently available for gas detection like potentiometric carbon dioxide (CO_2) sensors, amperometric oxygen sensors, metal oxide semiconductor field effect transistors, piezoelectric crystal sensors, and organic conducting polymers (Kress-Rogers, 2001). Many factors govern the possibility of marketable success, including selectivity, sensitivity, response and recovery times, and power consumption (Puligundla et al., 2012). Biosensors can be integrated into food packaging in various forms, such as patches, strips, or labels (Dincer et al., 2019; Vasilescu et al., 2020). Thin, adhesive patches that contain biorecognition elements to detect gases can be attached to food packaging. The color or appearance of the patch changes based on the gas levels detected, providing visual indications of freshness. Strip biosensors are similar to patch biosensors, and these are thin strips that can be applied to food packaging (D'Souza et al., 2017).

The strip changes color or displays a change in visual appearance to show the quality or freshness of the food, and the labels are biosensors incorporated into labels that are affixed to food packaging (Khan et al., 2024). The labels may contain indicators that change color based on gas levels or other factors (Turasan and Kokini, 2021). It provides a simple and visual way for consumers to assess the freshness and quality of the food. These forms of biosensors in food packaging allow for easy integration, providing consumers with convenient and reliable indicators of food freshness. However, there are some limitations, such as single-use, limited lifespan, specificity, packaging interference, and sometimes interpretation.

Biosensors can also monitor the moisture levels inside the packaging to prevent the growth of mold or bacteria, that is, microbial biosensors (Su et al., 2011). These biosensors are designed to detect the presence and activity of microorganisms by measuring certain biological markers or by monitoring changes in electrical signals or optical properties (Zahra et al., 2016). For example, a microbial biosensor may use specific enzymes or antibodies that react with the metabolic byproducts produced by mold or bacteria. When these microorganisms are present and actively growing, they release certain molecules or gases that can be detected by the biosensor. The biosensor then generates a signal that indicates the presence and level of microbial activity by continuously monitoring the growth of mold and bacteria (Zahra et al., 2016). These biosensors can provide an initial indication of potential spoilage or contamination in the packaged food, allowing for timely intervention and prevention of food-borne illnesses. For monitoring pH levels in food packaging, pH-sensitive biosensors can be used (Abedi-Firoozjah et al., 2022). These biosensors are designed to detect and measure the acidity or alkalinity of the food product inside the package, such as pH-sensitive dye or indicators (Luo et al., 2023). These dyes change color or fluorescence intensity in response to changes in pH. The biosensor incorporates these pH-sensitive dyes, and when the food comes into contact with the biosensor, the dye reacts to the pH of the food, causing a visible color change or a change in fluorescence intensity.

This color change or fluorescence signal can then be measured and correlated to the pH level of the food, so by using pH biosensors, and food producers can ensure that the pH of their products remains within the desired range for freshness, taste, and safety (Zhang et al., 2022). Bromothymol blue (Li et al., 2022), is commonly used in biosensors, which is a pH indicator that changes color depending on the acidity or alkalinity of the solution. At a lower pH (acidic conditions), bromothymol blue appears yellow, but as the pH increases and becomes more alkaline, the color transitions to green and eventually blue (Luo et al., 2022). This color change can be easily observed and measured, making it useful for pH detection in various applications, including food packaging. Since there are many dyes used as a pH indicator, like neutral blue (Kuswandi et al., 2020), alizarin yellow (Khan et al., 2023), and bromocresol Green (Wen et al., 2023), these pH-sensitive dyes offer different color changes that can be used in biosensors for pH monitoring in food packaging.

10.3.2 Nanosensors

Nanosensors are referred to as any chemical sensory elements used to transport information about nanoparticles to the microscopic world (Bogue, 2008). Nano sensors are the modified version of chemical or biosensor-engaging nanomaterial in an analytical device. However, biological and chemical sensors are precise recognition systems with physical or electrochemical signaling that may be incorporated into food packaging to deliver rapid, accurate "*in situ*" analysis. These devices need to be appropriate for printing technology for low-cost relative to the values of the food product or the mass production, easy to use, simple, reliable, and reproducible in their range of operation, and moderate to the environment as well as food contact safe (Kuswandi et al., 2011). In most of the situations, it is essential to detect multiple analytes or pathogens parallelly, which would not be probable with conventional sensors. Additionally, nanoscale sensors are more prioritized in current research on smart food packaging (Sobhan et al., 2021) and also due to the further advantages given by materials in nano dimensions such as sensitivity and selectivity. Hence, nanoparticles deliver a large surface area, and the diversity of biomolecules could be immobilized on these structures, and it also increases the number of sites able to interact with target analytes (Wu and Xu, 2009). By definition, nanosensors are nanotechnology-enabled sensors but are characterized by one of the following attributes: either the size of the sensor or the spatial interaction distance between the object given in nanometers and the sensors, or its selectivity is on the nanoscale (Khanna, 2008), whereas, three major reasons why nanoscale technologies signify the future generation of biosensors; (Abell et al., 2009)

1) Improved assay specificity, sensitivity, and limits of detection (LOD)
2) Increased sample throughput and
3) Reduced assay complexity and cost (when the detection is performed "*in situ*")

Biosensors can be integrated into food packaging using various nanomaterials, called nanosensors, such as quantum dots (Raul et al., 2022), (CNTs) (Ezati et al., 2022a; Riahi et al., 2022), or nanoparticles, nanofilms and so on. For instance, quantum dots can be used to detect the presence of pathogens or toxins in food by emitting a specific color when they come into contact with them. Moreover, the quantum dots can also be incorporated into food packaging to provide various functionalities. Quantum dots are tiny semiconductor particles that emit light of different colors when stimulated by an energy source (Raul et al., 2022). By incorporating specific quantum dots that react to certain gases or compounds, it can change color or emit light when the food inside the packaging undergoes spoilage or contamination. This visual indication can help consumers easily identify if the food is still safe to consume. Incorporating quantum dots that release antimicrobial agents or antioxidants can help extend the shelf life of the food by inhibiting the growth of bacteria or reducing oxidative reactions that lead to spoilage. This can help maintain the freshness and quality of the packaged food for a longer period. Additionally, quantum dots (Ezati et al., 2022b) can be used in packaging labels or QR codes to provide interactive package with information relating to the food quality. Quantum dots that emit different colors or patterns when exposed to specific light wavelengths can also be used to create unique packaging designs or interactive elements that enhance the consumer's experience.

CNTs are cylindrical structures made of carbon atoms (Zhao et al., 2023). It has unique electrical properties that can be utilized in nanosensors. By modifying the surface of CNTs with specific biomolecules, it can selectively bind to target analytes in food, causing changes in electrical conductivity. These changes can be measured to determine the presence or concentration of the analyte. CNTs can be incorporated into food packaging in different ways to enhance their properties. One common method is integrating CNTs into the packaging material itself (Bajpai et al., 2018). This can be achieved by dispersing CNTs within the polymer matrix during the manufacturing process. The resulting nanocomposite material exhibits improved mechanical strength, barrier properties, and even antimicrobial properties. The CNTs act as reinforcement, making the packaging material stronger and more resistant to damage, and can also enhance the barrier properties of the packaging, preventing the entry of oxygen or moisture, which helps preserve the freshness and quality of the food for a longer duration. Additionally, CNTs have been explored for their antimicrobial properties (Bajpai et al., 2018). It can inhibit the growth of bacteria and other microorganisms, reducing the risk of contamination and extending the shelf life of the packaged food, by incorporating CNTs into food packaging can improve its durability, enhance food preservation, and ensure safer consumption. Similarly, nanoparticles, such as gold (Paidari and Ibrahim, 2021) or silver (Carbone et al., 2016), can be functionalized with specific biomolecules to create nanosensors and incorporated into food packaging materials to provide various benefits. Similar to quantum dots, nanoparticles are tiny particles with unique properties that can enhance the functionality of packaging materials. The common application of nanoparticles in food packaging is as antimicrobial agents (Kumar et al., 2021).

Certain nanoparticles, such as silver, have antimicrobial properties that can inhibit the growth of bacteria and fungi on the surface of the packaging (Nakazato et al., 2017). This can help extend the shelf life of the food by reducing the risk of contamination and spoilage (Paul et al., 2017). Nanoparticles can also be used to enhance the barrier properties of food packaging. By incorporating nanoparticles with specific characteristics, such as clay nanoparticles (Deshmukh et al., 2023) or graphene oxide nanoparticles (Silva-Leyton et al., 2019), the packaging material can become more resistant to moisture, oxygen, and other gases (Deshmukh et al., 2023). This helps to maintain the freshness and quality of the food by preventing the entry of external elements that could lead to spoilage (Bratovčić et al., 2015). Furthermore, nanoparticles can be used to create intelligent or active packaging (Ghosh and Das, 2024). By incorporating nanoparticles that respond to specific stimuli, such as temperature or pH changes, the packaging can provide visual indicators or release active compounds to signal changes in the food's condition (Nakazato et al., 2017). This can help consumers easily determine if the food is still safe to consume. Overall, incorporating nanoparticles into food packaging offers exciting possibilities for improving food safety, extending shelf life, and enhancing the overall consumer experience.

Nanosensors are designed for controlling the internal as well as external conditions of food products, as well as they are applicable/classified for the detection of atmospheric impacts (external conditions) and microorganisms and chemical detection inside the food package (internal condition) (Ileš et al., 2011). For external conditions, decreasing dimensions, a good improvement in the development of computer technology, permits the possible application of food sensors, it is making to monitor all activities of atmospheric effects on the food product, from the manufacturing unit to the final customer. For instance, if macro and nanosensors for moisture, temperature, or other measurement devices are added to freshly picked fruit, then at the time of purchasing fruits, salespersons would be able to check and read data to check the freshness of any food product and also conditions of transportation or storage conditions (Ivanov et al., 2007), according to this data, they could decide whether to take or refuse the goods. These types of sensors are in the phase of research and development (Mao et al., 2006). For the transmission of the data, radio frequency identification sensors (RFID), as tags are used, which are very flexible, small, however, still greater than 1,000 nm. On the other hand, in internal conditions, there are two types of nanosensors in food, and they can be recognized in the nanoscale mainly depending on the type of detector used for detection: nanosensors and nano-biosensors. However, the chemical and biological sensors are fabricated similarly by putting chemical and bacterial detectors (reagents) on the probe in reaction with the targeted chemical element or bacteria, creating changes in the nanosensor's output (Jafarizadeh-Malmiri et al., 2019; Madhusudan et al., 2018; Pizzoni et al., 2014).

10.3.2.1 Bio-Nanosensors for Food Packaging

Sensors using biorecognition elements (in nanoscale dimension) are called biosensors. It enhances the sensing abilities for food packaging applications. Therefore, bio-nanosensors can be practical as coating the food packaging to easily detect conceivable leaking, higher moisture, and raised oxygen content, which can be the reason for microbial growth and spoilage. The quality and safety of the food can be ensured through the packaging of food (Caon et al., 2017). However, food packaging increases the shelf life of food products by ignoring unfavorable factors like oxygen, moisture, and pathogens. Novel bio-nanosensors have been advanced to utilize the concept of smart/intelligent bio-nanosensors and the various applications of bio-nanosensors in food packaging are shown in Table 10.2. The development of novel nanomaterials can be a possible approach to improving food quality and safety, and measuring environmental conditions (He et al., 2019). Newly, the demand for nanoparticle-based materials for diverse applications, e.g., the food industry, has increased. Many of these materials cover crucial elements, few of them nontoxic (Roselli et al., 2003), that can be stable at high temperatures and pressures (Mohamadian et al., 2019; Samadi et al., 2018).

The improved properties of nanomaterials have permitted and encouraged the development of technologies that extend the shelf life and freshness of products, the design of

Table 10.2 Application of bio-nanosensors in food packaging

Application	Nanomaterials	Bioreceptor	Analyte	Food Packaging	References
Detection of toxins	Tri-layer oxide (SiO_2 10 nm/Si_3O_4 10 nm/SiO_2 10 nm)	Monoclonal antibodies for aflatoxin-B_1, zearalenone and HT-2	Mycotoxins (aflatoxin-B_1 and zearalenone)	Bottles and jars, paperboard packaging, and caps and closures	Mak et al. (2010)
	Nanopipettes from quartz capillaries	Poly L-lysine, polyclonal antibody HPV_{16} E_6 ad monoclonal antibody for HT-2	HT-2	—	Actis et al. (2010)
	Colloid gold nanoparticles	Polyclonal antibody for botulinum neurotoxin type B and polyclonal antibody IgG, antibody PbTx Mab and polyclonal antibody IgG	Botulinum neurotoxin type B, brevetoxins (PbTx-1, PbTx-2, PbTx-3, PbTx-9)	Active packaging, sensor labels, antimicrobial coating	Chiao et al. (2004), Zhou et al. (2009)
	Carbon nanotubes	Antibody of microcystin-LR, bovine serum albumin, polyclonal antipalytoxin antibodies	Microcystin-LR, palytoxin	Active and intelligent packaging materials	Wang et al. (2009)
	Gold nanoparticles	Cysteamine, monoclonal antibody of aflatoxin B_1	Aflatoxin B_1	Smart packaging	Sharma et al. (2010)
	Carbon dots	Aflatoxin B_1 aptamer HS-AAA AAA GTT GGG CAC GTG TTG TCT CTC TGT GTC TCG TGC CCT TCG CTA GGC CCA CA	Aflatoxin B_1	Active packaging	Wang et al. (2016)
Detection of microbes	Single-walled carbon nanotubes	Polyclonal antibody for *Salmonella enterica*	*S. enterica* subsp. *enterica* serotype Infantis	Active packaging	Villamizar et al. (2008)
	Au nanoparticles	*Escherichia coli* O157 : H7-specific antibody, *E. coli* O157 : H7 intact cells and *E. coli* O157 : H7-specific antibody conjugated with horseradish peroxidase (HRP)	*E. coli O157 : H7*	Incorporation of AuNPs and grapheme oxide separately on PVA (polyvinyl alcohol) composite films	Lin et al. (2008)
	Carbon dots	Amino-modified aptamers of *Salmonella typhimurium*	*S. typhimurium*	Biopolymer-based active packaging	Wang et al. (2015)
	Graphene nanoplatelets	*E. coli* O157 : H7-specific antibody	*E. coli* O157 : H7	Flexible, stretchable electronic packaging materials	Pandey et al. (2017)
	Gold magnetic (Fe_3O_4) bifunctional nanobeads	Anti-*Salmonella choleraesuis* monoclonal antibodies (${}_{11}D_8$—D_4 as the detection antibody, ${}_{5}F_{11}$—B_{11} as the capture antibody)	*S. choleraesuis*	—	Xia et al. (2016)

tools and methods for rapid in situ analysis, and the production of environmentally safe and degradable food packaging. Radio frequency identification (RFID) technology takes this a step further to communicate the information regarding the quality of the packages and food can be traced/tracked from the manufacturer's website which uses the hardware with minimum dimensions (Caon et al., 2017). The incorporation of bio-nanosensors into food packaging, nanomaterials can be used as a key component. These nanomaterials can be integrated into the packaging material or applied as a coating on the surface. One way to incorporate bio-nanosensors is by blending nanomaterials, like CNTs or gold nanoparticles, into the packaging material itself (Saeed et al., 2022). This allows for the direct integration of the sensors, enabling real-time monitoring of food quality. Another approach is to apply a nanomaterial-based coating onto the packaging surface (Ashfaq et al., 2022).

This coating can contain bio-nanosensors that react to specific substances, such as gases released by spoilage or harmful bacteria. The sensors in the coating can then provide visual or electronic indicators of the food's freshness or safety. By incorporating bio-nanosensors with nanomaterials into food packaging, we can create smart packaging that actively monitors and protects the quality and safety of the food. Assessment of food packaging technologies certifies the safety of food, vigorous monitoring, and enhanced shelf life to fulfill the demands of consumers (Tumwesigye et al., 2016). Nevertheless, few nanomaterials protect food from moisture and spoilage. Moreover, titanium dioxide is a nontoxic and safer material used in food packaging (Dudefoi et al., 2018). There is another cost-effective food packaging with the least harmful effect on the environment is cellulose nanofibers (Abdal-Hay et al., 2016). Ensuring the food quality of the next generation involves food packaging because it detects, senses, and records changes in food products (Vanderroost et al., 2014), whereas smart packaging includes monitoring of food conditions. The changes arise due to detection through external and internal parameters along with customer feedback. Currently, present intelligent packaging involves the use of labels, dots, tags, and inks that work to ensure the safety and quality of the food (Ghaani et al., 2016). Labels function to monitor food conditions and alert consumers about freshness (Otles and Sahyar, 2016). However, the cost is the major challenge in developing robotic technology nowadays, and as related to food packaging it is around 50% of the whole cost compared to the fixed limit of 10% total cost.

The basic target of food packaging integration of nanomaterials into polymer packaging material is to raise the physical and mechanical properties of the packaging, such as resistance to temperature, gas barrier properties, mechanical strength, and flexibility. Currently, numerous nanocomposites or nanoparticles incorporated in polymers and nanostructures are manufactured for several oil and beverage industries. The composition of nanoparticles like clay nanoparticle composites is up to 5% (w/w), and it can help to improve the barrier properties of packaging material, like a decrease in oxygen and carbon dioxide permeation was up to 80–90% (Brody, 2007). There are studies (Kim and Cha, 2014) that reported excess clay loading in ethylene-vinyl alcohol (EVOH) copolymer-based nanocomposite films can lead to the decrease of tensile properties and optical transparency because of the formation of clay agglomerates. Nanomaterials appropriate for food packaging have more benefits than conventional packaging materials. The most prevalent nanotechnology method for food packaging properties is nano-coating. For providing a mass transfer barrier, there are different forms of food coating. Moreover, the edible food coating is appropriate for direct contact with the food, although nonedible coating can be used as the protection of the container with or without coming in contact with food (Joye et al., 2015).

The application of edible nano-coatings (~5 nm thin coatings) can be used in the meat processing industry, agriculture industry (for protection of fruit and vegetable products), cheese and bakery industry, etc., to provide flavor, color, enzymes, antioxidants, and antibrowning compounds to the products (Azeredo et al., 2009; Nile et al., 2020). However, nanomaterials such as nanoclays are also significantly studied and used for food packaging because of their thermal, mechanical, and barrier properties and cheap cost (Gabr et al., 2015). On the other hand, nanolaminates use various adsorbing substances like natural polyelectrolytes (proteins, polysaccharides), lipids (phospholipids and surfactants), and colloidal particles (micelles, vesicles, droplets) to form a different layer in nanolaminates for food packaging (Dasgupta et al., 2016). However, edible nano-coating are very easy to apply by immersion, rubbing, or spraying (Zambrano-Zaragoza et al., 2018) because they are mainly made up of eco-friendly, edible materials that do not harm the environment (Yousuf et al., 2018). Currently, these nano-coating films show nanostructures with natural ingredients with antioxidant and antimicrobial activity that can be integrated to increase their significant effects on fresh produce (Wu et al., 2012).

Nano-laminate films are made up of two or more layers, including a nanolayer, and the interlinkage might be physical or chemical, but layer by layer is the most common technique used for synthesis. This method enables surface lamination along multiple nanolayers as interfacial films based on different nanomaterials (Decher and Schlenoff, 2006). Different active compounds (e.g., antioxidants, antimicrobials, enzymes, antibrowning agents, odor, flavors, etc.) are integrated into the films (Galus et al., 2020) to increase the quality and extend the shelf life of packaged food products (Salgado et al., 2015) and with different adsorbing compounds like biobased or natural polyelectrolytes (polysaccharides and proteins), colloidal particles (vesicles, micelles, and droplets), and charged lipids (phospholipids, surfactants) could be used to provide and improve different layers properties (Kuswandi and Moradi, 2019).

Polymer/clay nanocomposites is significant in food packaging due to the increase in the barrier and mechanical properties of these packaging owing to the synergism between the matrix polymer and nano clay, as a small amount of nanofiller is integrated inside the polymer matrix (Decher and Schlenoff, 2006). Food packaging materials based on clay nanocomposite deliver improved shelf life, heat resistant, lightweight, and shatterproof. Currently, research is focused on using bio-composites strengthened with natural fibers, and the advantage depends on the behavior and properties of natural fibers as supporting fillers (Li et al., 2020). The most important characteristic of natural fibers is their positive impact on the environment (Rajak et al., 2019), but the poor interface quality between the fiber and matrix polymer leads to poor stress-transfer efficiencies and water absorption properties of natural fibers, thus represent a negative aspect for the possible industrial application of these materials in the food industry (Rajak et al., 2019).

10.4 Challenges and Future Trends

Biosensors, nanosensors, and bio-nanosensors have some limitations when it comes to food packaging because these sensors may not accurately detect all types of contaminants or spoilage indicators in the food. For example, a biosensor (enzyme-based biosensors such as glucose oxidase or urease, etc.) used to detect bacterial contamination in meat may not be sensitive enough to detect very low levels of bacteria (Kilic et al., 2023). However, researchers are working on improving the sensitivity and specificity of these sensors by developing new detection methods, using advanced materials, and exploring ways to enhance the sensors' ability to detect a wide range of contaminants and spoilage indicators, ensuring food safety. Another challenge is the cost and scalability of these sensors for large-scale food packaging applications. Currently, the production process for these sensors may be expensive and not easily scalable. However, scientists can invest in research and development to find cost-effective manufacturing techniques and explore innovative approaches to produce these sensors in larger quantities, making them more affordable and suitable for widespread use in food packaging. Additionally, there are concerns about the stability and compatibility of these sensors with different food packaging materials.

Some sensors may degrade over time or interact with the packaging, affecting their performance. To overcome this challenge, researchers can work and develop new sensor designs and materials that are more stable and compatible with various packaging materials. Protective coatings and encapsulation techniques can also be explored to ensure the longevity and reliability of these sensors in food packaging. The future of biosensors and nano biosensors is poised to be transformative, offering unique opportunities to revolutionize industries and enhance the quality of packed food. As technology continues to change, several trends are emerging that highlight the feasibility and economic potential of these advanced sensing devices.

The feasibility of bio-nanosensors majorly lies in advancements in biocompatibility, miniaturization potential, and wireless communication feasibilities, which are the key elements to make them suitable to seamlessly integrate into everyday life (Caon et al., 2017). In food packaging, nano biosensors are set to revolutionize food quality and food safety. These sensors can monitor food quality, detect freshness and spoilage, and identify pathogens in real time. By providing consumers with accurate data, nano biosensors enable the precise application of resources, minimizing waste, and maximizing yields. The feasibility of these applications lies in advancements in nanotechnology, DNA-based sensors, and remote data collection. The economic benefits include increased agricultural productivity, reduced environmental impact, and improved consumer trust in food products.

10.5 Conclusion

Biosensors, nanosensors, and bio-nanosensors have emerged as highly promising technologies for enhancing food packaging. These sensors offer real-time monitoring and quality control capabilities, enabling improved safety and freshness of perishable food products. Biosensors, with their biological recognition elements, provide high selectivity and sensitivity for detecting specific analytes, such as pathogens or spoilage markers. Nanosensors, on the other hand, leverage nanotechnology to detect and monitor parameters at the nanoscale, offering enhanced sensitivity, and fast response times. Integrating biological and nanoscale components in bio-nanosensors combines both advantages, resulting in increased sensitivity, improved selectivity, and enhanced durability. The application of these sensors in food packaging has the potential to revolutionize the industry by enabling improved traceability, reduced waste, and enhanced consumer confidence. Further research and development in this field will continue to drive innovation and advancements in food packaging technology, ensuring the safety and quality of food products throughout the supply chain.

References

Abdal-Hay, A., Hussein, K. H., Casettari, L., Khalil, K. A., & Hamdy, A. S. (2016). Fabrication of novel high performance ductile poly (lactic acid) nanofiber scaffold coated with poly (vinyl alcohol) for tissue engineering applications. *Materials Science Engineering, 60*, 143–150.

Abedi-Firoozjah, R., Salim, S. A., Hasanvand, S., Assadpour, E., Azizi-Lalabadi, M., Prieto, M. A., & Jafari, S. M. (2023). Application of smart packaging for seafood: a comprehensive review. *Comprehensive Reviews in Food Science and Food Safety, 22*(2), 1438–1461. https://doi.org/10.1111/1541-4337.13117

Abedi-Firoozjah, R., Yousefi, S., Heydari, M., Seyedfatehi, F., Jafarzadeh, S., Mohammadi, R., Rouhi, M., & Garavand, F. (2022). Application of red cabbage anthocyanins as pH-sensitive pigments in smart food packaging and sensors. *Polymers, 14*(8), 1629. https://doi.org/
10.3390/polym14081629

Abell, J., Driskell, J., Dluhy, R., Tripp, R., & Zhao, Y.-P. (2009). Fabrication and characterization of a multiwell array SERS chip with biological applications. *Biosensors Bioelectronics, 24*(12), 3663–3670. https://doi.org/10.1016/j.bios.2009.05.039

Abraham, J. (2022). Future of Food Packaging: Intelligent Packaging. In T. B. Annu, & S. Ahmed (Eds.), *Nanotechnology in Intelligent Food Packaging* (pp. 383–417). Wiley. https://doi.org/10.1002/9781119819011.ch16

Achatz, D. E., Meier, R. J., Fischer, L. H., & Wolfbeis, O. S. (2011). Luminescent sensing of oxygen using a quenchable probe and upconverting nanoparticles. *Angewandte Chemie International Edition, 50*(1), 260–263. https://doi.org/10.1002/anie.201004902

Actis, P., Jejelowo, O., & Pourmand, N. (2010). Ultrasensitive mycotoxin detection by STING sensors. *Biosensors and Bioelectronics, 26*(2), 333–337. https://doi.org/10.1016/j.bios.2010.08.016

Ashfaq, A., Khursheed, N., Fatima, S., Anjum, Z., & Younis, K. (2022). Application of nanotechnology in food packaging: pros and cons. *Journal of Agriculture Food Research, 7*, 100270. https://doi.org/10.1016/j.jafr.2022.100270

Aydınoğlu, D. (2020). Active food packaging technology as an application in the food industry. *Academic Studies in Engineering Sciences, 215*, 16–21.

Azeredo, H. M., Mattoso, L. H. C., Wood, D., Williams, T. G., Avena-Bustillos, R. J., & McHugh, T. H. (2009). Nanocomposite edible films from mango puree reinforced with cellulose nanofibers. *Journal of Food Science*, *74*(5), N31–N35. https://doi.org/10.1111/j.1750-3841.2009.01186.x

Bajpai, V. K., Kamle, M., Shukla, S., Mahato, D. K., Chandra, P., Hwang, S. K., Kumar, P., Huh, Y. S., & Han, Y.-K. (2018). Prospects of using nanotechnology for food preservation, safety, and security. *Journal of food drug analysis*, *26*(4), 1201–1214. https://doi.org/10.1016/j.jfda.2018.06.011

Bankar, S. B., Bule, M. V., Singhal, R. S., & Ananthanarayan, L. (2009). Glucose oxidase—an overview. *Biotechnology Advances*, *27*(4), 489–501. https://doi.org/10.1016/j.biotechadv.2009.04.003

Bogue, R. (2008). Nanosensors: a review of recent progress. *Sensor Review*, *28*(1), 12–17. https://doi.org/10.1108/02602280810849965

Bratovčić, A., Odobašić, A., Ćatić, S., & Šestan, I. (2015). Application of polymer nanocomposite materials in food packaging. *Croatian Journal of Food Science*, *7*(2), 86–94. https://doi.org/10.17508/CJFST.2015.7.2.06

Brody, A. L. X. (2007). Case studies on nanotechnologies for food packaging. *Food technology*, *61*(7).

Bueno, C. C., Garcia, P. S., Steffens, C., Deda, D. K., & de Lima Leite, F. (2017). 5—Nanosensors. In A. L. Da Róz, M. Ferreira, F. de Lima Leite, & O. N. Oliveira (Eds.), *Nanoscience and its Applications* (pp. 121–153). William Andrew Publishing. https://doi.org/10.1016/B978-0-323-49780-0.00005-3

Cammaroto, C., Diliberto, L., Ferralis, M., Manca, R., Sanna, A., & Giordano, M. (1998). Use of carbonic anhydrase in electrochemical biosensors for dissolved CO2. *Sensors and Actuators B: Chemical*, *48*(1), 439–447. https://doi.org/10.1016/S0925-4005(98)00082-3

Caon, T., Martelli, S., & Fakhouri, F. (2017). New Trends in the Food Industry: Application of Nanosensors in Food Packaging. In A. M. Grumezescu (Ed.), *Nanobiosensors* (pp. 773–804). Academic Press. https://doi.org/10.1016/B978-0-12-804301-1.00018-7

Carbone, M., Donia, D. T., Sabbatella, G., & Antiochia, R. (2016). Silver nanoparticles in polymeric matrices for fresh food packaging. *Journal of King Saud University-Science*, *28*(4), 273–279. https://doi.org/10.1016/j.jksus.2016.05.004

Chiao, D.-J., Shyu, R.-H., Hu, C.-S., Chiang, H.-Y., & Tang, S.-S. (2004). Colloidal gold-based immunochromatographic assay for detection of botulinum neurotoxin type B. *Journal of Chromatography B*, *809*(1), 37–41. https://doi.org/10.1016/j.jchromb.2004.05.033

Cova, C. M., Rincón, E., Espinosa, E., Serrano, L., & Zuliani, A. (2022). Paving the way for a green transition in the design of sensors and biosensors for the detection of volatile organic compounds (VOCs). *Biosensors*, *12*(2), 51. https://www.mdpi.com/2079-6374/12/2/51#

Dai, R., Sun, C., & Wu, G. (2001). Food active packagings and their application. *Storage Process*, *1*, 4–7.

Dalgaard, P. (1995). Modelling of microbial activity and prediction of shelf life for packed fresh fish. *International Journal of Food Microbiology*, *26*(3), 305–317. https://doi.org/10.1016/0168-1605(94)00136-T

Dasgupta, N., Ranjan, S., Patra, D., Srivastava, P., Kumar, A., & Ramalingam, C. (2016). Bovine serum albumin interacts with silver nanoparticles with a "side-on" or "end on" conformation. *Chemico-Biological Interactions*, *253*, 100–111. https://doi.org/10.1016/j.cbi.2016.05.018

Decher, G., & Schlenoff, J. B. (2006). Multilayer Thin Films: Sequential Assembly of Nanocomposite Materials. John Wiley & Sons.

Denizli, A., Nguyen, T. A., Rajendran, S., Yasin, G., & Nadda, A. K. (2021). Nanosensors for Smart Agriculture. Elsevier.

Deshmukh, R. K., Hakim, L., Akhila, K., Ramakanth, D., & Gaikwad, K. K. (2023). Nano clays and its composites for food packaging applications. *International Nano Letters4*, *13*(2), 131–153. https://doi.org/10.1007/s40089-022-00388-8

Dincer, C., Bruch, R., Costa-Rama, E., Fernández-Abedul, M. T., Merkoçi, A., Manz, A., Urban, G. A., & Güder, F. (2019). Disposable sensors in diagnostics, food, and environmental monitoring. *Advanced Materials*, *31*(30), 1806739. https://doi.org/10.1002/adma.201806739

D'Souza, A. A., Kumari, D., & Banerjee, R. (2017). Nanocomposite Biosensors for Point-of-Care—Evaluation of Food Quality and Safety. In *Nanobiosensors* (pp. 629–676). Elsevier. https://doi.org/10.1016/B978-0-12-804301-1.00015-1

Dudefoi, W., Villares, A., Peyron, S., Moreau, C., Ropers, M.-H., Gontard, N., & Cathala, B. (2018). Nanoscience and nanotechnologies for biobased materials, packaging and food applications: new opportunities and concerns. *Innovative food science emerging technologies*, *46*, 107–121. https://doi.org/10.1016/j.ifset.2017.09.007

Duncan, T. V. (2011). Applications of nanotechnology in food packaging and food safety: barrier materials, antimicrobials and sensors. *Journal of Colloid and Interface Science*, *363*(1), 1–24. https://doi.org/10.1016/j.jcis.2011.07.017

Ebrahimi Tirtashi, F., Moradi, M., Tajik, H., Forough, M., Ezati, P., & Kuswandi, B. (2019). Cellulose/chitosan pH-responsive indicator incorporated with carrot anthocyanins for intelligent food packaging. *International Journal of Biological Macromolecules*, *136*, 920–926. https://doi.org/10.1016/j.ijbiomac.2019.06.148

Ezati, P., Priyadarshi, R., & Rhim, J.-W. (2022a). Prospects of sustainable and renewable source-based carbon quantum dots for food packaging applications. *Sustainable Materials Technologies*, *33*, e00494. https://doi.org/10.1016/j.susmat.2022.e00494

Ezati, P., Rhim, J.-W., Molaei, R., & Rezaei, Z. (2022b). Carbon quantum dots-based antifungal coating film for active packaging application of avocado. *Food Packaging Shelf Life*, *33*, 100878. https://doi.org/10.1016/j.fpsl.2022.100878

Fazio, E., Spadaro, S., Corsaro, C., Neri, G., Leonardi, S. G., Neri, F., Lavanya, N., Sekar, C., Donato, N., & Neri, G. (2021). Metal-oxide based nanomaterials: synthesis, characterization and their applications in electrical and electrochemical sensors. *Sensors*, *21*(7), 2494. https://doi.org/10.3390/s21072494

Gabr, M. H., Okumura, W., Ueda, H., Kuriyama, W., Uzawa, K., & Kimpara, I. (2015). Mechanical and thermal properties of carbon fiber/polypropylene composite filled with nano-clay. *Composites Part B: Engineering*, *69*, 94–100. https://doi.org/10.1016/j.compositesb.2014.09.033

Gal-Oz, R., Gandhi, S., Ogungbile, A., Roy, D., Ghosh, M., & Vernick, S. (2023). Biocomposite-based electrochemical chip for ethylene detection. *Sensors Actuators B: Chemical*, *397*, 134652. https://doi.org/10.1016/j.snb.2023.134652

Galus, S., Arik Kibar, E. A., Gniewosz, M., & Kraśniewska, K. (2020). Novel materials in the preparation of edible films and coatings—a review. *Coatings*, *10*(7), 674. https://doi.org/10.3390/coatings10070674

Ge, L., Zhao, Y.-S., Mo, T., Li, J.-R., & Li, P. (2012). Immobilization of glucose oxidase in electrospun nanofibrous membranes for food preservation. *Food Control*, *26*(1), 188–193. https://doi.org/10.1016/j.foodcont.2012.01.022

Ghaani, M., Cozzolino, C. A., Castelli, G., & Farris, S. (2016). An overview of the intelligent packaging technologies in the food sector. *Trends in Food Science Technology*, *51*, 1–11. https://doi.org/10.1016/j.tifs.2016.02.008

Ghosh, S. K., & Das, N. C. (2024). Recent advances in nanoclay-and graphene-based thermoplastic nanocomposites for packaging applications. *Packaging Technology Science*. https://doi.org/10.1002/pts.2804

Givanoudi, S., Heyndrickx, M., Depuydt, T., Khorshid, M., Robbens, J., & Wagner, P. (2023). A review on bio- and chemosensors for the detection of biogenic amines in food safety applications: the status in 2022. *Sensors*, *23*(2).

Gómez, E., Martínez, A., & Laencina, J. (1995). Prevention of oxidative browning during wine storage. *Food Research International*, *28*(3), 213–217. https://doi.org/10.1016/0963-9969(95)93529-4

Govardhan, K., & Grace, A. N. (2016). Metal/metal oxide doped semiconductor based metal oxide gas sensors—a review. *Sensor letters*, *14*(8), 741–750. https://doi.org/10.1166/sl.2016.3710

Hanušová, K., Vápenka, L., Dobiáš, J., & Mišková, L. (2013). Development of antimicrobial packaging materials with immobilized glucose oxidase and lysozyme. *Open Chemistry*, *11*(7), 1066–1078. https://doi.org/10.2478/s11532-013-0241-4 (Open Chemistry)

He, X., Deng, H., & Hwang, H.-M. (2019). The current application of nanotechnology in food and agriculture. *Journal of Food and Drug Analysis*, *27*(1), 1–21. https://doi.org/10.1016/j.jfda.2018.12.002

Hochella, M. F., Jr. (2002). Nanoscience and technology: the next revolution in the earth sciences. *Earth Planetary Science Letters*, *203*(2), 593–605. https://doi.org/10.1016/S0012-821X(02)00818-X

Huang, X., Zhu, Y., & Kianfar, E. (2021). Nano biosensors: properties, applications, and electrochemical techniques. *Journal of Materials Research and Technology*, *12*, 1649–1672. https://doi.org/10.1016/j.jmrt.2021.03.048

Ileš, D., Martinović, G., & Kozak, D. (2011). Review of potential use, benefits, and risks of nanosensors and nanotechnologies in food. *Strojarstvo: časopis za teoriju i praksu u strojarstvu*, *53*(2), 127–136.

Ivanov, P., Gràcia, I., Sabatè, N., Fonseca, L., Figueras, E., Cané, C., Balanco, F., Vergara, A., Vilanova, X., & Correig, X. (2007). Spiral μ-pre Concentrator for Gas Sensor Detection in the ppb Range. In *Sensors* (pp. 151–154). IEEE.

Jafarizadeh-Malmiri, H., Sayyar, Z., Anarjan, N., Berenjian, A., Jafarizadeh-Malmiri, H., Sayyar, Z., Anarjan, N., & Berenjian, A. (2019). Nano-sensors in food nanobiotechnology. *Nanobiotechnology in Food: Concepts, Applications Perspectives*, 81–94. https://doi.org/10.1007/978-3-030-05846-3

Janssen, S., Schmitt, K., Blanke, M., Bauersfeld, M.-L., Wöllenstein, J., & Lang, W. (2014). Ethylene detection in fruit supply chains. *Philosophical Transactions of the Royal Society A: Mathematical, Physical Engineering Sciences*, *372*(2017), 20130311. https://doi.org/10.1098/rsta.2013.0311

Joo, H. J., Kim, G. R., Park, E.-C., & Jang, S.-I. (2020). Association between frequency of breakfast consumption and insulin resistance using a triglyceride-glucose index: a cross-sectional Study of the Korea National Health and Nutrition Examination Survey (2016–2018). *International Journal of Environmental Research and Public Health*, *17*(9), 3322.

Joye, I., Davidov-Pardo, G., & McClements, D. (2015). Nanotechnology in food processing. In *Encyclopedia of Food and Health*. Elsevier. https://doi.org/10.1016/B978-0-12-384947-2.00481-5

Kerry, J. P., O'Grady, M. N., & Hogan, S. A. (2006). Past, current and potential utilization of active and intelligent packaging systems for meat and muscle-based products: a review. *Meat Science*, *74*(1), 113–130. https://doi.org/10.1016/j.meatsci.2006.04.024

Khan, A., Ezati, P., & Rhim, J.-W. (2023). Alizarin: prospects and sustainability for food safety and quality monitoring applications. *Colloids Surfaces B: Biointerfaces*, *223*, 113169. https://doi.org/10.1016/j.colsurfb.2023.113169

Khan, S., Monteiro, J. K., Prasad, A., Filipe, C. D., Li, Y., & Didar, T. F. (2024). Material breakthroughs in smart food monitoring: intelligent packaging and on-site testing technologies for spoilage and contamination detection. *Advanced Materials*, *36*(1), 2300875. https://doi.org/10.1016/j.colsurfb.2023.113169

Khanna, V. K. (2008). New-generation nano-engineered biosensors, enabling nanotechnologies and nanomaterials. *Sensor Review*, *28*(1), 39–45. https://doi.org/10.1108/02602280810850017

Khatib, M., & Haick, H. (2022). Sensors for volatile organic compounds. *ACS nano*, *16*(5), 7080–7115. https://doi.org/10.1021/acsnano.1c10827

Kilic, N. M., Singh, S., Keles, G., Cinti, S., Kurbanoglu, S., & Odaci, D. (2023). Novel approaches to enzyme-based electrochemical nanobiosensors. *Biosensors*, *13*(6).

Kim, S. W., & Cha, S. H. (2014). Thermal, mechanical, and gas barrier properties of ethylene–vinyl alcohol copolymer-based nanocomposites for food packaging films: effects of nano clay loading. *Journal of Applied Polymer Science*, *131*(11). https://doi.org/10.1002/app.40289

Kim, J. J., Ryu, S., & Lee, H. (2018). Foodborne illness outbreaks in Gyeonggi province, Korea, following seafood consumption potentially caused by Kudoa septempunctata between 2015 and 2016. *Osong Public Health Research Perspectives*, *9*(2), 66. https://doi.org/10.24171%2Fj.phrp.2018.9.2.05

Kress-Rogers, E. (2001). 20—Chemosensors, Biosensors, Immunosensors, and DNA Probes: The Base Devices. In E. Kress-Rogers, & C. J. B. Brimelow (Eds.), *Instrumentation and Sensors for the Food Industry* (Second Edition, pp. 623–713). Woodhead Publishing. https://doi.org/10.1533/9781855736481.3.623

Kress-Rogers, E., & Brimelow, C. J. (2001). Instrumentation and Sensors for the Food Industry (Volume 65). Woodhead Publishing.

Kumar, S., Basumatary, I. B., Sudhani, H. P., Bajpai, V. K., Chen, L., Shukla, S., & Mukherjee, A. (2021). Plant extract mediated silver nanoparticles and their applications as antimicrobials and in sustainable food packaging: a state-of-the-art review. *Trends in Food Science Technology*, *112*, 651–666. https://doi.org/10.1016/j.tifs.2021.04.031

Kumar, T. S. M., Kumar, K. S., Rajini, N., Siengchin, S., Ayrilmis, N., & Varada Rajulu, A. (2019). A comprehensive review of electrospun nanofibers: food and packaging perspective. *Composites Part B: Engineering*, *175*, 107074. https://doi.org/10.1016/j.compositesb.2019.107074

Kuswandi, B., Asih, N. P., Pratoko, D. K., Kristiningrum, N., & Moradi, M. (2020). Edible pH sensor based on immobilized red cabbage anthocyanins into bacterial cellulose membrane for intelligent food packaging. *Packaging Technology Science. Wiley*, *33*(8), 321–332. https://doi.org/10.1002/pts.2507

Kuswandi, B., Maryska, C., Abdullah, A. J., & Heng, L. Y. (2013). Real-time on-package freshness indicator for guavas packaging. *Journal of Food Measurement and Characterization*, *7*(1), 29–39. https://doi.org/10.1007/s11694-013-9136-5

Kuswandi, B., & Moradi, M. (2019). Improvement of Food Packaging Based on Functional Nanomaterial. In *Nanotechnology: Applications in Energy, Drug-Food* (pp. 309–344). Springer. https://doi.org/10.1007/978-3-319-99602-8_16

Kuswandi, B., Moradi, M., & Ezati, P. (2022). Food sensors: off-package and on-package approaches. *Packaging Technology Science*, *35*(12), 847–862. https://doi.org/10.1002/pts.2683

Kuswandi, B., Wicaksono, Y., Abdullah, A. J., Heng, L. Y., & Ahmad, M. (2011). Smart packaging: sensors for monitoring of food quality and safety. *Sensing Instrumentation for Food Quality Safety*, *5*, 137–146. https://doi.org/10.1007/s11694-011-9120-x

Li, M., Pu, Y., Thomas, V. M., Yoo, C. G., Ozcan, S., Deng, Y., Nelson, K., & Ragauskas, A. J. (2020). Recent advancements of plant-based natural fiber–reinforced composites and their applications. *Composites Part B: Engineering*, *200*, 108254. https://doi.org/10.1016/j.compositesb.2020.108254

Li, N., Zhou, S., Yang, X., & Lin, D. (2022). Applications of natural polysaccharide-based pH-sensitive films in food packaging: current research and future trends. *Innovative Food Science Emerging Technologies*, *82*, 103200. https://doi.org/10.1016/j.ifset.2022.103200

Lin, Y.-H., Chen, S.-H., Chuang, Y.-C., Lu, Y.-C., Shen, T. Y., Chang, C. A., & Lin, C.-S. (2008). Disposable amperometric immunosensing strips fabricated by au nanoparticles-modified screen-printed carbon electrodes for the detection of foodborne pathogen *Escherichia coli* O157 : H7. *Biosensors and Bioelectronics*, *23*(12), 1832–1837. https://doi.org/10.1016/j.bios.2008.02.030

Luo, Q., Hossen, A., Sameen, D. E., Ahmed, S., Dai, J., Li, S., Qin, W., & Liu, Y. (2023). Recent advances in the fabrication of pH-sensitive indicators films and their application for food quality evaluation. *Critical Reviews in Food Science Nutrition*, *63*(8), 1102–1118. https://doi.org/10.1080/10408398.2021.1959296

Luo, X., Zaitoon, A., & Lim, L. T. (2022). A review on colorimetric indicators for monitoring product freshness in intelligent food packaging: indicator dyes, preparation methods, and applications. *Comprehensive Reviews in Food Science Food Safety*, *21*(3), 2489–2519. https://doi.org/10.1111/1541-4337.12942

Madhusudan, P., Chellukuri, N., & Shivakumar, N. (2018). Smart packaging of food for the 21st century—a review with futuristic trends, their feasibility and economics. *Materials Today: Proceedings*, *5*(10, Part 1), 21018–21022. https://doi.org/10.1016/j.matpr.2018.06.494

Mahalik, N. P., & Nambiar, A. N. (2010). Trends in food packaging and manufacturing systems and technology. *Trends in Food Science Technology*, *21*(3), 117–128. https://doi.org/10.1016/j.tifs.2009.12.006

Mak, A. C., Osterfeld, S. J., Yu, H., Wang, S. X., Davis, R. W., Jejelowo, O. A., & Pourmand, N. (2010). Sensitive giant magnetoresistive-based immunoassay for multiplex mycotoxin

detection. *Biosensors and Bioelectronics, 25*(7), 1635–1639. https://doi.org/10.1016/j.bios.2009.11.028
Mao, X., Yang, L., Su, X.-L., & Li, Y. (2006). A nanoparticle amplification based quartz crystal microbalance DNA sensor for detection of *Escherichia coli* O157 : H7. *Biosensors Bioelectronics, 21*(7), 1178–1185. https://doi.org/10.1016/j.bios.2005.04.021
Meng, X., Kim, S., Puligundla, P., & Ko, S. (2014). Carbon dioxide and oxygen gas sensors-possible application for monitoring quality, freshness, and safety of agricultural and food products with emphasis on importance of analytical signals and their transformation. *Journal of the Korean Society for Applied Biological Chemistry, 57*(6), 723–733. https://doi.org/10.1007/s13765-014-4180-3
Mills, A. (2005). Oxygen indicators and intelligent inks for packaging food [10.1039/B503997P]. *Chemical Society Reviews, 34*(12), 1003–1011. https://doi.org/10.1039/B503997P
Mohamadian, N., Ghorbani, H., Wood, D. A., & Khoshmardan, M. A. (2019). A hybrid nanocomposite of poly (styrene-methyl methacrylate-acrylic acid)/clay as a novel rheology-improvement additive for drilling fluids. *Journal of Polymer Research, 26*, 1–14. https://doi.org/10.1007/s10965-019-1696-6
Molina-Besch, K., & Olsson, A. (2022). Innovations in Food Packaging—Sustainability Challenges and Future Scenarios. In *Future Foods* (pp. 375–392). Elsevier.
Nakazato, G., Kobayashi, R. K., Seabra, A. B., & Duran, N. (2017). Use of nanoparticles as a potential antimicrobial for food packaging. In *Food Preservation* (pp. 413–447). Elsevier.
Nami, M., Taheri, M., Siddiqui, J., Deen, I. A., Packirisamy, M., & Deen, M. J. (2024). Recent progress in intelligent packaging for seafood and meat quality monitoring. *Advanced Materials Technologies, 2024*, 2301347. https://doi.org/10.1002/admt.202301347
Nguyen, H. H., Lee, S. H., Lee, U. J., Fermin, C. D., & Kim, M. (2019). Immobilized enzymes in biosensor applications. *Materials (Basel), 12*(1). https://doi.org/10.3390/ma12010121
Nile, S. H., Baskar, V., Selvaraj, D., Nile, A., Xiao, J., & Kai, G. (2020). Nanotechnologies in food science: applications, recent trends, and future perspectives. *Nano-Micro Letters, 12*, 1–34. https://doi.org/10.1007/s40820-020-0383-9
Olaniyi, O. D. (2015). Food Packaging Indicators and Sensors.
Olifant, G. E. (2023). Fabrication of Carbon Nanomaterial Solid-State Sensors with Metal/Polymer-Based Composites for the Detection of Ketones, Aldehydes, Chlorobenzenes, and Amines and the Study of their Sensing Mechanism at Room Temperature. *University of Johannesburg.*
Otles, S., & Sahyar, B. (2016). Intelligent Food Packaging. In *Comprehensive Analytical Chemistry* (Volume 74, pp. 377–387). Elsevier.
Paidari, S., & Ibrahim, S. A. (2021). Potential application of gold nanoparticles in food packaging: a mini review. *Gold Bulletin, 54*, 31–36. https://doi.org/10.1007/s13404-021-00290-9
Pandey, A., Gurbuz, Y., Ozguz, V., Niazi, J. H., & Qureshi, A. (2017). Graphene-interfaced electrical biosensor for label-free and sensitive detection of foodborne pathogenic *E. coli* O157 : H7. *Biosensors and Bioelectronics, 91*, 225–231. https://doi.org/10.1016/j.bios.2016.12.041
Park, Y. W., Kim, S. M., Lee, J. Y., & Jang, W. (2015a). Application of biosensors in smart packaging. *Biosensors Bioelectronics, 11*, 277–285.

Park, Y. W., Kim, S. M., Lee, J. Y., & Jang, W. (2015b). Application of biosensors in smart packaging. *Molecular Cellular Toxicology*, *11*, 277–285.

Patel, G., Pillai, V., Bhatt, P., & Mohammad, S. (2020). Application of Nanosensors in the Food Industry. In *Nanosensors for Smart Cities* (pp. 355–368). Elsevier.

Paul, I. E., Kumar, D. N., Rajeshwari, A., Alex, S. A., Karthiga, D., Raichur, A. M., Chandrasekaran, N., & Mukherjee, A. (2017). Detection of Food Contaminants by Gold and Silver Nanoparticles. In *Nanobiosensors* (pp. 129–165). Elsevier.

Perdomo, S. A., Marmolejo-Tejada, J. M., & Jaramillo-Botero, A. (2021). Review—bio-nanosensors: fundamentals and recent applications. *Journal of The Electrochemical Society*, *168*(10), 107506. https://doi.org/10.1149/1945-7111/ac2972

Pizzoni, D., Mascini, M., Lanzone, V., Del Carlo, M., Di Natale, C., & Compagnone, D. (2014). Selection of peptide ligands for piezoelectric peptide based gas sensors arrays using a virtual screening approach. *Biosensors Bioelectronics*, *52*, 247–254.

Prabha, H., Saradhambal, M., & Prathyusha, K. (2021). Intelligent packaging using sensor–an overview. *International Journal of Research in Engineering, Science Management*, *4*(2), 103–108.

Puligundla, P., Jung, J., & Ko, S. (2012). Carbon dioxide sensors for intelligent food packaging applications. *Food Control*, *25*(1), 328–333. https://doi.org/10.1016/j.foodcont.2011.10.043

Rajak, D. K., Pagar, D. D., Menezes, P. L., & Linul, E. (2019). Fiber-reinforced polymer composites: manufacturing, properties, and applications. *Polymers*, *11*(10), 1667. https://doi.org/10.1016/j.bej.2023.108861

Rath, R. J., Farajikhah, S., Oveissi, F., Dehghani, F., & Naficy, S. (2022). Chemiresistive sensor arrays for gas/VOC monitoring: a review. *Advanced Engineerimg Materials*, *25*(3).

Rathee, K., Dhull, V., Dhull, R., & Singh, S. (2016). Biosensors based on electrochemical lactate detection: a comprehensive review. *Biochemistry and Biophysics Reports*, *5*, 35–54. https://doi.org/10.1016/j.bbrep.2015.11.010

Raul, P. K., Thakuria, A., Das, B., Devi, R. R., Tiwari, G., Yellappa, C., & Kamboj, D. V. (2022). Carbon nanostructures as antibacterials and active food-packaging materials: a review. *ACS Omega*, *7*(14), 11555–11559.

Riahi, Z., Rhim, J.-W., Bagheri, R., Pircheraghi, G., & Lotfali, E. (2022). Carboxymethyl cellulose-based functional film integrated with chitosan-based carbon quantum dots for active food packaging applications. *Progress in Organic Coatings*, *166*, 106794.

Roselli, M., Finamore, A., Garaguso, I., Britti, M. S., & Mengheri, E. (2003). Zinc oxide protects cultured enterocytes from the damage induced by *Escherichia coli*. *The Journal of Nutrition*, *133*(12), 4077–4082.

Saeed, F., Afzaal, M., Niaz, B., Shah, Y. A., & Afzal, A. (2022). Nanomaterials-Based Biosensors for Packaging Application. In *Nanotechnology Interventions in Food Packaging and Shelf Life* (pp. 61–84). CRC Press.

Salgado, P. R., Ortiz, C. M., Musso, Y. S., Di Giorgio, L., & Mauri, A. N. (2015). Edible films and coatings containing bioactives. *Current Opinion in Food Science*, *5*, 86–92.

Samadi, A., Klingberg, H., Jauffred, L., Kjær, A., Bendix, P. M., & Oddershede, L. B. (2018). Platinum nanoparticles: a non-toxic, effective and thermally stable alternative plasmonic material for cancer therapy and bioengineering. *Nanoscale*, *10*(19), 9097–9107.

Sandhya. (2010). Modified atmosphere packaging of fresh produce: current status and future needs. *LWT- Food Science and Technology*, *43*(3), 381–392.

Sekhon, B. S. (2014). Nanotechnology in agri-food production: an overview. *Nanotechnology, Science Applications*, 31–53.

Sharma, A., Matharu, Z., Sumana, G., Solanki, P. R., Kim, C. G., & Malhotra, B. D. (2010). Antibody immobilized cysteamine functionalized-gold nanoparticles for aflatoxin detection. *Thin Solid Films, 519*(3), 1213–1218. https://doi.org/10.1016/j.tsf.2010.08.071

Sharma, A., Ranjit, R., Kumar, N., Kumar, M., & Giri, B. S. (2023). Nanoparticles based nanosensors: principles and their applications in active packaging for food quality and safety detection. *Biochemical Engineering Journal, 193*, 108861. https://doi.org/10.1016/j.bej.2023.108861

Shawon, Z. B. Z., Hoque, M. E., & Chowdhury, S. R. (2020). Nanosensors and Nanobiosensors: Agricultural and Food Technology Aspects. In *Nanofabrication for Smart Nanosensor Applications* (pp. 135–161). Elsevier.

Silva-Leyton, R., Quijada, R., Bastías, R., Zamora, N., Olate-Moya, F., & Palza, H. (2019). Polyethylene/graphene oxide composites toward multifunctional active packaging films. *Composites Science Technology, 184*, 107888. https://doi.org/10.1016/j.compscitech.2019.107888

Smolander, M., Hurme, E., & Ahvenainen, R. (1997). Leak indicators for modified-atmosphere packages. *Trends in Food Science & Technology, 8*(4), 101–106. https://doi.org/10.1016/S0924-2244(97)01017-0

Sobhan, A., Muthukumarappan, K., & Wei, L. (2021). Biosensors and biopolymer-based nanocomposites for smart food packaging: challenges and opportunities. *Food Packaging Shelf Life, 30*, 100745. https://doi.org/10.1016/j.fpsl.2021.100745

Su, L., Jia, W., Hou, C., & Lei, Y. (2011). Microbial biosensors: a review. *Biosensors Bioelectronics, 26*(5), 1788–1799. https://doi.org/10.1016/j.bios.2010.09.005

Takhistov, P. (2009). Biosensor Technology for Food Packaging Applications. *The Wiley Encyclopedia of Packaging Technology.*

Thakur, M. S., & Ragavan, K. V. (2013). Biosensors in food processing. *Journal of Food Science and Technology, 50*(4), 625–641. https://doi.org/10.1007/s13197-012-0783-z

Tichoniuk, M., Biegańska, M., & Cierpiszewski, R. (2021). Intelligent Packaging: Sustainable Food Processing and Engineering Challenges. In *Sustainable Food Processing and Engineering Challenges* (pp. 279–313). Elsevier.

Tumwesigye, K. S., Oliveira, J., & Sousa-Gallagher, M. J. (2016). New sustainable approach to reduce cassava borne environmental waste and develop biodegradable materials for food packaging applications. *Food Packaging Shelf Life, 7*, 8–19. https://doi.org/10.1016/j.fpsl.2015.12.001

Turasan, H., & Kokini, J. (2021). Novel nondestructive biosensors for the food industry. *Annual Review of Food Science Technology, 12*, 539–566. https://doi.org/10.1146/annurev-food-062520-082307

Valdés, M. G., Valdés González, A. C., García Calzón, J. A., & Díaz-García, M. E. (2009). Analytical nanotechnology for food analysis. *Microchimica Acta, 166*, 1–19. https://doi.org/10.1007/s00604-009-0165-z

Vanderroost, M., Ragaert, P., Devlieghere, F., & De Meulenaer, B. (2014). Intelligent food packaging: the next generation. *Trends in Food Science Technology, 39*(1), 47–62. https://doi.org/10.1016/j.tifs.2014.06.009

Vasilescu, A., Polonschii, C., Titoiu, A. M., Mishra, R., Peteu, S., & Marty, J.-L. (2020). Bioassays and Biosensors for Food Analysis: Focus on Allergens and Food Packaging. In *Commercial Biosensors and Their Applications* (pp. 217–258). Elsevier.

Villamizar, R. A., Maroto, A., Rius, F. X., Inza, I., & Figueras, M. J. (2008). Fast detection of *Salmonella infantis* with carbon nanotube field effect transistors. *Biosensors and Bioelectronics*, *24*(2), 279–283. https://doi.org/10.1016/j.bios.2008.03.046

Wang, B., Chen, Y., Wu, Y., Weng, B., Liu, Y., Lu, Z., Li, C. M., & Yu, C. (2016). Aptamer induced assembly of fluorescent nitrogen-doped carbon dots on gold nanoparticles for sensitive detection of AFB1. *Biosensors and Bioelectronics*, *78*, 23–30. https://doi.org/10.1016/j.bios.2015.11.015

Wang, L., Chen, W., Xu, D., Shim, B. S., Zhu, Y., Sun, F., Liu, L., Peng, C., Jin, Z., Xu, C., & Kotov, N. A. (2009). Simple, rapid, sensitive, and versatile SWNT—paper sensor for environmental toxin detection competitive with ELISA. *Nano Letters*, *9*(12), 4147–4152. https://doi.org/10.1021/nl902368r

Wang, R., Xu, Y., Zhang, T., & Jiang, Y. (2015). Rapid and sensitive detection of *Salmonella typhimurium* using aptamer-conjugated carbon dots as fluorescence probe [10.1039/C4AY02880E]. *Analytical Methods*, *7*(5), 1701–1706. https://doi.org/10.1039/C4AY02880E

Wanihsuksombat, C., Hongtrakul, V., & Suppakul, P. (2010). Development and characterization of a prototype of a lactic acid–based time–temperature indicator for monitoring food product quality. *Journal of Food Engineering*, *100*(3), 427–434. https://doi.org/10.1016/j.jfoodeng.2010.04.027

Wen, H., Ou, C., Tang, H., Sang, S., Du, Y., & Chen, J. (2023). Development, characterization and application of a three-layer intelligent ph-sensing indicator based on Bromocresol green (BCG) for monitoring fish freshness. *Journal of Ocean University of China*, *22*(2), 565–575. https://doi.org/10.1007/s11802-023-5326-x

Wu, Y., Luo, Y., & Wang, Q. (2012). Antioxidant and antimicrobial properties of essential oils encapsulated in zein nanoparticles prepared by liquid–liquid dispersion method. *LWT—Food Science and Technology*, *48*(2), 283–290. https://doi.org/10.1016/j.lwt.2012.03.027

Wu, C., & Xu, Q.-H. (2009). Stable and functionable mesoporous silica-coated gold nanorods as sensitive localized surface plasmon resonance (LSPR) nanosensors. *Langmuir*, *25*(16), 9441–9446. https://doi.org/10.1021/la900646n

Xia, S., Yu, Z., Liu, D., Xu, C., & Lai, W. (2016). Developing a novel immunochromatographic test strip with gold magnetic bifunctional nanobeads (GMBN) for efficient detection of *Salmonella choleraesuis* in milk. *Food Control*, *59*, 507–512. https://doi.org/10.1016/j.foodcont.2015.06.028

Yaparatne, S., Morón-López, J., Bouchard, D., Garcia-Segura, S., & Apul, O. G. (2024). Nanobubble applications in aquaculture industry for improving harvest yield, wastewater treatment, and disease control. *Science of The Total Environment*, *2024*, 172687. https://doi.org/10.1016/j.scitotenv.2024.172687

Yousuf, B., Qadri, O. S., & Srivastava, A. K. (2018). Recent developments in shelf-life extension of fresh-cut fruits and vegetables by application of different edible coatings: a review. *LWT*, *89*, 198–209. https://doi.org/10.1016/j.lwt.2017.10.051

Zahra, S. A., Butt, Y. N., Nasar, S., Akram, S., Fatima, Q., & Ikram, J. (2016). Food packaging in perspective of microbial activity: a review. *The Journal of Microbiology, Biotechnology, Food Sciences*, *6*(2), 752. https://doi.org/10.15414/jmbfs.2016.6.2.752-757

Zamani, F. G., Moulahoum, H., Ak, M., Demirkol, D. O., & Timur, S. (2019). Current trends in the development of conducting polymers-based biosensors. *Trends in Analytical Chemistry*, *118*, 264–276. https://doi.org/10.1016/j.trac.2019.05.031

Zambrano-Zaragoza, M. L., González-Reza, R., Mendoza-Muñoz, N., Miranda-Linares, V., Bernal-Couoh, T. F., Mendoza-Elvira, S., & Quintanar-Guerrero, D. (2018). Nanosystems in edible coatings: a novel strategy for food preservation. *International Journal of Molecular Sciences*, *19*(3).

Zhang, Z., Lou, Y., Guo, C., Jia, Q., Song, Y., Tian, J.-Y., Zhang, S., Wang, M., He, L., & Du, M. (2021). Metal–organic frameworks (MOFs) based chemosensors/biosensors for analysis of food contaminants. *Trends in Food Science Technology*, *118*, 569–588. https://doi.org/10.1016/j.tifs.2021.10.024

Zhang, Q., Yang, L., Han, Y., Wang, Z., Li, H., Sun, S., & Xu, Y. (2022). A pH-sensitive ESIPT molecule with aggregation-induced emission and tunable solid-state fluorescence multicolor for anti-counterfeiting and food freshness detection. *Chemical Engineering Journal*, *428*, 130986. https://doi.org/10.1016/j.cej.2021.130986

Zhao, L., Zhang, M., Mujumdar, A. S., & Wang, H. (2023). Application of carbon dots in food preservation: a critical review for packaging enhancers and food preservatives. *Critical Reviews in Food Science Nutrition*, *63*(24), 6738–6756. https://doi.org/10.1080/10408398.2022.2039896

Zhou, Y., Pan, F.-G., Li, Y.-S., Zhang, Y.-Y., Zhang, J.-H., Lu, S.-Y., Ren, H. L., & Liu, Z.-S. (2009). Colloidal gold probe-based immunochromatographic assay for the rapid detection of brevetoxins in fishery product samples. *Biosensors and Bioelectronics*, *24*(8), 2744–2747. https://doi.org/10.1016/j.bios.2009.01.034

11

Biodegradable Food Packaging and Additive Manufacturing Technology

Adity Bahndral[1], *Rafeeya Shams*[1], *and Pintu Choudhary*[2]

[1] *Department of Food Technology and Nutrition, School of Agriculture, Lovely Professional University, Phagwara, Punjab, India*
[2] *Department of Food Technology, Chaudhary Bansal Lal Government Polytechnic, Bhiwani, Haryana, India*

11.1 Introduction

Polymers are ubiquitous in modern society because they have many useful qualities and are simple to manufacture. The terms "biopolymer" and "bioplastic" are often used interchangeably. In line with European Bioplastics, "bioplastics are biobased, ecological, one and the other" (European Bioplastics 2018) whereas the term "biopolymer" is synonymous with "bioplastic." Both renewable plant materials and resources that are sourced from fossil fuels can be used to make biodegradable polymers. Polylactic acid or polylactide (PLA) is a biodegradable polymer derived from living organisms; poly (glycolic acid) or polyglycolide (PGA), which lacks resemblance from PLA by the presence of lateral CH_3 group, is an eco-friendly polymer derived from fossil fuels. In today's world, the vast majority of conventional polymers, including polyethylenes (PEs), polyvinyl chloride, (PVC), polyethylene terephthalate(PET), and polyamides, may also be produced from bio-derived resources. There is a lot of obscuring surrounding the use of terms "biodegradable" and "biobased," especially in the business world. Composting must take place according to a predetermined set of parameters for biodegradable polymers in order for them to be approved for use in accordance with international laws that have to be followed (ISO 17088:2012, EN 13432:2000, EN 14995:2006, ASTM D6400-12, AS 4736, ASTM D5338:15, ISO 14855-2:2018). Additionally, they degrade more slowly under typical settings than would be the case otherwise; therefore, they comply with other relevant requirements as well. There is no definitive proof to demonstrate that PLA breaks down biologically as quickly as likely in coastal habitats, and it takes a very long time to break down in natural ecosystems (significantly longer than the 180-day period that is anticipated by the applicable international standards). Bio-based polymers, which may or may not degrade, are approved in accordance with the standards EN 16640:2015, ISO 16620-4:2016, ASTM 6866-18, and EN 16785-1:2015. Specifically, the amount of biogenic carbon in a matter is

Nanotechnology for Sustainable Food Packaging, First Edition. Edited by C. Anandharamakrishnan, Jeyan A. Moses, and M. Maria Leena.

measured by 14C in EN 16640:2015, ISO 16620-4:2016, and ASTM 6866-18, but by radiocarbon and elemental scanning in EN 16785-1:2015 (Niaounakis, 2019).

In 2017, global biopolymer production was over 2.05 million metric tons. Production patterns are changing from biodegradable polymers to nonenvironment-friendly polymers derived from biomass. About 57.1% of the polymer product is made up of bio-derived, nonbiodegradable polymers, whereas biodegradable polymers account for just 42.9% of the output (European Bioplastics 2017, 2003). This runs counter to the common belief that biopolymers are harmless because they decompose in the environment. The packaging industry is a substantial consumer of biopolymers, representing around 58% of industry's market share. Nearly 58% of biopolymers' overall volume share is found in the packaging industry, making it the largest market for these materials. Less than one percent of the 3 billion tons of plastics manufactured in 2017 were PP-Fibers, which include coatings, adhesives, thermoplastics, thermosets, elastomers, and sealants. There was a rise to 359 million tons in 2018. The majority of suppliers come from Asia having 51%, followed by China up to 30%, Europe (17%), the MENA region nearly 7%, and other regions (Europe and EPRO 2019). India plays a major role in the global plastics industry. About 15 million metric tonnes of PE were produced in India in 2018–2019, mostly in terms of films and sheetings; this number is predicted to rise to 24 million metric tonnes by 2020 (Aryan et al., 2019). By 2022, it is expected that the worldwide biopolymer production capacity will have increased to 2.44 million tonnes.

Synthetic plastics, from which 95–99% of all plastics are made, are not a sustainable resource (Mangaraj et al., 2019)). The packaging industry utilizes 43% of India's yearly man-made polymer production. Synthetic plastics are widely employed in medical appliances, packaging, and building materials (FICCI, 2014). In contrast, synthetic plastics are not biodegradable and, hence, accumulate in landfills. Problems with the environment and human health are just two of the many outcomes of waste production. They aggregate in public spaces, causing drains to become clogged and overflow (Foolmaun and Ramjeeawon, 2012). The marine world is threatened by massive aggregates of plastic trash poured into oceans as well as rivers. Production of noxious fumes (for instance, both carbon mono and dioxide, 1, 3-butadiene, chlorine, furans, amine, dioxin, etc.) from incineration has a negative impact on atmospheric quality, increases the fear of climate crisis and can lead to a certain medical condition (Kennedy and Panesar, 2006). Many synthetic polymers cannot be broken down in landfills, which has led to increased waste management problems and harmful effects on the environment and human health, prompting calls for research into the development of environmentally preferable replacement materials (Luckachan and Pillai, 2011). In recent years, biodegradable polymers (sometimes called ecologically or environmentally degradable polymers) have gained popularity as a practical replacement for nonbiodegradable plastic in a wide variety of industrial settings. The European Bioplastics Organisation classifies polymers as bioplastics if they are either biobased (meaning they are [partially] generated from organic matter as sustainable resources) or ecological or alike. The bio-derived polymers used in food packages can be broken down into three distinct types: (i) synthetic polymers made from bio-derived monomers, (ii) polymers made by microbes, and (iii) natural bio-based polymers derived straight from raw sources (Cha and Chinnan, 2004).

Bio-based PET is an example of plastic partially derived from renewable resources and synthesized in the same way as conventional plastics (Landis, 2010; Suryanegara et al., 2018). Alternative plastics, such as PLA, can be made from renewable materials and cannot fit into the standard 1-6 classification system (Morschbacker, 2009; Tabone et al., 2010). Biopolymers can be thought of as a sustainable replacement for conventional plastics made from petroleum. They can be made from a wide range of feedstocks, such as algae or food waste, and agricultural products like corn or soybeans (Landis et al., 2007; Luo and Hu, 2017). Biopolymers are still being made, but they belong to a rising generation of materials, gaining a lot of attention as potential replacements in favor of traditional plastics and polymers made from petroleum. This is because more people are interested in environmentally responsible and sustainable development. Biobased and biodegradable materials often outperform those made from fossil fuels in terms of how well they block water vapor, how strong they are, how stable they are at high temperatures, and how well they can be processed (Platt, 2006). Biodegradation is the process by which a polymer, with the help of microorganisms, is broken down into its parts (water, carbon dioxide, biomass, and methane) (Meereboer et al., 2020). Biodegradation depends on temperature, humidity, and oxygen (aerobic or anaerobic circumstances) (Guzman et al., 2011). This means that their general acceptability and use are constrained by the difficulties of establishing enough barriers and mechanical qualities without sacrificing biodegradability (Nampoothiri et al., 2010). Polymer modification, coating, mixing, and the use of nanocomposites are all approaches that are likely to contribute to tackling these issues, as concluded by Nampoothiri et al. (2010). The degradable property of different biopolymers is compared and summarized in Table 11.1.

11.2 Emerging Concerns on the Usage of Synthetic Polymers

In its beginnings, the plastics industry had great promise as a positive development for society. Overuse has made this material extremely harmful to the environment and society as a whole. An increasing fraction of the polymer market is expected to be devoted to biopolymers shortly. Particularly in the fields of health, agriculture, engineering, and textiles—where biocompatibility and biodegradability are major means—because of their unique properties, these materials are better, easier to use, and likewise harmless for the ecology than manufactured polymers. Since the use of fossil fuels is discouraged because of the damage they cause to farmlands when they leak, using less of them and switching to biopolymers is recommended. As biopolymers continue to advance, new theories and the imperative to fully embrace their utilization are essential, yet this development comes at a price that must be examined and addressed objectively. Considering economic factors is important since the success of biopolymer products depends on their capacity to compete on price and on the willingness of consumers to pay for them. Because of the high cost of most biopolymers and the low cost of petroleum-based polymers, many industries have begun using the latter without first assessing the environmental impact of their actions (Pathak et al., 2014; Ibrahim et al., 2018). Polymerization carbon is extracted from

Table 11.1 The degradable property of different biopolymers

Polymer	Chemical Nature	Degradation Time	Degradation Mechanism	Reference
Cellulose	Repeating units of (1-4)-linked glucose	7–8 yr	Enzymatic: cellulase. Alkaline degradation	Pavasars et al. (2003), Hernández et al. (2014)
Starch	D-glucopyranose subunits α (1-4) and (1-6) glycosidic linkage	Approx. 174 d	Enzymatic: amylase. Bacterial Degradation	Hernández et al. (2014), Moon et al. (2020)
Gelatin	A large number of glycine, proline, and 4-hydroxy proline residues	5 d	Enzyme-induced decomposition	Benabid and Zouai (2016)
Chitin or chitosan	Polysaccharide with (1,4)-linked 2-acetamido-2-deoxy-D-glucose and 2-amino-2-deoxy-D-glucose units	>20 wk	Enzymatic: chitosanase and lysosome	Kean and Thanou (2010), Comblain et al. (2017)
Polycaprolactone	(1,7)-polyoxepan-2-one	2–3 yr	Radical interactions, temperature, pH, enzymatic activity, and cellular phagocytosis	Ramesh et al. (2018), Bartnikowski et al. (2019)
Poly (vinyl alcohol)	Consists of 1, 3-diol units or 1, 2-diol units	7 d	Bacterial Degradation—*Bacillus amyloliquefaciens*	Wu and Wang (2021)
Xanthan	Produced by the bacterium *Xanthomonas campestris*, containing glucose, mannose, and glucuronic acid in the molar ratio of 2:2:1.	5 yr	Temperature more than 80°	Seright and Henrlcl (1990), Shekarforoush et al. (2017)
Glucans	Glucose sugar together by using (1-3) glycosidic bonds and side-chain by (1, 6) connections	3–4 d	Three main enzymatic activities: endo-glucanase, exo-glucanase, and β-glucosidase activities	Nworu et al. (2015), Giovannoni et al. (2020)
Poly-g-glutamic acid	Polymerization of L- and/or D-glutamic acid via a g-amide link	4 hr	Proteolytic enzyme degradation	Yao et al. (2009), Hamman (2010)
Alginate	Polysaccharide with (1,4)-linked β-D-mannuronic and α-L-guluronic acid repeat units	About 80 d	Hydrolytic disintegration	Venkatesan et al. (2017)
Cyanophycin	Co-polymer of l-aspartic acid and l-arginine	25 d	Enzymatic: cyanophycinase	Burnat et al. (2014), Elias et al. (2015)
Poly (hydroxyl alkanoates)	Polyester consisting of hydroxyalkanoate repeat units	>24 mo	Bacterial fermentation	Cheng et al. (2017)

petroleum, where it has been for millions of years. The release of this carbon into the environment, whether through plastic incineration or some other kind of breakdown, would increase greenhouse gases generally (Gironi and Piemonte, 2011; Hottle et al., 2013).

There is a large variation in recycling rates depending on the sort of plastic along with the method used to dispose of it after its useful life has ended. In most countries, incineration is used to manage the 12% of plastic waste that is included in municipal solid waste that cannot be recycled. Plastic emits about 2.8 kg of carbon dioxide when burned (EPA, 2002). Although just 7% of plastics were recovered for recycling overall, the percent of particular plastic containers is much higher. Recycling rates for PET soda bottles and HDPE milk and water bottles were both around 28%. Only 9% of the 19.2 million metric tonnes of nondurable plastic packaging and packaging trash was recycled (EIA US, 2011). Biodegradability and compostability are two features of biopolymers that are required by the US standard particularization being biodegradable plastics. Bio-derived polymers attempt a way to cut back on the fossil fuels essential to generate the 21 million metric tonnes of plastic used per annum in packages and perishable objects, together with 16.7 million metric tonnes of plastic garbage that ends up in landfills (ASTM, 2004). Polypropylene (PP) films have multiple disposal options, the most common of which is landfilling. The disposal procedure accounts for 94% of the environmental load associated with carcinogens since not all landfill gas is collected for incineration and energy production. Furthermore, surface water is expected to be contaminated by the landfill's short-term emissions, and groundwater is expected to be contaminated by pollutants in the soil beneath the dump. However, 93% of the surrounding stress associated with fossil fuels is attributable to the extraction of fossil fuels along with the polymerization of propylene monomer by applying the heterogeneous Zieglere–Natta catalyst system. Since PP requires a lot of energy to produce and is made from fossil fuels, the effects in this area are significant. Carcinogens and fossil fuel impacts, which occur during raw material extraction and account for the vast majority of pollution, were found to be significantly greater in films made from PP than those made from chitosan (Goedkoop, 1999; Leceta et al., 2013). Nile-red tagging was used to check 259 bottles for microplastic contamination by Mason et al. (2018). Almost all of them (approximately 93%) were tainted with microplastics. There were more than 10.3 microplastic particles of 100 μm in size per liter of processed packed drinking water after controlling for any background (lab) contamination. FTIR spectroscopy indicated that 50% of these particles are polymeric in nature, with PP being the most frequent form of polymer (54%). Nile-red dye absorption suggests these particles are likely made of plastic. Akhbarizadeh et al. (2018) came to the conclusion that microplastics were present within the muscles of both benthic and pelagic fish species that were found in the Persian Gulf. Therefore, eating commercial fish species that have already consumed microplastics should be considered a potential pathway for their consumption by humans. This is because these fish have already digested the microplastics. It has been demonstrated that microplastics can be found in a variety of foods, including shellfish purchased from retail outlets (Li et al., 2018; Cho et al., 2019), canned fish (Karami et al., 2018), bottled water (Schymanski et al., 2018), salt (Renzi and Blašković, 2018), honey, and sugar (Liebezeit and Liebezeit, 2013). Humans are most vulnerable to microplastic exposure through breathing, digestion, and dermal exposer. Man-made fibers and rubber tires, dust are the most common sources of airborne microplastics (Prata, 2018); ingestion (through eating foul marine food or other food or polluted potable water) and inhalation are the most common routes by which microplastics

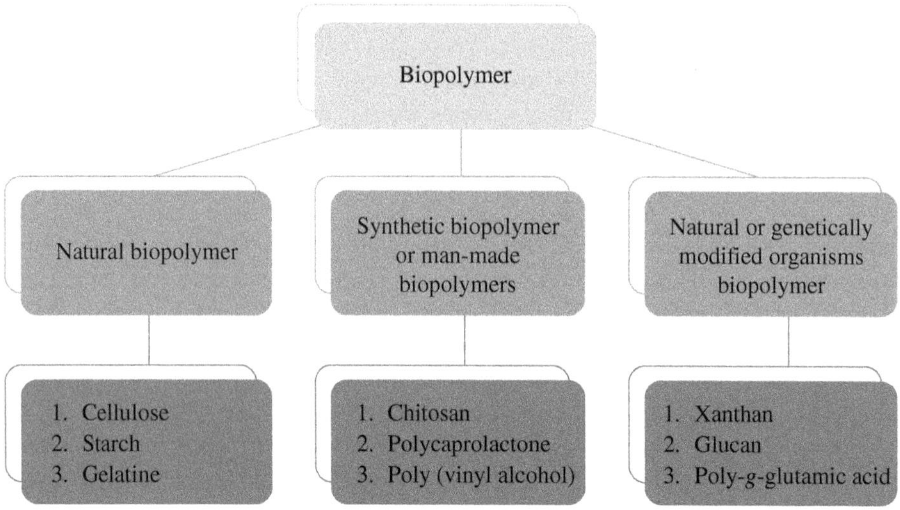

Figure 11.1 The classification of biopolymers on the basis of their structure.

enter the digestive tract. Completely digested organisms pose a greater threat than their partially digested counterparts (Carbery et al., 2018). Sweat glands, open skin injuries, and hair follicles are all potential entrance points for microplastics and other contaminants despite the skin's ability to prevent their direct passage (Schneider et al., 2009). Figure 11.1 shows the classification of biopolymers on the basis of their structure.

11.3 Biodegradable Materials Used for Food Packaging

Classifying polymers is a complex task, and biopolymers can be categorized based on various criteria. Data from several categories can be quite useful. Despite the fact that biopolymers can be categorized in a number of ways based on their genesis, chemical constitution, application, production, and processing (Abdul Khalil et al., 2018). On account of low C—C bond energy in bio-derived polymers, they are easily degraded by enzymes, water, heat, and light. Some common biopolymers include cellulose, starch, gelatin, keratin, DNA, RNA, lipids, and many more. Polysaccharides, lipids, and various eco-friendly materials are the most common types of biopolymer packaging. As opposed to synthetic polymers, they benefit more from the recycling process. Biodegradable polymers are useful as coating substances for sheets due to their great oxygen hindrance and low moisture-permeable rate (Avella et al., 1996).

11.3.1 Biopolymers Obtained from Natural Biomass Products

11.3.1.1 Cellulose

Plant-based cellulose was the earliest example of a thermoplastic polymer. Cellulose has the chemical formula $(C_6H_{10}O_5)_n$. It is a straight polysaccharide with repeating units having (1-4)-linked glucose in the range of 100–1,000. D-glucose is the form of glucose with the

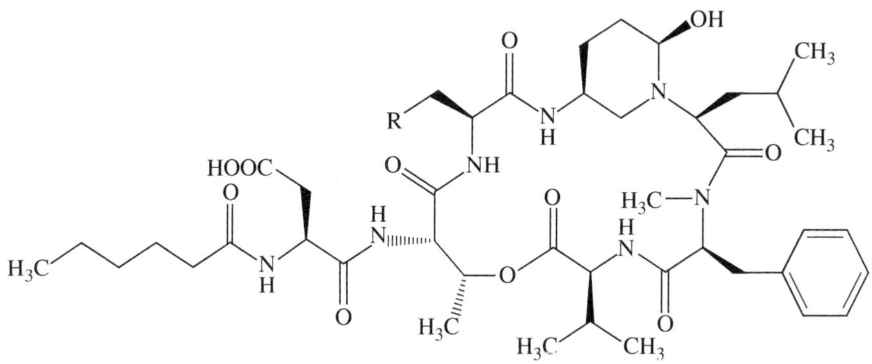

Figure 11.2 Representative structural formula of cyanophycin.

hydroxyl and carboxyl groups in the same plane (1–5) (Figure 11.2). D-hydroxyl glucose's group and oxygen atom form strong intermolecular and intramolecular H bonds. It is crystalline, has a high molecular weight (degree of polymerization is approximately between 8,000 and 10,000 dpi), and has an elevated molecular weight (Khosravi-Darani and Bucci, 2015). Chain length and Dp are two factors that influence the characteristics of cellulose-based polymers. The thermal properties of cellulose are low, but its chemical properties are high. It is a white, crystalline powder with no discernible flavor (Qiu et al., 2022). Some primary food packaging uses waxed or polyethene-coated paper; however, the vast majority of paper is employed in secondary packaging. With its hydrophilicity, insolubility, and crystalline structure, cellulose is a low-cost raw material with a steep learning curve. Baked foods, processed meat, cheese, and candies benefit from cellophane's improved barrier characteristics when layered using nitrocellulose wax or PVDC (polyvinylidene chloride).

Carboxy-methyl cellulose, methylcellulose, ethyl cellulose, hydroxyethyl cellulose, hydroxypropyl cellulose, and cellulose acetate are just a few of the many cellulose derivatives produced for commercial usage. Due to its crystalline structure, cellulose makes the initial derivatization processes time consuming and costly. Some examples of label names for cellulose-derived polymers include Tenite® (Eastman, USA), Bioceta® (Mazzucchelli, Italy), Fasal® (IFA, Austria), and Natureflex® (UCB, Germany) (Pawar and Purwar, 2013).

11.3.1.2 Starch

Starch is a long chain of D-glucopyranose sugars. For sugars, the pyranose ring structure is the most thermodynamically stable form to exist in solution. D-glucopyranose subunits are bound together in starch mostly by α (1-4) and (1-6) glycosidic connections (Weber, 2000a) (Figure 11.3). The energy in grains comes primarily from starch, which makes up between 60% and 75% of the total weight of most grains. It is the primary source of energy for humans and most other species, accounting for roughly 70–80% of daily caloric intake. It consists of the linear and crystalline polymer amylose (poly-1, 4-D-glucopyranoside) and the branched as well as amorphous polymer amylopectin (poly-1, 4-D-glucopyranoside with 1,6-D-glucopyranoside). Amylose makes up roughly 10% of starch, whereas amylopectin accounts for the remaining 80–90%; however, this percent varies widely from one starch source to the next. Amylose has a helical shape and dissolves in water.

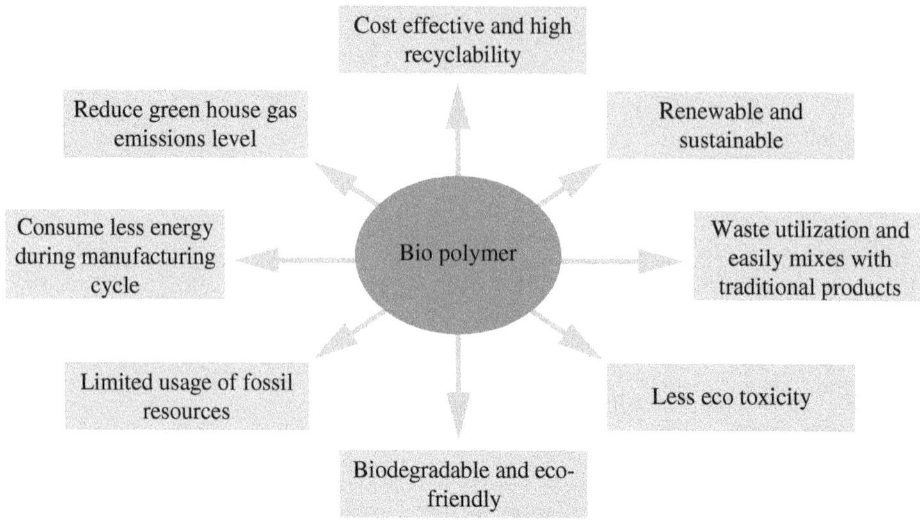

Figure 11.3 The merits of biopolymers are depicted schematically.

Different amounts and molar weights of amylose and amylopectin, which are in turn dependent on the starch supply, provide materials with distinct mechanical characteristics and biodegradability (Ratnayake et al., 2001). Thermoplasticity is a common application for starch. It undergoes destruction in the presence of a predetermined amount of solvent, plasticizers, and heat, allowing it to become plasticized so that it can be extruded. Due to its sensitivity to moisture, thermoplastic starch (TPS) is often used in low-humidity environments. As a result, the TPS that is formed is very close to becoming amorphous. Enzymes hydrolyze the acetal bond, allowing the starch to degrade (Chandra and Rustgi, 1998; Weber, 2000b). Amylases target the 1,4 linkage, while glucosidases target the 1,6 linkage.

11.3.1.3 Gelatin

Gelatin is a natural biopolymer obtained from the meat processing industry, where it is made from raw collagen (Dara et al., 2020). Collagen hydrolysate and denatured collagen are all names for gelatin (Fu et al., 2014). It's a staple of a healthy diet owing to its inclusion of 18 necessary amino acids (Auras et al., 2004). As a byproduct of hydrolyzing raw collagen found in animal skin, bone, and other connective tissues (Ahmed et al., 2020). Type-I collagen is used in its preparation, and it does not include cysteine. On the basis of production method and ingredient makeup, two distinct kinds of gelatin exist

In the presence of acid, raw collagen is hydrolyzed to generate type-A gelatin. There is nitrogen present in it at a percentage of 18.5% (Jaswir et al., 2020). In order to make type-B gelatin, collagen with an amide group (—CONH—) removed is hydrolyzed in an alkaline environment, releasing the 18% nitrogen atoms present in the collagen (Yu et al., 2020). At both low and high temperatures, the biopolymer exhibits unusual behavior. At high temperatures, it melts and forms coils. In contrast, it transforms from a coil to a helix at room temperature (Zambrano et al., 2020). Because of the existence of reactive groups, it is accessible for modification with composites and nanocomposites (such as —CONH—, —SH—, and —COOH—).

11.3.1.4 Chitin or Chitosan

Structured polysaccharide chitin (or chitosan) is made of several similar subunits of modified glucose units and has a nitrogen backbone. The outer walls of numerous algae, insects, fish, and vertebrates are composed of chitin. Unlike cellulose, chitin has amine components and a hydroxyl group on the respective monomer, with the methyl-amide substituent replacing the 2-hydroxyl group on each monomer. Chitin consists of N-acetylglucosamine monomers joined together through a glycosidic bond of the 1-4 type. Chitin shares several structural similarities with cellulose, including β (1-4) anhydrous glycosidic bonding. Like cellulose, it is indissoluble in water and many organic solvents; however, it can be dissolved in hexafluoro isopropanol, haloalcohols, and mineral acid solutions in water dimethylacetamide. In extreme circumstances, pure sugar and D-glucosamine are the by-products of its hydrolysis with concentrated mineral acids. Deacylated chitin (chitosan) is a cationic polysaccharide that combines with higher polysaccharides to generate polyelectrolyte complexes. In contrast to cellulose glucan, chitosan glucan is widely available (Weber, 2000b).

It is a weak basic that is indissoluble in water. However, it is dissolvable in dilute aqueous acidic solutions underneath its pKa (w6.3), where it has the ability to convert glucosamine units (eNH_2) into the dispersible protonated form (eNH_3). Because chitosan is solvable in diluted acid solutions, films with a thick and porous structure can be easily made by casting or dipping. Three basic processes involved in the chemical extrication of chitin from crab outerkeletons are demineralization, deproteinization, and deacetylation. The Chitin can be deacetylated in most cases by subjecting it to reacting with a concentrated sodium hydroxide solution (40–50%) at a temperature of 90–100 °C or higher. This can take out part or all of the acetyl groups that are present in the chitin (Abdou et al., 2008; Al Sagheer et al., 2009).

11.3.2 Synthetic or Man-made Biopolymers

11.3.2.1 Polycaprolactone

Biodegradable polymers can be synthetic, and polycaprolactone (PCL) is one such material. Caprolactone is polymerized into PCL by opening its ring when the catalyst is present like stannous octate. PCL is referred to as (1,7)-polyoxepan-2-one by the IUPAC and is utilized in biological applications. PCL is commonly utilized in the medical field, and it is also added to starch as an additive to decrease the price, increase biodegradability, and boost impact strength. The pellets of this polymer typically measure around 3 mm in diameter. This biopolymer has a melting temperature of 60–65 °C, a density of 1.145 g/cm^3, and a glass transition point of 60 °C (Weber, 2000b). PCL is a semirigid polymer at room temperature with modulus values between low and high-density PE, tensile strength of just 23 MPa, and an elongation at break with up to 700% (Pawar and Purwar, 2013). PCL is sold commercially under several brand names, including CAPA® (built by Solvay of Belgium), Tonc®

(made by Union Carbide of the United States), Celgreen® (produced by Daicel of Japan), and many others. The potential for its use in packaging has been explored (Pawar and Purwar, 2013).

11.3.2.2 Poly(vinyl alcohol)

Poly(vinyl alcohol), sometimes known as PVA, is a biodegradable thermoplastic polymeric substance that can be produced from polyvinyl acetate via hydrolysis. PVA can be found in pearl form or as a solution, and its chemical structure is often $[CH_2CH(OH)]_n$. It rapidly dissolves in water because of its improved crystallinity. Other man-made polymers, such as polyvinyl chloride, PE, PP, etc., are produced from their relevant monomers, while PVA is obtained by partially or entirely hydrolyzing polyvinyl acetate to take out its acetate groups. PVA is a flexible polymer, which means it is both extremely malleable as well as structurally sound. This is why PVA has found such wide application in the field of packaging. PVA, or polyvinyl alcohol, is a polymer that forms chemical bonds with surfaces and is, therefore, useful for floating objects that need to be kept above the water (Weber, 2000b). Many other biodegradable polymers have been made for packaging applications.

11.3.3 Biopolymers Obtained from Natural or Genetically Modified Organisms

Scientists now have access to state-of-the-art methods to investigate and control biological systems, all thanks to modern biotechnology. When it comes to regulating when, where, how much, and what kind of "gene expression" occurs, genetic engineering provides unprecedented precision. Protein polymers constitute the simplest example. One can change the manufacturing system and the molecular makeup of a protein polymer if they have access to its corresponding gene. Uniform in length and with the same stereochemistry (or spatial orientation). Polymer chains can be synthesized using recombinant DNA techniques (Rao et al., 2014). The Food and Drug Administration does not require labeling or approval for food products that are structurally and functionally similar to those already on the market. However, the United States Department of Agriculture and Environmental Protection Agency monitors field tests of genetically modified plants to ensure their safety (EPA). More than 700 permits to test GM crops and livestock in the wild had been issued (Steinbüchel and Lütke-Eversloh, 2003).

11.3.3.1 Xanthan

Xanthan gum, a bacterially generated complex copolymer, was one of the first fermented bacterial polysaccharides to find commercial success. *Xanthomonas campestris* bacteria use five distinct sugar groups to construct the "repeat units" that makeup xanthan polymers. Polymerization releases xanthan gum from the bacterium, which can then be precipitated out using alcohol, and the biomass of the bacteria can be discarded. It is used in many different industries, both food and nonfood. Provides viscosity control in drilling mud fluids for oil recovery; acts as a biocide in mineral ore processing; changes the way paper is made; helps plants grow in agriculture; is being tested for use in prolonged drug release; and is used in cosmetics (controls dust release). In the food processing industry, it is used as a thickener in many sweets like cheese spreads, ice creams, puddings, and whipped cream (Kennedy and Bradshaw, 1984; Hassler and Doherty, 1990; Prakash et al., 2013).

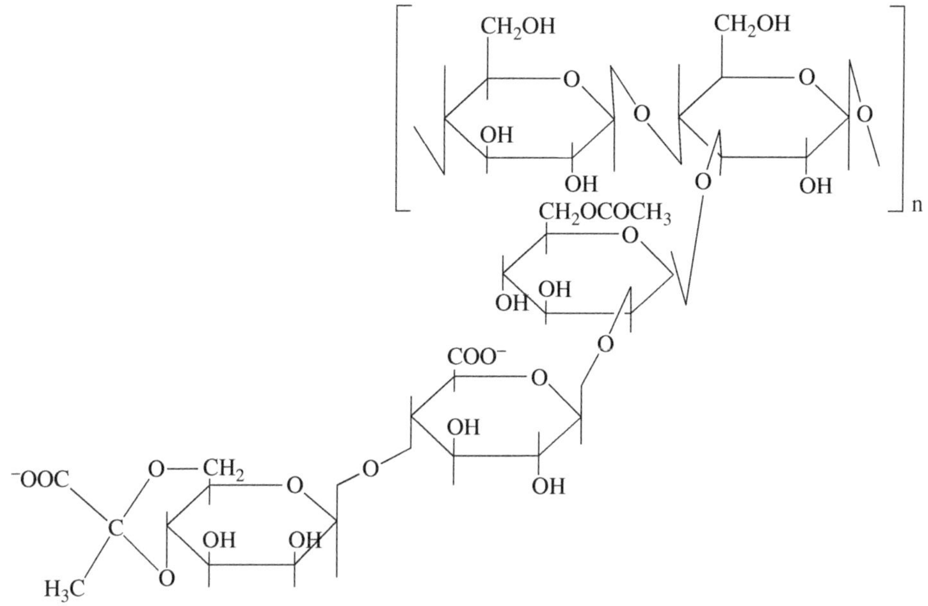

However, due to its exceptional mechanical and physical qualities, this biopolymer has uses in a variety of industries, including food (which accounts for 60% of the market) and nonfood (which accounts for 40%). It is used as a biocide in the mining industry, as a paper industry conditioner, for controlled and delayed drug delivery in the pharmaceutical industry, and in the cosmetics industry for regulated spray release. It's also used in the creation of transparent gel toothpaste. It's safe to say that microbial xanthan is a significant player in the polysaccharide or biopolymer production market in terms of total volume (Becker 2015).

11.3.3.2 Glucans

Glucans are molecules that are entirely made up of glucose sugar. The word "glucan" is frequently used to refer to the glucan component that is present within yeast cell walls. *Saccharomyces cerevisiae* is one of the most prevalent sources of this glucan; however, it can also be found in other sources. Glucans are the most prevalent type of polymer found in yeast and account for around 12–14% of the organism's total dry cell mass. By subjecting yeast cells to a hot alkali treatment, one may easily remove all other cellular elements and recover the insoluble glucan material. This method of yeast glucan purification is known as "hot alkali precipitation." Glucans are radioprotective, which is yet another fascinating quality of these remarkable molecules. This increases the likelihood of the test animals surviving amounts of radiation that would otherwise be fatal. Glucans consist of a backbone constituted of glucose monomers that are bonded together using (1-3) glycosidic bonds. They also have affixed alternate chain glucose residues that are bonded together via (1, 6) connections (McIntosh et al., 2005). The (1, 3)-1, 6-glucans lentinan, schizophyllan, and grifolan all originate from *Lentinus edodes*, *Schizophyllum commune*, and *Grifola frondosa*, respectively. All three glucans have been reported to branch at different rates. Different doses of each are effective against the same cancer model (Di Luzio et al., 1979; Demleitner et al., 1992; McIntosh et al., 2005).

11.3.3.3 Poly-g-glutamic acid

Poly-g-glutamic acid (PGGA) is a new anionic polypeptide molecule formed by the polymerization of L- and/or D-glutamic acid via a g-amide link (Cao et al., 2018). PGGA is optically active in its unmodified form because each glutamate unit contains a chiral molecule. When D-glutamate homo-polymer as well as L-glutamate homo-polymer are merged irregular units of D and L glutamate are produced (Inbaraj et al., 2006). There are four main stages in the PGA process: racemization, polymerization, regulation, and degradation (Bajaj and Singhal, 2011). Bacillus is a most popular genus for producing PGA through microbial fermentation, and other strains of Bacillus, like *B. licheniformis* as well as *B. subtilis*, are employed to produce PGA. Bacteria engineered for PGA production can be split into two types, L-glutamic acid-dependent and separated, based on nutrient restrictions for development. The first category includes *Bacillus subtilis* and *Bacillus licheniformis* 9945. *B. subtilis* strains CGMCC 0833 (Verma et al., 2020), *B. subtilis* strain ATCC 15,245 (Bhat et al. 2013), *B. subtilis* strain C10 (Zhang et al., 2012), and *Bacillus amyloliquefaciens* LL3 (Cao *et al.*, 2011). The amount of L-glutamic acid absorbed in this method of PGA generation directly affects the amount of PGA produced.

11.3.3.4 Alginate

In contrast to other polymeric compounds, alginate does not exhibit branching patterns or monomeric repeats (Anderson et al., 2018)). It is comprised of dual separate substances: a-L-guluronic acid (C5-epimer) and b-D-mannuronic acid. Typically, brown algae are used to make it. Bacteria such as *Azotobacter* and *Pseudomonas* also have the necessary metabolic pathways to produce alginate. Polymerization of the cytosolic precursor guanosine diphosphate-mannuronic acid yields the membrane-passing polymer poly-mannuronic acid, the first step in the biosynthetic pathway. The enzyme used by bacteria to produce alginate undergoes a change known as epimerization or acetylation before it is released into the external environment. Numerous industries find value in the specific properties of tailored alginate biopolymers. Alginates, which are negatively charged bio-based polymers, assist in material qualities such as solution viscosity and the gel structure of particles in the prescnce of cations (Barcelos et al., 2020).

11.3.3.5 Cyanophycin

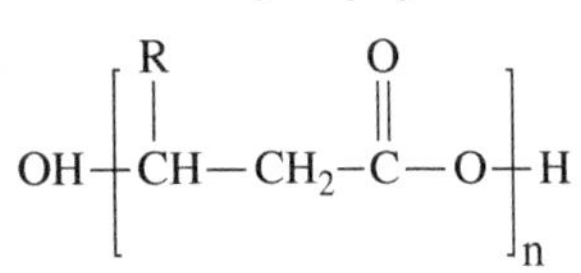

Linking amine group of arginine groups to the —COOH of aspartic acid results in the formation of the poly (aspartic acid) backbone that is present in the polyamide molecule cyanophycin (Altun et al., 2018). Cyanophycin is a chemical that accumulates nutrients and provides the host organism with carbon and nitrogen (Diniz et al., 2006). The majority of the effort that goes into the production of the cyanophycin polymer is done by the enzyme known as cyanophycin synthetase. It is possible to successfully transfer the gene for this enzyme into prokaryotic organisms, which enables genetic alteration that enhances the organism's capacity to make the biopolymer in issue (Lippi et al., 2018).

In addition to bacteria, this biopolymer can also be synthesized in sufficient amounts by yeast and higher plant species, as well as by a number of other organisms. These biopolymers are broken down both within and outside of cells by microbial cyanophycinases known as CGPase, intracellular cyanophycinases known as CphB, and extracellular cyanophycinase, in that order (CphE). The biodegradation process produces dipeptides, which are then broken down even further by dipeptidase inside the cell to produce amino acids (Sallam et al., 2009).

11.3.3.6 Poly (hydroxyl alkanoates) (PHAs)

Polyhydroxyalkanoates are natural bio-friendly plastics made by Gram-positive and Gram-negative bacteria like *Cuprividusnecator*, *Pseudomonas putida*, *Azotobacter*, *Aeromonas*, *Clostridium*, *Methylobacterium*, *Syntrophomonas*, *Ralstonia*, *Alcaligenes latus*, and *Alcaligenes eutrophus* (Abd alFattah Amara, 2008). Cyanobacteria, also known as blue-green algae among the prokaryotic species that, by nature, make the homopolymer of PHB (Banerjee et al., 2016). PHAs have a low water vapor permeability, which is a huge plus for food packaging (Weber, 2000b). PHB, a biotechnologically created and extensively studied biodegradable polyester, has been in existence since 1925 (Zhang et al., 1997).

The amorphous phase undergoes secondary crystallization while being stored at ambient temperature. Consequently, the polymer becomes more brittle and rigid, with increased stress along with elongation modulus (E = 1.7 GPa). Failure elongation is only 10% after that (El-Hadi et al., 2002). It has a smaller process ability window than regular plastics. In the neighborhood of its melting point, PHB is prone to heat deterioration (Grassie et al., 1984). A natural component of blood, 3-hydroxybutyric acid, is produced via hydrolytic breakdown, albeit at a slow rate. Biodegradable polymers for use in wastewater treatment have been prepared by grafting various monomers onto PHB. Depending on their hydrophobicity or hydrophilicity, the grafted monomers were either styrene, methyl acrylate, or acrylic acid (Hsieh et al., 2009).

The term "reactive blending" has been applied to these techniques. Notably, various chemical methods for producing PHB have been established in addition to the bacterial synthesis method. It is also possible to synthesize PHB by polymerizing butyrolactone in a ring-opening fashion (Hori et al., 1994; Juzwa and Jedliñski, 2006). Framework can be made in a number of various ways depending on the synthesis technique used. Polyhydroxy butyrate (PHB) and related materials (e.g., Biopol) have been adopted for usage in many different industries.

11.4 Advantages and Limitations of Biodegradable Polymer

Many countries are gradually coming around to the idea of using biodegradable materials. It is believed that by minimizing trash, these materials will aid the environment. According to Mohanty, "the emerging concern of garbage causes in the limitation of dump yard accessibility and the necessity for the ecologically liable usage of resources" are two primary motivations for employing biodegradable materials. One could find clear benefits in manufacturing bio-friendly plastics more of a reality, as the government and many groups are fighting to conserve the environment. Due to their high price, bio-friendly polymers are not extensively utilized in the packaging business; therefore, conventional plastics have a monopoly on that market. The key to lowering prices is for several businesses to collectively purchase a sizable quantity of biodegradable materials. According to the theory of supply and demand, an increase in demand should result in lower prices. Biodegradable plastics must behave and perform in the same ways as their nonbiodegradable counterparts, and they must share the same structural and functional features. Biodegradable materials must be capable of "bacterial as well as ecological deterioration over disposition, unless any harmful environmental consequences" (Mohanty et al., 2005). Since biopolymers are so readily available, inexpensive, lightweight, strong, and robust, they have mostly been studied as a reinforcement for polymer composites.

Various derivatives of plant and animal-based polysaccharides and proteins are explored to improve the properties of nanocomposites. Micro/nanofibrillated cellulose (Tingaut et al., 2010), cellulose nanocrystals (González et al., 2015), nanowhiskers (Sucinda et al., 2021), microbial cellulose nanofibers (Woehl et al., 2010), carboxymethyl cellulose (Ghanbarzadeh et al., 2013), and lignin nanoparticles (Yang et al., 2015), are few of the examples of bio-based reinforcement fillers used in the preparation of bionanocomposites. All these reinforcement fillers are the derivatives or processed nanofibers/nanoparticles or the derivatives of plant or animal-based carbohydrates or proteins. Apart from nanomaterials, it is based on the functionality and applicability of various dyes/pigments (Micó-Vicent et al., 2020), essential oils, and nanoliposomes (Ghasempour et al., 2022). Metal-based nanoparticles are inexpensive and inert sources of reinforcement, and these are used in biodegradable films as antiradiation or antimicrobial agents due to their photocatalytic activity. Nanocrystals, on the other hand, exhibit good mechanical properties with a good network-forming ability due to their high crystalline structure with a high aspect ratio (Noshirvani et al., 2018).

11.5 Bionanocomposites for Food Packaging

Biopolymers possess poor mechanical and barrier attributes owing to their biodegradability and water-loving nature. Making them adequate for food package uses, biopolymers are reinforced with nanoparticles to improve their barrier and mechanical properties, which are known as bionanocomposites. The introduced nanoparticles bind with biopolymers, reducing the gas and water vapor permeability due to the formation of a complex diffusion path.

Biomaterial-based nanocomposites are a novel type of biohybrid material that consists of biopolymers reinforced with inorganic molecules or solids that possess various morphological properties. On a bionanocomposite matrix, the biopolymers are the continuous phase, while the reinforcement agent is a dispersed phase. Bionanocomposites have better structural and barrier properties than conventional biopolymers. Based on the morphological characteristics of reinforcement materials, these are classified as particulate bionanocomposites (isodimensional reinforcement particles), elongated particle (like nanotubes and nanofibrils) bionanocomposites, and layered polymer nanocomposites. These are likewise biodegradable, and the raw materials are plentiful from renewable sources. This property of bio nanocomposites makes them suitable for food packaging applications. Despite being an environmentally clean, safe, and eco-friendly method for food packaging applications, they are not suited for ingestion, unlike edible films, due to their higher strength and barrier qualities. Biopolymer, bio nanofiller, or both are used in bio-based nanocomposites.

Bionanocomposites can be amended by mixing nanofillers such as MMT (montmorillonite), TiO_2, AgO, ZnO, SiO_2, and bio-derived polymers comprising PHB, polybutylene succinate (PBS), PLA also poly-caprolactone (PCL), in addition to natural polymers like starch and chitosan. The property of biopolymers can be improved via nanocomposite formations as discussed in detail in Chapters 3 and 4.

The reinforcement agents incorporated into the biopolymeric films not only further improve their mechanical attributes but also have a significant impact on various physical and structural properties, which enables them to be suitable materials for food packaging. Several metal oxide-based reinforcement agents have also been proven to provide potential antimicrobial activity. However, incorporation of several other agents has proved to be beneficial in improving the shelf life of products to be packed, which is also explored by researchers, where the release of active compounds is evidently controlled by the reinforcement agents.

11.5.1 Gas and Water Vapor Permeability

Among the main purposes of comprising mineral particles as fillers in a polymer matrix is to modify their barrier qualities for definite uses. The addition of fillers has the main focus of decreasing the transmission rate of water vapor across the matrix (Castillo et al., 2015). Several biopolymers, such as starch and cellulose derivatives, possess good film-forming properties but have poor mechanical and water vapor permeability properties. These can be improved by two approaches. One is to mix the polymers together to produce biodegradable blended films with modification in their properties. Another approach is the production of nanocomposites by adding a filler to improve their properties.

The preparation of bionanocomposites with suitable reinforcement agents tend to decrease water vapor permeability. The addition of fillers into biopolymeric materials results in a physical blockage to water molecule transportation by forming a complex route. This is primarily influenced by the concentration of the nanofiller and the chemical correlation between the nanofillers and the matrix. The incorporation of nanofillers into the biopolymer matrices also resulted in a notable decease in the elongation at break. The use of nanofibers as reinforcement in bionanocomposites has been reported to possess improved tensile and puncture strength along with reduced water vapor permeability of the developed materials (Carvalho et al., 2020). Incorporation of cellulose nanocrystals has decreased water vapor permeability due to its higher water resistance as a result of its crystalline structure and hydrophobic character and increase in tortuosity (Noshirvani *et al.*, 2018).

11.5.2 Mechanical Properties (Tensile Strength, Elongation at Break, and Work of Fracture)

Various mechanical features like tensile strength, elongation at break, Young's modulus, etc. were the determining factors representing the impact of added fillers or reinforcement agents on the mechanical properties, which subsequently results in their suitability in food packaging applications. These mechanical properties are derived from the tensile stress–strain curves in standard texture profile analysis method. Most of the biopolymeric materials are reported to improve their tensile strength and other mechanical characteristics owing to the addition of reinforcement agents. In short, the mechanical strength of the bionanocomposites increases as a result of the incorporation of nanofillers. Bionanocomposites made of biopolymeric materials and loaded with nanoparticles can sustain large forces and are not brittle (Murali et al., 2019).

11.5.3 Microstructure

It was noted that the levels of nanoparticle (nanofiller) incorporation affect the uniformity of nanoparticle distribution in the formed film. The study that incorporated TiO_2 nanoparticles in the starch biopolymer film has shown that the film was homogeneous and smooth until the addition of 1% of the nanoparticle, where the high content of nanoparticles resulted in a lack of affinity between the polymer matrix and nanoparticles. As a result, the aggregation of nanoparticles took place and inhomogeneity in surface morphology was evident. Wu et al. (2019) studied the incorporation of chitosan-gallic acid nanoparticles incorporation at lower levels, resulting in a relatively compact and smoother surface without the presence of cracks, whereas an increase in the nanoparticle concentration to 15% results in the development of visible microcracks and discontinuous void space. This would have been due to the agglomeration of nanoparticles inside the film matrix when incorporated at high concentrations (Wu et al., 2019). A similar phenomenon was also observed in several other studies of bionanocomposites production.

11.5.4 Antibacterial Properties

Nanoparticles exhibiting antibacterial properties have been explored to be incorporated in packaging films and proven to be effective against food spoilage bacteria. The direct incorporation of metallic NPs into bionanocomposite films would pose health risks due to their toxicity to eukaryotic cells and their possible migration from the food package to the food product. Surface modification or functionalization of inorganic NPs such as TiO_2 and SiO_2 is a strategy to improve the antimicrobial characteristics of bionanocomposite matrix (Hajizadeh et al., 2020). In another study, ZnO-reinforced chitosan bionanocomposite film has shown intrinsic antimicrobial properties for the storage of fresh meat, when used as primary packaging material (Souza et al., 2020). Another study has incorporated Silver nanoparticles into agar films, which have shown higher levels of antibacterial activity as a result of interaction between the positively charged silver ions and negatively charged phosphorus or sulfur-containing biomacromolecule compounds (proteins and nucleic acids), which can cause structural changes and deformation in bacterial cell walls and membranes that leads to disruption of metabolic processes followed by cell death (Butkus et al., 2003; Rhim et al., 2014). Apart from metal oxides, the nanocomposites prepared with the incorporation of essential oils have also been shown to be beneficial over the antimicrobial activity. The reinforcement agents in these studies often hinder the release of active compounds over time (Souza et al., 2019). Metal nanoparticles also have proven the potential to treat bacteria, including both Gram-positive and Gram-negative strains, with a less-toxic effect. In a study, chitosan film was embedded with silver nanoparticles to improve the sustained release kinetics of the Itraconazole drug and has been shown to provide a controlled drug delivery profile (Saruchi et al., 2022). In another study, Magnesium oxide was reinforced in chitosan bio nanocomposite, which was incorporated with clove oil for the active food packaging application (Sanuja et al., 2014).

11.5.5 Thermal Properties

Thermal properties of the nanocomposites have been explored by numerous researchers, as the change in thermal properties impacts the suitability of bionanocomposites films as food packaging materials. The addition of reinforcement fillers in general increases the glass transition temperature of the nanocomposites. It has also been reported that the addition of nanofibers has improved the thermal resistance of the polymer (PVA) matrix, as a result of the increasing crystalline structure results. This in turn results in the increase in temperature required for the thermal decomposition (Abral et al., 2020). Similar results of improved thermal stability were also reported in other studies. Apart from thermal decomposition kinetics, studies have also revealed thermal stability at high temperatures as in terms of mass loss. It was observed that the mass loss was lesser for pure films as compared to the bionanocomposites (Wu et al., 2019). At higher thermal range, mass loss takes place as a result of depolymerization, evaporation of water, and volatile molecules. The preparation of nanocomposites results in lowering the mass loss, which is indicative of improved thermal stability (Safaei et al., 2019).

11.6 Biodegradation and Waste Valorization of Eco-friendly and Sustainable Biopolymers

Wheat straw, rice husk, corn straw, and sugarcane bagasse are only some examples of agricultural wastes that have been the subject of several studies aimed at perfecting the technique for cellulose extraction through chemical, physical, biological, or biochemical means (Das et al., 2014; Menon et al., 2017; Lubis et al., 2018). Corn cobs and blue agave are two examples of plentiful agricultural trash. When it comes to getting cellulose out of corn straw in a sustainable way, Song et al. (2020) reported on a method they called a low-temperature laccase system (LTLS). The extraction of cellulose to a high purity level was accomplished using the system that was used in the study. Some of the benefits of this system include its easy operation, great yield, and the reusing of solutions that are used in the process of cellulose recovery. Grapes are one of the most frequently farmed fruit crops in the world, and their skins are one of the best sources of cellulose (Ju and Song, 2020).

Cellulose may be extracted from grape skins using organic solvent and acid-base treatment. Isolated cellulose was utilized in the synthesis of nanostructured cellulose, which is also referred to as nanocrystals (Lu and Hsieh, 2012). For the production of nanofibers, several manufacturing enterprises have turned to the cellulose extracted from Nypa fruit peel (Fauzee and Othaman, 2013). It has been investigated that acid hydrolysis can be used to extract cellulose from pineapple leaves, which is another type of agricultural waste that is readily available. Additionally, it has been reported that this technique has been used in industrial settings for the yield of cellulose nanocrystals (Dos Santos et al., 2013). The usual feedstock used for pectin extraction includes a banana peel, jackfruit peel, orange peel, grapefruit peel, lemon peel, passion fruit peel, cocoa peel, fig seed, guava pulp, mango peel, beet pulp, and apple pomace. It has been stated that TPS (a starch-and-plasticizer combination that is heated to plasticize it) may be produced from a variety of plant species using a variety of ways (Fabunmi et al., 2006). *Cupriavidus necator* H16 yields an ideal 83% PHB when grown on waste palm oil and lard waste from an animal source. This same microbe was studied by Obruca et al. (2010); they observed the highest yields of 67.9% and 89.10% for PHB synthesis from waste-fried rapeseed oil and waste oil from coffee grounds, respectively (Obruca et al., 2010). *Pseudomonas sp.* was used in another investigation (Chaudhry et al., 2011) to transform used frying corn oil into PHA with a maximum efficiency of 35.63%.

Concerns have been raised about the viability of recycling biopolymers. Right now, we are talking about the benefits and downsides of reusing biopolymers. Some people believe that recycling wastes the potential of biodegradable polymers. Furthermore, bio-derived polymers highlight a serious concern when they end up in the same recycling streams as conventional plastics or the same composting streams as food scraps. PET recyclers, recycling associations, and government authorities have expressed growing concern regarding the possibility of PLA bottle contamination of the PET recycling stream, yield loss, increased sorting costs, and detrimental effects on recovered PET quality and processing. These issues have prompted a rise in the level of anxiety felt by these groups. It is important to have a critical mass of PLA (at least 200 million kg created annually) in order for autonomous rigid packaging recycling to be able to function as a financially viable business (Cornell, 2007). However, proponents of recycling argue that the majority of biodegradable polymers, such as PLA, dissolve easily in normal settings, even in the presence of microbes. The disposal of significant quantities of PLA may in the future, cause problems for the surrounding ecosystem. The disposal of biodegradable plastics in landfills can lead to the production of methane, which is harmful to the environment and the climate. Another issue is that the disposal of biopolymer products results in the loss of valuable raw resources (e.g., hydroxy acids). If you want to manufacture biopolymer monomers while using a minimum amount of resources that can be renewed, recycling biopolymers is a crucial step to take. Waste biopolymers that have been degraded over an extended period of time can be used to produce monomers and intermediates. At this time, there is a significant disparity between the supply of postconsumer PLA and the demand for compost. It is possible to recycle bio-based polymers that do not biodegrade, such as bio-PET, by mixing them with their fossil fuel-based equivalents, such as industrial PET, and then recycling the mixture. Last but not least, biopolymer recycling complies with the guidelines set forth by the circular economy.

11.7 Market Trends in Biodegradable Food Packaging

11.7.1 Smart Food-Packaging System

Intelligent packaging (IP) is a type of packaging technology that guarantees to send along only the best and safest goods. Put simply, "smart packaging" is when a product's quality and shelf life are improved by the addition of certain components to the packaging.

During transit and storage, the quality of packaged food can be monitored by an intelligent system. These innovations were developed in response to the rising need for longer-lasting, more hygienic food products. The use of packaging materials that incorporate active and IP systems or systems is anticipated to create a lucrative market including, but not limited to, temperature, luminosity, moisture, atmospheric pressure, microbial life, gaseous emissions, and so on. It features products of varying sizes and forms for the consumer's convenience and simplicity of use (Yam et al., 2005; Marsh and Bugusu, 2007). Alternatively, active and IP concepts, which are at the heart of smart packaging systems, rely on the user's engagement with the package and its contents to offer the food additional layers of protection. Table 11.2 shows a summary of the intelligent response displayed by various formulations of biodegradable film.

Taking into account the primary packaging, it is imperative to create novel food-packaging (NFP) systems by utilizing a variety of biomaterials and techniques, embedding

Table 11.2 Activity of different compounds in biodegradable packaging films

Biodegradable Packaging Material	Antimicrobial Activity Toward	Application	References
Poly (vinyl alcohol) (PVA)	*Staphylococcus aureus and Escherichia coli*	Active food packaging	Sarwar et al. (2018)
Chitosan	*Escherichia. coli, Staphylococcus aureus, Aspergillus niger, and Penicillium. Citrinum*	Active food packaging for litchi fruits	Wu et al. (2018)
PVA	*S. Typhimurium and S. aureus*	Active food packaging for chicken meat	Mathew et al. (2019)
Cellulose nanofibrils	*E. coli and Listeria monocytogenes*	Active food packaging	Yu et al. (2019)
Chitosan	*E. coli, Salmonella, S. aureus, and L. monocytogenes*	Active and intelligent food packaging	Qin et al. (2019)
Cellulose/keratin	*S. aureus (MRSA) and Enterococcus (VRE)*	Active food packaging	Tran et al. (2018)
Gelatin/furcellaran	*E. coli, S. aureus, and S. aureus*	Active food packaging for kiwi fruits	Jamróz et al. (2019)
Starch	*Mucor sp., Mucor circinelloides*	Active food packaging for mushrooms	Zhang et al. (2019)
Chitosan/polylactic acid (PLA)	*E. coli, S. aureus, B. cinerea, and Rhizopus*	Active food packaging for fruit and vegetable	Yun'an Qing et al. (2018)
Alginate	*L. monocytogenes*	Active food packaging for smoked salmon meat	Vizzini et al. (2020)
Polylactide/polyethylene glycol/ polycaprolactone	*S. aureus and E. coli*	Active food packaging for eggs	Ahmed et al. (2019)
Chitosan/PVA	*S. aureus, P. aeruginosa, E. coli (O157 : H7), and Candida albicans*	Active food packaging for soft white cheese	Youssef et al. (2015)

sensors and indicators, functions, biodegradable materials, nanotechnology, essential oils, and plant extracts, while also retaining the quality and nutrition of food consistent and extending its shelf life; the environmental effects of the package restricted the need for food shortages. The absence of complacency and negative inactivity between packaging and food components, long-term performance, preventative measures of food spoilage, and improvement of human wellbeing are the strengths of NFP, which is also known as active, smart, and green technologies, and these can be deemed as the ultimate objective for the future of food packaging technology. In addition, NFP is also known as active, smart, and green technologies. A number of other names have been used to refer to NFP, including active technologies, smart technologies, and green technologies (Lone et al., 2016).

11.7.2 Biodegradable Active Packaging

A package is considered "active" if it is active in some way beyond simply protecting the goods from the environment (Rooney, 1995). Active packaging is a process in which the product, the container, and the surroundings work together to improve the product or increase its shelf life. It is also known as "packaging that alters to preserve the quality of the food by extending its shelf life, preventing it from going bad or enhancing its sensory attributes" (Ahvenainen, 2003). Using active packaging methods such as coating, microperforation, lamination, coextrusion, or a polymer blend, the ambient concentration of gaseous chemicals inside the container can be altered. Extrusion and coextrusion are two additional examples of active packing methods. Following is a summary of the use of active function in biodegradable films, as outlined in Table 11.1.

11.7.2.1 Using Plant Extract

Polyphenol-rich plant extracts have been used as an antimicrobial ingredient in biodegradable active films, according to multiple studies. Phenolic substances, acting as ionophores, can lower the pH difference across microbial membranes. As a result, the proton loses some of its propulsion. This ultimately leads to nutrient malabsorption and bacterial cell death (Xu et al., 2021). As an example of the incorporation of natural extracts in functional packaging, Wu (2019) created films with pomelo peel powder using polyphenols from tea extract at 5%, 10%, 15%, and 20% amounts. The polyphenols from tea enhanced the films' antioxidant and antimicrobial action against *Escherichia coli* and *Staphylococcus aureus* as their concentration was raised. Moreover, the tea-infused film showed improved efficacy against Gram-positive bacteria in the current study (*S. aureus*). Antimicrobial and antioxidant effects of hop extract (0 and 0.15 wt% in the film-forming solution) in chitosan and gelatin films were studied by Xu et al. (2021). Hop extract not only increased the films' antioxidant activity, but it also inhibited the development of Gram-positive (*B. subtilis* and *Listeria monocytogenes*) and Gram-negative (*E. coli* and *Shigella sonnei*) microorganisms. Though Gram-positive microbes were hit harder than Gram-negative ones, the former were the primary targets. The antioxidant potential of chitosan films was significantly bolstered by the addition of extracts from pine nutshells, peanut shells, and jujube leaves (Zhang *et al.*, 2020). The highest levels of antioxidant activity were found in films that included jujube leaf extract. A similar level of antioxidant activity was found in Khorasan wheat starch sheets containing 1% moringa leaf extract (Ju *et al.*, 2019). According to Ju and Song (2020), yellow onion peel extract has great free radical scavenging ability and was used in funoran films derived from red algae (Ju and Song, 2020). Camucamu extract was used in teff starch films at ratios of 0.3%, 0.6%, 1%, and 2% by Ju and Song (2019). The 2,2′-azino-bis (3-ethylbenzothiazoline-6-sulfonic acid) radical scavenging activity was greatest in films having a 2% camucamu extract, while the 2,2-diphenyl-1-picrylhydrazyl (DPPH) radical scavenging activity was increased from 42.64% to 89.19% at concentrations of 0.3% and 1%, respectively (Ju and Song, 2019).

11.7.2.2 Using Essential Oils

Essential oils (EO) are a type of subsidiary plant metabolite that is added to food because of their antimicrobial and antioxidant qualities. The antimicrobial activity of essential oils results from the presence of numerous compounds, many of which are hydrophobic (Pateiro et al., 2021). The influence of lavender essential oil concentration on the characteristics of biodegradable starch, furcellaran, and gelatin sheets was investigated by Jamroz et al. (2018). These authors found that the antioxidant activity and the inhibition zone against Gram-negative (*E. coli*) and Gram-positive (*S. aureus*) bacteria increased with higher concentrations (Jamróz et al., 2018). Influences of thyme, clove, and rosemary essential oils (1.5% concentration) in poultry protein films were studied by Sarcaoglu and Turhan (2020). Essential oils of clove, thyme, and rosemary all performed similarly, but the essential oil of clove performed the best. Essential oil-enriched films were also tested for their antibacterial effects against *B. subtilis, S. aureus, E. coli,* and *L. monocytogenes*. Thyme essential oil was more effective resistance to *L. monocytogenes* than clove oil was, but clove oil was more successful contrary to *B. subtilis, S. aureus, and E. coli*. The antibacterial effects of rosemary essential oil were the weakest (Sarıcaoglu and Turhan, 2020). It was discovered by Lian et al. (2020) that the growth performance of E. coli was reduced when thyme essential oil was present in active films made from chitin and pullulan (Lian et al., 2020). Likewise, the use of rosemary EO (Abdollahi et al., 2012), apricot kernel EO (Priyadarshi et al., 2018), *Thymus daenensis* EO (Moghimi et al., 2017), ginger EO (Alexandre et al., 2016), and rockrose EO (Luís et al., 2020) in biodegradable films has been investigated and reported.

11.7.3 Biodegradable Intelligent Packaging

The term "intelligent packaging" refers to a modern concept that has emerged in response to the growing concerns over the internal movement of preservatives into food and the packaging's emerging function as a communication tool to aid in consumer decision-making. According to the regulations set forth in EC/450/2009, "intelligent materials and articles" are those that monitor the storage conditions of food. Food storage and cooking instructions can be communicated to the customer via "intelligent packaging" (temperature, pH). Conventional packaging's communication function is extended because of its ability to notice, sense, and record changes in the product's environment and then transmit this data to the customer. (Restuccia et al., 2010; Realini and Marcos, 2014). Foods packaged in containers do not improve in quality or last longer. It can, however, monitor the food's quality and give quantitative information about changes in pigment or physical characteristics in real time (Li et al., 2019; Chen et al., 2020). Active packaging is designed to keep things in top shape, while IP conveys data about quality. "Smart packaging" can also be made by combining intelligent and active container technologies (Jamróz et al., 2019). The consumer can immediately tell that the product is safe to consume because of the packaging's shift in color. We explain and summarize in Table 11.3 how natural dyes can be incorporated into biodegradable films.

Table 11.3 A summary of the intelligent response displayed by various formulations of biodegradable film

Biodegradable Packaging Material	Intelligent Compound	Intelligent Response	References
Cellulose-polyvinyl alcohol	Acidochromic dye (1-hydroxy-4-[4-(ethylsulfurate sulfonyl)-phenylazo]—naphthalene)	• At pH 7, 10, and 12, the film color changed to yellow, brick red, and purple, respectively; when applied to shrimp, the shrimp changed from yellow to brown after 24 hours when stored at 25 °C.	Ding et al. (2020)
Chitosan	Alizarin	• At pH 4, the film's color was yellow; at pH 5, it was khaki; after that, it remained pale until neutral pH; at alkaline pH, it turned purple. • Applying it to fish, transformation from green to yellow was observed the fish went bad (after 48 hr at 25°C).	Ezati and Rhim (2020)
Konjac glucomannan and oxidized chitin nanoparticles	Red cabbage anthocyanins	• When exposed to a pH range of 2.0–4.0, films with low RCA looked yellow, while those with high red cabbage anthocyanins became red. • When the pH was increased, blue or light brown films turned green.	Wu et al. (2020)
Alginate and polyvinyl alcohol	Red beetroot extract	• Seven days of storage at 4 °C resulted in no noticeable color change, while storage at 40 °C and 60 °C resulted in visible color change.	Tekin et al. (2020)
CO_2 indicators Polystyrene and branched polyethyleneimine	Bromothymol blue and tetra butyl ammonium	• The color of the indicator shifted from blue (0.3% CO_2) to yellow (65.8% CO_2) as the CO_2 concentration increased during fermentation; this was observed in kimchi storage (3 d at 20 °C).	Lyu et al. (2019)

Table 11.3 (Continued)

Biodegradable Packaging Material	Intelligent Compound	Intelligent Response	References
pH-based freshness indicators starch-polyvinyl alcohol, nutmeg oil	Jamun extract	• A film's violet jamun extract turned a bright cherry crimson when exposed to acid. • Under alkaline circumstances, its color shifted to a brownish yellow.	Jayakumar et al. (2019)
Furcellaran, gelatin	Green tea extract	• Under acidic circumstances (pH 3.0), the chlorophyll in green tea extract degraded into pheophytin and pheophorbide, which caused the films to turn white. • Tested in carp fillets for 24 hr at 22 °C, chlorogenic acid (CA) (found in green tea) oxidized to CA quinines, causing a shift in color to brown, yellow, or green.	Jamroz et al. (2019)
Succinylated chitosan and hydroxypropyl chitosan	Bromocresol blue and methyl red (1:1)	• Evaluation of banana freshness (104 hr at 10 °C) using a linear color connection to CO_2 concentration in packaging atmosphere (10–50%).	Wan et al. (2021)
Chitosan and agarose	Anthocyanins	• After 10 hr of storing, the applied dye on fish turned pink, and after 20 hr, it turned purple.	Wu et al. (2018)
Starch/polyvinyl alcohol	Red pitaya peel betalains	• As the TVBN concentration increased to 19.38 mg/100 g (after 24 hr), the film's color changed to a pale yellow. • The generation of volatile amines caused the film to turn an absolute yellow at 35.18 mg TVBN/100 g (after 48 hr).	Qin et al. (2020)
Glucomannan polyvinyl alcohol	Betacyanin	• When applied to a fish, the TVBN concentration increased to 39.74 mg/100 g after 8 d, causing the fish to change color from purple to yellow at a chilling temperature.	Ardiyansyah et al. (2018)

11.7.3.1 pH-based Freshness Indicators

Many foods, including fruits, veggies, meat, and dairy products, have their pH changed by respiration and microbial development while they are being stored. With the addition of pH-sensitive compounds to food packaging, it is feasible to track the safety of these products. Color shifts occur as a consequence of chemical and structural changes in these substances. Anthocyanins (obtained from purple potatoes) were investigated for their potential as a smart component in chitosan- and chitin-based biodegradable films by Li et al. (2019). Pigment variation as an outcome of pH was then tracked after exposing the resulting images to acidic and alkaline conditions (Li et al., 2019). The researchers demonstrated a connection between the chemical structure of anthocyanins and the color of films at different pH levels. Specifically, it indicates that at a pH between 1 and 3, the anthocyanin molecules predominantly exist in the form of the flavylium cation, which is responsible for the crimson (red) coloration. Flavylium cation hydration and following conversion to methanol and chalcone base made films colorless in the pH range of 3.0–6.0. The quinone compound turned blue when the pH was made more alkaline. Similar color changes occurred when anthocyanins were incorporated into packaging sheets made of cellulose nanofibers, novel konjac glucomannan, fish gelatin, chitosan, and k-carrageenan. (Liu et al., 2019; Chen et al., 2020; Wu et al., 2020). pH-sensitive green tea extracts have been shown to be useful for monitoring pH variations in furcellaran and gelatin-based films (Jamróz et al., 2019), complementing the use of anthocyanins in this application. As the pH was raised from 3.0 to 12.0, the color of the films (in buffer solution) changed from white to orange. Rosemary extracts in turbellarian-gelatin (Jancikova et al., 2019), jamun extract in starch polyvinyl alcohol (Jayakumar et al., 2019), betalains in starch polyvinyl alcohol (Qin et al., 2020), and alizarin in chitosan all show pH-dependent color shifts in compostable films (Ezati and Rhim, 2020).

11.7.3.2 CO_2 Limit Detectors

For example, when microorganisms spoil foods like fruits, veggies, and meat products, carbon dioxide gas is constantly being produced and accumulating in the container (Kimbuathong et al., 2020). Numerous reports have detailed the creation of biodegradable films and coatings sensitive enough to identify CO_2 levels inside a container. For example, Saliu and Della Pergola (2018) made a lysine and, poly-lysine and anthocyanins aqueous indicator to be used in labeling an ethyl cellulose film. When exposed to air, labels quickly changed colors even at low CO_2 levels (below 2.5%). The acidification of the film and the generation of a carbamic acid derivative formed the basis of the color change mechanism, which was based on the reaction between the -amino group of lysine and CO_2. This caused a shift in the coloration of anthocyanins. Further testing revealed that CO_2 production from microbial decomposition at low temperatures (0–5°C) was extremely successful as an active coating for preserving fresh chicken breast (Saliu and Della Pergola, 2018). Lyu et al. (2019) designed a packaging method for the monitoring of kimchi fermentation using bromothymol blue and tetrabutylammonium as indicators for CO_2 detection. This CO_2-sensitive packaging was able to adapt to variations in its surrounding atmosphere (from bluish to yellowish color as the concentration of CO_2 increased) (Lyu et al., 2019). Wan et al. (2021) looked into the feasibility of using a biodegradable film (containing bromocresol blue and methyl red as markers) to track a banana's development while it was stored at 10 °C (Wan et al., 2021).

11.8 Emerging Applications of Novel Techniques for the Development of Biodegradable Packaging Materials

11.8.1 3D Food Printing

3D printing (3DP) technology has been used to create foods for dysphagic patients (Pant et al., 2021), nutrient-enriched snacks (Wilson et al., 2023), food waste utilization (Muthurajan et al., 2021), food packaging (Nida et al., 2022), and other products. Layer-by-layer food product deposition is used in 3D printing to create a three-dimensional edible object. Rheological properties play a significant role in the layer-by-layer deposition process used in food 3D printing. When printing, it's important to know how the input material behaves in terms of viscosity, elasticity, and consistency. The printed structure's tactile and visual qualities are also impacted by its rheological properties. Furthermore, additives like hydrocolloids, lipids, carbohydrates, salts, etc., have been used to increase the printing material's flowability. Materials with additives not only have better rheology, but they also make printed food taste, smell, and feel better (Chen et al., 2022). When a 3D-printed object's physical and chemical state shifts over time (the fourth dimension) in response to environmental factors like pH, temperature, water, light, etc., the process is known as "4D printing." Changing the form of 3D purple sweet potatoes (He et al., 2020), starch-based gel (Liu et al., 2021) 2021), and pumpkin paper are just some of the culinary uses of 4D printing (Chen et al., 2021b). Anthocyanin-potato starch gel and sago starch colored with turmeric were both investigated as potential colorants for 4D printing (Ghazal et al., 2019). While 3D and 4D printing has shown promising uses in the food industry, there are still some obstacles to overcome. In 3D printing, additional support material is required for manufacturing complex structures, such as those with curved surfaces. The flat shape required for 3D printing provides a weak spot in the construction. In the early days of 3D food printing, chocolate, and sugar powder were the primary printing substrates due to their favorable properties for 3D printing. In addition to printing, blending, tempering, and mixing are essential first steps in food manufacture. The composition of the meal, especially the macronutrients, is also crucial. An effectively printed food item will feature all of these characteristics (Shanthamma et al., 2021).

11.8.1.1 3D Printing in Packaging Material Design

The currently available 3DP processes can be categorized into seven major groups: vat photo polymerizations, material jetting, binder jetting, powder bed fusion, material extrusion, directed energy deposition, and sheet lamination (Ian Gibson, 2015; Calignano et al., 2017). However, based on material nature and price, extrusion is the most suited technique for food applications. Notably, research has also been done on the practice of powder bed fusion, material jetting, and binder jetting for the production of specific types of consumables. Binder jetting can also be employed to print food using eatable powdered materials (such as sugar, starch, or corn flour) and edible liquid binders (such as sugar solutions or xanthan-based binders) that attach the powder particles inside the 3D structure (Holland et al., 2018). Consequently, the properties of the pulverized material, the liquid binder, and the operating conditions have a substantial influence on the nature of the materials produced by this method.

A second approach is the powder bed fusion procedure. Lasers or electron beams are used to dissolve and fuse powdered materials. Due to its high precision, this printing technology can make complicated material geometries. PBF printers can be time consuming and costly because of the necessity for preheating, vacuum generation, and cooling times before every print. PBF procedures include a wide variety of techniques, the most common of which is selective laser melting, but other methods, such as electron beam melting and selective heat sintering, also fall into this category. Powdered materials like sugar, fat, or starch granules can be used in 3D printers that are equipped for PBF to create edible items (Liu et al., 2017). However, neither its research nor its applications in the commercial sector are particularly ubiquitous at this time. The degree of precision and accuracy that can be achieved in food production using the PBF technique is directly proportional to the particle size, flowability, bulk density, and wettability of the materials being pulverized (Godoi et al., 2016). Printing with pure sugar utilizing a technique that involves particular hot air sintering and melting was one of the important ways that CandyFab, a company that operates for profit, evaluated the potential of PBF as a food printing technology. The sugar crystals liquefied due to the heated air, and after that, they merged with one another to create edible three-dimensional objects.

Material extrusion is the most favored 3DP method for both academic and marketable use because of its low cost and ease of implementation. To create an object with a material extrusion 3DP system, a nozzle deposits substance onto a platform, heat bonds each layer, and finally fused together to make the final product. Because of their potential to be extruded effortlessly using a syringe along with maintaining their shape after printing, edible materials like hydrogels, sugar frostings, cheeses, and chocolate have received the most attention from researchers interested in food inks for culinary applications (Sun et al., 2015). Pureed potatoes with the appropriate viscous could be made by adding potato starch to the mixing (Liu et al., 2018). Adding NaCl to fish and surimi gels facilitated protein crosslinking, which improved the gels' textural qualities (Wang et al., 2018). Further, by adding pectin to the formulation, the texture along with the durability of 3D-printed fruit-purée-based treats have been improvised (Derossi et al., 2018).

11.8.2 4D Food Printing

Additive manufacturing, or "4D printing," is an expansion of the 3D printing method. The field of four-dimensional printing (4DP) used to create edible items is a young one. This idea was first developed in 2013 by a group of researchers at MIT (Campbell et al., 2014). Differences between fourth-generation printing (4DP) and third-generation printing (3DP) center on the use of "smart design" and "smart materials," which allow 4DP-printed buildings to undergo "morphological evolution" (Pei and Loh, 2018). Intelligent materials with self-assembly, adaptability, and self-healing characteristics are used in 4D printing (Liu et al., 2021). Materials used in 4DP are typically either monomeric or polymeric (Kuang et al., 2019). It consists of liquid crystal elastomers, hybrid hydrogels, and shape-memory polymers (Singholi and Sharma, 2020). MIT scientists were the first to introduce 4DP to the food service business. Using glucose, cellulose, and proteins as well as water as a stimulant, they produced a 2D picture. The addition of water transformed the printed 2D material into a 3D object (Gadelha et al., 2017). The 4D printing of edibles will make it easier for companies to better cater to their customer's needs and create new and exciting flavor combinations. Ink made from different combinations of food materials, tailored to each individual food's structure and formula, is used to create 4D-printed food samples that alter in response to external stimuli (Teng et al., 2021). The color, texture, aroma, and shape of food can all change in response to environmental factors like temperature and pH. As a result of the stimulation provided by microwave heating, a 4D metamorphosis occurred in 3D-printed buckwheat pastries and lotus stem powder gel. (Chen et al., 2021a). Likewise, it has been observed that 3D-printed blends of soy protein isolate, pumpkin, and beetroot exhibit pH-induced color variations (Phuhongsung et al., 2020). Food-based hydrocolloids (Pant et al., 2021), chocolate (Rando and Ramaioli, 2021), soybeans (Phuhongsung et al., 2020), beef (Wilson et al., 2023), and carbohydrates (Zheng et al., 2021).

11.8.2.1 4D Printing in Packaging Material Design

The stimulus-impelled alterations in the 3D printed specimen are referred to as the 4DP. As a result, canonical designs are crucial in the 4D printing procedure. An efficacious software resolution is critical in 4DP because it allows for the production of the intended product. A number of software resolutions and management software were used in the 4D model design and imprinting procedures. It contains one answer for each of the five phases, including simulation, modelling, slicer, host/firmware, monitoring, and printing management software. Simulation is a way of studying an object's or system's behavior by simulating it on a computer application. Simulation software is typically used prior to the scheme and production of a product; therefore, it lowers the threat of failing before real development applications. It is founded on the mathematical modeling of physical-world phenomena. It allows the user to comprehend the complete task except having to perform it. The simulation technology used in 4DP is classified into two different types: simulation utilizing finite element analysis (FEA) software and simulation function development software (Chung et al., 2017). After the simulation is completed, the 3D design of the printing item is created using modeling software. Modeling generates an object model with all of its qualities, which could be comparable to the physical object. It creates a 3D model of the item. Rhino 6 software and Rhinoceros 5.0 (Liu et al., 2021), Sketch Up Pro2015, and others are some popular computer-based graphics and computer-aided design software used for design augmentation in food imprinting. The properties of a synthesized 3D object alter gradually over time in 4DP. These modifications form when the object is placed in a suitable environment. Monitoring software allows for the monitoring and prompting of this stimulus-induced self-transformation of 4D-printed objects. This monitoring software can be installed on user devices and constantly checked. The 4DP procedure is made up of a number of unique steps and software. The output of one stage may use the input of subsequent steps. As a result, the software in each stage serves a certain purpose and may influence the next stage; all stages are interconnected and function as a network. As a result, the user monitors and controls all software at each step of processing. This is accomplished by utilizing a printing management procedure that is linked to every software at each phase and is intended to improve the usage of printers (Chung et al., 2017).

11.8.3 5D and 6D Food Printing

Despite the fact that 3DP and 4DP have demonstrated numerous and certain restrictions must be overcome for successful applications in the food business. Convoluted structure, such as arched surfaces, cannot be imprinted using 3DP without the benefit of additional support material. The flat shape of 3D-printed structures creates a vulnerable point within them. In contrast, 4DP also requires additional material and postprocessing techniques to alter the shape of an object and remove the supporting material. This increases the expense of printing by reason of the prolonged run time (5DP) required to build the structures. Because it does not require the use of fillers or further assisting framework, the 5DP method can be used to build structures that have complicated geometries. Additionally for greater mechanical strengththis enables it to be utilized in the process of overcoming the challenges that have been presented. The 5DP manufacturing process is a type of hybrid production method that incorporates subtractive and additive processes. It includes the printing of an extremely intricate framework utilizing triplicate movement axes (*X*, *Y*, and *Z*) and two rotational axes as part of the process. The *X*, *Y*, and *Z* coordinates stand in for the printing head of the printer, while the rotational axis denotes the printing bed. The year 2016 witnessed the debut of 5DP, which was developed by Mitsubishi Electric Research Laboratories (MERL). The MERL investigation group attempted to manufacture pressure covers by using a 5D printer as their tool of choice. It was found that the mechanical strength of the cap that was printed employing 5D printing (3.7 MPa) was significantly farther up than that of the structure that was manufactured using 3D printing (0.1 MPa) (Reddy and Devi, 2018). Ethereal Machines, based in Bengaluru, India, has created a hybrid "Halo 5D printer" that is suitable for producing jewelry as well as components for automobiles (Sathe, 2017). Therefore, a product that is more durable could be produced in numerous dimensions using 25% less material than 3DP (Reddy and Devi, 2018). Previous work on 5DP has shown that this technology has the potential to be used in the 5D imprinting of food and food containers. The technology has shown that this possibility is realized to some degree. Candies in the shape of balls, chips with concave shapes, and snacks are just some of the edible products that could be printed using the 5DP technology. The technology also allows for the printing of complex designs. Cutlery, for instance glassware and dishes that must have added support material in order to be impression using 3DP, could instead be printed through 5DP technology. Objects that have been produced using 6D printing technology are possible thanks to the combination of smart materials and fifth-generation additive manufacturing (5DP). The concept was initially presented in 2021 by Georgantzinos et al., and it is regarded as a successor to the 5DP technology. They have provided evidence to show that the 6DP technology is more advanced than the 4DP technology. 6DP has the ability to potentially save both time and money if it uses the appropriate processing specifications. Therefore, 6DP could be used in the food industry to produce robust materials with increased sensitivity to stimuli. This would be an advantage for the industry.

11.9 Conclusion and Future Perspective

Within the packaging sphere, bio-friendly plastics are among the most cutting-edge innovations. Businesses are unable to keep up with the demand for this revolutionary technology. The level of public support for and commitment to environmental protection will determine the scope of biodegradable plastics' adoption. There is no shortage of raw materials or financial backing to develop and support more applications for biodegradable polymers. There has been a dramatic surge in biodegradable technology in recent years, and signs suggest that the public is quite interested in learning more about these items. Many common polymers are made from petroleum and other fossil fuels like coal, gas, and natural gas. They are expected to live in the ecosystem for much longer than humans do (Ingrao et al., 2015). The development of man-made polymer in 2013 attained 299 million tonnes, up 3.9% from the previous year and 9.7% from 2010, referenced from (2016, The Statistics Portal). Despite decades of trial and error to boost the efficiency along with effectiveness of conventional plastics, the market share of bio-based plastics is still quite small (Organisation for Economic Co-operation and Development, 2013). Nevertheless, it is anticipated that the generation capacity of bio-based polymers will expand from 5,132,000 tonnes in 2013 to 17,000,000 tonnes in 2020; this still accounts for 2% and 4% of the global polymer output in those years. The market for biodegradable polyesters (PLA and PHA), which are comprised of bio-based polymers, has grown at the quickest rates lately. Although there existed an upsurge in the worldwide production of bio-based polymers, it is important to note that not all bio-driven polymers are Eco friendly, just as not entirely biomaterials are created using sustainable resources. Both naturally occurring polymers like starch and cellulose, as well as biodegradable synthetics and polyesters derived from biological sources, are capable of being hydrolyzed and broken down quickly in the environment (water, soil, and others). Polyesters, on the other hand, exhibit a broad spectrum of biodegradation periods. Unlike poly (L-lactide), which dissolves hydrolytically in a few weeks, PCL's abiotic hydrolysis takes much time hingeing on molar mass and crystallinity. However, the mechanical and processing properties of aromatic synthetic polyesters are superior, notwithstanding their sluggish rate of degradation. Accordingly, by contrasting the resulting controlled qualities and tailoring them to the right customer requirements, by 2016, the volume of municipal garbage that can be composted or recycled should be 35% lower than it was in 1995, as mandated by EU directives. Because they can be made from renewable resources, compostable plastics are a step toward a greener future by reducing the demand for nonbiodegradable petrochemicals. Green polymers are widely utilized for food packaging purposes because they provide a safe, biodegradable barrier without harming the environment. Though relatively modest in size, biodegradable packaging has become increasingly important as a result of rising consumer awareness of the environmental and social benefits it provides. Changes in composting infrastructure, newly developed technology, increased financial capacity, and policies regarding bioplastics and biodegradable polymers are required before the widespread adoption of bio-friendly thermosetting plastic in the packaging domain and the satisfaction of the rising expectations of environmentally conscious consumers. On the other hand, it is anticipated that more biodegradable packaging will be used in the years to come (Organisation for Economic Co-operation and Development, 2013). Hence, only a few nations have achieved considerable headway in the production of biodegradable polymers.

References

Abd alFattah Amara, A. (2008). Polyhydroyalkanoates: from basic research and molecular biology to application. *IIUM Engineering Journal*, *9*(1), 37–73.

Abdollahi, M., Rezaei, M., & Farzi, G. (2012). Improvement of active chitosan film properties with rosemary essential oil for food packaging. *International Journal of Food Science & Technology*, *47*(4), 847–853.

Abdou, E. S., Nagy, K. S. A., & Elsabee, M. Z. (2008). Extraction and characterization of chitin and chitosan from local sources. *Bioresource Technology*, *99*(5), 1359–1367.

Abdul Khalil, H. A., Banerjee, A., Saurabh, C. K., Tye, Y. Y., Suriani, A., Mohamed, A., Karim, A., Rizal, S., & Paridah, M. T. (2018). Biodegradable films for fruits and vegetables packaging application: preparation and properties. *Food Engineering Reviews*, *10*, 139–153.

Abral, H., Ariksa, J., Mahardika, M., Handayani, D., Aminah, I., Sandrawati, N., Sapuan, S. M., & Ilyas, R. A. (2020). Highly transparent and antimicrobial PVA based bionanocomposites reinforced by ginger nanofiber. *Polymer Testing*, *81*, 106186. https://doi.org/10.1016/J.POLYMERTESTING.2019.106186

Ahmed, M. A., Al-Kahtani, H. A., Jaswir, I., AbuTarboush, H., & Ismail, E. A. (2020). Extraction and characterization of gelatin from camel skin (potential halal gelatin) and production of gelatin nanoparticles. *Saudi Journal of Biological Sciences*, *27*(6), 1596–1601.

Ahvenainen, R. (2003). *Novel Food Packaging Techniques*. Elsevier.

Akhbarizadeh, R., Moore, F., & Keshavarzi, B. (2018). Investigating a probable relationship between microplastics and potentially toxic elements in fish muscles from northeast of Persian Gulf. *Environmental Pollution*, *232*, 154–163.

Al Sagheer, F. A., Al-Sughayer, M. A., Muslim, S., & Elsabee, M. Z. (2009). Extraction and characterization of chitin and chitosan from marine sources in Arabian Gulf. *Carbohydrate Polymers*, *77*(2), 410–419.

Alexandre, E. M. C., Lourenço, R. V., Bittante, A. M. Q. B., Moraes, I. C. F., & do Amaral Sobral, P. J. (2016). Gelatin-based films reinforced with montmorillonite and activated with nanoemulsion of ginger essential oil for food packaging applications. *Food Packaging and Shelf Life*, *10*, 87–96.

Altun, M., Wiefel, L., & Steinbüchel, A. (2018). Cyanophycin production from feather hydrolysate using biotechnological methods. *Preparative Biochemistry and Biotechnology*, *48*(7), 589–598.

Anderson, L. A., Islam, M. A., & Prather, K. L. J. (2018). Synthetic biology strategies for improving microbial synthesis of "green" biopolymers. *Journal of Biological Chemistry*, *293*(14), 5053–5061.

Ardiyansyah, Apriliyanti M.W., Wahyono A., Fatoni M., Poerwanto B., Suryaningsih W. (2018) 'The Potency of Betacyanins Extract from a Peel of Dragon Fruits as a Source of Colourimetric Indicator to Develop Intelligent Packaging for Fish Freshness Monitoring', in *IOP Conference Series: Earth and Environmental Science*. IOP Publishing, p. 012038. https://iopscience.iop.org/article/10.1088/1755-1315/207/1/012038/pdf

Aryan, Y., Yadav, P., & Samadder, S. R. (2019). Life cycle assessment of the existing and proposed plastic waste management options in India: a case study. *Journal of Cleaner Production*, *211*, 1268–1283.

Astm, D. (2004). Standard Specification for Compostable Plastics. *ASTM International*, West Conshohocken, PA.

Auras, R., Harte, B., & Selke, S. (2004). An overview of polylactides as packaging materials. *Macromolecular Bioscience*, *4*(9), 835–864.

Avella, M., Immirzi, B., Malinconico, M., Martuscelli, E., & Volpe, M. G. (1996). Reactive blending methodologies for Biopol. *Polymer International*, *39*(3), 191–204.

Bajaj, I., & Singhal, R. (2011). Poly (glutamic acid)–an emerging biopolymer of commercial interest. *Bioresource Technology*, *102*(10), 5551–5561.

Banerjee, A., Singh, M., Das, K., & Sharma, S. (2016). Study of biodegradable polyesters from algal sources for use in future textile fiber applications. *AATCC Journal of Research*, *3*(1), 1–6.

Barcelos, M. C. S., Vespermann, K. A., Pelissari, F. M., & Molina, G. (2020). Current status of biotechnological production and applications of microbial exopolysaccharides. *Critical Reviews in Food Science and Nutrition*, *60*(9), 1475–1495.

Bartnikowski, M., Dargaville, T. R., Ivanovski, S., & Hutmacher, D. W. (2019). Degradation mechanisms of polycaprolactone in the context of chemistry, geometry and environment. *Progress in Polymer Science*, *96*, 1–20.

Becker, A. (2015). Challenges and perspectives in combinatorial assembly of novel exopolysaccharide biosynthesis pathways. *Frontiers in Microbiology*, *6*, 687. https://doi.org/10.3389/fmicb.2015.00687

Benabid, F. Z., & Zouai, F. (2016). Natural polymers: cellulose, chitin, chitosan, gelatin, starch, carrageenan, xylan and dextran. *Algerian Journal of Natural Products*, *4*(3), 348–357.

Bhat, A. R., Irorere, V. U., Bartlett, T., Hill, D., Kedia, G., Morris, M. R. et al. (2013). Bacillus subtilis natto: a non-toxic source of poly-γ-glutamic acid that could be used as a cryoprotectant for probiotic bacteria. *Amb Express*, *3*, 1–9.

Burnat, M., Herrero, A., & Flores, E. (2014). Compartmentalized cyanophycin metabolism in the diazotrophic filaments of a heterocyst-forming cyanobacterium. *Proceedings of the National Academy of Sciences*, *111*(10), 3823–3828.

Butkus, M. A., Edling, L., & Labare, M. P. (2003). The efficacy of silver as a bactericidal agent: advantages, limitations and considerations for future use. *Journal of Water Supply: Research and Technology-AQUA*, *52*(6), 407–416. https://doi.org/10.2166/AQUA.2003.0037

Calignano, F., Manfredi, D., Ambrosio, E. P., Biamino, S., Lombardi, M., Atzeni, E., Salmi, A., Minetola, P., Iuliano, L., & Fino, P. (2017). Overview on additive manufacturing technologies. *Proceedings of the IEEE*, *105*(4), 593–612.

Campbell, T. A., Tibbits, S., & Garrett, B. (2014). The next wave: 4D printing. *Atlantica*, 1–16. https://www.jstor.org/stable/pdf/resrep03594.pdf

Cao, M., Geng, W., Liu, L., Song, C., Xie, H., Guo, W., Jin, Y., & Wang, S. (2011). Glutamic acid independent production of poly-γ-glutamic acid by *Bacillus amyloliquefaciens* LL3 and cloning of pgsBCA genes. *Bioresource Technology*, *102*(5), 4251–4257.

Cao, X., Zhang, M., Mujumdar, A. S., Zhong, Q., & Wang, Z. (2018). *Effects of ultrasonic pretreatments on quality, energy consumption and sterilization of barley grass in freeze drying*. *Ultrasonics Sonochemistry*, *40*, 333–340. https://doi.org/10.1016/j.ultsonch.2017.06.014

Carbery, M., O'Connor, W., & Palanisami, T. (2018). Trophic transfer of microplastics and mixed contaminants in the marine food web and implications for human health. *Environment International*, *115*, 400–409.

Carvalho, R. A., De Oliveira, A. C. S., Santos, T. A., Dias, M. V., Yoshida, M. I., Borges, S. V. et al. (2020). WPI and cellulose nanofibres bio-nanocomposites: effect of thyme essential oil on the morphological, mechanical, barrier and optical properties. *Journal of Polymers and the Environment*, *28*(1), 231–241. https://doi.org/10.1007/S10924-019-01598-6/METRICS

Castillo, L. A., López O.V., Ghilardi J., Villar M.A., Barbosa S.E., García M.A. (2015) Thermoplastic starch/talc bionanocomposites. Influence of particle morphology on final properties, *Food Hydrocolloids*, 28: 51, pp. 432–440. doi: https://doi.org/10.1016/J.FOODHYD.2015.05.030.

Cha, D. S., & Chinnan, M. S. (2004). Biopolymer-based antimicrobial packaging: a review. *Critical Reviews in Food Science and Nutrition*, *44*(4), 223–237.

Chandra, R., & Rustgi, R. (1998). Biodegradable polymers. *Progress in Polymer Science*, *23*(7), 1273–1335.

Chaudhry, W. N., Jamil, N., Ali, I., Ayaz, M. H., & Hasnain, S. (2011). Screening for polyhydroxyalkanoate (PHA)-producing bacterial strains and comparison of PHA production from various inexpensive carbon sources. *Annals of Microbiology*, *61*, 623–629.

Chen, S., Wu, M., Lu, P., Gao, L., Yan, S., & Wang, S. (2020). Development of pH indicator and antimicrobial cellulose nanofibre packaging film based on purple sweet potato anthocyanin and oregano essential oil. *International Journal of Biological Macromolecules*, *149*, 271–280.

Chen, C., Zhang, M., Guo, C., & Chen, H. (2021a). 4D printing of lotus root powder gel: color change induced by microwave. *Innovative Food Science & Emerging Technologies*, *68*, 102605.

Chen, F., Zhang, M., Liu, Z., & Bhandari, B. (2021b). 4D deformation based on double-layer structure of the pumpkin/paper. *Food Structure*, *27*, 100168.

Chen, Y., Zhang, M., Sun, Y., & Phuhongsung, P. (2022). Improving 3D/4D printing characteristics of natural food gels by novel additives: A review. *Food Hydrocolloids*, *123*, 107160.

Cheng, J., Hu, Y. N., Luo, Z. G., Chen, W., Chen, H. M., & Peng, X. C. (2017). Preparation and properties of octenyl succinate beta-cyclodextrin and its application as an emulsion stabilizer. *Food Chemistry*, *218*, 116.

Cho, Y., Shim, W. J., Jang, M., Han, G. M., & Hong, S. H. (2019). Abundance and characteristics of microplastics in market bivalves from South Korea. *Environmental Pollution*, *245*, 1107–1116.

Chung, S., Song, S. E., & Cho, Y. T. (2017). Effective software solutions for 4D printing: a review and proposal. *International Journal of Precision Engineering and Manufacturing-Green Technology*, *4*, 359–371.

Comblain, F., Rocasalbas, G., Gauthier, S., & Henrotin, Y. (2017). Chitosan: A promising polymer for cartilage repair and viscosupplementation. *Bio-medical Materials and Engineering*, *28*(s1), S209–S215.

Cornell, D. D. (2007). Biopolymers in the existing postconsumer plastics recycling stream. *Journal of Polymers and the Environment*, *15*(4), 295–299.

Dara, P. K., Raghavankutty, M., Sebastian, N., Chatterjee, N. S., Mathew, S., Ravishankar, C. N., & Anandan, R. (2020). Rheological, physico-chemical, and surface-active properties of gelatin extracted from bigeye tuna (*Thunnus obesus*) skin waste. *Journal of Aquatic Food Product Technology*, *29*(5), 428–444.

Das, A. M., Ali, A. A., & Hazarika, M. P. (2014). Synthesis and characterization of cellulose acetate from rice husk: eco-friendly condition. *Carbohydrate Polymers*, *112*, 342–349.

Demleitner, S., Kraus, J., & Franz, G. (1992). Synthesis and antitumour activity of sulfoalkyl derivatives of curdlan and lichenan. *Carbohydrate Research*, *226*(2), 247–252.

Derossi, A. et al. (2018). Application of 3D printing for customized food. A case on the development of a fruit-based snack for children. *Journal of Food Engineering*, *220*, 65–75.

Di Luzio, N. R., Williams, D. L., RB, M. N., Edwards, B. F., & Kitahama, A. (1979). Comparative tumor-inhibitory and anti-bacterial activity of soluble and particulate glucan. *International Journal of Cancer*, *24*(6), 773–779.

Ding, L., Li, X., Hu, L., Zhang, Y., Jiang, Y., Mao, Z., Xu, H., Wang, B., Feng, X., & Sui, X. (2020). A naked-eye detection polyvinyl alcohol/cellulose-based pH sensor for intelligent packaging. *Carbohydrate Polymers*, *233*, 115859.

Diniz, S. C., Voss, I., & Steinbüchel, A. (2006). Optimization of cyanophycin production in recombinant strains of *Pseudomonas putida* and *Ralstonia eutropha* employing elementary mode analysis and statistical experimental design. *Biotechnology and Bioengineering*, *93*(4), 698–717.

Dos Santos, R. M., Neto, W. P. F., Silvério, H. A., Martins, D. F., Dantas, N. O., & Pasquini, D. (2013). Cellulose nanocrystals from pineapple leaf, a new approach for the reuse of this agro-waste. *Industrial Crops and Products*, *50*, 707–714.

EIA, US (2011). *Voluntary Reporting of Greenhouse Gases Program Fuel Emission Coefficients*. EIA.

El-Hadi, A., Schnabel, R., Straube, E., Müller, G., & Henning, S. (2002). Correlation between degree of crystallinity, morphology, glass temperature, mechanical properties and biodegradation of poly (3-hydroxyalkanoate) PHAs and their blends. *Polymer Testing*, *21*(6), 665–674.

Elias, L. M., Silva-Stenico, M. E., Alvarenga, D. O., Rigonato, J., Fiore, M. F., & de Lira, S. P. (2015). Molecular and chemical analyses of cyanobacterial blooms in tropical lagoons from Southeast Brazil. *Journal of Water Resource and Protection*, *7*(01), 50.

EPA (2002). Municipal Solid Waste in the United States: 2000 Facts and Figures. In *Office of Solid Waste and Emergency Response (5305W). US EPA*. https://elibrary.asabe.org/abstract.asp?aid=24179

European Bioplastics 2017 (2003). Facts and figures. *Folding Carton Industry*, *30*(5), 12.

Ezati, P., & Rhim, J.-W. (2020). pH-responsive chitosan-based film incorporated with alizarin for intelligent packaging applications. *Food Hydrocolloids*, *102*, 105629.

Fabunmi, O. O., Tabil, L. G., Chang, P. R., & Panigrahi, S. (2006). Developing Biodegradable Plastics from Starch. In *ASABE/CSBE North Central Intersectional Meeting* (p. 1). American Society of Agricultural and Biological Engineers.

Fauzee, S. N., & Othaman, R. (2013). Extraction and Dissolution of Cellulose from Nypa Fruit Husk for Nanofibers Fabrication. In *AIP Conference Proceedings* (pp. 904–910). American Institute of Physics.

FICCI. (2014). Potential of the Plastics Industry in Northern India with a Special Focus on Plastic Culture and Food. Available at: https://ficci.in/public/storage/SPDocument/20607/Kowledge-Paper-Plastic.pdf.

Foolmaun, R. K., & Ramjeeawon, T. (2012). Comparative life cycle assessment and life cycle costing of four disposal scenarios for used polyethylene terephthalate bottles in Mauritius. *Environmental Technology*, *33*(17), 2007–2018.

Fu, W., Liu, Z., Feng, B., Hu, R., He, X., Wang, H., Yin, M., Huang, H., Zhang, H., & Wang, W. (2014). Electrospun gelatin/PCL and collagen/PLCL scaffolds for vascular tissue engineering. *International Journal of Nanomedicine*, *9*, 2335–2344.

Gadelha, M., Maji, S., & Wang, R. (2017). 3d Shape Induction from 2d Views of Multiple Objects. In *2017 International Conference on 3d Vision (3DV)* (pp. 402–411). IEEE.
1. https://arxiv.org/abs/1612.05872
2. https://www.semanticscholar.org/paper/3D-Shape-Induction-from-2D-Views-of-Multiple-Gadelha-Maji/e0485ca5fc4600740edd1372fe81b419d1409ad4

Ghanbarzadeh, B., Almasi, H., & Oleyaei, S. A. (2013). A novel modified starch/carboxymethyl cellulose/montmorillonite bionanocomposite film: structural and physical properties. *International Journal of Food Engineering*, *10*(1), 121–130. https://doi.org/10.1515/IJFE-2012-0197/MACHINEREADABLECITATION/RIS

Ghasempour, Z., Khodaivandi, S., Ahangari, H., Hamishehkar, H., Amjadi, S., Kia, E. M., & Ehsani, A. (2022). Characterization and optimization of Persian Gum/whey protein bionanocomposite films containing betanin nanoliposomes for food packaging utilization. *Journal of Polymers and the Environment*, *30*(7), 2800–2811. https://doi.org/10.1007/S10924-021-02367-0/METRICS

Ghazal, A. F., Zhang, M., & Liu, Z. (2019). Spontaneous color change of 3D printed healthy food product over time after printing as a novel application for 4D food printing. *Food and Bioprocess Technology*, *12*, 1627–1645.

Giovannoni, M., Gramegna, G., Benedetti, M., & Mattei, B. (2020). Industrial use of cell wall degrading enzymes: the fine line between production strategy and economic feasibility. *Frontiers in Bioengineering and Biotechnology*, *8*, 356.

Gironi, F., & Piemonte, V. (2011). Bioplastics and petroleum-based plastics: strengths and weaknesses. *Energy Sources, Part A: Recovery, Utilization, and Environmental Effects*, *33*(21), 1949–1959.

Godoi, F. C., Prakash, S., & Bhandari, B. R. (2016). 3d printing technologies applied for food design: status and prospects. *Journal of Food Engineering*, *179*, 44–54.

Goedkoop, M. (1999). A damage oriented method for life cycle impact assessment. *The Eco-indicator*, *99*.

González, K., Retegi, A., González, A., Eceiza, A., & Gabilondo, N. (2015). Starch and cellulose nanocrystals together into thermoplastic starch bionanocomposites. *Carbohydrate Polymers*, *117*, 83–90. https://doi.org/10.1016/J.CARBPOL.2014.09.055

Grassie, N., Murray, E. J., & Holmes, P. A. (1984). The thermal degradation of poly (-(d)-β-hydroxybutyric acid): part 2—changes in molecular weight. *Polymer Degradation and Stability*, *6*(2), 95–103.

Guzman, A., Gnutek, N. and Janik, H. (2011) 'Biodegradable Polymers for Food Packaging–Factors Influencing Their Degradation and Certification Types–a Comprehensive Review'. Academic Journals and Conferences https://science.lpnu.ua/jcct/all-volumes-and-issues/volume-5-number-1-2011/biodegradablepolymers-food-packing-factors

Hajizadeh, H., Peighambardoust, S. J., Peighambardoust, S. H., & Peressini, D. (2020). Physical, mechanical, and antibacterial characteristics of bio-nanocomposite films loaded with Ag-modified SiO_2 and TiO_2 nanoparticles. *Journal of Food Science*, *85*(4), 1193–1202. https://doi.org/10.1111/1750-3841.15079

Hamman, J. H. (2010). Chitosan based polyelectrolyte complexes as potential carrier materials in drug delivery systems. *Marine Drugs*, *8*(4), 1305–1322.

Hassler, R. A., & Doherty, D. H. (1990). Genetic engineering of polysaccharide structure: production of variants of xanthan gum in Xanthomonas campestris. *Biotechnology Progress*, *6*(3), 182–187.

He, C., Zhang, M., & Devahastin, S. (2020). Investigation on spontaneous shape change of 4D printed starch-based purees from purple sweet potatoes as induced by microwave dehydration. *ACS Applied Materials & Interfaces*, *12*(34), 37896–37905.

Hernández, N., Williams, R. C., & Cochran, E. W. (2014). The battle for the "green" polymer. Different approaches for biopolymer synthesis: bioadvantaged vs. bioreplacement. *Organic & Biomolecular Chemistry*, *12*(18), 2834–2849.

Holland, S., Foster, T., MacNaughtan, W., & Tuck, C. (2018). Design and characterisation of food grade powders and inks for microstructure control using 3D printing. *Journal of Food Engineering*, *220*, 12–19.

Hori, Y., Takahashi, Y., Yamaguchi, A., & Nishishita, T. (1994). Ring-Opening Copolymerization of Optically Active β-Butyrolactone with Several Lactones: Synthesis of New Biodegradable Polyesters. In G. Henrici-Olive, H.-J. Cantow, G. Dall'Asta, S. Olive, J. D. Ferry, H. Fujita, et al. (Eds.), *Studies in Polymer Science* (Volume 12, pp. 549–556). Elsevier.

Hottle, T. A., Bilec, M. M., & Landis, A. E. (2013). Sustainability assessments of bio-based polymers. *Polymer Degradation and Stability*, *98*(9), 1898–1907.

Hsieh, W.-C., Wada, Y., & Chang, C.-P. (2009). Fermentation, biodegradation and tensile strength of poly (3-hydroxybutyrate-co-4-hydroxybutyrate) synthesized by Delftia acidovorans. *Journal of the Taiwan Institute of Chemical Engineers*, *40*(2), 143–147.

Ian Gibson, I. G. (2015). Additive Manufacturing Technologies 3D Printing, Rapid Prototyping, and Direct Digital Manufacturing. Springer.

Ibrahim, M. S., Sani, N., Adamu, M., & Abubakar, M. K. (2018). Biodegradable polymers for sustainable environmental and economic development. *MOJ Bioorganic & Organic Chemistry*, *2*, 192–194.

Inbaraj, B. S., Chien, J. T., Ho, G. H., Yang, J., & Chen, B. H. (2006). Equilibrium and kinetic studies on sorption of basic dyes by a natural biopolymer poly (γ-glutamic acid). *Biochemical Engineering Journal*, *31*(3), 204–215.

Ingrao, C., Tricase, C., Cholewa-Wójcik, A., Kawecka, A., Rana, R., & Siracusa, V. (2015). Polylactic acid trays for fresh-food packaging: a carbon footprint assessment. *Science of the Total Environment*, *537*, 385–398.

Jamróz, E., Juszczak, L., & Kucharek, M. (2018). Investigation of the physical properties, antioxidant and antimicrobial activity of ternary potato starch-furcellaran-gelatin films incorporated with lavender essential oil. *International Journal of Biological Macromolecules*, *114*, 1094–1101.

Jamróz, E., Kulawik, P., Krzysciak, P., Talaga-Cwiertnia, K., & Juszczak, L. (2019). Intelligent and active furcellaran-gelatin films containing green or pu-erh tea extracts: Characterization, antioxidant and antimicrobial potential. *International Journal of Biological Macromolecules*, *122*, 745–757.

Jancikova, S., Jamroz, E., Kulawik, P., Tkaczewska, J., & Dordevic, D. (2019). Furcellaran/gelatin hydrolysate/rosemary extract composite films as active and intelligent packaging materials. *International Journal of Biological Macromolecules*, *131*, 19–28.

Jaswir, I., Al-Kahtani, H. A., Octavianti, F., Lestari, W., & Yusof, N. (2020). Camel Gelatin Composition, Properties, Production, and Applications. In *Handbook of Research on Health and Environmental Benefits of Camel Products* (pp. 306–327). IGI Global.

Jayakumar, A., Heera, K. V., Sumi, T. S., Joseph, M., Mathew, S., Praveen, G., Nair, I. C., & Radhakrishnan, E. K. (2019). Starch-PVA composite films with zinc-oxide nanoparticles and phytochemicals as intelligent pH sensing wraps for food packaging application. *International Journal of Biological Macromolecules, 136*, 395–403.

Ju, A., & Song, K. B. (2019). Development of teff starch films containing camu-camu (*Myrciaria dubia* Mc. Vaugh) extract as an antioxidant packaging material. *Industrial Crops and Products, 141*, 111737.

Ju, A., & Song, K. B. (2020). Incorporation of yellow onion peel extract into the funoran-based biodegradable films as an antioxidant packaging material. *International Journal of Food Science & Technology, 55*(4), 1671–1678.

Ju, A., Baek, S. K., Kim, S., & Song, K. B. (2019). Development of an antioxidative packaging film based on khorasan wheat starch containing moringa leaf extract. *Food Science and Biotechnology, 28*, 1057–1063.

Juzwa, M., & Jedliñski, Z. (2006). Novel synthesis of poly (3-hydroxybutyrate). *Macromolecules, 39*(13), 4627–4630.

Karami, A., Golieskardi, A., Choo, C. K., Larat, V., Karbalaei, S., & Salamatinia, B. (2018). Microplastic and mesoplastic contamination in canned sardines and sprats. *Science of the Total Environment, 612*, 1380–1386.

Kean, T., & Thanou, M. (2010). Biodegradation, biodistribution and toxicity of chitosan. *Advanced Drug Delivery Reviews, 62*(1), 3–11.

Kennedy, J. F., & Bradshaw, I. J. (1984). Production, properties and applications of xanthan. *Progress in Industrial Microbiology, 19*, 319–371.

Kennedy, J. F., & Panesar, P. S. (2006). Biodegradable Polymers for Industrial Application. In R. Smith (Ed.),. Woodhead Publishing Ltd/CRC Press, Cambridge, UK/Boca Raton, FL. (2005),(xvi+ 531 pp.,£ 135.00, ISBN 0-8493-3466-7)

Khosravi-Darani, K., & Bucci, D. Z. (2015). Application of poly (hydroxyalkanoate) in food packaging: Improvements by nanotechnology. *Chemical and Biochemical Engineering Quarterly, 29*(2), 275–285.

Kimbuathong, N., Leelaphiwat, P., & Harnkarnsujarit, N. (2020). Inhibition of melanosis and microbial growth in Pacific white shrimp (Litopenaeus vannamei) using high CO2 modified atmosphere packaging. *Food Chemistry, 312*, 126114.

Kuang, X., Roach, D. J., Wu, J., Hamel, C. M., Ding, Z., Wang, T., Dunn, M., & Qi, H. (2019). Advances in 4D printing: materials and applications. *Advanced Functional Materials, 29*(2), 1805290.

Landis, A. E. (2010). Cradle to Gate Environmental Footprint and Life Cycle Assessment of Poly (Lactic Acid). In Rafael Auras, Loong-Tak Lim, Susan E. M. Selke, Hideto Tsuji, *Poly (lactic acid) Synthesis, Structures, Properties, Processing, and Applications* (pp. 431–441). John Wiley & Sons, Elsevier.

Landis, A. E., Miller, S. A., & Theis, T. L. (2007). Life cycle of the corn – soybean agroecosystem for biobased production. *Environmental Science & Technology, 41*(4), 1457–1464.

Leceta, I., Guerrero, P., Cabezudo, S., & de la Caba, K. (2013). Environmental assessment of chitosan-based films. *Journal of Cleaner Production, 41*, 312–318.

Li, J., Green, C., Reynolds, A., Shi, H., & Rotchell, J. M. (2018). Microplastics in mussels sampled from coastal waters and supermarkets in the United Kingdom. *Environmental Pollution, 241*, 35–44.

Li, Y., Ying, Y., Zhou, Y., Ge, Y., Yuan, C., Wu, C., & Hu, Y. (2019). A pH-indicating intelligent packaging composed of chitosan-purple potato extractions strength by surface-deacetylated chitin nanofibers. *International Journal of Biological Macromolecules*, *127*, 376–384.

Lian, H., Shi, J., Zhang, X., & Peng, Y. (2020). Effect of the added polysaccharide on the release of thyme essential oil and structure properties of chitosan based film. *Food Packaging and Shelf Life*, *23*, 100467.

Liebezeit, G., & Liebezeit, E. (2013). Non-pollen particulates in honey and sugar. *Food Additives & Contaminants: Part A*, *30*(12), 2136–2140.

Lippi, L., Bähr, L., Wüstenberg, A., Wilde, A., & Steuer, R. (2018). Exploring the potential of high-density cultivation of cyanobacteria for the production of cyanophycin. *Algal Research*, *31*, 363–366.

Liu, X., Chi, B., Zhiwei, J., Tan, J., Liu, F., & Yang, W. (2017). A large-scale double-stage-screw 3 D printer for fused deposition of plastic pellets. *Journal of Applied Polymer Science*, *134*(31), 45147.

Liu, Z., Zhang, M., & Yang, C. (2018). Dual extrusion 3D printing of mashed potatoes/ strawberry juice gel. *Lwt*, *96*, 589–596.

Liu, J., Yong, H., Liu, Y., Qin, Y., Kan, J., & Liu, J. (2019). Preparation and characterization of active and intelligent films based on fish gelatin and haskap berries (*Lonicera caerulea* L.) extract. *Food Packaging and Shelf Life*, *22*, 100417.

Liu, Z., He, C., Guo, C., Chen, F., Bhandari, B., & Zhang, M. (2021). Dehydration-triggered shape transformation of 4D printed edible gel structure affected by material property and heating mechanism. *Food Hydrocolloids*, *115*, 106608.

Lone, A., Anany, H., Hakeem, M., Aguis, L., Avdjian, A. C., Bouget, M., Atashi, A., Brovko, L., Rochefort, D., & Griffiths, M. W. (2016). Development of prototypes of bioactive packaging materials based on immobilized bacteriophages for control of growth of bacterial pathogens in foods. *International Journal of Food Microbiology*, *217*, 49–58.

Lu, P., & Hsieh, Y.-L. (2012). Cellulose isolation and core–shell nanostructures of cellulose nanocrystals from chardonnay grape skins. *Carbohydrate Polymers*, *87*(4), 2546–2553.

Lubis, M., Harahap, M. B., Ginting, M. H. S., Sartika, M., & Azmi, H. (2018). Production of bioplastic from avocado seed starch reinforced with microcrystalline cellulose from sugar palm fibers. *Journal of Engineering Science and Technology*, *13*(2), 381–393.

Luckachan, G. E., & Pillai, C. K. S. (2011). Biodegradable polymers-a review on recent trends and emerging perspectives. *Journal of Polymers and the Environment*, *19*, 637–676.

Luís, Â., Ramos, A., & Domingues, F. (2020). Pullulan films containing rockrose essential oil for potential food packaging applications. *Antibiotics*, *9*(10), 681.

Luo, Y., & Hu, Q. (2017). Food-Derived Biopolymers for Nutrient Delivery. In *Nutrient Delivery* (pp. 251–291). Elsevier.

Lyu, J. S., Choi, I., Hwang, K.-S., Lee, J.-Y., Seo, J., Kim, S. Y., & Han, J. (2019). Development of a BTB−/TBA+ ion-paired dye-based CO2 indicator and its application in a multilayered intelligent packaging system. *Sensors and Actuators B: Chemical*, *282*, 359–365.

Mangaraj, S., Yadav, A., Bal, L. M., Dash, S. K., & Mahanti, N. K. (2019). Application of biodegradable polymers in food packaging industry: a comprehensive review. *Journal of Packaging Technology and Research*, *3*, 77–96.

Marsh, K., & Bugusu, B. (2007). Food packaging—roles, materials, and environmental issues. *Journal of Food Science*, *72*(3), R39–R55.

Mason, S. A., Welch, V. G., & Neratko, J. (2018). Synthetic polymer contamination in bottled water. *Frontiers in Chemistry*, *6*, 389699.

McIntosh, M., Stone, B. A., & Stanisich, V. A. (2005). Curdlan and other bacterial (1→ 3)-β-D-glucans. *Applied Microbiology and Biotechnology*, *68*, 163–173.

Meereboer, K. W., Misra, M., & Mohanty, A. K. (2020). Review of recent advances in the biodegradability of polyhydroxyalkanoate (PHA) bioplastics and their composites. *Green Chemistry*, *22*(17), 5519–5558.

Menon, M. P., Selvakumar, R., & Ramakrishna, S. (2017). Extraction and modification of cellulose nanofibers derived from biomass for environmental application. *RSC Advances*, *7*(68), 42750–42773.

Micó-Vicent, B., Viqueira, V., Ramos, M., Luzi, F., Dominici, F., Torre, L., Jiménez, A., Puglia, D., & Garrigós, M. C. (2020). Effect of lemon waste natural dye and essential oil loaded into laminar nanoclays on thermomechanical and color properties of polyester based bionanocomposites. *Polymers*, *12*(7), 1451. https://doi.org/10.3390/POLYM12071451

Moghimi, R., Aliahmadi, A., & Rafati, H. (2017). Antibacterial hydroxypropyl methyl cellulose edible films containing nanoemulsions of *Thymus daenensis* essential oil for food packaging. *Carbohydrate Polymers*, *175*, 241–248.

Mohanty, A. K., Misra, M., & Drzal, L. T. (2005). Natural Fibers, Biopolymers, and Biocomposites: An Introduction. In Amar K. Mohanty, Manjusri Misra, Lawrence T. Drzal, *Natural Fibers, Biopolymers, and Biocomposites* (pp. 17–51). CRC Press.

Moon, H. C., Han, S., Borges, J., Pesqueira, T., Choi, H., Han, S. Y., Cho, H., Park, J. H., Mano, J. F., & Choi, I. S. (2020). Enzymatically degradable, starch-based layer-by-layer films: application to cytocompatible single-cell nanoencapsulation. *Soft Matter*, *16*(26), 6063–6071.

Morschbacker, A. (2009). Bio-ethanol based ethylene. *Journal of Macromolecular Science® Part C: Polymer Reviews*, *49*(2), 79–84.

Murali, S., Kumar, S., Koh, J., Seena, S., Singh, P., Ramalho, A., & Sobral, A. J. (2019). Bio-based chitosan/gelatin/Ag@ZnO bionanocomposites: synthesis and mechanical and antibacterial properties. *Cellulose*, *26*(9), 5347–5361. https://doi.org/10.1007/S10570-019-02457-2/METRICS

Muthurajan, M., Veeramani, A., Rahul, T., Gupta, R. K., Anukiruthika, T., Moses, J. A., & Anandharamakrishnan, C. (2021). Valorization of food industry waste streams using 3D food printing: A study on noodles prepared from potato peel waste. *Food and Bioprocess Technology*, *14*(10), 1817–1834.

Nampoothiri, K. M., Nair, N. R., & John, R. P. (2010). An overview of the recent developments in polylactide (PLA) research. *Bioresource technology*, *101*(22), 8493–8501.

Niaounakis, M. (2019). Recycling of biopolymers–the patent perspective. *European Polymer Journal*, *114*, 464–475.

Nida, S., Moses, J. A., & Anandharamakrishnan, C. (2022). Emerging applications of 5D and 6D printing in the food industry. *Journal of Agriculture and Food Research*, *10*, 100392.

Noshirvani, N., Hong, W., Ghanbarzadeh, B., Fasihi, H., & Montazami, R. (2018). Study of cellulose nanocrystal doped starch-polyvinyl alcohol bionanocomposite films. *International Journal of Biological Macromolecules*, *107*, 2065–2074. https://doi.org/10.1016/J.IJBIOMAC.2017.10.083

Nworu, C. S., Ihim, S. A., Okoye, F. B., Esimone, C. O., Adikwu, M. U., & Akah, P. A. (2015). Immunomodulatory and immunorestorative activities of β-D-glucan-rich extract and polysaccharide fraction of mushroom, Pleurutus tuberregium. *Pharmaceutical Biology, 53*(11), 1555–1566.

Obruca, S., Marova, I., Snajdar, O., Mravcova, L., & Svoboda, Z. (2010). Production of poly (3-hydroxybutyrate-co-3-hydroxyvalerate) by *Cupriavidus necator* from waste rapeseed oil using propanol as a precursor of 3-hydroxyvalerate. *Biotechnology letters, 32,* 1925–1932.

Pant, A., Lee, A. Y., Karyappa, R., Lee, C. P., An, J., Hashimoto, M., Tan, U. X., Wong, G., Chua, C. K., & Zhang, Y. (2021). 3D food printing of fresh vegetables using food hydrocolloids for dysphagic patients. *Food Hydrocolloids, 114,* 106546.

Pateiro, M., Munekata, P. E. S., Sant'Ana, A. S., Domínguez, R., Rodríguez-Lazaro, D., & Lorenzo, J. M. (2021). Application of essential oils as antimicrobial agents against spoilage and pathogenic microorganisms in meat products. *International Journal of Food Microbiology, 337,* 108966.

Pathak, S., Sneha, C. L. R., & Mathew, B. B. (2014). Bioplastics: its timeline based scenario & challenges. *Journal of Polymer and Biopolymer Physics Chemistry, 2*(4), 84–90.

Pavasars, I., Hagberg, J., Borén, H., & Allard, B. (2003). Alkaline degradation of cellulose: mechanisms and kinetics. *Journal of Polymers and the Environment, 11,* 39–47.

Pawar, P. A., & Purwar, A. H. (2013). Biodegradable polymers in food packaging. *American Journal of Engineering Research, 2*(5), 151–164.

Pei, E., & Loh, G. H. (2018). Technological considerations for 4D printing: an overview. *Progress in Additive Manufacturing, 3,* 95–107.

Phuhongsung, P., Zhang, M., & Devahastin, S. (2020). Influence of surface pH on color, texture and flavor of 3D printed composite mixture of soy protein isolate, pumpkin, and beetroot. *Food and Bioprocess Technology, 13,* 1600–1610.

Platt, D. K. (2006). *The Future of Biodegradable Films for Flexible Packaging.* Pira International Limited.

Prakash, U., Singh, L., & Sharma, V. (2013). Role of xanthan gum (*Xanthomonas compestris*) in gastroretentive drug delivery system: an overview. *International Research Journal of Pharmacy, 4*(4), 35–37.

Prata, J. C. (2018). Airborne microplastics: consequences to human health? *Environmental Pollution, 234,* 115–126.

Priyadarshi, R., Kumar, B. S., Deeba, F., Kulshreshtha, A., & Negi, Y. S. (2018). Chitosan films incorporated with apricot (*Prunus armeniaca*) kernel essential oil as active food packaging material. *Food Hydrocolloids, 85,* 158–166.

Qin, Y., Liu, Y., Zhang, X., & Liu, J. (2020). Development of active and intelligent packaging by incorporating betalains from red pitaya (*Hylocereus polyrhizus*) peel into starch/polyvinyl alcohol films. *Food Hydrocolloids, 100,* 105410.

Qiu, L., Zhang, M., Mujumdar, A. S., & Liu, Y. (2022). Recent developments in key processing techniques for oriental spices/herbs and condiments: a review. *Food Reviews International, 38*(8), 1791–1811.

Ramesh, N., Moratti, S. C., & Dias, G. J. (2018). Hydroxyapatite–polymer biocomposites for bone regeneration: a review of current trends. *Journal of Biomedical Materials Research Part B: Applied Biomaterials, 106*(5), 2046–2057.

Rando, P., & Ramaioli, M. (2021). Food 3D printing: effect of heat transfer on print stability of chocolate. *Journal of Food Engineering, 294,* 110415.

Rao, M. G., Bharathi, P., & Akila, R. M. (2014). A comprehensive review on biopolymers. *Scientific Reviews & Chemical Communications*, *4*(2), 61–68.

Ratnayake, W. S., Hoover, R., Shahidi, F., Perera, C., & Jane, J. (2001). Composition, molecular structure, and physicochemical properties of starches from four field pea (*Pisum sativum* L.) cultivars. *Food Chemistry*, *74*(2), 189–202.

Realini, C. E., & Marcos, B. (2014). Active and intelligent packaging systems for a modern society. *Meat Science*, *98*(3), 404–419.

Reddy, P. R., & Devi, P. A. (2018). Review on the advancements of additive manufacturing-4D and 5D printing. *International Journal of Mechanical and Production Engineering Research and Development*, *8*(4), 397–402.

Renzi, M., & Blašković, A. (2018). Litter & microplastics features in table salts from marine origin: Italian versus croatian brands. *Marine Pollution Bulletin*, *135*, 62–68.

Restuccia, D., Spizzirri, U. G., Parisi, O. I., Cirillo, G., Curcio, M., Iemma, F., Puoci, F., Vinci, G., & Picci, N. (2010). New EU regulation aspects and global market of active and intelligent packaging for food industry applications. *Food Control*, *21*(11), 1425–1435.

Rhim, J. W., Wang, L. F., Lee, Y., & Hong, S. I. (2014). Preparation and characterization of bio-nanocomposite films of agar and silver nanoparticles: laser ablation method. *Carbohydrate Polymers*, *103*(1), 456–465. https://doi.org/10.1016/J.CARBPOL.2013.12.075

Rooney, M. L. (1995). *Overview* of Active Food Packaging. In M.L. Rooney, *Active Food Packaging* (pp. 1–37). Springer.

Safaei, M., Taran, M., & Imani, M. M. (2019). Preparation, structural characterization, thermal properties and antifungal activity of alginate-CuO bionanocomposite. *Materials Science and Engineering: C*, *101*, 323–329. https://doi.org/10.1016/J.MSEC.2019.03.108

Saliu, F., & Della Pergola, R. (2018). Carbon dioxide colorimetric indicators for food packaging application: applicability of anthocyanin and poly-lysine mixtures. *Sensors and Actuators B: Chemical*, *258*, 1117–1124.

Sallam, A., Kast, A., Przybilla, S., Meiswinkel, T., & Steinbüchel, A. (2009). Biotechnological process for production of β-dipeptides from cyanophycin on a technical scale and its optimization. *Applied and Environmental Microbiology*, *75*(1), 29–38.

Sanuja, S., Agalya, A., & Umapathy, M. J. (2014). Studies on magnesium oxide reinforced chitosan bionanocomposite incorporated with clove oil for active food packaging application. *International Journal of Polymeric Materials and Polymeric Biomaterials*, *63*(14), 733–740. https://doi.org/10.1080/00914037.2013.879445

Sarıcaoglu, F. T., & Turhan, S. (2020). Physicochemical, antioxidant and antimicrobial properties of mechanically deboned chicken meat protein films enriched with various essential oils. *Food Packaging and Shelf Life*, *25*, 100527.

Saruchi, Kaur, M., Kumar, V., Ghfar, A.A. and Pandey, S. (2022) A green approach for the synthesis of silver nanoparticle-embedded chitosan bionanocomposite as a potential device for the sustained release of the itraconazole drug and its antibacterial characteristics, *Polymers* 2022, *Vol. 14*(9), 1911. doi: https://doi.org/10.3390/POLYM14091911.

Sathe, G.2018 Award, 2017.

Schneider, M., Stracke, F., Hansen, S., & Schaefer, U. F. (2009). Nanoparticles and their interactions with the dermal barrier. *Dermato-Endocrinology*, *1*(4), 197–206.

Schymanski, D., Goldbeck, C., Humpf, H. U., & Fürst, P. (2018). Analysis of microplastics in water by micro-Raman spectroscopy: release of plastic particles from different packaging into mineral water. *Water Research*, *129*, 154–162.

Seright, R. S., & Henrlcl, B. J. (1990). Xanthan stability at elevated temperatures. *SPE Reservoir Engineering*, *5*(01), 52–60.

Shanthamma, S., Preethi, R., Moses, J. A., & Anandharamakrishnan, C. (2021). 4D printing of sago starch with turmeric blends: a study on pH-triggered spontaneous color transformation. *ACS Food Science & Technology*, *1*(4), 669–679.

Shekarforoush, E., Faralli, A., Ndoni, S., Mendes, A. C., & Chronakis, I. S. (2017). Electrospinning of xanthan polysaccharide. *Macromolecular Materials and Engineering*, *302*(8), 1700067.

Singholi, A. K. S., & Sharma, A. (2020). Recent advancement and research possibilities in 4D printing technology. *Materialwissenschaft und Werkstofftechnik*, *51*(10), 1332–1340.

Song, H., Jia, H., Wang, Q., Zhao, X., Yang, G., Zhang, M. et al. (2020). A new environmentally-friendly system for extracting cellulose from corn straw: the low temperature laccase system. *Materials*, *13*(2), 437.

Souza, V. G. L., Rodrigues, C., Ferreira, L., Pires, J. R. A., Duarte, M. P., Coelhoso, I., & Fernando, A. L. (2019). In vitro bioactivity of novel chitosan bionanocomposites incorporated with different essential oils. *Industrial Crops and Products*, *140*, 111563. https://doi.org/10.1016/J.INDCROP.2019.111563

Souza, V. G. L., Rodrigues, C., Valente, S., Pimenta, C., Pires, J. R. A., Alves, M. M., Santos, C. F., Coelhoso, I. M., & Fernando, A. L. (2020). Eco-friendly ZnO/Chitosan bionanocomposites films for packaging of fresh poultry meat. *Coatings*, *10*(2), 110. https://doi.org/10.3390/COATINGS10020110

Steinbüchel, A., & Lütke-Eversloh, T. (2003). Metabolic engineering and pathway construction for biotechnological production of relevant polyhydroxyalkanoates in microorganisms. *Biochemical Engineering Journal*, *16*(2), 81–96.

Sucinda, E. F., Majid, M. A., Ridzuan, M. J. M., Cheng, E. M., Alshahrani, H. A., & Mamat, N. (2021). Development and characterisation of packaging film from Napier cellulose nanowhisker reinforced polylactic acid (PLA) bionanocomposites. *International Journal of Biological Macromolecules*, *187*, 43–53. https://doi.org/10.1016/J.IJBIOMAC.2021.07.069

Sun, J., Peng, Z., Zhou, W., Fuh, J. Y., Hong, G. S., & Chiu, A. (2015). A review on 3D printing for customized food fabrication. *Procedia Manufacturing*, *1*, 308–319.

Suryanegara, L., Kurniawan, Y. D., Syamani, F. A., & Nurhamiyah, Y. (2018). Mechanical Properties of Composites Based on Poly (Lactic Acid) and Soda-Treated Sugarcane Bagasse Pulp. In *Sustainable Future for Human Security: Society, Cities and Governance* (pp. 277–285).

Tabone, M. D., Cregg, J. J., Beckman, E. J., & Landis, A. E. (2010). Sustainability metrics: life cycle assessment and green design in polymers. *Environmental Science & Technology*, *44*(21), 8264–8269.

Tekin, E., Mazı, I. B., & Türe, H. (2020). Time temperature indicator film based on alginate and red beetroot (*Beta vulgaris* L.) extract: in vitro characterization. *Ukrainian Food Journal*, *9*(2), 344–360.

Teng, X., Zhang, M., & Mujumdar, A. S. (2021). 4D printing: recent advances and proposals in the food sector. *Trends in Food Science & Technology*, *110*, 349–363.

Tingaut, P., Zimmermann, T., & Lopez-Suevos, F. (2010). Synthesis and characterization of bionanocomposites with tunable properties from poly(lactic acid) and acetylated microfibrillated cellulose. *Biomacromolecules*, *11*(2), 454–464. https://doi.org/10.1021/BM901186U/SUPPL_FILE/BM901186U_SI_001.PDF

Venkatesan, J., Anil, S., Singh, S. K., & Kim, S. K. (2017). Preparations and Applications of Alginate Nanoparticles. In Jayachandran Venkatesan, Sukumaran Anil and Se-Kwon Kim, *Seaweed Polysaccharides* (pp. 251–268). Elsevier.

Verma, M. L., Kumar, S., Jeslin, J., & Dubey, N. K. (2020). Microbial production of biopolymers with potential biotechnological applications. In *Biopolymer-based formulations* (pp. 105–137). Elsevier.

Wan, X., He, Q., Wang, X., Liu, M., Lin, S., Shi, R., Tian, J., & Chen, G. (2021). Water-soluble chitosan-based indicator label membrane and its response behavior to carbon dioxide. *Food Control*, *130*, 108355.

Wang, L., Zhang, M., Bhandari, B., & Yang, C. (2018). Investigation on fish surimi gel as promising food material for 3D printing. *Journal of Food Engineering*, *220*, 101–108.

Weber, A. (2000a). Production of nanoscale powders from sols using supercritical fluids. *German Patent DE, 19904990.*

Weber, C. J. (2000b) *Biobased Packaging Materials for the Food Industry: Status and Perspectives, a European Concerted Action*. KVL https://agris.fao.org/search/en/providers/122465/records/647235d053aa8c896302456a

Wilson, A., Anukiruthika, T., Moses, J. A., & Anandharamakrishnan, C. (2023). Preparation of fiber-enriched chicken meat constructs using 3D printing. *Journal of Culinary Science & Technology*, *21*(1), 127–138.

Wilson, A., Anukiruthika, T., Moses, J. A., & Anandharamakrishnan, C. (2023). Preparation of fiber-enriched chicken meat constructs using 3D printing. *Journal of Culinary Science & Technology*, *21*(1), 127–138.

Woehl, M. A., Canestraro, C. D., Mikowski, A., Sierakowski, M. R., Ramos, L. P., & Wypych, F. (2010). Bionanocomposites of thermoplastic starch reinforced with bacterial cellulose nanofibres: Effect of enzymatic treatment on mechanical properties. *Carbohydrate Polymers*, *80*(3), 866–873. https://doi.org/10.1016/J.CARBPOL.2009.12.045

Wu, R., & Wang, L. (2021). Vinyl alcohol formation via catalytic β-dehydrogenation of ethanol on Ir (100). *Chemical Physics Impact*, *3*, 100040.

Wu, Z., Huang, X., Li, Y. C., Xiao, H., & Wang, X. (2018). Novel chitosan films with laponite immobilized Ag nanoparticles for active food packaging. *Carbohydrate Polymers*, *199*, 210–218.

Wu, C., Li, Y., Du, Y., Wang, L., Tong, C., Hu, Y., Pang, J., & Yan, Z. (2019). Preparation and characterization of konjac glucomannan-based bionanocomposite film for active food packaging. *Food Hydrocolloids*, *89*, 682–690. https://doi.org/10.1016/J.FOODHYD.2018.11.001

Wu, C., Li, Y., Sun, J., Lu, Y., Tong, C., Wang, L., Yan, Z., & Pang, J. (2020). Novel konjac glucomannan films with oxidized chitin nanocrystals immobilized red cabbage anthocyanins for intelligent food packaging. *Food Hydrocolloids*, *98*, 105245.

Xu, D., Chen, T., & Liu, Y. (2021). The physical properties, antioxidant and antimicrobial activity of chitosan–gelatin edible films incorporated with the extract from hop plant. *Polymer Bulletin*, *78*, 3607–3624.

Yam, K. L., Takhistov, P. T., & Miltz, J. (2005). Intelligent packaging: concepts and applications. *Journal of Food Science*, *70*(1), R1–R10.

Yang, W., Fortunati, E., Dominici, F., Kenny, J. M., & Puglia, D. (2015). Effect of processing conditions and lignin content on thermal, mechanical and degradative behavior of lignin nanoparticles/polylactic (acid) bionanocomposites prepared by melt extrusion and solvent casting. *European Polymer Journal*, *71*, 126–139. https://doi.org/10.1016/J.EURPOLYMJ.2015.07.051

Yao, J., Jing, J., Xu, H., Liang, J., Wu, Q., Feng, X., & Ouyang, P. (2009). Investigation on enzymatic degradation of γ-polyglutamic acid from *Bacillus subtilis* NX-2. *Journal of Molecular Catalysis B: Enzymatic*, *56*(2–3), 158–164.

Yu, F., Xu, R. C., Huang, L., Luo, M. L., Li, J., Tay, F. R., Niu, L. N., & Chen, J. H. (2020). Isocyanate-terminated urethane-based methacrylate for in situ collagen scaffold modification. *Materials Science and Engineering: C*, *112*, 110902.

Zambrano, M. C., Pawlak, J. J., Daystar, J., Ankeny, M., Goller, C. C., & Venditti, R. A. (2020). Aerobic biodegradation in freshwater and marine environments of textile microfibers generated in clothes laundering: Effects of cellulose and polyester-based microfibers on the microbiome. *Marine Pollution Bulletin*, *151*, 110826.

Zhang, L., Deng, X., Zhao, S., & Huang, Z. (1997). Biodegradable polymer blends of poly (3-hydroxybutyrate) and starch acetate. *Polymer international*, *44*(1), 104–110.

Zhang, H., Zhu, J., Zhu, X., Cai, J., Zhang, A., Hong, Y. et al. (2012). High-level exogenous glutamic acid-independent production of poly-(γ-glutamic acid) with organic acid addition in a new isolated Bacillus subtilis C10. *Bioresource Technology*, *116*, 241–246.

Zhang, X., Lian, H., Shi, J., Meng, W., & Peng, Y. (2020). Plant extracts such as pine nut shell, peanut shell and jujube leaf improved the antioxidant ability and gas permeability of chitosan films. *International Journal of Biological Macromolecules*, *148*, 1242–1250.

Zheng, L., Liu, J., Liu, R., Xing, Y., & Jiang, H. (2021). 3D printing performance of gels from wheat starch, flour and whole meal. *Food Chemistry*, *356*, 129546.

12

Nanoscale Surface Modification by Fatty Acid Grafting Technologies

Corina L. Reichert, Lisa-Marie Dietz, and Markus Schmid

Sustainable Packaging Institute SPI, Albstadt-Sigmaringen University, Sigmaringen, Germany

12.1 Introduction

An increase in the use of biopolymer-based materials for packaging applications leads to new challenges. For instance, bio-based and biodegradable materials often lack sufficient water repellence or low water vapor barrier properties. This functionality was announced to be one of the main limiting parameters for the application of biobased and/or biodegradable materials for food packaging (Wu et al., 2021).

An eco-friendly technique to improve the water barrier function is called chromatogenic chemistry (Schmid et al., 2012), also named the fatty acid grafting procedure, patented by Samain (2002). Using this technique, a surface modification is done to increase the barrier properties to reach the requirements of the packaged good (Pellis et al., 2021). The technique can be used to create a polymeric substrate, with the treated side being water-repellant and the other untreated side being water-permeable. Furthermore, the fatty acid grafting procedure, which will be outlined in detail in this chapter, can be used to protect moisture-sensitive biopolymers and induce a hydrophobic character of the treated substrates.

In this chapter, we want to give a detailed description of the fundamentals as well as hands-on advice for the application of the fatty acid grafting procedure. First, we want to give an overview of the chemical process of the fatty acid grafting procedure (Chapter 3), the techniques possible to do the fatty acid grafting (Chapter 4), an overview of the industrial fields of application (Chapter 5), legal aspects (Chapter 6), occupational safety (Chapter 7), and outlining the parameters that are important when using the fatty acid grafting for an industrial application. At last, we want to give an assessment regarding the sustainability aspects of this eco-friendly procedure and future scenarios for the application of this technique (Chapter 8).

Nanotechnology for Sustainable Food Packaging, First Edition. Edited by C. Anandharamakrishnan, Jeyan A. Moses, and M. Maria Leena.

12.2 Chemical Process

By using the fatty acid grafting process an increase in hydrophobicity shall be reached. By increasing the hydrophobic character of a surface, the barrier properties of this material shall increase, for instance, against water, oil, or water vapor. The hydrophobic effect is a result of the attachment of fatty acids onto the surface of a substrate. In the patent filed by Samain (2002), the grafting process was described as the production of "covalent grafting of hydrophobic groups R onto hydrophilic functions with the formation of a volatile compound HX under the reactions conditions" (Figure 12.1). This technology can be classified as a nanoscale surface functionalization technology, as the fatty acids are nanoscale in length.

In detail, the fatty acid grafting reaction describes the esterification between the acyl group of the fatty acid chloride and the functional groups of a substrate, which is schematically shown in (Figure 12.1). The grafting reactions take place via a nucleophilic acyl substitution, and as a byproduct, hydrochloric acid is formed when using fatty acid chlorides as reagents (Lee et al., 2018; Samain, 2002). The formation of an ester linkage between the functional groups of the substrate and the fatty acid (chloride) was confirmed by infrared spectroscopy (ATR-FTIR), which was done before and after grafting and showed a decrease in the hydroxy peak and an increase in the carbonyl peak (Schmid et al., 2014).

The fatty acid grafting is a surface reaction, which means that the depth of the fatty acid layer is very low in nonporous materials. The effect of the fatty acid grafting was shown to be just on the surface and not the inner layers (Schmid et al., 2014). Schmid et al. (2014) studied the penetration depth of the fatty acids by IR spectroscopy and concluded that the esterification reaction takes place at the surface, but the thickness and density of the fatty acid grafting are not clear and pointed out that there is a need for more research.

When applying the fatty acid grafting procedure, a hydrophobic zone is formed, which is greater than the point that was directly in contact with the fatty acid solution. This effect was explained in the patent from (Samain, 2002), who pointed out that this technique is very material efficient and the fatty acids diffuse a little bit into the material (Samain, 2002).

Grafting parameters were extensively studied for PVA polymer or films (Haas et al., 2017; Shaikh et al., 1998), PVA-coated paper (Lee et al., 2018; Schmid et al., 2012, 2014),

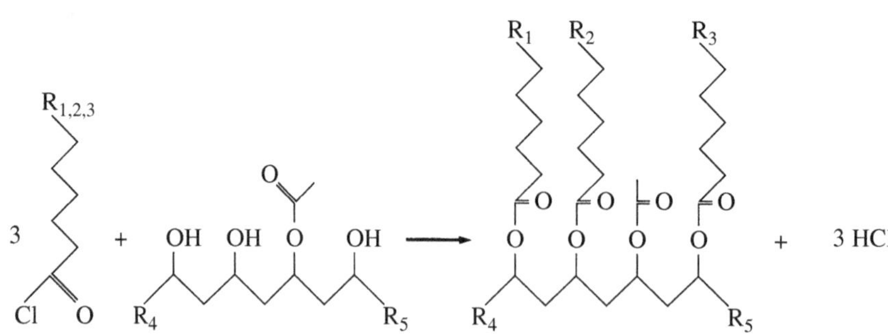

Figure 12.1 Schematic illustration of the chemical reaction used for the fatty acid grafting process, R_{1-3} represent fatty acid residues; R_4 and R_5 represent the remaining parts of the substrate molecule (based on *Source:* Schmid et al. (2014)/MDPI/C-BY-4.0)

PVA-coated PET and glassine (Germek et al., 2020), cellulose fibers (Berlioz et al., 2008; Stinga, 2008), and cotton fabrics (Zhong and Netravali, 2016).

The grafting parameters include:

- Grafting temperature
- Time; reaction duration, reaction kinetics
- Type of fatty acid
- Concentration of fatty acid
- Type of substrate
- Type of grafting process/technique

12.2.1 Grafting Temperature

The grafting temperature describes the temperature used, which influences two main phenomena. The first describes the temperature used for the transfer of the fatty acids via the transmitter paper to the substrate (important for the transfer method; see Chapter 4.1.). This means the temperature that is needed for the fatty acids to be transferred via the gas phase to the substrate.

The second phenomenon describes the temperature needed for the esterification reaction; in detail, the reaction kinetics of the grafting process are influenced by the temperature.

In the patent filed by Samain (2002), it is described that the amount of stearic acid chloride in pentane did not vary in weight with varying grafting temperatures (20 and 150 °C) tested. The authors concluded that the temperature does not influence the yield of the reaction. However, this information is part of the patent, and there is no detailed methodology to evaluate this conclusion.

In the literature, it is indicated that the effect of the grafting temperature depends on the other grafting parameters, such as time and fatty acid concentration. There was no effect on the solubility of PVA films when increasing the grafting temperature from 90 °C to 150 °C when using a fatty acid concentration of 1%. However, there was a "major influence" on the solubility of the PVA film with increasing temperature when using 5% fatty acid concentration and a grafting time of two minutes. The highly soluble PVA films (using a grafting temperature of 90 °C) changed to an unsoluble PVA film when increasing the grafting temperature to 150 °C (within the solubility time tested of a maximum 1,800 seconds). An increase in the grafting time to 20 minutes; however, the grafting temperature had no effect (Haas et al., 2017).

Similar results were found by Lee et al. (2018). Using a low fatty acid concentration (0.1% fatty acid concentration), the grafting temperature showed minor effects on the water drop contact angle of grafted PVA-coated paper. With the increase in fatty acid concentration (0.5% and 1.0%), an increase in the grafting temperature (from 140 °C to 160 °C or 180 °C) led to an increase in contact angle (88°–91° at 160 °C and 96°–100° at 180 °C).

Besides this, the grafting temperature may also lead to changes in the substrate, e.g., an increase in the crystallinity of PVA films through heat treatment that also changes the surface energy (Byron and Dalby, 1987) referenced by Haas et al. (2017). This means the heat treatment may change the properties of the substrate itself and thus also gas barrier and water repellence properties. If the effect of the fatty acid grafting comes from esterification or heat treatment of the substrate, it needs to be evaluated as the case arises.

12.2.2 Time

The grafting time refers to the time of the fatty acid chloride exposure under heat treatment, so for the direct and transfer method, the grafting time refers to the (reaction) time of the substrate covered with fatty acids and left in an oven or heated by a heating device for the esterification reaction to occur.

Schmid et al. (2012) plotted the dependency of the grafting parameters temperature and time against the contact angle of palmitoyl chloride grafted PVA: Both grafting parameters temperature (90–170 °C) and time (2–10 minutes) influenced the hydrophobic character of the PVA surface. An increase in contact angle was reached by chemical grafting using temperatures up to approximately 150 °C, whereafter a plateau was reached and even subsequently decreased. The results indicate that grafting temperature and grafting time need to be chosen carefully, and in many cases, these two parameters depend on each other. However, the authors pointed out that the results of the contact angle measurements are not "significant enough to draw general conclusions."

Similar results for the grafting temperature were measured for the grafting time. The effect of the grafting time depends on the other grafting parameters, and depending on the time used, it has a minor or a huge effect. For instance, Haas et al. (2017) measured a small increase in solubility time when increasing the grafting time from 2 to 20 minutes at a constant fatty acid chloride concentration of 1% and a constant temperature of 90 °C. In contrast, the solubility time of the PVA film increased to a great extent from 250 seconds to a maximum of 1,800 seconds when increasing the grafting time from 2 to 20 minutes at a constant fatty acid concentration of 5% and constant temperature of 90 °C (Haas et al., 2017).

Haas et al. (2017) varied the grafting parameters in several experiments and measured the solubility time of the PVA film. The authors concluded that the grafting time had the largest influence on the film solubility.

12.2.3 Types of Fatty Acids

Fatty acids, in general, are composed of a polar carboxyl group and a nonpolar carbon chain. The grafting reagent fatty acids can be used in different types. Fatty acids themselves are not suitable as they are not reactive (Berlioz et al., 2008). Amongst the fatty acid halides, fatty acid chlorides have the highest reactivity. Fatty acid chlorides are present in different types; they can vary in physical state (liquid or solid at room temperature) depending on the chain length. They can be either saturated or unsaturated, and there are different carbon lengths from commonly 6 to 24 carbon atoms. This variation in fatty acids is represented by the residues marked with R_1, R_2, and R_3 in Figure 12.1.

For the fatty acid grafting, commonly fatty acid chlorides and, to a lower extent, fatty acid anhydrides were tested. Fatty acid chlorides have several advantages as they obtain high reactivity, and there are some commercially available. For instance, palmitoyl and stearyl chloride are commonly tested in the literature (Germek et al., 2020; Haas et al., 2017; Lee et al., 2018; Schmid et al., 2012, 2014). Palmitoyl chloride can be purchased, for instance, from Sigma-Aldrich at a price of about €84.90/100 mL (MerckGaA, 2022).

However, also shorter fatty acid chlorides such as decanoic acid (Shaikh et al., 1998) and longer fatty acid chlorides have been tested, such as behenic chloride, which was found to induce a "very significantly greater" hydrophobic effect than that provided by stearic acid

(Samain, 2002). However, behenic acid chloride is quite expensive with about $414 per gram as offered by A2 Chem (A2 Chem LLC, 2022).

Furthermore, melissic acid, having 30 carbons in its fatty acid chain, was also tested in the patent, providing an even greater effect than using behenic acid (Samain, 2002); however, it is not realistic for industrial application mainly due to a very high price and limited commercial availability.

Overall, it was postulated that the chain length of the fatty acid determined the degree of hydrophobicity; the longer the chain length, the higher the hydrophobic effect. This was found by Germek et al. (2020) and Schmid et al. (2014), who measured decreased surface energy (and increased contact angle against water) for longer fatty acid chlorides; however, the authors tested only palmitoyl and stearoyl chloride.

In general, the deposition of fatty acids onto the surface depends on the grafting conditions. For instance, this process depends on the amount of fatty acids being transferred via the gas phase which is used in the transfer method and determines the grafting density (see Chapter 4.1).

12.2.4 Concentration of Fatty Acid

Besides the type of fatty acid used for the grafting process, also their concentration was indicated to influence the fatty acid grafting efficiency and, thus the hydrophobic effect.

Commonly, a concentration of 1–5% was used (Germek et al., 2020; Haas et al., 2017; Schmid et al., 2012, 2014) and fatty acid chlorides have been diluted often using petroleum ether (Germek et al., 2020; Haas et al., 2017; Schmid et al., 2012, 2014). But also smaller concentrations have been tested, i.e., 0.1–1% palmitoyl chloride in petroleum ether (Lee et al., 2018).

The effect of the fatty acid concentration can have a remarkable effect, meaning an enhanced grafting efficiency with increasing fatty acid concentration (Haas et al., 2017; Lee et al., 2018). An increase in fatty acid concentration from 1% to 5% (at a grafting time of 20 minutes) led to an increase in the grafting efficiency and solubility time from 550 to the maximum of 1,800 seconds, which points out that the film solubility was reduced by the increase in fatty acid chloride concentration (Haas et al., 2017).

In several studies, an optimization of the grafting parameters was tested, which was measured by an increase in hydrophobicity. For grafting PVA films, Haas et al. (2017) tested grafting temperature between 90 °C and 150 °C, grafting time between 2 and 20 minutes, and fatty acid chloride concentrations of 1%, 2%, and 3%. The authors found optimized process parameters, which are a grafting temperature of 130 °C, a grafting time of 14 minutes, and 3% of palmitoyl chloride in petroleum ether (Haas et al., 2017).

12.2.5 Types of Substrate

The type of substrate that is grafted was indicated to also strongly influence the grafting efficiency.

For instance, PVA with different degrees of hydrolysis have been grafted with fatty acids. The results showed that an increase in the degree of hydrolysis led to an increase in fatty acid grafting efficiency. This was shown by the higher contact angle of grafted PVA with a higher degree of hydrolysis compared to the lower contact angle for grafted PVA films with a lower degree of hydrolysis (Schmid et al., 2014).

The reason for this was related to the availability of functional groups on the surface which are needed for the fatty acid grafting reaction, specifically the esterification. This was also found by Germek et al. (2020), who observed the surface energy of PET and glassine, which were coated with PVA and grafted with fatty acids. The results revealed that the amount of functional groups in terms of hydroxyl groups was essential for the efficiency of the grafting. The higher the number of hydroxyl groups, the lower the surface energy, and the higher the grafting density was observed for both glassine and PET coated with PVA and grafted with fatty acids (Germek et al., 2020).

Besides functional groups, other parameters of the substrate, such as the surface smoothness (Lee et al., 2018) and the technique applied (see following Chapter 4), seem to influence the grafting efficiency.

12.2.6 Type of Grafting Process/Techniques

There are some techniques tested in laboratory and in-part pilot scale that will be outlined in the following chapters (Chapter 12.2.6.1 – 12.2.6.4) with individual advantages and challenges summarized in Table 12.1.

12.2.6.1 Transfer Method

The transfer method was one of the first methods used for the chemical fatty acid grafting technique, enabling a nanoscale layer of monomolecular fatty acids attached to a substrate. For instance, this method was used for grafting pure PET, pure paper, PVA-coated paper, and PVA-coated PET with fatty acid chlorides (Germek et al., 2020), PVA films (Haas et al., 2017; Schmid et al., 2012), and cellulose fibers (Stinga, 2008)

The transfer method is performed by the following steps, which were described in the literature as follows (Schmid et al., 2012):

- If needed: Modification of the substrate to obtain a surface with many functional groups available for the esterification process
- Preparation of the fatty acid solution (commonly fatty acid chloride in petroleum ether)
- Immersion of a transmitter paper with the fatty acid solution
- Drying of the transmitter paper to get rid of the solvent (commonly petroleum ether)
- The transmitter paper is put on top of the substrate and is usually fixed
- The transmitter paper on top of the substrate is heated to induce the chemical esterification reaction

Heating can be done in an oven (Germek et al., 2020; Schmid et al., 2012, 2014), with a hair dryer (Samain, 2002), microwave heating (Zhong and Netravali, 2016), or in principle with any heating device.

The success of fatty acid grafting can be determined by contact angle measurements, which was described to be a useful validation tool for the successful transfer method (Schmid et al., 2012). Alternatively, the water vapor permeability measurements can be used as validation of the success of the transfer method/chemical grafting (Schmid et al., 2012).

An enormous reduction in water vapor transmission rate (WVTR) was determined by Schmid et al. (2012), showing a reduction of the WVTR from 312 g/d*m^2 for untreated

PVA-coated paper to $48\,g/d{*}m^2$ for grafted PVA-coated papers when using either palmitoyl or stearoyl chloride applied using the transfer method (Schmid et al., 2012).

Strong effects on the water contact angle have been studied by the PVA-coated substrates showing an increase in contact angle from 38° (PVA Mowiol 6-98) to 104° and even 110° by chemical fatty acid grafting by the transfer method using palmitoyl acid chloride (C16) or stearoyl acid chloride (C18), respectively (Schmid et al., 2012).

In the literature, the transfer method was mainly used at the lab scale. However, there is one study describing the transfer method on PVA-coated paper at a pilot scale using a pilot machine with a twin-roll system (Lee et al., 2018).

Overall, using the transfer method the solvent does not get in contact with the substrate because the transfer method uses a transmitter paper for the transfer of the fatty acids. This is a big advantage, especially when you think of food packaging applications as mostly the fatty acid chlorides are diluted in solvents such as petroleum ether. The great advantages of the transfer method are the efficiency in modifying the surface of a substrate to a hydrophobic one and increasing the water repellency as described earlier. The fatty acid grafting technique using the transfer method leads to a nanoscale surface modification, as the fatty acids transferred via the gas phase attach as a monomolecular layer. However, there are also disadvantages such as the limited upscalability, possible changes through the heat treatment, and the procedure needs several steps as described before. In addition, the fatty acid grafting using the transfer method may result in changes in the color. For instance, a change in whiteness to a more yellow color was studied for PVA-coated papers after grafting, which may be a defect of the final product. Strong discoloration effects occur, for instance, if the hydrochloric acid formed during grafting is not removed quickly, which can be limited by neutralization of the hydrochloric acid directly upon its formation (Lee et al., 2018).

12.2.6.2 Direct Method

The direct method of the fatty acid grafting process describes the direct immersion of a substrate into a fatty acid solution, either pure or diluted in a solvent. This process can be performed for substrates that have functional groups on their surface, such as paper (Germek et al., 2020) or are used for hydrophobic PVA polymers (Shaikh et al., 1998). In the method described by Shaikh et al. (1998), fatty acid chlorides were dissolved in dimethylformamide and added dropwise to dissolved urethanized PVA polymers.

In principle, the direct method needs fewer steps compared to the transfer method; in detail, the transmitter paper is not needed, and the substrate is directly immersed in the fatty acid solution (Germek et al., 2020; Stinga, 2008).

12.2.6.3 Gravure Printing

Gravure printing represents a direct application process to transfer the fatty acid chloride solution, pure or diluted in a solvent, to a substrate. The process is comparable to the gravure roller application in the printing sector (Blechschmidt, 2013). Gravure printing is realized by a pair of rollers, in which the lower roller is the applicator roller. A relief structure consisting of small grid cells filled with liquid reagent is applied to the lower roller. The number and volume of the grid cells determine the amount of reagent that can be transferred to the substrate (Eichholz et al., 2015). Excess reagents can be removed by a doctor blade or controlled with an intermediate transfer roller (Eichholz et al., 2015). When the applicator (or transfer) roller is rotated against the upper substrate-carrying roller, the grafting reagent is applied to the substrate at specific points in the roller nip. Samain et al. (2013) and Eichholz et al. (2015) have applied for patents on this type of application. According to Eichholz et al. (2015), the parameters that influence the transfer of the grafting reagent to the substrate during gravure printing are the diameter, distances, and contact forces between the dossier (or transfer) roller, as well as the circumferential speed. In addition, the wetting properties of the printing substrate, the surface properties, the viscosity of the ink as well as the shape, volume, and filling level of the cells within the gravure printing device affect the gravure printing process (Kipphan, 2001).

The main challenge in the gravure printing application consists of dosing the grafting reagent in the grid cells in such a way that no excessive application results and the material efficiency of the process is maintained. Like the transfer method, gravure roller application is limited to a planar application, so this method can only be used on substrate sheets. In contrast, the required work steps are much simpler than with the transfer method, which makes the scalability of gravure printing on an industrial scale more attractive.

Within the EU project BIOnTop (grant agreement no. 837761), a gravure printing device was tested for the upscaling of fatty acid grafting using a Labratester (nsm Norbert Schläfli AG).

To analyze the efficiency of the gravure printing device, whey protein films have been grafted with stearoyl chloride either using the transfer method as explained in Chapter 12.2.6.1 or by the gravure printing technique using different grafting temperatures. After adding the fatty acid chlorides directly (without a solvent) onto the whey protein films through the gravure printing device, the films were put in an oven at different temperatures for the chemical reaction to occur. The grafting efficiency was determined by the WVTR. The results illustrated in Figure 12.2 revealed different grafting effects. Using the

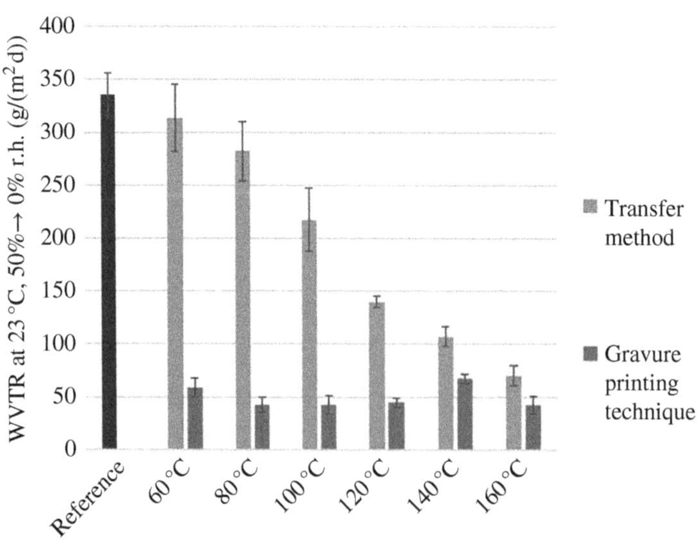

Figure 12.2 Water vapor transmission rate (WVTR) of fatty acid grafted whey protein films using either the transfer or the gravure printing technique

transfer method, the temperature highly influences the final WVTR, whereas, for the gravure printing technique, the temperature barely affects the grafting efficiency. The authors suggest that this is based on the transfer of the fatty acid chlorides through the vapor using the transfer method, and with increasing temperature, more fatty acid chlorides can be transferred.

Although gravure printing represents a promising technique for the upscalability of the fatty acid grafting process, the amount of material applied may be much more compared to the monomolecular nanoscale fatty acid layer reached by the transfer method.

12.2.6.4 Spraying

Spraying of fatty acid chlorides represents an application method for fatty acid grafting, which was already mentioned in the patent of Samain (2002). The application by spraying is indirect, as the liquid jet or film is first converted by an atomizer into a dispersion into the gaseous atmosphere (Lefebvre, 2017). In this way, the fatty acids are transferred through the gas phase to the substrate surface, on which the droplets are deposited (Goldschmidt and Streitberger, 2014).

The atomization during spraying is accompanied by an increase in surface area, for which energy is required. In spray applications, this energy is provided by mechanical forces or electric field forces. Mechanical atomization relies on the impact of dynamic pressure forces acting on the liquid due to the high relative velocity between the liquid and the surrounding atmosphere. This is achieved by high-velocity air or gas flow (pneumatic atomization), by increasing the kinetic energy of the fluid (airless atomization), or by the action of inertial forces through rotation (high rotation process) (Lefebvre, 2017; Westkämper and Warnecke, 2011). Thereby, the resulting spray is coupled to the internal geometry of the atomizer, the properties of the surrounding gaseous medium, and the liquid (Lefebvre, 2017).

In general, indirect application by spraying offers the advantage of overcoming the limitations of a planar application, such as those that exist in the gravure printing process. This makes spraying important for the functionalization of objects with complex three-dimensional geometries. Nevertheless, it should be noted that atomization also has some challenges. For instance, based on the high overspray, atomization is considered problematic from an ecological point of view. The reasons for this are the loss of reagents caused by overspray and the high solvent emissions, especially when volatile organic compounds are used (Goldschmidt and Streitberger, 2014).

The need for using a volatile solvent to atomize fatty acid chlorides depends on the liquid flow rate of the atomizer, the spray time, and the application efficiency. However, if the process parameters cannot be adjusted to produce a coating with a small excess of fatty acid chlorides, a solvent for dilution is required to maintain material efficiency. Depending on the aggregate state of the fatty acid chlorides at room temperature, a solvent is also needed to liquefy the fatty acid chlorides and ensure sprayability. Tempering the fatty acid chlorides as an alternative to liquefaction is unsuitable for spraying since this would facilitate the early loss of reagent during atomization by evaporation (Samain, 2002).

The potential of spraying as an application method for fatty acid chlorides covers both laboratory and industrial scales. However, it must be taken into account that the dispersion during spraying favors the formation of an explosive atmosphere, especially when volatile

Table 12.1 Advantages and challenges of grafting techniques

Method	Advantages	Challenges
Transfer method (Chapter 4.1)	• Very low reagent weight on the substrate is possible (in comparison to other methods described in this chapter) (Stinga, 2008; Zhong and Netravali, 2016) • No contact between solvent and substrate (is evaporated prior to contact with the substrate) (Germek et al., 2020; Haas et al. 2017; Schmid et al., 2012; Schmid et al., 2014; Stinga, 2008). • Adjustment of solubility of each side of films (Haas et al., 2017; Schmid et al., 2014)	• Solvent needed (Germek et al., 2020; Haas et al., 2017; Schmid et al., 2014) • Some reagents remain unused in the transmitter paper (own experiments, EU project BIOnTop project)
Direct method (Chapter 4.2)	• No transmitter needed (Germek et al., 2020; (own experiments, EU project BIOnTop project)) • Easy process (Germek et al., 2020; (own experiments, EU project BIOnTop project) • Grafting temperature may be lower as no transfer via the gas phase is necessary (own experiments, EU project BIOnTop project)	• Often, solvents are used (Germek et al., 2020) • High reagent weight (higher than needed for intended function) (own experiments, EU project BIOnTop project) • Remaining of not covalently bond reagents (own experiments, EU project BIOnTop project)
Gravure printing (Chapter 4.3)	• No solvent is needed, but possible (Samain et al., 2013; Eichholz et al., 2015) • Adjustable application weight (Eichholz et al., 2015; own experiments, EU project BIOnTop project) • Uniform and continous distribution of reagent possible (Eichholz et al., 2015) • Amount of reagent can be adjusted by using different printing plates (own experiments, EU project BIOnTop project)	• Planar application (Eichholz et al., 2015) • High reagent weight (higher than needed for intended function) (Eichholz et al., 2015)
Spraying (Chapter 4.4)	• Functionalization of objects/substrates with three-dimensional geometries possible (Samain, 2002) Solvent not necessarily required	• Overspray • Reagent losses due to premature evaporation • Explosive atmosphere with the use of solvent for dilution (subsequent step with gas stream free of grafting reagent needed) (Samain, 2002) • Liquid aggregate state of the reagent is necessary (e.g. stearoyl chloride at ambient temperature is solid (Sigma-Aldrich (2021b))

organic compounds are used (Sigma-Aldrich, 2022; *TRGS 721*, 2020). A reliable application of spraying in fatty acid grafting requires a high level of safety precautions. The implementation of explosion protection is required in the design of the spraying system and its peripherals (supply and exhaust air system, filter systems, liquid supply, and conveying equipment). It is recommended that spraying is performed in closed systems with suitable ventilation systems, in the best case, an object exhaust system. In the event that this cannot be ensured, comprehensive personal protective equipment is required (see also Chapter 12.6). Measures to ensure process safety are listed in the Technical Rules for Hazardous Substances (*TRGS 722*, 2021).

12.3 Upscaling Options and Requirements

The development of an industrial device for grafting is the focus of the patent filed by Samain et al. (2013). Lee et al. (2018) published a scientific article using a pilot machine with a twin-roll system for the upscaling of the fatty acid grafting technology. Furthermore, Eichholz et al. (2015) addressed the application of grafting on an industrial scale in his patent.

Basically, the following elements are required for the application of fatty acid grafting on an industrial scale (Samain et al., 2013). In Figure 12.3, the elements are combined on a generic basis.

1) Conveyor system
   To transport the substrate in an industrial process, a conveyor system is required.
2) Pretreatment module
   Depending on the substrate, a pretreatment, e.g., tempering, can help to avoid loss of reagent. For this purpose, Eichholz et al. (2015) preferred convection drying in order to achieve a moisture content of less than 5% for cellulose-based materials such as paper.

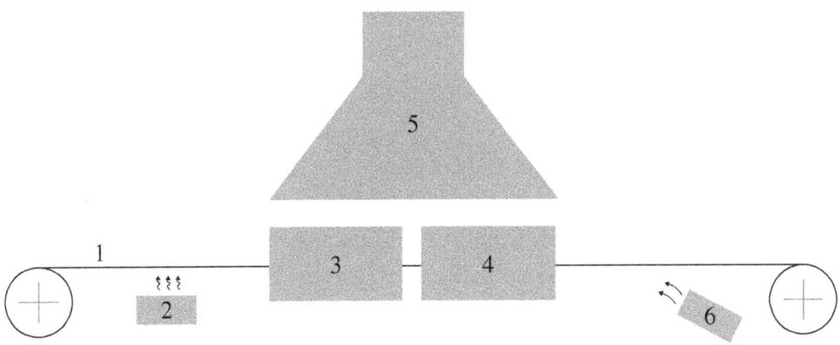

Figure 12.3 Generic application device for fatty acid grafting on an industrial scale with the elements: 1. Convoyer system, 2. Pretreatment module, 3. Application unit, 4. Tempering unit, 5. Exhausting unit, 6. Module to eliminate residual grafting reagent (*Source:* own illustration adapted from Samain et al. (2013))

3) Application unit
   Based on the application method, the application of the fatty acid chlorides must be designed in a way that the material efficiency is maintained. The amount of material required is dependent on the type of substrate; however, 0.1–2 g/m^2 is assumed to be sufficient (Samain et al., 2013). The challenge here is to apply a controlled amount of reagent to the substrate surface. For this, solvent may be necessary. For environmental conditions, Eichholz et al. (2015) specified that the reagent should be applied in a controlled atmosphere and preferably at less than 5% relative humidity.
   On the pilot plants of Eichholz et al. (2015), Lee et al. (2018), and Samain et al. (2013), types of direct anilox roller application are used to apply the reagent on the substrate.
4) Tempering unit
   Reaction parameters for temperature control must be defined so that the grafting reaction takes place within the specified reaction time. In industrial processes, it is necessary to reduce the reaction time to a minimum (Samain et al., 2013).
5) Exhausting module to remove hydrogen chloride
   When hydrogen chloride is formed as a byproduct of the covalent reaction of the fatty acid chlorides with the functional groups of the substrates, it shall be removed to prevent corrosion of the machine components, to maintain the kinetics of the reaction, and to avoid substrate degradation (Samain et al., 2013).
6) Module to eliminate the residual grafting reagent
   It is necessary to remove the reagent that was not covalently bonded to the substrate surface to avoid hydrochloric acid development downstream. This may especially be problematic when a substrate web is grafted and rolled up for storage. For the pilot plant, an air knife is used for this purpose (Lee et al., 2018).
   As already mentioned, the design of the reagent application unit in the industrial process depends on the application technique used. The following table gives an overview of the advantages and challenges of the different techniques.

12.4 Industrial Fields of Application

The effect of fatty acid grafting is mainly based on hydrophobization. For applications needing a hydrophobic effect, the fatty acid grafting technology may be an option. In the patent, Samain (2002) describes the application of fatty acid grafting onto windshields in the automotive industry, maintaining a hydrophobic effect for 6 months. In addition, fatty acid grafting technology can replace the wax coating of fibers to induce a hydrophobic surface (Eichholz et al., 2015). Other possible applications may be based on the fat absorption of the fatty acids when grafted onto a substrate. For instance, wood chips or sawdust composition can be hydrophobized by the fatty acid grafting technique being able to absorb hydrocarbons and oil while still maintaining the ability to float on the surface of water (Samain, 2002). In addition, the induced fat absorption property of grafted Whatman filter paper can be used to filter in terms of separating an emulsion (Samain, 2002).

The mentioned applications are theoretically possible and described in the mentioned references; however, there is not a lot of information available. In the following, some more fields of application that have been researched more extensively are outlined.

12.4.1 Partial Water-Soluble Packaging

Packaging having one side which is untreated and the other side fatty acid grafted, provides a slight hydrophobic character but dissolves upon flushed with water, providing useful packaging for detergents for dishwashers or washing machines. For instance, Haas et al. (2017) studied the water reactivity time, which was defined as the time needed for 1 mL of water to start disrupting the PVA surface as well as the solubility of palmitoyl chloride grafted PVA films. The authors concluded that by adjusting the grafting parameters (time, temperature, fatty acid concentration), the water reactivity, as well as the solubility of PVA films, can be controlled. Depending on the grafting parameters chosen, the PVA film can be tuned in the degree of hydrophobicity remaining for more than 30 minutes in water without dissolution compared to completely cold-water (20 °C) soluble PVA films without fatty acid grafting (Haas et al., 2017).

Similar effects have also been described for PVA-coated paper (Schmid et al., 2014). For instance, the WVTR of grafted PVA (PVA type Mowiol 28–99) coated paper was reduced from 1,046 g/m^2day to 60 g/m^2day when grafting with palmitoyl chloride using the transfer method. By grafting the PVA-coated paper, the water sensitivity was strongly reduced, which is important to maintain the oxygen barrier effect of PVA. This is because the oxygen barrier of PVA depends on the relative humidity, which decreases at high relative humidity (Schmid et al., 2014). Furthermore, fatty acid grafting of PVA-coated paper led to a high reduction in Cobb60 values compared to untreated PVA-coated papers (Schmid et al., 2012). These articles demonstrate the hydrophobic character of grafted layers, which initially had a hydrophilic character. Further possible fields of application are outlined in Chapter 12.7.

12.4.2 Flexible Packaging

Another field of application was outlined by Germek et al. (2020), who studied fatty acid grafting of the substrate polyethylene terephthalate (PET) or glassine, which were previously coated with PVA as application as release liner (e.g., address labels). Using the fatty acid grafting technique, PVA coating was hydrophobized with either stearoyl or palmitoyl acid chloride. The peel strength of labels onto differently grafted PVA-coated glassine and PET were analyzed. It was shown that by varying the coating and grafting conditions, the peel strength could be adjusted in part to similar peel strength as the benchmarks tested (Germek et al., 2020).

Another application was described by Lee et al. (2018) pointing out that fatty acid grafting of PVA-coated paper may be an alternative to laminating synthetic polymers on the surface of paper substrates. Herewith, the fatty acid grafting technique is used for hydrophobic paper to prevent the disintegration of water, thus allowing the paper to be suitable for liquid-based material or food packaging (Lee et al., 2018).

12.4.3 Paper-based Packaging

In its natural form, paper has no diffusion barrier against water due to the porous composite and the hydrophilic properties of the fibers (Blechschmidt, 2013). Considering that the strength properties of paper are sensitive to moisture, the lack of water-resistant properties is particularly constrictive for the applicability of paper in packaging (Emblem and Emblem, 2012; Hauptmann, 2017; Hauptmann et al., 2015; Rastogi and Samyn, 2015).

In order to enhance the hydrophobicity of paper, functionalization is necessary. Therefore, fatty acid grafting represents a material-efficient method for hydrophobizing and improving the water-repellent properties of paper and paperboard (Stinga, 2008). In chemical terms, wood fibers represent a biopolymer consisting of 38–55% cellulose, 17–35% hemicelluloses, up to 30% lignin, and up to 6% other extractives, including waxes, resins, and fats, depending on the type of wood (Blechschmidt, 2013). Both cellulose and hemicelluloses have hydroxyl groups in their chemical structure, which enable esterification and, thus, the grafting of fatty acids.

The use of fatty acid grafting to hydrophobic paper is already proposed in the patent of Samain (2002) to obtain water-impermeable papers for packaging purposes or to provide an alternative to sizing for printing papers. In the first experiments, Samain (2002) confirmed the grafting ability of paper with stearic acid chloride and palmitic acid chloride and estimated the amount of reagent required to hydrophobize a paper at a ratio of 1/1,000.

Stinga (2008) gives a comprehensive overview and discussion about the influence of paper composition on the grafting of fatty acids onto paper. In her PhD thesis, the influence of fiber type, fiber milling, and starch content on the hydrophobic effect was investigated. Regardless of these factors, Stinga (2008) determined a contact angle in the range of 120°–130° for grafted papers and a water absorption capacity according to Cobb60 of 14–20 g/m^2. Higher water resistance of paper has been achieved by fatty acid grafting. However, to further improve the hydrophobicity, Stinga (2008) applied both a coating of PVA and a coating with starch. While grafting PVA was successful, starch-coated papers also revealed improved grafting compared to uncoated paper. However, the durability of the hydrophobic effect is limited with starch-coated paper due to the mobility of the molecules. Research is needed to find a suitable starch coating for durable hydrophobicity by fatty acid grafting.

In a patent on fatty acid grafting with gravure printing application, paper was also used as a substrate. It was shown that predrying the substrate to a moisture content of less than 5% improved the hydrophobic effect of paper. This concerns, in particular, the hydrophobic result of the reverse side. With regard to the strength properties and stability of packaging, the patent stated that the strength properties of grafted papers could be maintained to a sufficient extent after exposure to moisture (Eichholz et al., 2015).

12.4.4 Textile

The use of fatty acid grafting in the field of textiles can be seen in the study of Zhong and Netravali (2016) to provide water repellency to cotton or even superhydrophobic surfaces of cotton textiles modified by a complex coating and stearic acid (Xue et al., 2009). Similar to paper, cotton consists of cellulose fibers and is hydrophilic in its natural form. In contrast to previous studies in the field of fatty acid grafting, Zhong and Netravali (2016) used fatty anhydride instead of fatty acid chlorides as a grafting reagent. Fatty anhydrides have a higher reactivity to fatty acids and a lower reactivity than fatty acid chlorides. This has been the main reason why fatty acid chlorides have normally been used as reagents (Samain, 2002; Stinga, 2008). While grafting with fatty acid chlorides leads to the formation of hydrogen chloride as a byproduct, free fatty acids are formed as a byproduct when fatty anhydrides are used as reagents (Zhong and Netravali, 2016).

To produce fatty acid esters with cellulose, acetic anhydride is used as a nontoxic coreagent. By using this coreagent, fatty acids are converted to more reactive fatty anhydrides under the influence of microwave heating. The success of the conversion to fatty anhydrides, as well as the grafting of fatty acids onto the substrate, was followed by Zhong and Netravali (2016) using high-performance liquid chromatography and attenuated total reflection Fourier transform infrared spectroscopy (ATR-FTIR).

For grafting the fatty anhydrides onto cellulose, microwave heating was applied. The effect of microwave power level, heating duration as well as the influence of different fatty acid chain lengths are investigated. In order to evaluate the grafting success with fatty anhydrides, the study will be used to measure the contact angle, tensile strength, and duration of water repellency of the laundry (Zhong and Netravali, 2016).

Study results by Zhong and Netravali (2016) provide promising results that grafting is also successful with fatty anhydrides. Results of ATR-FTIR spectroscopy showed that with increasing duration of tempering from 3 minutes to 15 minutes, a higher amount of grafted fatty acids onto the substrate can be detected. The contact angle measurement indicated that the hydrophobicity is positively correlated with both heating duration and power level: While untreated cotton has a contact angle of 0°, the contact angle with 100% microwave power (equivalent to 900 W) and 15-minute tempering time resulted in an increase in contact angle above 137° (Zhong and Netravali, 2016).

The durability of water resistance is considered by means of laundry cycles. Also, in this case, the results showed that the water resistance of grafted textile samples was improved with increasing microwave power and heating duration. At 100% power and 15-minute heating, a contact angle of >105° was achieved over 36 ± 2 washes. Each cycle in this study is equivalent to five hand or household washes, which means that the 15-minute treatment is sufficient to provide hydrophobicity for over 175 household washes. For the decrease in contact angle over the washes, Zhong and Netravali (2016) have several assumptions. On the one hand, the detergent could cause hydrolysis of ester bonds; on the other hand, cellulase enzymes in the detergent could have attacked the top, grafted layer of the textile.

In addition to hydrophobicity, the influence of microwave treatment during the grafting reaction on the tensile strength properties of cotton in both warp and weft was studied. While untreated cotton has a tensile breaking strength of 733 MPa (warp) and 630 MPa (weft), this was reduced by 20% (warp) and 27% (weft) after 10 minutes of grafting and microwave treatment at maximum power. After 15 minutes of annealing, the tensile breaking strength is about half of the original value. The reduction was suggested to be linked to the degradation of cellulose fibers (Zhong and Netravali, 2016).

12.5 Legal Aspects

Legal aspects to be considered in the implementation of fatty acid grafting technology are related to the control of physicochemical, health, and environmental hazards. A selection of regulations to be considered when performing fatty acid grafting is outlined subsequently.

Regulation (EC) No 1907/2006 of the European Parliament and of the Council of 18 December 2006 concerning the Registration, Evaluation, Authorisation, and Restriction of Chemicals (REACH) and establishing a European Chemicals Agency is generally applied to companies using chemicals. Those are defined as downstream users under REACH. For substances with an extended safety data sheet, downstream users are obliged to check whether the use of the chemical in the company is covered by the exposure scenarios described in the safety data sheet. If not, they have to check which measures they have to take according to REACH (ECHA, 2014). The solvent that is often used to dilute fatty acid chloride is petroleum ether (commonly with a boiling range of 100–140 °C) has an extended safety data sheet, pointing out that screening can be required depending on the intended use (Sigma-Aldrich, 2022).

During fatty acid grafting, hazards are essentially dependent on the substances used or produced (grafting reagent, solvent, and byproduct). The corresponding safety data sheets provide a basic overview of identified hazards in accordance with the international Globally Harmonized System classification, which is implemented within the Regulation (EC) No 1272/2008 of the European Parliament and of the Council of 16 December 2008 on classification, labeling, and packaging of substances and mixtures (Stark and Koch-Jugl, 2014). Hazardous substance classifications according to Regulation (EC) No 1272/2008 are available for fatty acid chlorides, petroleum ether, and hydrogen chloride (Sigma-Aldrich, 2021a, 2021b, 2022).

Depending on a quantity threshold classified as hazardous, Directive 2012/18/EU of the European Parliament and of the Council of 4 July 2012 on the control of major-accident hazards involving dangerous substances also apply to the operational use of fatty acid chlorides, petroleum ether, and hydrogen chloride. The directive deals with the control of major accident hazards caused by dangerous substances. On the basis of this directive, companies are required to submit a written accident prevention concept if the quantity threshold is exceeded.

In Germany, hazardous substances must be assessed in a hazard assessment based on §5 of the Occupational Health and Safety Act. During the hazard assessment, the conditions at the workplace are evaluated, hazards are determined, and protective measures are defined and tested for their effectiveness. The structure of the risk assessment is described in §6 of the Hazardous Substances Ordinance and in detail in the Technical Rules for Hazardous Substances (Ermer et al., 2021).

When using organic solvents for fatty acid grafting, the branch-specific emission limits of the 31stFederal Commission Control Ordinance must also be complied with.

12.6 Occupational Safety

In general, it must be taken into account that fatty acid grafting is performed with substances that require safety precautions during the process and for the operators. According to §7–12 of the Hazardous Substances Ordinance, the STOP hierarchy for protective measures needs to be considered. Options for implementation are listed in Table 12.2 (Ermer et al., 2021):

Overall, it should be noted that fatty acid grafting will only be controllable on a laboratory and industrial scale with a high safety effort.

Table 12.2 STOP protection measures concept

	Measures	Possibilities of Implementation for Fatty Acid Grafting
S	Substitution	The top priority is the substitution of hazardous substances for less hazardous substances (Ermer et al., 2021): effective grafting by substitution of fatty acid chlorides has not been achieved under the appropriate grafting conditions (Samain, 2002; Stinga, 2008). However, it is possible: This is shown by the results of Zhong and Netravali (2016) using fatty anhydrides.
T	Technical measures	If no substitution is applicable, collective protection must be implemented using technical protection measures (Ermer et al., 2021). A technical protective measure is to conduct the fatty acid grafting in a controlled atmosphere. This must be coupled with an extraction system to remove solvent emissions, vaporized reagent as well as the byproduct hydrogen chloride. In order to prevent the hydrolysis of the fatty acid chlorides on the one hand and the formation of hydrochloric acid on the other, a low relative humidity of less than 20% is useful in a controlled atmosphere. The temperatures should also be kept below the boiling temperature of the fatty acid chlorides (Eichholz et al., 2015). Depending on the application method (spraying) and the use of solvents, it should be considered that an explosive atmosphere can occur. So far, the periphery of the fatty acid grafting must be designed to be explosion-proof. This means that any kind of flammable sources must be avoided. In this context, tempering to improve the grafting reaction cannot be combined with the application and requires physical separation (TRGS 721, 2020; TRGS 722, 2021).
O	Organizational measures	Organizational measures also serve collective protection and regulate, for example, the exposure time or the reduction of the number of persons in the exposure area (Ermer et al., 2021).
P	Personal measures	If exposure is not adequately prevented by other protective measures and an open system is given (also on a laboratory scale), personal protective equipment becomes necessary. What type of personal protective equipment to use should be determined by considering the safety data sheets of the reagent, solvent, and byproduct (Sigma-Aldrich, 2021a, 2021b, 2022).

12.7 Sustainability Aspects and Future Scenarios

According to the recent state of knowledge, food packaging is in a field of tension due to the numerous, partly conflicting requirements. Packaging effectiveness, packaging efficiency, and cost-effective production and disposal have to be weighed against each other (Licciardello et al., 2017). In addition to packaging concepts tailored to the specific needs of users, product-specific considerations are a priority. Wikström et al. (2019) postulated the relevance of packaging functions such as reseal ability and especially the protective function for unopened packaging. Furthermore, food packaging contributes significantly to efficient logistics in extended transport chains and enables voluntary communication and legally mandatory declaration (Food packing materials 2017). These multiple functions are likely to be one reason for the high volume of primary packaging in Europe. Oversized packaging can still be found on the market due to the sometimes limited knowledge of how to achieve packaging functionalities and the advances in material development. Small and medium-sized companies, in particular, are affected by this, although the use of packaging with improved intrinsic performance and, thus, reduced weight can have positive ecological effects and, at the same time, economic net balances (Licciardello, 2017). This is confirmed in application-specific studies. In the case of the two-layer film bags considered for vacuum food packaging applications, a layer thickness thinning by 20 μm from 85 μm to 65 μm is possible without impairing functionality (Siracusa et al., 2014). Comparative shelf-life studies for industrially produced sliced durum wheat bread show the potential for packaging weight reduction by about 20% without compromising shelf-life standards (Licciardello et al., 2017). In modified atmosphere packaging (MAP) for fresh meat, the barrier properties of polypropylene (PP) are sufficient to keep the oxygen and carbon dioxide in the package during the shelf life. Multilayer barrier films, whose oxygen permeability rate is several times lower than the oxygen permeability rate of the PP tray, are additionally used as lidding films. Consequently, the barrier performance of the lid is over-dimensioned for this application (Müller und Gibis, D., Schmid, M.). Flexible plastic packaging is also often an alternative to semi-rigid/rigid plastic packaging, which allows thinner and more material-efficient packaging concepts. Pauer et al. (2020) also confirm that the weight reduction of plastic packaging can lead to greater environmental relief than increased recyclability.

As described in the previous chapters, the grafting technology can lead to a highly material-efficient surface functionalization on a nanoscale with interesting functional and packaging-relevant properties for different sectors. This material-efficient approach can therefore, contribute to sustainability depending on its final applications in many ways. Three of these potential approaches are described as example scenarios in this chapter. It must be mentioned, that these grafting approaches are potential future scenarios, which are—according to the author's knowledge—not on the market yet. However, these approaches could contribute to sustainability in the future.

12.7.1 Release Liners for Pressure-Sensitive Adhesives (PSAs)

Release liners are mandatory to protect the topliner with PSA during storage and transportation (e.g., self-adhesive labels). These release liners are single-use materials and are usually discarded after removal from the topliner with PSA (see Figure 12.4). As release liner material recycling is economical but still not beneficial, silicone-based release liners are partially incinerated or landfilled (Germek et al., 2020).

Germek et al. (2020) evaluated the effect of different grafting parameters on PVA-coated paper (see Figure 12.4 and Chapter 3). As PVA is a water-soluble polymer, it can be assumed that such coatings do not negatively affect the recyclability (reliability) of release liner papers and therefore could contribute to sustainability. In addition, and as described in this chapter above, chemical grafting is able to produce hydrophobic and low surface energy surfaces. This can positively affect the peel strength toward pressure-sensitive adhesives (PSAs) on top liners. Therefore, the grafting technology could potentially be an interesting treatment for hydrophilic surfaces (such as PVA-coated paper) to substitute silicone-based release liner substrates. However, as introduced above, also the functionality needs to be

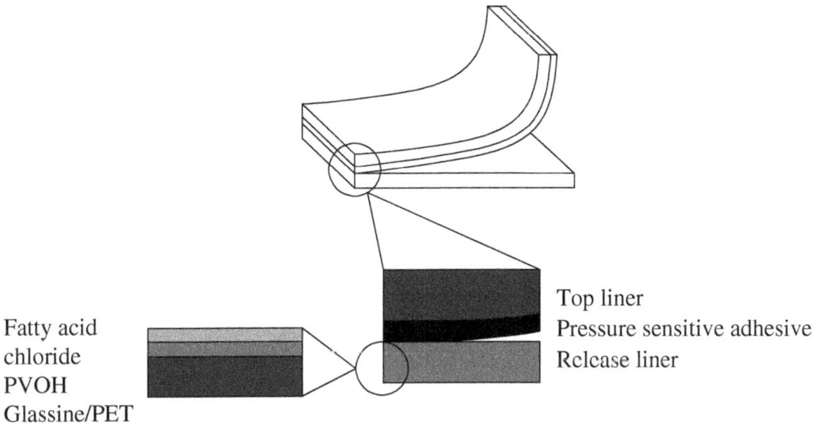

Figure 12.4 Simplified illustration of a construction showing a top liner, pressure-sensitive adhesive (PSA), and a release liner. The illustration on the left shows a magnification of the new release liner concept (*Source:* Germek et al. (2020)/Frontiers Media S.A./CC-BY-4.0)

given to a similar extent as silicon-based release liner materials. Therefore, the functional properties, such as surface energy and adhesion strength of acrylic PSAs, were tested.

The authors postulated that they were able to demonstrate that the grafting technology with nanoscale fatty acid chlorides produced potential alternative release liner materials that could be used for self-adhesive labels with acrylate-based PSAs. Based on these findings, it can be concluded that the grafting technology with fatty acid chlorides could potentially be used to produce paper-based release liners with similar functionalities compared to silicon-based release liners. In theory, these novel-release liner materials can be highly material efficient (Schmid et al., 2014) and significantly increase the recyclability of such liner materials (Germek et al., 2020).

12.7.2 Paper and Paperboard-based Packaging

In order to use paper and paperboard as transport packaging material, they need to be more resistant to moisture than other fiber-based materials. In fact, paper and paperboard offer protection against mechanical stress (pressure, shock, and vibration) due to their good mechanical strength and flexibility, but their porous structure and the hydrophilic nature of the fibers limit their ability to maintain these properties in highly humid conditions. Considering that the strength properties of paper are sensitive to moisture, the lack of water-resistant properties is particularly constrictive for the applicability of paper in packaging (Emblem and Emblem, 2012; Hauptmann, 2017; Hauptmann et al., 2015; Rastogi and Samyn, 2015). To ensure that paper and aperboard meet these requirements, sizing agents such as alum rosin, alkenyl succinic anhydride (Hubbe, 2007), or alkyl ketene dimer (Lindström and Larsson, 2008) are widely used. However, these cannot completely prevent water absorption, so functionalization of the surface is recommended (Lee et al., 2018). In the packaging industry, it is known to coat paper and paperboard with fossil polymers or metals for this purpose. Impregnations with waxes are also used (Eichholz et al., 2015; Rastogi and Samyn, 2015; Stinga, 2008). With such coatings, the functional properties are achieved, whereas the material combinations have limited recyclability in existing recycling streams, and biodegradability is lost. In this context, fatty acid grafting offers a possible alternative to the hydrophobization of paper while retaining its ecological advantages.

By retaining the strength properties of paper under the effects of moisture, grafted papers have an extended application potential as shipping bags or water-repellent packaging boxes since the focus of this type of packaging is on protection against mechanical effects and protection against moisture (Eichholz et al., 2015; Kotschik, 2015). According to Eichholz et al. (2015), the intended use of fatty acid grafting for paper is in wax replacement. Eichholz et al. (2015) consider that grafted papers can also be used in corrugated board production and in the processing of fruit/vegetable trays. In addition to packaging applications, Samain considers the potential of fatty acid grafting as an alternative to sizing in papers (Samain, 2002). According to (Lee et al., 2018), another advantage of the grafting process is that the appearance of the paper does not significantly change after the grafting process.

12.7.3 Repellent Properties of Textiles and Easy Emptying of Packaging

As already described in a previous chapter (Chapter 12.4.4), the application of fatty acid grafting in the field of textiles can provide water repellency to cotton fabrics for improved water repellent properties in a highly material-efficient way (Zhong and Netravali, 2016). This concept of reducing the surface energy by the fatty acid grafting process is hydrophobizing the material surface and therefore could be used to significantly improve the easy emptying behavior of hydrophilic packed goods (e.g., ketchup).

In addition to packaging design, food residues in packaging can negatively affect the recycling of plastic packing materials (Wohner et al., 2019). Furthermore, food or other packed goods residues in packaging must be classified as waste. These residues would still be fit for consumption but are lost due to the fact that they are disposed of together with the packaging. Since there are usually significantly more resources bound in the packaged goods than in the packaging itself, food or other goods residues in packaging and the resulting waste have a very unfavorable impact on the environment since the resources and energy associated with the production of the food or other goods are lost unused.

In certain applications with hydrophilic fillings, a hydrophobization of the packaging's inner surface could lead to improved emptying behavior and, therefore, lead to a reduction of residues in the packaging. This could reduce food waste therefore would significantly improve the sustainability of such packaging concepts.

Acknowledgments

This project has received funding from the Bio-based Industries Joint Undertaking (JU) under the European Union's Horizon 2020 research and innovation program under grant agreement No 837761. The JU receives support from the European Union's Horizon 2020 research and innovation program and the Bio-based Industries Consortium.

References

A2B Chem LLC. (2022). Docosanoyl Chloride. MFCD00056354. *Available online* https://www.a2bchem.com/21132-76-3.html (accessed 09 January 2024).

Berlioz, S., Stinga, C., Condoret, J., & Samain, D. (2008). SFGP 2007-investigation of a novel principle of chemical grafting for modification of cellulose fibers. *International Journal of Chemical Reactor Engineering*, *6*(1). https://doi.org/10.2202/1542-6580.1672

Byron, P. R., & Dalby, R. N. (1987). Effects of heat treatment on the permeability of poly(vinyl alcohol) films to a hydrophilic solute. *Journal of Pharmaceutical Science*, *76*, 65–67.

Blechschmidt, J. (Ed.). (2013). *Taschenbuch der Papiertechnik*. Carl Hanser Verlag GmbH & Co. KG. https://doi.org/10.3139/9783446437012

ECHA European Chemicals Agency. (2014). Guidance for Downstream Users. Version 2.1, October 2014, https://echa.europa.eu/documents/10162/2324906/du_en.pdf/9ac65ab5-e86c-405f-a44a-190ff4c36489, last checked on 09.01.2024

Eichholz, C., Kröner, H., Schmidt, H. B., Wengeler, L., Gattermayer, J., & Lawrenz, N. (2015). *Method for Continuous Coating of a Cellulose-based Substrate Web with Fatty Acid Chloride*. Google Patent Index EP 3177408 B1 20200429.

Emblem, A., & Emblem, H. (2012). *Packaging Technology: Fundamentals, Materials and Processes*. Woodhead Pub (Woodhead Publishing, Cambridge, Philadelphia. in materials). http://search.ebscohost.com/login.aspx?direct=true&scope=site&db=nlebk&db=nlabk&AN=679488

Ermer, A., Bender, H., et al. IVSS Sektion Chemie (publisher) (2021). *Vorbereitet für den Umgang mit Gefahrstoffen? Ein Einstieg zur sicheren Handhabung von gefährlichen Chemikalien*. Chapter 2 Gefährliche Eigenschaften von Chemikalien (pp. 11–25) & Chapter 4.4 Schutzmaßnahmen treffen (pp. 41–49). https://safety-work.org/fileadmin/safety-work/articles/Vorbereitet_fuer_den_Umgang_mit_Gefahrstoffen/issa-10_Gesamtdokument.pdf; ISBN 978-92-843-4233-4. ISSA.

Germek, E., Schmid, M., & Lindner, M. (2020). Grafting of fatty acids on polyvinyl alcohol: effects on surface energy and adhesion strength of acrylic pressure sensitive adhesives. *Frontiers in Materials*, *6*, 335. https://doi.org/10.3389/fmats.2019.00335

Goldschmidt, A., & Streitberger, H.-J. (2014). *BASF-Handbuch Lackiertechnik. [Überarb. Aufl.]. BASF; Vincentz Network (Farbe und Lack Bibliothek)*, Münster, Hannover.

Haas, A., Schlemmer, D., Grupa, U., & Schmid, M. (2017). Effect of chemical grafting parameters on the manufacture of functionalized PVOH films having controlled water solubility. *Frontiers in Chemistry*, *5*, 38. https://doi.org/10.3389/fchem.2017.00038

Hauptmann, M. (2017). Neue Einsatzpotentiale naturfaserbasierter Materialien in der Konsumgüterproduktion durch die technologische Entwicklung des Ziehverfahrens am Beispiel der Verpackung. *(Doctoral Dissertation, Habilitationsschrift, Dresden, Technische Universität Dresden)*.

Hauptmann, M., Wallmeier, M., Erhard, K., Zelm, R., & Majschak, J.-P. (2015). The role of material composition, fiber properties and deformation mechanisms in the deep drawing of paperboard. *Cellulose*, *22*(5), 3377–3395. https://doi.org/10.1007/s10570-015-0732-x

Hubbe, M. (2007). Paper's resistance to wetting - A review of internal sizing chemicals and their effects. *BioResources*, *2*, 106–145.

Kipphan, H. (Ed.). (2001). *Handbook of Print Media: Technologies and Production Methods/ Helmut Kipphan*. Springer (Springer Electronic Media), Berlin, Heidelberg.

Kotschik, G. (2015). *Leitfaden für umweltgerechte Versandverpackungen im Onlinehandel.* https://www.umweltbundesamt.de/sites/default/files/medien/378/publikationen/fachbroschuere_leitfaden_fuer_umweltgerechte_versandverpackungen_im_versa.pdf, Chapter 1 Funktion der Versandverpackung (p5). ISSN 2363-832X

Lee, T.-J., Lee, D.-J., Seo, J.-H., Lee, K.-S., Guerin, D., Martinez, P., Lee, M.-K., & Ryu, J.-Y. (2018). Preventing discoloration of poly(vinyl alcohol)-coated paper hydrophobized by gas grafting with palmitoyl chloride. *ACS Sustainable Chemistry & Engineering*, *6*(2), 1702–1707. https://doi.org/10.1021/acssuschemeng.7b02925

Lefebvre, A. H. (2017). *Atomization and Sprays* In collaboration with Vincent G. McDonell; McDonell V.G. (Ed.); 2nd Edition. Chapter 1 General Considerations (p- 1-16) Taylor & Francis CRC Press (Combustion).

Licciardello, F. (2017). Packaging, blessing in disguise. Review on its diverse contribution to food sustainability. *Trends in Food Science & Technology*, *65*, 32–39. https://doi.org/10.1016/j.tifs.2017.05.003

Licciardello, F., Giannone, V., Del Nobile, M. A., Muratore, G., Summo, C., Giarnetti, M., Caponio, F., Paradiso, V. M., & Pasqualone, A. (2017). Shelf life assessment of industrial durum wheat bread as a function of packaging system. *Food Chemistry*, *224*, 181–190. https://doi.org/10.1016/j.foodchem.2016.12.080

Lindström, T., & Larsson, P. (2008). Alkyl ketene dimer (AKD) sizing—a review. *Nordic Pulp and Paper Research Journal*, *23*(S), 202–209.

MerckGaA (2022). Palmitoyl chloride. *MERCK, 98. 112-67-4*

Pauer, E., Tacker, M., Gabriel, V., & Krauter, V. (2020). Sustainability of flexible multilayer packaging: environmental impacts and recyclability of packaging for bacon in block. *Cleaner Environmental Systems*, *1*, 100001. https://doi.org/10.1016/j.cesys.2020.100001

Pellis, A., Malinconico, M., Guarneri, A., & Gardossi, L. (2021). Renewable polymers and plastics: performance beyond the green. *New Biotechnology*, *60*, 146–158. https://doi.org/10.1016/j.nbt.2020.10.003

Rastogi, V., & Samyn, P. (2015). Bio-based coatings for paper applications. *Coatings*, *5*(4), 887–930. https://doi.org/10.3390/coatings5040887

Samain, D. (2002). *US 6342268 B1* (Patent No. US 6342268 B1).

Samain, D.; Guerin, D.; Lemaitre, A.; Lyannaz, L.; Guillouty, G. (2013). *US 20130236647A1*. Machine and treatment process via chromatogenous grafting of a hydroxylated substrate (Patent No. US 20130236647A1).

Schmid, M., Benz, A., Stinga, C., Samain, D., & Zeyer, K. P. (2012). Fundamental investigations regarding barrier properties of grafted PVOH layers. *International Journal of Polymer Science*, *2012*, 1–6. https://doi.org/10.1155/2012/637837

Schmid, M., Sängerlaub, S., Miesbauer, O., Jost, V., Werthan, J., Stinga, C., Samain, D., Stramm, C., Noller, K., & Müller, K. (2014). Water repellence and oxygen and water vapor barrier of PVOH-coated substrates before and after surface esterification. *Polymers*, *6*(11), 2764–2783. https://doi.org/10.3390/polym6112764

Shaikh, S., Asrof Ali, S. K., Hamad, E. Z., Al-Nafaa, M., Al-Jarallah, A., & Abu-Sharkh, B. (1998). Synthesis, characterization, and solution properties of hydrophobically modified poly(vinyl alcohol). *Journal of Applied Polymer Science*, *70*(12), 2499–2506. https://doi.org/10.1002/(SICI)1097-4628(19981219)70:12<2499::AID-APP23>3.0.CO;2-9

Sigma-Aldrich. (2021a). Safety data sheet. *Palmitinsäurechlorid zur Synthese*. https://www.sigmaaldrich.com/DE/de/sds/mm/8.00510 (accessed 09 January 2024)

Sigma-Aldrich. (2021b). Safety data sheet. *Stearoyl chloride. Version 6.2*. https://www.sigmaaldrich.com/DE/en/product/aldrich/171158 (accessed 09 January 2024)

Sigma-Aldrich. (2022). Safety data sheet. *Petroleumbenzin Siedebereich 100-140 GRAD C (Naphthabenzin) EMPLURA®. Version 8.6*. https://www.sigmaaldrich.com/DE/de/sds/mm/1.01770 (accessed 09.01.2024)

Siracusa, V., Ingrao, C., Lo Giudice, A., Mbohwa, C., & Dalla Rosa, M. (2014). Environmental assessment of a multilayer polymer bag for food packaging and preservation: an LCA approach. *Food Research International*, *62*, 151–161. https://doi.org/10.1016/j.foodres.2014.02.010

Stark, C.; Koch-Jugl, J. (2014). *Das neue Einstufungssystem für Chemikalien nach GHS. Leitfaden zur Anwendung der CLP-Verordnung*. Chapter 3 Umsetzung des GHS in Europa. Umweltbundesamt

Stinga, N. C. (2008). Utilisation de la chimie chromatogénique pour la conception et la réalisation de matériaux cellulosiques barrières à l'eau, aux graisses et aux gaz. *(Doctoral Dissertation, Université Joseph-Fourier-Grenoble I), Français. Chimie.*

TRGS 721. (2020). Technische Regeln Für Gefahrstoffe. Gefährliche Explosionsfähige Gemische—Beurteilung Der Explosionsgefährdung. Hg. v. Bundesanstalt Für Arbeitsschutz Und Arbeitsmedizin. http://gaa.baden-wuerttemberg.de/servlet/is/16495/5_721.pdf (accessed 31 January 2022).

TRGS 722. (2021). Hnische Regeln Für Gefahrstoffe. Vermeidung Oder Einschränkung Gefährlicher Explosionsfähiger Gemische.

Westkämper, E., & Warnecke, H.-J. (2011). *Einführung in die Fertigungstechnik. 8., aktualisierte und erw. Aufl., korrigierter Nachdr*. Vieweg + Teubner (Fertigung), Wiesbaden. https://link.springer.com/content/pdf/10.1007%2F978-3-8348-9798-5.pdf

Wikström, F., Verghese, K., Auras, R., Olsson, A., Williams, H., Wever, R., Grönman, K., Kvalvåg Pettersen, M., Møller, H., & Soukka, R. (2019). Packaging strategies that save food: a research agenda for 2030. *Journal of Industrial Ecology*, *23*(3), 532–540. https://doi.org/10.1111/jiec.12769

Wohner, B., Pauer, E., Heinrich, V., & Tacker, M. (2019). Packaging-related food losses and waste: an overview of drivers and issues. *Sustainability*, *11*(1), 264. https://doi.org/10.3390/su11010264

Wu, F., Misra, M., & Mohanty, A. K. (2021). Challenges and new opportunities on barrier performance of biodegradable polymers for sustainable packaging. *Progress in Polymer Science*, *117*, 101395. https://doi.org/10.1016/j.progpolymsci.2021.101395

Xue, C.-H., Jia, S.-T., Zhang, J., & Tian, L.-Q. (2009). Superhydrophobic surfaces on cotton textiles by complex coating of silica nanoparticles and hydrophobization. *Thin Solid Films*, *517*(16), 4593–4598. https://doi.org/10.1016/j.tsf.2009.03.185

Zhong, Y., & Netravali, A. N. (2016). "Green" surface treatment for water-repellent cotton fabrics. *Surface Innovations*, *4*(1), 3–13. https://doi.org/10.1680/jsuin.15.00022

13

Applications of Nanotechnology in the Packaging of Special/Space Foods

Vidushi Kapoor, Vijayakumar Raja, Ishita Neogi, and C. Anandharamakrishnan

CSIR–National Institute for Interdisciplinary Science and Technology (NIIST), Ministry of Science and Technology–Government of India, Industrial Estate PO, Thiruvananthapuram, Kerala - 695019, India

13.1 Introduction

Food accessibility is a significant concern when humans are at war, hit by a calamity, or on space exploration. In these exceptional cases, food must be made accessible to them for their survival and optimum health. Foods for special applications are meant for a selected group of humans under certain circumstances and are termed "special foods" (Long et al., 2022). These meals differ in composition, packaging material, nutritional value, storage needs, and food consumption. Unlike ordinary foods, special foods need to be safer, nutritionally dense, easy to carry, portable, and lightweight, apart from having ultralong shelf stability to sustain harsh environments. Depending upon the requirements, special foods need appropriate processing techniques and packaging systems to extend their shelf life. Shelf life can be defined as the microbiological, physicochemical, enzymatic, nutritional stability, and optimal sensory characteristics of foods (Sánchez-García et al., 2021). Survival in special situations like aerospace, military, and disaster relief situations requires stable prepackaged foods with long shelf life durations (Long et al., 2022).

For space settings, as dependence on Earth is reduced, maintaining the shelf life of the packaged food is crucial to provide adequate nutrition to the space crew. The quality of food available for consumption directly impacts the success rate of the missions, especially long-duration space missions. Therefore, shelf-life-extending processing technologies are necessary to produce high-quality space foods. In addition to this, whole supply and processing chains need to be optimized, considering the raw materials, formulations, packaging techniques, and storage facilities required (Pereira and Vicente, 2010). Space food packaging is essential as it helps maintain the nutritional value, acceptability, and, most importantly, the shelf life of foods for space crew members.

Space exploratory missions have gained popularity in recent years with advancing technology and success in achieving milestones. With the advent of technology in the development of spacecraft, well-known space exploration agencies like National Aeronautics and

Nanotechnology for Sustainable Food Packaging, First Edition. Edited by C. Anandharamakrishnan, Jeyan A. Moses, and M. Maria Leena.

Space Administration (NASA) pay 22.6 billion dollars annually for research and development under the space food category (Raut et al., 2021).

The microgravity environment causes a range of physiological changes associated with weightlessness. These changes lead to detrimental health effects like calcium and potassium losses, red blood cell mass loss, dehydration, space motion sickness, and intestinal disorders. In addition to this, bone loss, loss of muscle mass, weaker immune function, slower gastrointestinal mobility, and intestinal transit time, leading to constipation among astronauts during long-term manned missions, have also been reported (Chaloulakou et al., 2022). These changes in physiological factors must be considered when designing or selecting processing technologies and packaging systems for space foods, as processed foods must exhibit good nutrient bioaccessibility to maintain astronauts' physical health (Pandith et al., 2023). In addition to meeting the nutritional needs of astronauts, foods consumed by the space crew affect social psychology during long-duration missions (Douglas et al., 2020). The food system in space is not just about providing sustenance but is also a socializing tool that helps astronauts bond with one another and feel a sense of connection with Earth. Therefore, keeping the food nutritious and appetizing is the key to the survival of astronauts in manned spaceflights.

Food quality attributes such as appearance, color, texture, flavor, and aroma need to be manipulated in order to avoid the physiological and psychological disturbances faced by the space crew (McPhee and Charles, 2009). With atmospheric and gravitational changes in the space environment, the requirement for food and nutrition also changes. To prevent health deterioration in space and ensure sufficient dietary intake during long-duration missions, packaged space food must ensure not only the palatability and freshness of the food but also the bioavailability of nutrients (Tang et al., 2021). Besides being nutritionally adequate and palatable, the food must counteract harmful effects like oxidative stress and radiation exposure and be safe to consume in microgravity conditions. Therefore, packaging for space foods must be designed in such a way that they can overcome the adverse effects of vibration, galactic cosmic rays, radiation, and harsh environmental factors, such as low pressure in outer space (Grover et al., 2021). Considering all these factors, specialized packaging materials for space foods are essential to sustain space crews in manned spaceflight missions (Douglas et al., 2021). A variety of packaged space foods are exclusively designed and formulated for astronauts to use in the microgravity environment. The chapter provides an overview of current packaging systems, limitations, and challenges. It aims to provide insights into the application of nanotechnology to present packaging technologies to address the present limitations of space food packaging.

13.2 Criteria for Packaging of Space Foods

Packaging of space foods is one of the most crucial factors as it is directly linked with the shelf life of processed food products. Food packaging material acts as a barrier that safeguards internal contents from external damage while also reducing the exchange of water vapor and gases. The packaging materials are desired to protect the food and prevent loss of flavor and nutrients while being nonreactive with the food ingredients (Jiang et al., 2020a). Packaging materials should also be able to maintain the pH and nutrient content of foods with enhanced shelf life and palatability of crew members (Long et al., 2022). Contact of food products with moisture could be detrimental to microbial spoilage, and oxidation of food by oxygen gas can reduce the palatability and acceptability of space foods (Cooper et al., 2011). Therefore, factors like maintaining weight-to-volume ratios, improving shelf life, protection from harmful radiation, convenient storage, and waste disposal capacity must be considered when preparing packaging suitable for space applications. In addition, it is preferable for packaging materials to be heat-sealable, puncture-resistant, and flexible (Cooper et al., 2011; Crisp et al., 2015).

NASA has revised requirements for packaging for space food applications. According to recent guidelines, packaging must have barrier properties such as a thermostabilized pouch, oxygen permeability lower than 0.0003 cc/100 inch^2/day, and water vapor permeability lower than 0.0004 cc/100 inch^2/day. It must also be able to accommodate the microwave-assisted thermal sterilization (MATS) or pressure-assisted thermal sterilization (PATS) process as well as vacuum packaging and be foil-free, transparent, and flexible so that broken pieces are visible (DeVieneni et al., 2012).

The quality of space packaging materials in terms of physical, mechanical, and functional properties can be improved by the incorporation of nanomaterials in packaging. The incorporation of nanoparticles into the polymer matrix can improve barrier characteristics, elasticity, and temperature stability (Sharma et al., 2017). To adhere to the revised packaging rules, nanomaterials have high-prospect applications in aerospace food packaging. Composite nanomaterial packaging is very promising with its high barrier, high flame retarding, and thermally stable characteristics, which can extend the shelf life up to 3–5 years (Sun et al., 2016).

13.3 Evolution of Space Foods and Packaging Technologies

Space and extraterrestrial planet exploration was initiated in the mid-20th century to discover other habitable planets in outer space. Most notable missions like Mercury, Gemini, Apollo, and Skylab have contributed to the evolution of space foods. Previous and current space explorations and their contribution to the evolution of packaging systems have been summarized in Table 13.1.

13.4 Exploration Agencies and Their Packaging Trends

India's space exploration started with the support of the Union of Soviet Socialist Republics (USSR) in 1970 and has advanced to independently launching men and machines. Indian Space Research Organization (ISRO) has successfully launched men on the moon's surface. Such manned missions demand necessary technologies for designing foods to cater to the special needs of the space crew. Facilities to formulate special foods adhering to adequate safety evaluation and packaging requirements are available at the Defense Food Research Laboratory (DFRL), Mysore. Indian cuisine foods with spices are processed, packed, and stored on spaceflights. Regional Indian meals, including rice and noodle varieties with curries, were dehydrated and packaged in a multilayered retort flexible pouch.

Table 13.1 Space missions and evolution of space food packaging systems

Timeline	Space Program	Discoveries in the Realm of Space Food	Specifications	References
1962–1964	Mercury	• Novel bite-sized cubes and powder using freeze-drying technology were engineered. • Semi-liquid foods packed in aluminum tubes were introduced.	The texture and flowability of food materials were modified to comply with suitable organoleptic properties.	Cooper et al. (2011), Gupta and Gupta (2010), Perchonok and Bourland (2002),
1965–1967	Gemini	• In order to tackle the crumbling challenges of dehydrated, powdered and bite-sized foods, oil coating was provided.	The quality of food products was improved with innovative packaging.	Cooper et al. (2011), Douglas et al. (2020)
1968–1975	Apollo	• Irradiated foods and flexible pouches for thermostabilized foods were introduced for the first time in this mission.	Instead of using wet packs to keep the food moist, astronauts used water to rehydrate the food.	Douglas et al. (2020), Pometto and Bourland (2003),
1971–1973	Skylab	• The spacecraft was equipped with a refrigerator, freezer, warming trays, food containers, and a table. Most of the foods were packed in aluminum cans and hydratable packages.	A simple heating device was used in space for heating foods through conduction. In addition, frozen foods could also be stored in this mission.	Douglas et al. (2020), Gupta and Gupta (2010), Pometto and Bourland (2003)
1972–present	Space Shuttle Program (ISS)	• A standard special menu consisting of three familiar balanced meals with appetizing snacks was incorporated every 7 d. • Food items like peach yogurt, butterscotch pudding (thermostabilized food), shrimp and cocktail, coffee and cream (rehydratable food and beverage), intermediate moisture foods like dried apricots, mushrooms and beef tips, irradiated steak, and natural form foods are some examples. • Space farming and 3D food printing in space was also introduced.	This mission presented various ways to ensure a food experience similar to Earth. For long-duration space missions, National Aeronautics and Space Administration's (NASA's) Advanced Food Technology Project initiated new ways to ensure adequate food supply. Astronauts could design their own menus based on their unique homely cuisine. A food preparation was also installed on the orbiter's mid-deck.	Gupta and Gupta (2010), Perchonok and Bourland (2002), Pometto and Bourland (2003)
Mid–2030s	Mars Mission	A mechanical centrifuge was used to produce artificial gravity to carry out research on space gardening. This could provide thirty years of shelf life to foods. In addition, due to weight and space limitations, a variety of ready-to-eat bars were developed. Novel packaging techniques to extend shelf-life for long-duration missions are being explored.	NASA collaborated with a private company, for long-duration space missions. Further, a vegetable production system was developed at ISS.	Perchonok and Bourland (2002), Pometto and Bourland (2003), Tang et al. (2021), Pandith et al., (2023)

Multiple layers of various polymers and aluminum foils are used in the packaging materials that exhibit better moisture and gas barrier properties and improved aroma retention. Powdered fruit juices and beverages were stored in flexible rehydratable pouches (Pandith et al., 2023; Singh, 2021).

Japan Aerospace Exploration Agency (JAXA) provides a variety of delicious Japanese meals in freeze-dried forms. Freeze-dried products like rice, noodles, green tea, and soft drinks are stored in multilayered flexible pouches. The food is rehydrated with hot or cold water before consumption. Thermostabilized retort pouches are used to pack heat-processed Japanese yakitori, fish, and curry. Candy, biscuits, and yokan (sweet bean paste) are processed and packed in transparent plastic boxes, and multilayered flexible pouches act as secondary packaging. Similarly, seasonings and condiments are stored in small plastic bottles and then packed in multilayered flexible pouches (HoJeon, 2007)

NASA uses food packaging materials such as bite-size food pouches and rehydratable pouches made of nylon, ethylene vinyl alcohol, and polyethylene, which were optimized for specific dimensions and thickness. Combitherm paxx materials were used for designing multilayered packaging and overwrap for additional barrier properties. Packaging materials like nylon (polyamide), ethylene vinyl alcohol (EVOH), and polyethylene were used to pack foods in bite-size food pouches with a thickness of 0.0045 inch and 7.5-inch × 3.563-inch dimensions. Beverage pouches were made with low-density polyethylene (LDPE), linear LDPE, polyethylene terephthalate (PET), and aluminum foil. The dimensions of these pouches are 8.8-inch × 3.84-inch (length × width) with 0.00498-inch thickness. A multilayered rehydratable pouch made of nylon, EVOH, and polyethylene with a length of 6 inch and width of 5 inch is used to store rehydratable food products. Secondary packaging or overwraps with laminates of PET, LDPE white, aluminum foil, and surlyn are used to improve the barrier properties of the rehydratable pouch. Thermostabilized retort pouches are laminated with layers of PET, stretched nylon, aluminum foil, and cast polypropylene. Sixty-five thermostabilized prepackaged products were tested by NASA at room temperature for palatability after five years, out of which only seven qualified (DeVieneni et al., 2012).

13.5 Applicability of Nano Packaging to Processed Space Foods

The main aim of processing the food is to extend the shelf life and provide adequate nutrients to the space crew during their stay in space. Due to limited facilities available on spacecraft, the space crew has access to only ready-to-eat packaged foods, which require simple rehydration or heating (Long et al., 2022). It is, therefore, imperative to understand the processing techniques and adequate packaging applied to such foods.

13.5.1 Packaging for Freeze-Dried Foods

Freeze drying, also referred to as lyophilization or cryodesiccation, is a method of dehydration conducted at low temperatures. This process involves the prefreezing of wet materials, followed by sublimation of frozen solvents under a vacuum environment, and the removal of residual bound water from the material matrix (Ratti, 2008; Zhang et al., 2017).

Various combinations of time-temperature coefficients and pressure are used in this process (Nowak and Jakubczyk, 2020). The resulting product features a porous dried particle structure, retaining the natural odor and color of the original food while having a lower density (Huang and Zhang, 2012; Wang et al., 2010). In addition to this, the food's original nutrition, flavor, and taste remain intact without the need for preservatives, making it an ideal choice for processing space foods. Multi-layered co-extruded materials are majorly used for freeze dried space foods.

13.5.2 Beverage Pouch

The current scenario in space does not allow for refrigerated storage of food. Foods dried with conventional approaches are generally prone to crystallization or collapse upon storage, which reduces consumer acceptability. Freeze-drying technology enables the dried product to exhibit good color retention while being nutritionally more stable and retaining shape amid space environment. A variety of foods could be included in space missions with this technology. China's green and black tea, Korea's seaweed soup and Thai's durian dried fruit, NASA's ice cream, shrimp cocktail, chicken, and vegetables were among the most notable freeze-dried products utilized as space foods for spaceflight missions. Beverages like fruit juices, tea, and coffee are the most commonly freeze-dried foods, which are stored in flexible rehydratable pouches. Beverages are freeze-dried and packed in multilayered flexible pouches as dry powders. After placing the powder in the package, it is flushed with nitrogen gas to reduce oxygen gas availability and thus reduce oxidation reaction in foods, making them shelf stable. The packages are then sealed after inserting a septum made of silicon rubber (Perchonok, 2014).

Other freeze-dried products like rice, noodles, soft drinks, and green tea are stored in multilayered flexible pouches, which are rehydrated using hot or cold water in space. Primary packaging material usually does not provide adequate barrier properties to freeze-dried foods and is, therefore, overwrapped with an additional packaging layer to offer an extended shelf life of up to 18 months. To minimize the weight burden on spacecraft, packaging materials with nanocomposite films instead of metal foil layers could be designed, which would reduce the total weight of the packaging and enhance its functionality (Kumar and Gaikwad, 2023). The use of nanomaterials in the packaging of freeze-dried foods can provide high barrier properties with better biodegradability and reusability, which could provide better shelf life for foods and reduce the solid waste burden on land.

13.5.3 Packaging for Irradiation Sterilized Foods

Ionizing radiation, which was earlier used in medicine and nuclear power, was patented in 1905 as a microbial inactivation technique. This radiation disrupts the normal cell function of the bacteria by damaging its nucleic acid through direct or indirect function. After its first evaluation against trichinae in pork in 1921, it has been explored for many food applications, including space foods. It has simple operation, offers high efficiency, and maintains color, flavor, and nutrients without chemical additives. In the context of space foods, this is especially advantageous as reheating irradiated foods is not required, ensuring maximum retention of nutritional properties. Food irradiated at high doses, even though mostly sensorially acceptable, may have some undesirable characteristics such as strange smell, low viscosity, discoloration, or darkening, lipid oxidation, and altered texture. Upon studying the minimum electron beam dose required for space food sterilization, it was found that a dose of 15 kGy is sufficient to achieve the safety and long-term shelf life targets for space foods (Bhatia et al., 2017). Using irradiation techniques in combination with other processing techniques like freezing, refrigeration, vacuum packaging, and the use of calcium, vitamin C, and other antioxidants as additives can overcome the adverse effects of high radiation doses. Further, irradiation dose can be reduced when combined with heat treatment without affecting the quality of food (Chen et al., 2016; Kirkin et al., 2014; Ouattara et al., 2001). It is essential to choose packaging materials with careful consideration of these factors related to irradiation processing.

13.5.4 Retort Pouches/Thermostabilized Packaging

Developed for battlefield use by the United States Army Natick Research and affiliates, the retort pouch was made from flexible plastic laminates and metal foils to save on volume and mass for space applications. Retort packaging technology is widely used because of its greatest potential to achieve a shelf life of 3–5 years. Foods like grilled ham, chicken, tuna fish, and puddings are heat processed and stored in the retort pouches to reduce microbial spoilage. These flexible pouches can also be used to pack thermostabilized and irradiated food products to ensure food's acceptability, safety, and nutritional value. Also known as ready-to-eat pouches, they are laminated with polyolefin, aluminum foil, polyamide, and polyester combinations (Prisco, 2014). These quad-laminated pouches have very low oxygen and water permeability. The storage period of retort-processed products can extend for 3–5 years while maintaining food's sensory acceptability (Jiang et al., 2020a). For enhanced oxygen barrier properties, the aluminum foil layer could be swapped with a polymer nanocomposite, especially boron nitride nanosheets, in combination with polymer layers for better oxygen barrier properties and efficient waste disposal.

13.5.5 Metal Cans

Metal cans, mainly constituted of tinplate and aluminum alloy, provide a shelf life of 3 years due to better barrier properties. Frozen and heat-stabilized foods were packed in aluminum cans during the Skylab mission to withstand changes in temperature and pressure and maintain two years of shelf life (Klicka and Smith, 1982; Perchonok and Bourland, 2002). The headspace volume in the aluminum cans was maintained by flushing with nitrogen gas at least three times before sealing. Metal cans are currently used in Russian manned spaceflight missions. A major drawback of metal can packaging is that they are bulky and due to garbage disposal problems and weight constraints in spacecrafts, they are unsuitable for long-term manned missions. Alternatively, a cooking bag made of metal materials and flexible plastic is highly suitable as it is easy to use, has high shelf stability, and can be used in many forms (Sun et al., 2016). Nanocomposite packaging materials could achieve this as they provide thermal resistance, low density, mechanical strength, efficient gases and vapors barrier efficiency, and transparency, providing both flexibility to packages and high shelf-stability to foods.

13.5.6 Packaging of High-Pressure-Processed (HPP) Foods

High-pressure processing is one of the emerging technologies that was first adopted by NASA for space applications. To achieve a goal of 5-year shelf life for all food formulations (Perchonok and Catauro, 2009). Conditions like high pressure ranging from 200 MPa and 600 MPa cause changes in the structure of microbes and food enzymes, leading to inactivity (Khan et al., 2017). The use of high-pressure sterilization began in the 1990s in Japan for jams, jellies, and juice processing (Pandrangi et al., 2014). After a study performed by the United States on the processing of avocado, it was found that high-pressure conditions did not alter taste, color, or texture and were able to significantly increase the shelf life from 3 to 30 days (Brown, 2011).

HHP technology, when used in combination with traditional thermal treatment, can effectively overcome the cost and technical constraints of HHP technology alone. Successfully developed by the National Center for Food Safety and Technology in the United States, PATS is one such example that employs the features of both high temperature and pressure (Balasubramaniam et al., 2016).

13.5.7 High-Barrier Flexible Packaging

The products are preheated to attain the required temperature, followed by compression heating to reach the pressure-temperature. The temperature of the product is lowered to the ambient temperature during the decompression period (Barbosa-Cánovas and Juliano, 2008). PATS processed low-acid foods are vacuum packaged in high-barrier flexible containers or pouches and stored. This technology, along with reducing the sterilization time, improves the quality of low-acid food. PATS-treated fruits are more shelf stable than retort processed fruits. In combination with refrigeration, PATS-processed fruits can achieve a 5-year long shelf life. The applicability of HPP processed foods is limited to some foods because it can significantly alter the texture and organoleptic properties of most food products (Jiang et al., 2020a; Long et al., 2022).

13.5.8 Packaging for MATS (Microwave-Assisted Thermal Sterilization) Foods

A sterilization procedure that uses 915 MHz, as opposed to the domestic frequency of 2,450 MHz, offers an alternative to conventional technology (Barbosa-Cánovas et al., 2014). MATS features in-container processing of food and provides better nutrient retention. Microwave sterilization uses water as a heating medium, making it an effective processing technology and can avoid uneven heating and edge effects, as found in traditional methods. However, it can result in modification of some ingredients in a few cases, which might be undesirable. According to research conducted by NASA, when MATS and conventional sterilization were applied on sweet and sour pork and coined carrot, MATS processed products showed better sensory and nutritional characteristics with a superior shelf life of more than five years (Douglas et al., 2016). However, aluminum-coated pouches cannot be used for MATS sterilization as they might interfere with the electromagnetic waves, which may negatively affect the water and gas barrier properties of the packaging. A comparison study between MATS and retort sterilization was conducted by NASA to ascertain if lower heat can preserve nutrients and extend shelf life. It was learned that despite the MATS product being able to retain color, texture, and vitamin stability, the advantages were unable to be sustained because of the nonmetallic packaging film, which could not act as an appropriate oxygen barrier (Cooper and Douglas, 2015). High-quality canned food can be provided for astronauts using microwave-assisted sterilization techniques by optimizing the dielectric properties formula of packaging. In order to achieve optimum storage and solve problems of alteration of food characteristics in canned food and traditional thermal treatment, China developed a novel compound technology vacuum low-temperature seasoning-microwave sterilization-flexible packaging (Jiang et al., 2020b). Nanocellulose and graphene polymer nanocomposite films can be used instead of metal for packaging films to improve the oxygen barrier properties of MATS-sterilized products.

13.6 Emerging/Novel Space Food Packaging Technologies

Orion spacecraft engineered for NASA's deep space missions has weight limitations due to the extended duration of space missions. Therefore, only limited supplies can be carried on board (Kumar and Gaikwad, 2023). In order to overcome challenges related to aerospace packaging, NASA's Deep Space Food Challenge bought many innovative packaging materials that may be considered for maintaining safe, nutritious, and palatable foods for future long-duration space missions Beehex company proposed drying plant and meat powder using a Universal Food Fabricator (UFF) and then storing it in hermetically sealed cartridges to elongate shelf life for more than five years. One group from Hawthorne came up with an innovative idea of Space Bread, described as a food system comprising multifunctional plastic bag that allows storage of ingredients that could be baked to create a ready-to-eat-yeast-risen-roll (Evans, 2023). Another group proposed the idea of a flexible graphite food packaging material that is not only lightweight but also compatible with NASA's sterilization and disposal requirements that could maintain the food's shelf life and quality for up to 3 years. However, it was later found to fragment too easily, making it inappropriate for use in space (Evans, 2023). The following section highlights the prospective application of nano packaging in novel and emerging packaging systems (Figure 13.1).

13.6.1 Edible Films

Food films have been used in space food packaging for their ability to isolate bacteria, retain moisture content, and minimize food residue, but their poor heat resistance and barrier properties make them unsuitable for long-term space missions. Edible films comprising starches, protein, polysaccharides, and fat have high utility and can be modified to design edible functional films by adding phenolic substances and organic acids. It serves as a protection layer that prevents alteration in color, flavor, and texture of fresh fruits and vegetables, frozen, baked, and meat products during storage and transportation (Zhang and Zhang, 2011). Further, the film maintains good food quality and ensures enhanced shelf life by preventing the migration of gases, water vapor, fragrance components, and solute contents. A process of coating with nontoxic and biodegradable materials like resins, beeswax,

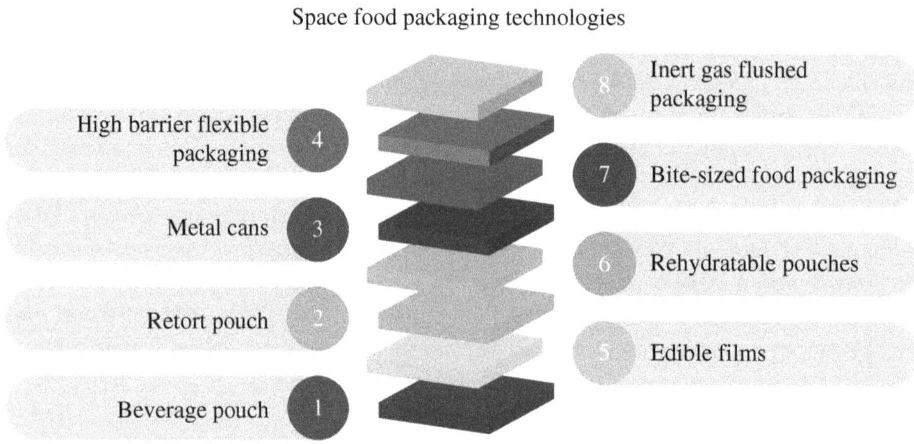

Figure 13.1 Space food packaging technologies

paraffin, or titanium dioxide can further enhance nutrient stability (Hu and Zhou, 2010). However, the developed technology is not very suitable for long-duration space missions yet as factors like tensile strength, sealing capacity, water, and heat resistance need to be optimized for space conditions to withstand the requirement of 3–5 years of shelf life (Hartung et al., 2006). Coating with food-grade nanomaterials could provide optimum strengthening and enhanced barrier properties to edible films while also being sustainable.

13.6.2 Modified Atmospheric Packaging (MAP) for Rehydratable Foods

Rehydratable pouch packaging is for food products, which are dried into powders and rehydrated with cold or hot water in space to retain their physical and chemical properties. It is designed by a co-extrusion process, and the thermoformed base comprises multiple layers of nylon, medium-density polyethylene layers of EVOH, and linear LDPE to provide sufficient water and gas barriers. Both the base and lid of this packaging are flexible, which makes them easily compressible as trash. The oxygen and water vapor permeability of the thermoformed base is about $0.053\,cc/(100\,inch^2 \times day)$ and $0.1784\,g/(100\,inch^2 \times day)$, respectively (Cooper et al., 2011). The food, after being placed inside the base, is flushed with nitrogen gas three times, followed by vacuum sealing. The texture of food largely depends upon the vacuum pressure used, which is different for different types of food. Cereal, toast, cornbread dressing, and spicy beans are examples of food products packed with modified air packaging with rehydratable pouches. The food is consumed after the rehydration process by injection of water with a needle (Jiang et al., 2020a; Perchonok, 2014). However, multilayered overwrapping with polyester lamination, polyethylene, and aluminum foil as secondary packaging may be required to enhance the water vapor, gas, and aroma barrier properties for space conditions.

Compared to the three-piece thermoformed rehydratable packaging, a gusseted pouch is an emerging technology that is mass, weight, volume, and cost-efficient. The one-piece packaging design reduces production time and cost and has resulted in an overall mass reduction of 66% (Raut et al., 2021).

Inert gas-flushed packaging developed by NASA is a novel method specially designed to maintain the freshness and quality of bread for consumption in space. The procedure first involved cleaning the package with 70% ethyl alcohol, followed by nitrogen flushing of the bread and package three times, and then packing each piece of bread individually in a sterile environment. This method allows for a 14-week storage period of bread without mold (Hartung et al., 2006).

Alternatively, polymer films with 2D nanomaterials like graphene, tungsten disulfide, and hexagonal boron nitride could be investigated because of their interesting attributes, such as high oxygen barrier properties. Being lightweight materials, considerable mass reduction and less oxygen transmission rate (OTR) can be achieved. Oxygen permeability reduction of 42% with just 0.017 vol% could be achieved with nanosheets (Xie et al., 2015).

13.6.3 Bite-Size Food Packaging

Similar to the packaging materials and techniques used for rehydratable packaging, bite-size food packaging involves co-extrusion of nylon, EVOH, nylon, LF adhesive, HV polyethylene, and linear LDPE. It is usually used for the storage of low-moisture foods like peanuts, almonds, nuts, and granola bars. After the bite-size food is placed inside multilayered packages, the package is completely sealed after flushing nitrogen gas to remove oxygen gas. This modified atmospheric packaging technique enhances the shelf life of foods and reduces oxidation. However, they do not provide sufficient barrier properties and are hence unsuitable for long-duration missions. Overwrapping with a combination of polyester, polyethylene, and aluminum foil can strengthen barrier properties, providing an extended shelf-life for space foods (Cooper et al., 2011; Lane et al., 2006). The addition of an aluminum foil layer improves barrier ability but makes packaging difficult to incinerate, dispose of, and recycle. Therefore, nanomaterials that provide enhanced barrier properties may add value to existing limitations.

A comparison between the conventional quad-plastic laminate and plastic laminate and a novel nano-packaging material, aluminum-oxide-coated plastic laminate, was used to assess the barrier ability of the packaging materials toward food products susceptible to air and moisture exposure in a simulated space environment. The results revealed that over a period of two years, the quad plastic laminate materials did not provide a sufficient barrier to oxygen and moisture and may require additional barrier support, while the plastic laminates with aluminum materials provided sufficient barrier independently. Current operations at the International Space Station (ISS) packaging system could benefit from the novel single-packaging materials (Cooper et al., 2011).

13.6.4 Polymer Composite Packaging

Packaging with polymer nanocomposites is another efficient packing system with a high barrier effect on oxygen and water, which helps preserve food flavor and increases food stability. Developed during the Gemini mission in 1965–1966, high-barrier materials like EVOH copolymer are primarily used alongside metal oxides like SiO*x*, alumina, and titanium oxide. EVOH exhibits good barrier performance and high physical strength that guarantees stability under harsh environmental conditions like high temperature and humidity.

A variety of foods, like acidic foods, intermediate moisture foods, and dehydrated foods, can be effectively packed using this technology (Sun et al., 2016).

Synthesis of polyethylene, polypropylene (PP), polystyrene, poly (caprolactone), polycarbonate, PAS, EVOH, and PET is the main focus of the study of polymer nanocomposites for application in packaging. They not only have superior physical properties but also better gas resistance, with the ability to reduce oxygen, water vapor diffusion, and food's volatile aroma. Nanocomposite packaging of foods provides excellent functions and improves safety and reliability, which makes it more suitable for use in long-term manned missions (Nguyen, 2006). Polymer nanocomposite packaging is a lightweight material and biodegradable in nature that has the potential to cater to emerging needs. The safety and reliability of polymer nanocomposite applications in aerospace packaging can be explored to understand its efficacy.

13.6.5 Hurdle Technology

This technology enhances the shelf stability and quality of foods by controlling water activity, pH, and the amount and type of preservatives and antioxidants to be added to preserve foods. It can also be used in aerospace food packaging. Factors such as color change, texture, nutritional value, sensory quality, rate of water replenishment of food products, and evaluation of packaging film barrier performance and its mechanical integrity are identified, and stabilization technologies are applied to maintain optimum quality and minimize nutrition degradation. Satisfactory fulfillment of these hurdles would lead to unique identification for processed food formulation to achieve a 5-year shelf-life goal (Cooper et al., 2017). Processing foods can reduce nutritional content; therefore, sensitive nutrients like vitamin C need encapsulation as a strengthening method. In order to provide adequate nutrition to astronauts, NASA has launched projects to microencapsulate spices, probiotics, health foods, and nutrients like omega-3 fatty acids, vitamins, and trace elements to supplement the diet in space and ensure their sustained release.

13.7 Challenges Faced in Space Food Packaging

Nutrient-dense foods with enhanced shelf-life are appropriately processed and packaged on land for consumption in space. Due to the storage of space foods in harsh microgravity environments, the overall nutrient composition and acceptability reduce over a period of time, especially in the case of thermostabilized packaging (Kumar and Gaikwad, 2023). Furthermore, in order to provide high barrier properties, multilayer packaging is used, which adds to the weight of the spacecraft. An overwrap of aluminum foil is also utilized to increase barrier ability, which adds to the weight. The foods must be designed to occupy minimum space and be lightweight because of the spacecraft's mass and volume limitations. This is a significant challenge where the minimization of mass and volume contradicts the ability of foods to be barrier-free. Further, even though barrier properties are provided, not all foods are in the desired form as some foods tend to become sticky in space due to high surface tension and stick on the sides of the packaging material, while some liquid items must be consumed from a small opening, which reduces their acceptability. The development of appropriate packaging material that meets these limitations of space foods is of prime importance. Furthermore, the packaging must ensure that it creates a barrier to the radiation environment of space, preventing the generation of reactive oxygen species, degradation of nutritional content, and alteration of the taste quality of food products (Perchonok et al., 2012).

Current packaging systems include multiple layers of protection to achieve good barrier properties and longer shelf life of foods, which adds to the weight and volume burden on spacecraft. This is one of the major drawbacks of the current space food packaging system, as the amount of waste generated in the space shuttle post consumption of food is enormous. The waste is stored until the next supply vehicle arrives and then burns in the Earth's atmosphere with the return of the vehicle (Hanford 2006). This is highly unsustainable, with a substantial environmental impact and demands for an urgent sustainable solution. Scientists are currently working on developing biodegradable and reusable packaging solutions to minimize solid waste processing in space or on Earth. The following section highlights the novel and upcoming packaging techniques for space applications.

13.8 Use of Nanotechnology for Quality Detection of Space Foods

Protein analysis and immunological enzymology are effective traditional methods for detecting food adulteration but are unsuitable for large-scale applications due to expensive operations and requirements. Due to similar disadvantages, techniques like High-Performance Liquid Chromatography (HPLC), mass spectroscopy, and immunoassay are unfit for food quality detection at a large scale. Therefore, simple and faster imaging technologies should be used. For detecting adulterated meat mince, multispectral imaging technology (MSI) has excellent potential and suitability for large-scale applications (Ropodi et al., 2016). Near-infrared hyperspectral imaging technology is one such technology that is used to detect adulterants in foods, such as the detection of melamine in milk.

Ma et al. (2018) developed an accurate, easy-to-use food spoilage detection system using a nanostructured conductive gas sensor made of polymer. This technology is based on near field communication (NFC) function, detects food spoilage by monitoring amines, and can track food quality on a smartphone. In another development, a silver nanorod-mediated SPR sensor, based on traditional surface plasmon resonance, can effectively detect if pathogenic bacteria exceed in food due to its increased detection range and sensitivity (Yi et al., 2020).

13.9 Future Scope of Packaging Materials for Space Foods

Upcoming space missions for Mars exploration last more than three to five years. Safe and nutritious food throughout the duration of the mission is crucial. Food quality factors like shelf-life, nutrition and vitamin concentration, and maintaining weight-to-volume ratios in the spacecraft depend a lot on the efficiency of packaging. Any limitation in the packaging aspect can negatively affect the health of astronauts and the mission's success. The main drawbacks of the existing packaging technologies are the increase in weight and difficulty in disposability, while trying to achieve better barrier properties.

A suitable packaging material for the packaging of space foods must be environment-friendly and sustainable. Bacterial cellulose, for example, is a naturally occurring food-grade nanomaterial cultivated by the bacteria belonging to the *Komagataeibacter* genus and has been previously explored for food applications due to its high purity and yield. It has been used in packaging films and coatings of fruit purees, sushis, and sandwiches because of its unique properties like web-like network structure, high water holding capacity, and high porosity, showing promising potential as a packaging material for space applications (Azeredo et al., 2019; Rajwade et al., 2015; Ruka et al., 2012). However, it has not been explored for space application yet.

Scientists have previously incorporated algae and chloroplasts into biodegradable polymers via 3D printing to complement the properties of the polymer with the photosynthetic properties of the plant cells (Balasubramanian et al., 2021; Yu et al., 2021). Applying the same logic, nanomaterial packaging materials like bacterial cellulose could be incorporated with radiosynthetic organisms (that convert high-energy radiation to chemical energy) to provide protection from radioactive and cosmic rays of the space environment while also being food-grade in nature. One such material is melanin-producing *Cladosporium sphaerospermum* fungus, which is halotolerant, cosmopolitan and found to degrade industrial gas emissions. Apart from being able to thrive in radioactive environments like in the cooling pools of the Chernobyl Nuclear Power plant and in low Earth orbit, it can also bind with other substances (Shunk et al., 2020). Therefore, NASA could implement such strategies involving the incorporation of organisms into sustainable polymers using 3D printing which could strengthen the barrier properties of packaging materials, providing enhanced protection to food materials in space.

Even though there have been tremendous advancements in space food technology, the lack of resources for human survival in the space environment is a major limitation. Considering cultural differences and food preferences can potentially improve the quality of space missions and ensure better health of astronauts. Many challenges related to shelf-life extension while maintaining food quality, weight, and volume restrictions in space remain unresolved to date, which could be addressed using nanomaterials in space food packaging.

13.10 Conclusion and Outlook

Nutrition requirements in space differ from those on Earth. It is challenging to meet nutrition requirements while sustaining harsh microgravity environments and mass and volume constraints. To combat this, various advancements in technology have been observed over the years, which have made it possible to survive in space for an extended duration. An abundance of processing and packaging technologies for space foods has been studied, as discussed in this chapter. However, there is a broad scope for improvement and novel innovations that could help astronauts safeguard their health, thus improving the success rate of space missions. Packaging of space foods plays an integral role in providing nutritional adequacy, safety, and food acceptability, which is crucial for sustainable missions. The high barrier properties of packaging can provide safe food for astronauts for a long duration. Advanced food packaging materials are currently being explored to solve current limitations. Leveraging the unique properties of nanomaterials can help in meeting the demanding requirements of packaging in the space food sector. More explorations in this area can help capture the potential to comply with in-demand properties in space microgravity environments while maintaining shelf stability.

References

Azeredo, H. M. C., Barud, H., Farinas, C. S., Vasconcellos, V. M., & Claro, A. M. (2019). Bacterial cellulose as a raw material for food and food packaging applications. *Frontiers in Sustainable Food Systems, 3*. https://doi.org/10.3389/fsufs.2019.00007

Balasubramaniam, V. M., Barbosa-Cánovas, G., & Lelieveld, H. (2016). High Pressure Processing of Food-Principles, Technology and Application. Springer. https://doi.org/10.1007/978-1-4939-3234-4

Balasubramanian, S., Yu, K., Meyer, A. S., Karana, E., & Aubin-Tam, M.-E. (2021). Bioprinting of regenerative photosynthetic living materials. *Advanced Functional Materials, 31*(31), 2011162. https://doi.org/10.1002/adfm.202011162

Barbosa-Cánovas, G. V., & Juliano, P. (2008). Food Sterilization by Combining High Pressure and Thermal Energy. In G. F. Gutiérrez-López, G. V. Barbosa-Cánovas, J. Welti-Chanes, & E. Parada-Arias (Eds.), *Food Engineering: Integrated Approaches* (pp. 9–46). Springer. https://doi.org/10.1007/978-0-387-75430-7_2

Barbosa-Cánovas, G. V., Medina-Meza, I., Candoğan, K., & Bermúdez-Aguirre, D. (2014). Advanced retorting, microwave assisted thermal sterilization (MATS), and pressure assisted thermal sterilization (PATS) to process meat products. *Meat Science, 98*(3), 420–434. https://doi.org/10.1016/j.meatsci.2014.06.027

Bhatia, S., Wall, K., Kerth, C., & Pillai, S. (2017). Benchmarking the minimum electron beam (eBeam) dose required for the sterilization of space foods. *Radiation Physics and Chemistry, 143*, 72–78. https://doi.org/10.1016/j.radphyschem.2017.08.007

Brown, A. C. (2011). *Understanding Food: Principles and Preparation* (4th Edition). Wadsworth Pub Co.

Chaloulakou, S., Poulia, K. A., & Karayiannis, D. (2022). Physiological alterations in relation to space flight: the role of nutrition. *Nutrients, 14*(22), 4896. https://doi.org/10.3390/nu14224896

Chen, Q., Cao, M., Chen, H., Gao, P., Fu, Y., Liu, M., Wang, Y., & Huang, M. (2016). Effects of gamma irradiation on microbial safety and quality of stir fry chicken dices with hot chili during storage. *Radiation Physics and Chemistry, 127*, 122–126. https://doi.org/10.1016/j.radphyschem.2016.06.022

Cooper, M. R., & Douglas, G. L. (2015). Integration of Product, Package, Process, and Environment: A Food System Optimization. *NASA Human Research Program (HRP) Investigators Workshop, Galveston, TX*. https://ntrs.nasa.gov/citations/20140012829

Cooper, M., Douglas, G., & Perchonok, M. (2011). Developing the NASA food system for long-duration missions. *Journal of Food Science*, *76*(2), R40–R48. https://doi.org/10.1111/j.1750-3841.2010.01982.x

Cooper, M., Perchonok, M., & Douglas, G. L. (2017). Initial assessment of the nutritional quality of the space food system over three years of ambient storage. *Npj Microgravity*, *3*(1), 1. https://doi.org/10.1038/s41526-017-0022-z

Crisp, N. H., Smith, K., & Hollingsworth, P. (2015). Launch and deployment of distributed small satellite systems. *Acta Astronautica*, *114*, 65–78. https://doi.org/10.1016/j.actaastro.2015.04.015

DeVieneni, A., Velez, C. A., Benjamin, D., & Hollenbeck, J. (2012). Cost-Effective Additive Manufacturing in Space: HELIOS Technology Challenge Guide. Marshall Space Flight Center. https://ntrs.nasa.gov/citations/20130001751

Douglas, G. L., Cooper, M., Bermudez-Aguirre, D., & Sirmons, T. (2016). Risk of Performance Decrement and Crew Illness Due to an Inadequate Food System. Johnson Space Center. https://ntrs.nasa.gov/citations/20160011582

Douglas, G. L., Wheeler, R. M., & Fritsche, R. F. (2021). Sustaining astronauts: resource limitations, technology needs, and parallels between spaceflight food systems and those on Earth. *Sustainability*, *13*(16), 16. https://doi.org/10.3390/su13169424

Douglas, G. L., Zwart, S. R., & Smith, S. M. (2020). Space food for thought: challenges and considerations for food and nutrition on exploration missions. *The Journal of Nutrition*, *150*(9), 2242–2244. https://doi.org/10.1093/jn/nxaa188

Evans, R. (2023). Space food packaging: A review of its past, present and future materials and technologies. *Packaging Technology and Science: An International Journal*, *36*(8), 617–627. https://doi.org/10.1002/pts.2752

Grover, Y., Bhasin, J., Dhingra, B., Nandi, S., Hansda, M., Sharma, R., Paul, V., Idrishi, R., Tripathi, A., & Agarwal, A. (2021). Developments and scope of space food. *Current Nutrition & Food Science*, *18*, 248–258. https://doi.org/10.2174/1573401317666210809113956

Gupta, C., & Gupta, S. (2010). Food for space. *International Journal of Biological Technology*, *1*, 121–123.

Hanford, A., (2006). Advanced Life Support—Baseline Values and Assumptions Document. Technical Reports. 3.

Hartung, T., Bullerman, L., Arnold, R., & Heidelbaugh, N. (2006). Application of low dose irradiation to fresh bread system for space flight. *Journal of Food Science*, *38*, 129–132. https://doi.org/10.1111/j.1365-2621.1973.tb02795.x

HoJeon, C. J. (2007). *Development of packaging material for space Japanese food. The Monthly Packaging World*, *11*, 68–73.

Hu, & Zhou (2010). Application of novel preservation technology in the preservation of food. *Food Research and Development*, *31*, 242–246.

Huang, L., & Zhang, M. (2012). Trends in development of dried vegetable products as snacks. *Drying Technology*, *30*(5), 448–461. https://doi.org/10.1080/07373937.2011.644648

Jiang, J., Zhang, M., Bhandari, B., & Cao, P. (2020a). Current processing and packing technology for space foods: a review. *Critical Reviews in Food Science and Nutrition*, *60*(21), 3573–3588. https://doi.org/10.1080/10408398.2019.1700348

Jiang, J., Zhang, M., Bhandari, B., & Cao, P. (2020b). Development of chinese yam/chicken semi-liquid paste for space foods. *LWT*, *125*, 109251. https://doi.org/10.1016/j.lwt.2020.109251

Khan, I., Tango, C. N., Miskeen, S., Lee, B. H., & Oh, D.-H. (2017). Hurdle technology: a novel approach for enhanced food quality and safety—a review. *Food Control, 73*, 1426–1444. https://doi.org/10.1016/j.foodcont.2016.11.010

Kirkin, C., Mitrevski, B., Gunes, G., & Marriott, P. J. (2014). Combined effects of gamma-irradiation and modified atmosphere packaging on quality of some spices. *Food Chemistry, 154*, 255–261. https://doi.org/10.1016/j.foodchem.2014.01.002

Klicka, M. V., & Smith, C. M. (1982). Food for US Manned Space Flight. *United States Army Natick Research & Development Laboratories Natick*, Massachusetts.

Kumar, L., & Gaikwad, K. K. (2023). Advanced food packaging systems for space exploration missions. *Life Sciences in Space Research, 37*, 7–14. https://doi.org/10.1016/j.lssr.2023.01.005

Lane, H. W., Kloeris, V., & Perchonok, M. (2006). *Changes in Space Food Over the Last 45 Years*. NASA.

Long, Y., Zhang, M., Devahastin, S., & Cao, P. (2022). Progresses in processing technologies for special foods with ultra-long shelf life. *Critical Reviews in Food Science and Nutrition, 62*(9), 2355–2374. https://doi.org/10.1080/10408398.2020.1853034

Ma, Z., Chen, P., Cheng, W., Yan, K., Pan, L., Shi, Y., & Yu, G. (2018). Highly sensitive, printable nanostructured conductive polymer wireless sensor for food spoilage detection. *Nano Letters, 18*(7), 4570–4575. https://doi.org/10.1021/acs.nanolett.8b01825

McPhee, J. C., & Charles, J. B. (2009). *Human Health and Performance Risks of Space Exploration Missions: Evidence*. Government Printing Office, Washington, DC. https://books.google.co.in/books/about/Human_Health_and_Performance_Risks_of_Sp.html?id=avR4A9l80W0C&redir_esc=y

Nguyen, Q. T. (2006). Preparation of Polymer–Clay Nanocomposites and Their Properties—Nguyen—2006—Advances in Polymer Technology—Wiley Online Library. Retrieved November 15, 2023, from https://onlinelibrary.wiley.com/doi/10.1002/adv.20079

Nowak, D., & Jakubczyk, E. (2020). The freeze-drying of foods-the characteristic of the process course and the effect of its parameters on the physical properties of food materials. *Foods (Basel, Switzerland), 9*(10), 1488. https://doi.org/10.3390/foods9101488

Ouattara, B., Sabato, S. F., & Lacroix, M. (2001). Combined effect of antimicrobial coating and gamma irradiation on shelf life extension of pre-cooked shrimp (Penaeus spp.). *International Journal of Food Microbiology, 68*(1–2), 1–9. https://doi.org/10.1016/s0168-1605(01)00436-6

Pandith, J. A., Neekhra, S., Ahmad, S., & Sheikh, R. A. (2023). Recent developments in space food for exploration missions: a review. *Life Sciences in Space Research, 36*, 123–134. https://doi.org/10.1016/j.lssr.2022.09.007

Pandrangi, S., Balasubramaniam, V. M., Tao, Y., & Sun, D.-W. (2014). High-Pressure Processing of Salads and Ready Meals. In Y. Tao (Ed.), *Emerging Technologies for Food Processing* (pp. 25–34). *Academic Press*. https://doi.org/10.1016/B978-012676757-5/50004-9

Perchonok, M. (2014). The Challenges of Developing a Nutritious Food System for a Mars Mission. *(No. JSC-CN-31374).*

Perchonok, M., & Bourland, C. (2002). NASA food systems: past, present, and future. *Nutrition (Burbank, Los Angeles County, Calif), 18*(10), 913–920. https://doi.org/10.1016/s0899-9007(02)00910-3

Perchonok, M. H., & Catauro, P. M. (2009). *Thermostabilized Shelf-Life Study*. HRP Investigator's Workshop (No. JSC-17649).

Perchonok, M. H., Cooper, M. R., & Catauro, P. M. (2012). Mission to mars: food production and processing for the final frontier. *Annual Review of Food Science and Technology*, *3*, 311–330. https://doi.org/10.1146/annurev-food-022811-101222

Pereira, R. N., & Vicente, A. A. (2010). Environmental impact of novel thermal and non-thermal technologies in food processing. *Food Research International*, *43*(7), 1936–1943. https://doi.org/10.1016/j.foodres.2009.09.013

Pometto, A. L., & Bourland, C. (2003). NASA Food Technology Commercial Space Center Mission and Activities. SAE Technical paper 2003-01-2375,2003, https://doi.org/10.4271/2003-01-2375

Prisco, J. (2014). "Foodini" Machine Lets You Print Edible Burgers, Pizza|CNN Business. https://edition.cnn.com/2014/11/06/tech/innovation/foodini-machine-print-food/index.html

Rajwade, J. M., Paknikar, K. M., & Kumbhar, J. V. (2015). Applications of bacterial cellulose and its composites in biomedicine. *Applied Microbiology and Biotechnology*, *99*(6), 2491–2511. https://doi.org/10.1007/s00253-015-6426-3

Ratti, C. (Ed.). (2008). *Advances in Food Dehydration*. CRC Press. https://doi.org/10.1201/9781420052534

Raut, S., Hegde, S., Modak, S., & Bhande, R. (2021). Advancements in space food processing technologies. *International Journal of Recent Scientific Research*, *12*(6), 42033–42037.

Ropodi, A., Panagou, E., & Nychas, G.-J. (2016). Multispectral imaging (MSI): a promising method for the detection of minced beef adulteration with horsemeat. *Food Control*, *73*, 57–63. https://doi.org/10.1016/j.foodcont.2016.05.048

Ruka, D. R., Simon, G. P., & Dean, K. M. (2012). Altering the growth conditions of *Gluconacetobacter xylinus* to maximize the yield of bacterial cellulose. *Carbohydrate Polymers*, *89*(2), 613–622. https://doi.org/10.1016/j.carbpol.2012.03.059

Sánchez-García, F., Hernández, I., Palacios, V. M., & Roldán, A. M. (2021). Freshness quality and shelf life evaluation of the seaweed *Ulva rigida* through physical, chemical, microbiological, and sensory methods. *Foods*, *10*(1), 1. https://doi.org/10.3390/foods10010181

Sharma, C., Dhiman, R., Rokana, N., & Panwar, H. (2017). Nanotechnology: an untapped resource for food packaging. *Frontiers in Microbiology*, *8*, 1735. https://doi.org/10.3389/fmicb.2017.01735

Shunk, G., Gomez, X., & Averesch, N. (2020). *A self-replicating radiation-shield for human deep-space exploration: radiotrophic fungi can attenuate ionizing radiation aboard the international space station*. *BioRxiv*, *2020*, 2020–20207. https://doi.org/10.1101/2020.07.16.205534

Singh, J. (2021). *Enjoying Food in Space–it's a Matter of Taste!*

Sun, J. C., Qu, W. L., & Dong, H. S. (2016). Requirement analysis of development in space food packaging. *Space Medicine and Medical Engineering*, *29*, 451–456.

Tang, H., Rising, H. H., Majji, M., & Brown, R. D. (2021). Long-term space nutrition: a scoping review. *Nutrients*, *14*(1), 194. https://doi.org/10.3390/nu14010194

Wang, R., Zhang, M., & Mujumdar, A. S. (2010). Effect of food ingredient on microwave freeze drying of instant vegetable soup. *LWT—Food Science and Technology*, *43*(7), 1144–1150. https://doi.org/10.1016/j.lwt.2010.03.007

Xie, S., Istrate, O. M., May, P., Barwich, S., Bell, A. P., Khan, U., & Coleman, J. N. (2015). Boron nitride nanosheets as barrier enhancing fillers in melt processed composites. *Nanoscale*, *7*(10), 4443–4450.

Yi, R.-M., Zhang, Z., Liu, C.-X., & Qi, Z.-M. (2020). Gold-silver alloy film based surface plasmon resonance sensor for biomarker detection. *Materials Science & Engineering. C, Materials for Biological Applications*, *116*, 111126. https://doi.org/10.1016/j.msec.2020.111126

Yu, K., Feng, Z., Du, H., Xin, A., Lee, K. H., Li, K., Su, Y., Wang, Q., Fang, N. X., & Daraio, C. (2021). Photosynthesis-assisted remodeling of three-dimensional printed structures. *Proceedings of the National Academy of Sciences*, *118*(3), e2016524118. https://doi.org/10.1073/pnas.2016524118

Zhang, M., Chen, H., Mujumdar, A. S., Tang, J., Miao, S., & Wang, Y. (2017). Recent developments in high-quality drying of vegetables, fruits, and aquatic products. *Critical Reviews in Food Science and Nutrition*, *57*(6), 1239–1255. https://doi.org/10.1080/10408398.2014.979280

Zhang, Y. B., & Jiang, J (2011). Research progress of edible films. *China Food Additives 1*, 191–198.

14

Life Cycle Analysis in Food Packaging

Shweta Deotale, Divyajyoti Biswal, and Sachin A. Mandavgane

Department of Chemical Engineering, Visvesvaraya National Institute of Technology (VNIT), Nagpur, Maharashtra, India

14.1 Introduction

Food deterioration is a serious global concern because of the lack of suitable packing technologies. The food industry uses a variety of packaging materials, such as plastic (polyethylene, polypropylene, and polyethylene terephthalate) glass, metal, flexible packaging, edible coating, aluminum foils, active packaging, intelligent packaging, and biodegradable polymeric film. Out of these, some are not environment friendly, but according to the modern lifestyle and eating trends of generations (ready-to-eat, ready-to-cook, ready-to-serve, and ready-to-use), would not function without packaging; Hence, the packaging is necessary for containment, convenience, communication and protection of food (Marsh and Bugusu, 2007; Sundqvist-Andberg and Åkerman, 2021). Food packaging can make food last longer by giving it an ideal environment, such as by keeping away from dirt, light, moisture, oxygen, microorganisms, and other environmental factors. The packing materials not only protect the food from physical damage, but they also create the right physical and chemical conditions that are needed to keep food fresh for longer. Ideally, it also has thermal, mechanical, and optical properties. The main function of packaging material is to protect the food from food-borne microorganisms (Rhim et al., 2013; Wesley et al., 2014). Because of the rising number of diseases that are transmitted by food, antimicrobial packaging materials have been implemented (Appendini and Hotchkiss, 2002). For such functions as antimicrobial properties, a barrier against gases and aromatic compounds and more, nanotechnology is turning out to be very beneficial for the packaging industry. Nanotechnology is a new science that focuses on fabricating, manipulating, characterizing, and creating material at the nanoscale (1–100 nm). Recent investigations showed that the application of nanotechnology enhances food packaging. Nanoparticles are able to modify the physical and mechanical characteristics of polymeric material by making it stronger, more durable, more flexible, effective barrier properties, and easier to reuse

Nanotechnology for Sustainable Food Packaging, First Edition. Edited by C. Anandharamakrishnan, Jeyan A. Moses, and M. Maria Leena.

(Bumbudsanpharoke et al., 2015; Cerqueira et al., 2018). Innovative nano-based food packaging has distinctive properties such as oxygen scavenging, antimicrobial, and gas/moisture barriers, which help to enhance the product-keeping qualities without any undesirable changes (Rai et al., 2019). Although the benefits of nanomaterials as novel packaging agents have been extensively studied, their widespread use in the food packaging industry may pose health and environmental risks that have still not been fully comprehended.

It has been reported that municipal solid waste (MSW) consists of materials that are frequently thrown away, such as food parcels, leftover food, yard garbage, used packaging materials including metal cans, glass, plastic, multilayer packages, paper, thin films, and paperboard. Food is the only product category that every person consumes three times every day. During the COVID-19 Pandemic, the use of plastic packaging and single-use plastic has increased due to hygienic perception. Furthermore, National lockdowns and home quarantine upsurge the demand for packaging materials for groceries, food deliveries, etc. This is also one of the plausible reasons for increasing the food packaging waste in MSW. As a result, by volume, food packaging makes up close to two-thirds of all packaging waste, which has a low rate of recyclability (Scaraboto et al., 2020; Vanapalli et al., 2021). Further, the Food and Agriculture Organization of the United Nations estimates that about one-third of food is lost or wasted. The problem gets worse in nations that are more developed. In the meantime, the world packaging market will be worth more than $1 trillion by 2023, and another $150 billion will be added to it by 2028 (Abejón et al., 2020; Pira, 2019). Therefore, the assessment and resultant reduction of ecological impacts associated with the production, use, and disposal of packaging materials is presently one of the most important priorities of the packaging and food industries.

Recently, food packaging has benefited from numerous technical advancements. To ensure that the following characteristics are maintained in the new types of packaging materials:

- It should be helpful all over the entire life cycle.
- It should be designed in such a way that it can satisfy the market requirements for cost and performance.
- Should produce with the help of best practices and eco-friendly production methods.
- It should be made from materials that are appropriate for all end-of-life conditions.
- It should be physically engineered to optimize material and energy consumption.
- It should be capable of being adequately recovered through either mechanical, biological, or energetic means.
- Should abide by the rules for disposing of packing materials that are periodically established worldwide.

Despite the numerous advantages of packaging food products, it is crucial to pay attention to the safety of packaging materials, their carbon footprints and environmental impacts, waste management, and the recycling of packaging materials. It is possible to address these issues with the help of technical breakthroughs. An environmental assessment method based on a life cycle thinking (LCT) approach is very important for the evaluation of the environmental burden of food packaging since packaging has an impact on the environment not only during manufacture or disposal but also throughout packing and distribution. There is a steady increase in LCT research activity. Life cycle assessment (LCA), life cycle costing, social LCA, eco-design, carbon footprint, water footprint, and ecological footprint are some of the various LCT tools and approaches. LCA is the most extensively used approach for evaluating the environmental performance of products. LCA is a system for identifying and thoroughly evaluating the potential environmental impact of a material, product, service, or process over the course of its full life cycle, from raw material extraction and processing to manufacture, transit, usage, and final disposal (Guinee, 2002). In the present chapter, the LCA approach has been discussed to evaluate the environmental implications of traditional and novel food packaging techniques in order to establish environmental guidelines for the selection of packaging materials for foods as a means of developing more effective and sustainable food items over their whole life cycle.

14.2 Sustainable Food Packaging

To reduce both quantitative and qualitative losses of food products, a variety of organic, inorganic, and mixed nanoparticles are utilized in the development of effective food packaging. They aid in extending the shelf life of food products by enhancing the mechanical and barrier qualities of packaging materials. A lot of research is being done on nanotechnology to enhance food packaging. Functional nanoparticles have been discovered to enhance packaging material's barrier qualities, thermal stability, strength, and durability, resulting in extending the shelf life of food items (Ashfaq et al., 2022; Shukla et al., 2019). In general, there are two ways to create nano-based packaging materials for preserving food quality and extending shelf life: either by adding nanoparticles to conventional food packaging materials like films and containers or by creating nanocomposite multilayer packaging materials and organic, inorganic, and combined nanocoatings by immersion, spraying, or rubbing techniques (Tsagkaris et al., 2018). Nanotechnology aids in the fabrication of improved packaging (having better physical and mechanical properties), active packaging (having antibacterial, antioxidative, and UV absorption properties), and smart or intelligent packaging (having monitored/controlled food conditions) explained in Figure 14.1. Several types of active nanomaterials have been used to achieve effective food packaging (Primožič et al., 2021). A recent review elaborated on the utilization of food waste as a raw material for manufacturing nanocomposite materials, which possess potential barrier and tensile properties. Various food scraps, such as the leftovers from the production and preparation of fruits, vegetables, grains, and other foods, consist of essential bioactive components like polyphenols, dietary fiber, proteins, lipids, vitamins, organic acids, and minerals, some of which are concentrated in higher levels in the parts that are thrown away than in the parts that are sold. Such bioactives/active components like antibrowning and antioxidant agents, colorants, nutrition, taste, and antibacterial compounds are able to be transported by biopolymer frameworks. The bionanocomposite materials' capabilities were enhanced by these bioactive components, making them appropriate for use as proactive food packaging. The ethylene absorbers, moisture absorbers, ethanol emitters, and taste-absorbing/releasing components can all be manufactured from biopolymer-based nanofiber materials. The utilization of such biopolymer-based nanocomposites in nanofood packaging is effective in enhancing the shelf

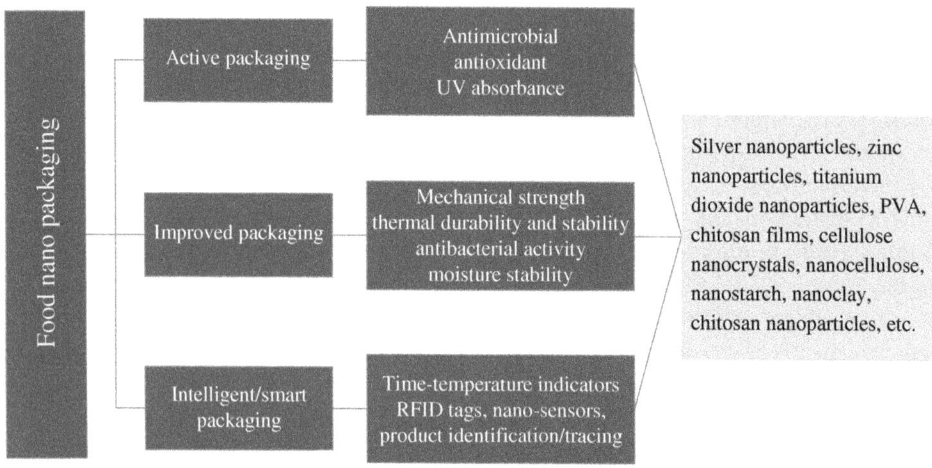

Figure 14.1 Type of food nanopackaging, its principle functions, and nanomaterials used for it Ashfaq et al. (2022), Hernández-Muñoz et al. (2018), Kuswandi (2016)

life of food products as well as are ecofriendly and biodegradable (Basumatary et al., 2022; Tripathi et al., 2023). Such biodegradable materials include biopolymers, commonly referred to as green nanocomposites or bioplastics (Riaz et al., 2024).

14.3 Life Cycle Assessment

Packaging needs to add value for customers as well as business entities. In addition, it should be inexpensive, sustainable, and able to effectively store the product for the specified duration of time. It is crucial to evaluate the overall influence of packaging material on the environment; that's why the life cycle approach must be employed. LCA is sometimes referred to as a cradle-to-grave analysis, eco-balance, and life cycle analysis.

LCA is a comprehensive method that provides a systematic approach to assessing the environmental impacts of a product or service throughout its entire life cycle. The life cycle includes all stages, from the extraction of raw materials through manufacturing and distribution to usage, repair and maintenance, disposal, and recycling. By examining each phase, LCA aims to identify potential environmental hotspots and help decision makers make informed choices to reduce their environmental footprint. LCA has gained increasing prominence as sustainability and environmental concerns have become critical global issues. The method is governed by the ISO 14040 International Series of Standards, which ensures standardized and consistent procedures are followed during the assessment process. Designers, engineers, and manufacturers use LCA as a powerful tool to evaluate their work, as it allows them to understand the environmental implications of their products and identify opportunities for improvement. The primary objective of LCA is to compare the environmental impacts of different products or services. By quantifying all inputs and outputs of material flows throughout a product's life cycle, LCA provides a comprehensive understanding of how each phase contributes to environmental consequences. This includes assessing resource consumption, energy usage, emissions, waste generation, and potential impacts on ecosystems and human health.

At the core of LCA lies the concept of the life cycle inventory (LCI), which involves compiling detailed data on the resource inputs and emissions associated with each life cycle stage. These inventories help create a holistic understanding of the product's environmental profile, enabling comparisons between different alternatives. One of the key advantages of LCA is that it goes beyond traditional assessments that often focus solely on direct impacts and neglect indirect effects. For example, while a product's manufacturing stage might have immediate environmental implications, a more in-depth LCA can uncover the hidden environmental costs associated with its supply chain, including the extraction of raw materials or transportation of components from distant locations. LCA data plays a crucial role in supporting policy development and sustainability initiatives. Policymakers use LCA results to inform regulations, set eco-labeling standards, and develop incentives for environmentally friendly products and services. Industries also leverage LCA to guide eco-design efforts, identifying areas where changes can be made to minimize environmental impacts. Additionally, LCA results are used to foster eco-innovation, encouraging the development of more sustainable technologies and processes. By identifying the "hotspots" of environmental impact within a product's life cycle, LCA helps prioritize areas for improvement and supports the creation of greener and more efficient alternatives. Moreover, LCA serves as a valuable tool in educating consumers about the environmental impacts of their choices. Eco-labels and environmental product declarations based on LCA data enable consumers to make informed purchasing decisions that align with their values and contribute to a more sustainable economy. (ISO, 2006). LCA aids in the quantification of environmental impacts occurring throughout the entire life cycle of a product/process/service across a plethora of indicators, signifying environmental implications across a multitude of dimensions. It provides a comprehensive outlook into the emission profile generated due to the material and energy balance of the system under investigation. It perfectly quantifies all emissions in terms of greenhouse gases, hazardous materials, and much more. Selecting relevant mid-point indicators allows LCA practitioners to assess the potential environmental burdens that occur at specific points in a product's life cycle. For example, analyzing the impact of greenhouse gas emissions (such as CO_2) provides crucial insights into the contribution of a product or process to global climate change. By identifying the major contributors, decision makers can devise targeted measures to reduce emissions and combat climate change effectively. Endpoint indicators, on the other hand, enable the assessment of the ultimate consequences of a product or process on specific environmental areas, such as human health, ecosystem quality, or resource depletion. By evaluating impacts on these broader endpoints, LCA can provide a more comprehensive understanding of a product's overall environmental performance (Vignali, 2016).

In general, this method follows step-by-step analysis to find out the environmental impact associated with a product, process, or activity (Figure 14.2); those steps are

I) Goal and scope definition
II) Inventory analysis
III) Impact assessment
IV) Interpretation and conclusion

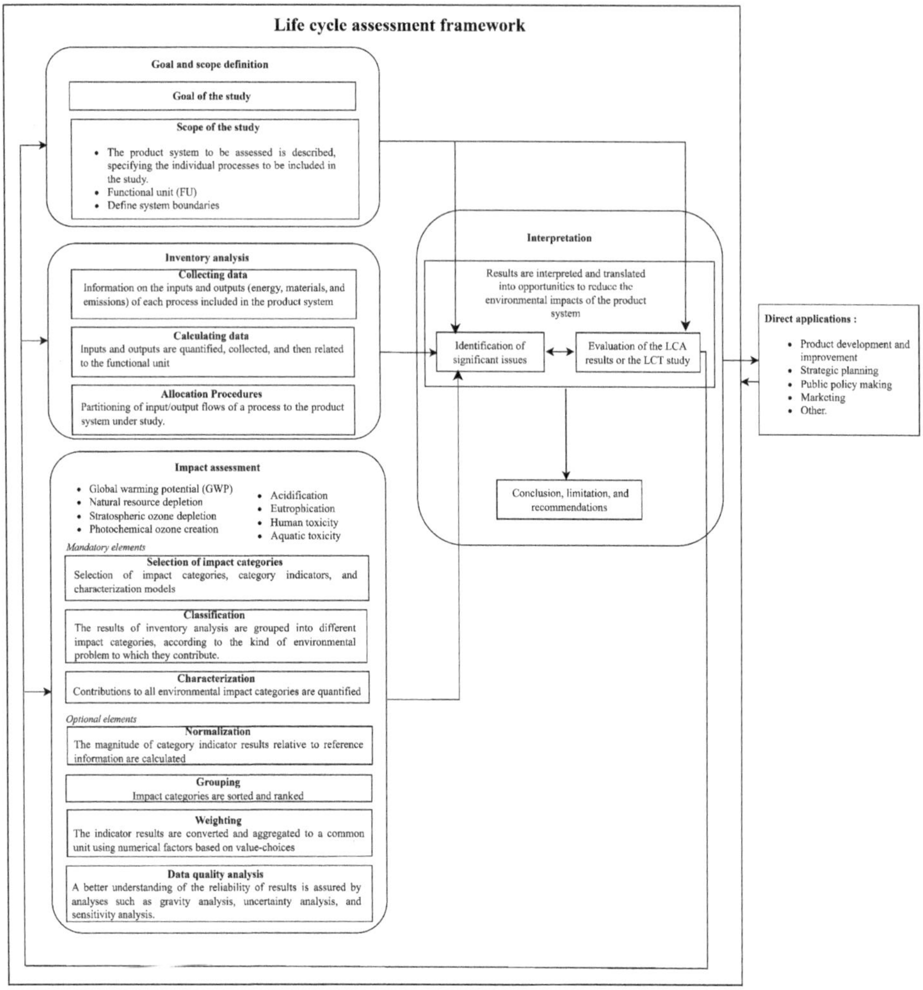

Figure 14.2 Methodological representation of LCA (*Source:* Restuccia et al. (2016)/with permission of Elsevier)

14.3.1 Goal and Scope Definition

The goal and scope definition is a critical starting point in the LCA process. It lays the foundation for the entire analysis by setting clear objectives, establishing the system boundaries, and defining the scope of the study. This initial step ensures that the assessment is conducted in a transparent, consistent, and meaningful manner, enabling accurate comparisons and reliable conclusions about the environmental impacts of a product, process, or service. One of the primary elements of the goal and scope definition is determining the system boundary. The system boundary delineates the extent of the life cycle that will be considered in the assessment. It includes all the operations and phases that are involved in the entire life cycle of the product, process, or service, from cradle to grave. This means that all stages, from the extraction of raw materials and resource acquisition to production, distribution, use, and end-of-life scenarios, are evaluated. Siracusa and co-workers also explained the system boundaries to be considered during the LCA of packaging material (Figure 14.3[a]).

By defining the system boundary comprehensively, LCA aims to avoid the problem of burden shifting, where improving one stage of the life cycle might inadvertently worsen the environmental impact in another phase. For instance, shifting production to an eco-friendly location might reduce emissions during manufacturing but increase transportation-related impacts if materials are now transported from a farther distance. Setting the system boundary also enables a holistic perspective on the environmental impacts, preventing cherry-picking of data or focusing only on specific stages to achieve biased results. This comprehensive approach provides decision makers with a complete picture of the product's sustainability and helps avoid unintended consequences in environmental management. In defining the goal and scope, LCA practitioners must also make assumptions about data availability and precision. These assumptions are essential to ensure that the study remains feasible and cost-effective while delivering meaningful results. Data availability can vary significantly depending on the industry, location, and specific product being assessed. Hence, the analyst must carefully consider what data is available, what data gaps exist, and how to address these gaps while maintaining a scientifically rigorous approach. Transparency is key in the goal and scope definition, as all the assumptions made must be documented and justified. This transparency allows for the reproducibility of the study and enhances its

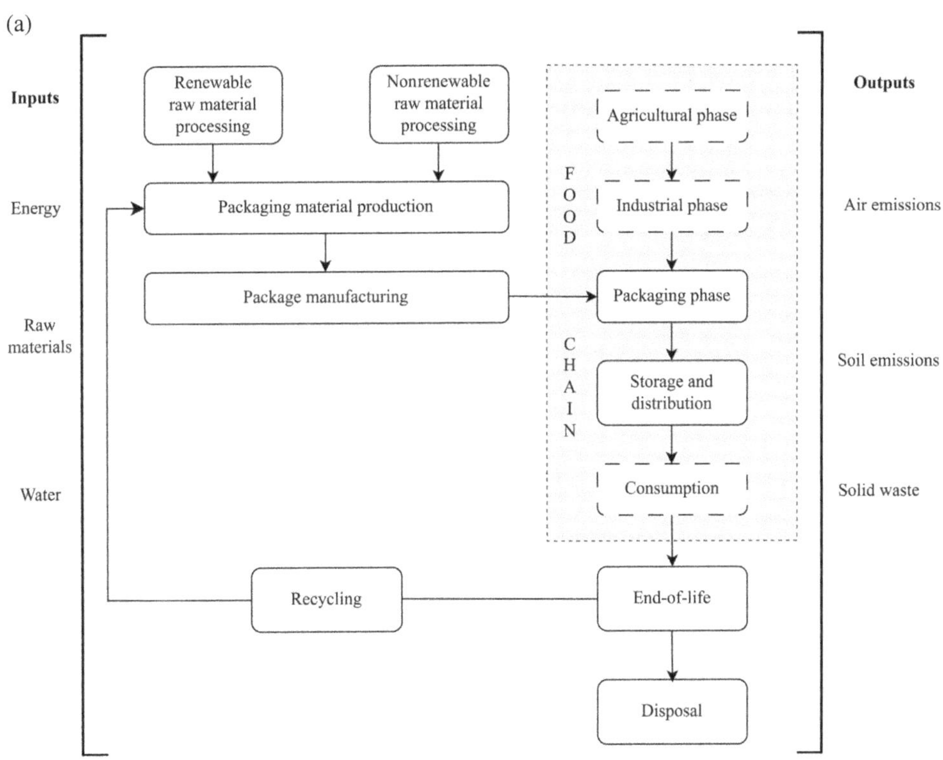

Figure 14.3 (a) LCA for nanopackaging of food (*Source:* Restuccia et al. (2016)/with permission of Elsevier), (b) System boundary for nanopackaging of food (*Source:* WebPAINTER-Std / Adobe Stock)

(b)

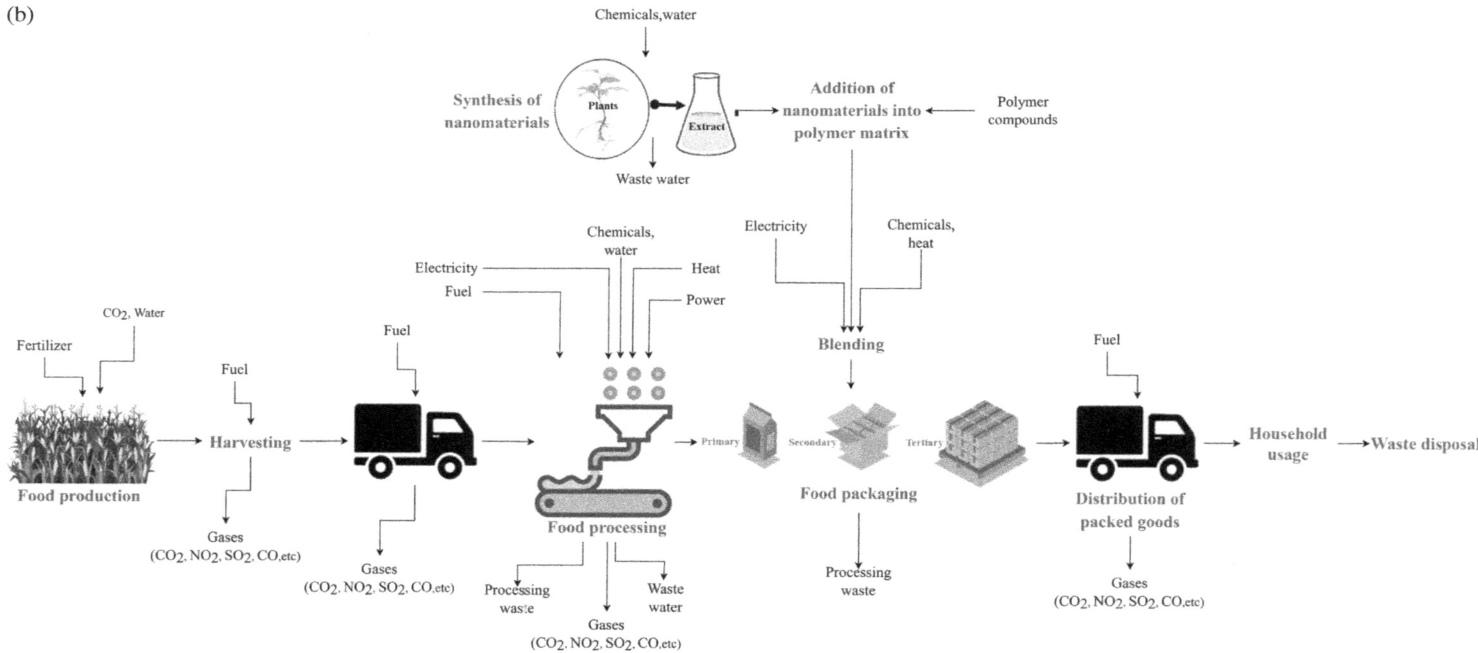

Figure 14.3 (Continued)

credibility. It enables stakeholders and peer reviewers to evaluate the study's validity and understand the potential limitations in interpreting the results. The limitations of the study must also be explicitly addressed in the goal and scope definition. These limitations could be related to data availability, time constraints, budget constraints, or the complexity of the system being assessed. Acknowledging these limitations upfront helps manage expectations and ensures that the results are interpreted with appropriate caution. In addition to system boundaries and assumptions, the goal and scope definition may include other elements, such as the functional unit and the intended application of the study. The functional unit defines the quantifiable measure that represents the function or service provided by the product, process, or service. It serves as the basis for normalizing the results and allows for comparisons between different products or alternatives that fulfill the same function. The intended application of the study refers to the purpose for which the LCA results will be used. This could be for eco-design, product development, policy-making, environmental labeling, or marketing. The intended application influences the focus and depth of the analysis, ensuring that the LCA provides the most relevant information for the specific decision-making context. Overall, the goal and scope definition is a crucial step in conducting a robust and meaningful LCA. It sets the stage for a comprehensive evaluation of the environmental impacts of a product, process, or service. By defining the system boundary, making appropriate assumptions, and addressing limitations, LCA practitioners can provide valuable insights to support sustainable decision making and drive continuous environmental improvement. The transparent and well-documented nature of the goal and scope definition also enhances the credibility and usefulness of the LCA results, making it a valuable tool for stakeholders in various industries and sectors seeking to assess and improve their environmental performance. The initial steps that should be taken in an LCA are to (i) identify the functional unit (i.e., the subject of the analysis), (ii) specify the research motivation (i.e., the intended assessment), and (iii) choose the actual recipients of the findings (i.e., the people who will use the assessment) (Gahane et al., 2022). For nanopackaging materials, the main goal was to calculate their net carbon footprint to reduce food waste (Zhang et al., 2019). It helps in assessing reduced environmental impact by avoiding food loss due to the use of nanopackaging and additional resources consumption and emission resulting from its production. This data will be helpful in the manufacture of nanopackaging materials.

14.3.2 Functional Unit

For the purpose of normalization, all problem data must be correlated with a functional unit. For example, in the subject of food-packaging systems, a functional unit may be 1 kg of packed food product or 1 kg of packaging materials.

14.3.3 Life Cycle Inventory

Inventory analysis is a rigorous and systematic approach that relies on extensive data to assess the resource consumption and environmental impacts of a product throughout its entire life cycle. By meticulously quantifying the energy and material requirements, as well as the pollution emissions released into the air, water, and soil, inventory analysis provides valuable insights into the product's overall sustainability. This method aids decision makers, researchers, and industries in identifying opportunities for optimizing resource efficiency, minimizing environmental burdens, and promoting environmentally friendly practices to achieve a more sustainable and responsible approach to product development and consumption (Bohlmann, 2004). When choosing life cycle inventories for a nanopackaging system, the focus primarily centers on the nanomaterials utilized, the polymer matrix employed, and the specific food products involved. These critical factors play a pivotal role in the overall environmental assessment of the packaging system. Nanomaterials and polymer matrices are essential components that can significantly influence the environmental impacts, resource consumption, and potential risks associated with nanopackaging. Considering the specific food products is vital as different products have distinct characteristics and requirements that can impact the overall LCA. By analyzing these elements in-depth, researchers and industries can make informed decisions to develop more sustainable and eco-friendly nanopackaging solutions for various food items. LCA relies heavily on data to provide accurate and meaningful results. Therefore, ensuring data quality and declaring its reliability before conducting the analysis is of paramount importance. The process starts with defining the system boundary and inventory, which determines the scope of data collection. A well-defined boundary ensures that all relevant life cycle stages are included, while an appropriately designed inventory targets specific data points aligned with the study's objectives. By adopting these measures, LCA practitioners can gather comprehensive and pertinent data, enhancing the precision of the assessment and enabling informed decision making for achieving environmental sustainability and effective resource management.

14.3.4 Life Cycle Impact Assessment (LCIA)

Life cycle impact assessment (LCIA) is a crucial phase of LCA that evaluates the potential environmental impacts associated with nanopackaging throughout its entire life cycle. In the context of nanopackaging, LCIA is particularly significant due to the unique properties and potential risks of nanomaterials, which may have distinct environmental implications compared to conventional packaging materials. During LCIA, various environmental impact categories are considered, such as global warming potential (GWP), ozone depletion, eutrophication, acidification, and human toxicity, among others. The analysis seeks to quantify the potential adverse effects that nanopackaging may have on the environment, human health, and ecosystems across all life cycle stages, from raw material extraction to disposal or recycling. When assessing nanopackaging, special attention is given to the nanomaterials employed and their specific properties. Nanomaterials have unique characteristics, such as high surface area to volume ratio and increased reactivity, which can potentially lead to altered environmental behaviors. The release of nanoparticles during production, use, or disposal stages may introduce new risks and require careful evaluation. LCIA for nanopackaging also considers potential benefits. Nanotechnology can enhance packaging properties, such as barrier performance, mechanical strength, and antimicrobial properties, which can lead to reduced food waste and extended shelf life. A thorough assessment is needed to weigh these benefits against the potential environmental risks and determine the overall sustainability of nanopackaging solutions.

It is essential to conduct LCIA with accurate and reliable data, considering uncertainties and limitations inherent in the assessment of nanomaterials' impacts. Researchers and practitioners should collaborate closely to address data gaps and develop standardized methodologies for assessing nanomaterials' environmental impact, promoting transparency, and credibility in the analysis. By considering the unique properties and potential risks of nanomaterials, LCIA provides valuable insights into the overall environmental performance of nanopackaging throughout its life cycle. Careful evaluation of the benefits and potential drawbacks of nanotechnology in packaging allows for the development of more sustainable and responsible packaging solutions that balance environmental considerations with functional requirements and economic viability.

14.4 LCA of Nanofood Packaging

The nanopackaging of food includes processing of raw materials, nanomaterial production, addition into polymer matrix followed by blending and the final packaged food products (Figure 14.3[b]). Incorporation of nanoparticles into the polymer matrix was found to be effective in food preservation, increasing storage life, which results in the reduction of food waste. A study reported that the utilization of nanomaterials for packing fresh fruits and cooked ham showed a reduction in food waste generation and shelf life extension. The GWP decreased due to reduced food waste generation. Utilizing nanopackaging has the potential to significantly improve the net carbon footprint of the food supply chain, particularly for animal-based goods that have a high environmental impact during their life cycles (Zhang et al., 2019). Table 14.1 shows the summary of LCA of different nanomaterials used for food packaging of fresh fruits and vegetables and animal products. LCA offers a way to measure the potential reduction of packaging's environmental impact based on specific indicators, and improved package designs may decrease that impact (Laura et al., 2013).

Among the pool of researchers in the field of food packaging, (Siracusa et al. 2011) have discussed the LCA of multilayer polymer film used on food packaging. The primary objective of this study is to comprehensively comprehend and address the mounting environmental challenges posed by plastic packaging. The focus centers on devising effective strategies for mitigating these challenges, with a specific emphasis on sustainable solutions like material recovery and innovative recycling technologies. By delving into the intricacies of these alternatives, the research endeavors to identify pathways that can alleviate the environmental strain attributed to plastic packaging. Furthermore, the investigation probes into the feasibility of incorporating recycled polymer pellets into the packaging industry. By employing recycled materials, the research aims to substantially curtail the overall environmental repercussions traditionally associated with plastic packaging. This endeavor aligns with the broader objective of transitioning towards a circular economy, where resources are conserved, waste is minimized, and environmental pressures are alleviated.

A LCA comparing the environmental impact of polylactide (PLA) and polypropylene (PP) in food packaging applications is presented by Bohlmann (2004). The LCA methodology comprehensively assesses the complete journey of products, encompassing material extraction, manufacturing, usage, and end-of-life management. Through an inventory analysis, essential data regarding energy and material requisites, alongside greenhouse gas emissions, are collated for both Polylactic Acid (PLA) and Polypropylene (PP) systems. A pivotal focal point of the study is elucidating the significance of functional units within the LCA

Table 14.1 Summary of LCA of different nanomaterials used for food nanopackaging

Type of Food Item	Polymer Matrix	Nanomaterials Used for Packaging (%)	Functional unit	System Boundaries	Impact Assessment	Reference
Apricot	LLDPE + PE-g-MA	Nanoclay (5.72)	1 kg	Cradle-to-grave	GWP	Zhang et al. (2019)
Tomato paste	LLDPE + PET	Nanoclay (4.0)	1 kg	Cradle-to-grave	GWP	
Orange juice	LDPE	P105 (95% TiO_2 + 5% nano-Ag) (10.0)	1 kg	Cradle-to-grave	GWP	
Cooked ham	PA6 + LDPE	Nanoclay (2.0)	1 kg	Cradle-to-grave	GWP	
Fresh cut lettuce	Polypropylene	ZnO nanoparticles	130 g	Cradle-to-grave	ReCiPe indicators (18 midpoint indicators)	Vigil et al. (2020)
Fresh cut vegetables	Polylactic acid	ZnO nanoparticles	130 g	Cradle-to-grave		
Fresh cut vegetables	PLA	Nanoclay (Na-cloisite 20 and Na-cloisite 30B)	100,000 kg	Cradle to grave	IMPACT 2002+	Lorite et al. (2017)
Food packaging film	Chitosan	Cellulose nanofbril (CNF)	1 kg	Cradle-to-gate	GWP	Ponnusamy and Mani (2022)
Tomato	PP	Zein nanofibers	1 kg	Cradle-to-grave	Environmental footprint (EF)	Tsouti et al. (2023)

LLDPE, linear low-density polyethylene; PP, polypropylene; PLA, polylactic acid; LDPE, low-density polyethylene; PE, polyethylene; PA, polyamide; PET, polyethylene terephthalate; PE-g-MA, polyethylene glycol methacrylate; GWP, global warming potential.

framework, enabling a direct, meaningful juxtaposition of disparate systems. Particular emphasis is placed on recognizing the significance of end-of-life treatment, an aspect vital for evanescent consumer goods such as packaging, to accurately gauge the holistic environmental load. The study's fundamental objective revolves around unraveling insights into the ecological merits inherent in biodegradable polymers like PLA vis-à-vis their conventional counterpart, PP. This exploration specifically hones in on greenhouse gas emissions and the overarching sustainability quotient. In summation, this research employs the LCA methodology to meticulously scrutinize the entire lifecycle of products, underscoring the vitality of functional units and the need for a comprehensive appraisal of end-of-life stages. By evaluating the environmental advantages offered by biodegradable polymers like PLA in contrast to conventional options like PP, the study aims to contribute substantively to the discourse on mitigating ecological impacts and fostering sustainable practices.

Pardo and Zufía (2012) undertook a comprehensive comparison of the ecological repercussions stemming from diverse food preservation techniques while simultaneously pinpointing avenues for enhancement. Through a systematic analysis, the findings illuminated the potential of innovative methodologies, such as Modified Atmosphere Packaging (MAP), to yield diminished environmental footprints in contrast to conventional thermal methods. Notably, the outcomes underscored that MAP, in particular, exhibited superior performance across a spectrum of impact categories, primarily attributed to its minimal energy demands during both packaging and gas injection phases. However, a salient caveat remains pertinent: the temporal extension of shelf life through MAP remains limited when juxtaposed with alternative preservation modalities. The study's overarching significance stems from its role as a compass, guiding the quest for sustainable food preservation practices. By dissecting the environmental implications inherent in various approaches, the research endeavors to illuminate pathways that reconcile food security with ecological integrity. Of paramount interest are the ramifications of MAP, which, despite its commendable environmental profile, must contend with constraints related to longevity in shelf life extension. In essence, this study contributes substantively to the discourse surrounding ecologically responsible food preservation. It not only accentuates the comparative advantages of emergent techniques like MAP over thermal processes but also underscores the necessity of a holistic perspective that encompasses ecological, economic, and practical considerations. In doing so, the research lays a cornerstone for advancing the paradigm of sustainable food preservation, fostering a future where both nourishment and environmental equilibrium coalesce harmoniously (Pardo and Zufía, 2012).

In 1969, the Coca-Cola Company conducted one of the earliest studies on the environmental impact of food packaging (and most likely the first study on packaging to use what would later be termed LCA). The study quantified the energy, material, and environmental impacts of a glass bottle's life cycle and compared it to plastic and refillable bottles (Hunt et al., 1996). Since then, many LCAs on food packaging have been conducted and published, thanks to the Society of Environmental Toxicology and Chemistry (and eventually the ISO) improving and consolidating the LCA approach. As a result, Food packaging systems have been extensively studied from an LCA standpoint (Siracusa et al., 2014). Earlier literature review reveals that most of the LCA studies have been conducted on food and drink products, their packaging, packaging materials, comparisons among alternative packaging materials, food preservation technologies and food waste management systems

(Restuccia et al., 2016). According to Flanigan and co-workers, they critically reviewed the studies based on LCA of packaging, and they found that out of 110 studies, 23 were concentrated on food packaging, 23 on beverage packaging, whereas the remaining studies were addressed on packaging materials or its application related to food and beverage industries. Thirty-seven studies showed the comparison among different and alternative packaging materials; 20 studies were carried out on food or beverage products with a major focus on packaged materials to evaluate their impact during assessment. The remaining studies are concerned with end-of-life scenarios or LCA methodology. Among these, all 58 studies considered cradle-to-grave impact of packaging and/or cradle-to-table impact of the food. Further they concluded that the application of LCA can be an effective tool to evaluate the ecological impact of packaging (Laura et al., 2013).

14.5 Environmental Impacts of Nanofood Packaging Materials

The packaging sector is constantly looking for alternative approaches that maintain a balance between several challenges, such as energy and material costs, environmental awareness, pollution control, and MSW, in order to meet regulatory requirements. Additionally, it should prevent food from microbial contamination, light, physical damage, and chemical changes. Plastic, polystyrene, cardboard, paper, glass, and aluminum are the most common materials used to package food since they are inexpensive, lightweight, and sturdy. The production rate of waste has increased quickly, along with a rise in the global population. Humans, animals, and the environment are all negatively impacted by this. When food packaging material deteriorates, it emits some hazardous substances into the air and also leads to pollution by disposing inappropriately when it does not deteriorate properly. Nonbiodegradable packaging materials pollute soil because they remain in the environment for a long time. They also deplete soil nutrients that are essential for plant growth, agriculture, and the grazing of livestock, which ultimately results in food insecurity for humans (Basavegowda et al., 2020; Guillard et al., 2018).

Manufacturing of nanomaterials involves molecular-level chemical/biochemical synthesis as well as the physical or chemical breakdown of bigger particles. In order to create packaging materials, the designed nanomaterials must first be extracted and purified further before being distributed and combined with polymers. As a result, the synthesis of nanomaterials may result in a number of environmental burdens, such as the consumption of energy, water, and other resources and the emissions that accompany them. The environmental costs associated with the production of nanomaterials, as well as the possible advantages of nanopackaging in terms of reducing food waste, must be taken into account when assessing the sustainability of nanomaterial application for food packing (Silvestre et al., 2011; Zhang et al., 2019). In most of the nanotechnology applications, silver nanoparticles (Ag NPs), zinc nanoparticles (ZnO NPs), titanium dioxide nanoparticles (TiO_2 NPs), and other nanomaterials are used due to their potential antimicrobial, antibacterial, etc. activities; its global environmental impact varies based on synthesis methods, raw material used for its synthesis, scale, and desired product application (Table 14.2) (Papadaki et al., 2017; Temizel-Sekeryan and Hicks, 2020; Zhang et al., 2017).

Table 14.2 Environmental impact assessment of nanoparticles used in food packaging using LCA

Nanomaterials Used in Food Packaging	Functional Unit	System Boundaries	Software Used	Impact Assessment	Results Reported	Reference
Ag NPs	1 kg	Cradle-to-gate	SimaPro 7.3.3	IPCC GWP 100a (CO_2 eq. kg) Energy, nonrenewable, fossil (MJ)	291.8 CO_2 eq. kg 3,531 MJ	Zhang et al. (2017)
TiO_2 NPs	1 kg	Cradle-to-gate	SimaPro 7.3.3	IPCC GWP 100a (CO_2 eq. kg) Energy, nonrenewable, fossil (MJ)	8.24 CO_2 eq. kg 133.47 MJ	
ZnO NPs	1 m^2/g	Cradle-to-gate	SimaPro 7.3	ReCiPe impact assessment	Synthesis temperature and microwave power are inversely proportional to environmental sustainability	Papadaki et al. (2017)
Cellulose nanocrystals	1 kg	Cradle-to-grave	SimaPro	Eco-Indicator 99	0.454 eco Pts as environmental score	Petrucci et al. (2018)

When we talked about analyzing the end-of-life stage of nanomaterials, it is critical since it affects environmental as well as human health. Asmatulu and the team worked on different nano products used in various industrial segments in order to understand the end-of-life of these materials. They distributed 1,014 nano products into nine end-of-life categories such as recycling, ingestion, absorption by skin then public sewer or water body, public sewer or water body, burning then landfill, landfill, air release, initial air release then public sewer or water body and others based on their final usage. The most probable end-of-life outcome has been identified to be recyclability. Absorption through the skin, ingestion, and then landfilling occur less frequently. For existing nanoparticles, exposure to air and water has a far reduced risk of end-of-life consequences (Asmatulu et al., 2012).

14.6 Major Challenges and Future Perspective

Recently, Turkcu and Tura expounded on the "Dark side of sustainable packaging," which explains the major challenges faced by industries in order to produce sustainable packaging materials. They mentioned that the majority of conflicts are found to be caused by high operational and investment expenses, a lack of standards and guidelines, different legislation in different nations and regions, evolving regulations, and a lack of awareness among customers, policymakers, and end-of-life management organizations. One of the most common ways to resolve such conflicts is to join associations, collaborate with NGOs, get certifications and labels, and carry out LCAs (Turkcu and Tura, 2023). In such cases, the application of LCA methodology helps to assess the environmental performance of alternative product packaging systems, with the goal of demonstrating how this technique could assist industry and packaging makers in their management activities to make environmentally driven decisions with awareness (De Monte et al., 2005). For a better understanding of 'proof-of-concept' case studies on net carbon footprint, GWP from a life cycle perspective can help people and industries, too. Such emerging technologies will help food industry professionals in decision making for crucial aspects. Depending on the goals and ideals of the user, LCA is a very useful tool for promising environmentally sustainable packaging. It can be used in combination with other methods to measure other significant economic, technical, or social attributes.

14.7 Conclusion

The concept that producers, products, and packaging users bear some degree of accountability for their environmental impact is the foundation of the relatively new scientific field known as packaging LCA. The cycle was formerly thought to start when raw materials are brought to the manufacturer and finish when the finished product "comes out" of the production and is delivered to the customer. The environmental impact is especially important because consumers and retailers frequently view packing as trash. Packaging is being handled more and more in compliance with laws these days. Additionally, LCA gives firms the ability to select safe production methods with less hazardous substances and risks that exist, as well as to enhance recycling, waste management, and energy efficiency. LCA can be utilized as a tool to estimate the environmental impact of different processes; using LCA, it has been clearly indicated that the incorporation of nanoparticles into active packaging materials proved to be superior compared to traditional packaging techniques in terms of environmental burdens generated. Nanopackaging served as an innovative packaging technique for extending shelf life and reducing environmental impact and GWP. The application of LCA in this area needs to be explored soon.

References

Abejón, R., Bala, A., Vázquez-Rowe, I., Aldaco, R., & Fullana-i-Palmer, P. (2020). When plastic packaging should be preferred: life cycle analysis of packages for fruit and vegetable distribution in the Spanish peninsular market. *Resources, Conservation and Recycling, 155*, 104666. https://doi.org/10.1016/j.resconrec.2019.104666

Appendini, P., & Hotchkiss, J. H. (2002). Review of antimicrobial food packaging. *Innovative Food Science & Emerging Technologies, 3*(2), 113–126. https://doi.org/10.1016/S1466-8564(02)00012-7

Ashfaq, A., Khursheed, N., Fatima, S., Anjum, Z., & Younis, K. (2022). Application of nanotechnology in food packaging: Pros and Cons. *Journal of Agriculture and Food Research, 7*, 100270. https://doi.org/10.1016/j.jafr.2022.100270

Asmatulu, E., Twomey, J., & Overcash, M. (2012). Life cycle and nano-products: end-of-life assessment. *Journal of Nanoparticle Research, 14*(3), 720. https://doi.org/10.1007/s11051-012-0720-0

Basavegowda, N., Mandal, T. K., & Baek, K. H. (2020). Bimetallic and trimetallic nanoparticles for active food packaging applications: a review. *Food and Bioprocess Technology, 13*(1), 30–44. https://doi.org/10.1007/s11947-019-02370-3

Basumatary, I. B., Mukherjee, A., Katiyar, V., & Kumar, S. (2022). Biopolymer-based nanocomposite films and coatings: recent advances in shelf-life improvement of fruits and vegetables. *Critical Reviews in Food Science and Nutrition, 62*(7), 1912–1935. https://doi.org/10.1080/10408398.2020.1848789

Bohlmann, G. M. (2004). Biodegradable packaging life-cycle assessment. *Environmental Progress—Wiley Online Library, 23*(4), 342–346.

Bumbudsanpharoke, N., Choi, J., & Ko, S. (2015). Applications of nanomaterials in food packaging. *Journal of Nanoscience and Nanotechnology, 15*(9), 6357–6372. https://doi.org/10.1166/jnn.2015.10847

Cerqueira, M. A., Vicente, A. A., & Pastrana, L. M. (2018). Chapter 1—Nanotechnology in Food Packaging: Opportunities and Challenges. In M. Â. P. R. Cerqueira, J. M. Lagaron, L. M. Pastrana Castro, & A. A. M. B. T.-N. for F. P. de Oliveria Soares Vicente (Eds.), *Micro and Nano Technologies (pp. 1–11)*. Elsevier. https://doi.org/10.1016/B978-0-323-51271-8.00001-2

De Monte, M., Padoano, E., & Pozzetto, D. (2005). Alternative coffee packaging: an analysis from a life cycle point of view. *Journal of Food Engineering, 66*(4), 405–411. https://doi.org/10.1016/j.jfoodeng.2004.04.006

Gahane, D., Biswal, D., & Mandavgane, S. A. (2022). Life cycle assessment of biomass pyrolysis. *BioEnergy Research, 15*(3), 1387–1406. https://doi.org/10.1007/s12155-022-10390-9

Guillard, V., Gaucel, S., Fornaciari, C., Angellier-Coussy, H., Buche, P., & Gontard, N. (2018). The next generation of sustainable food packaging to preserve our environment in a circular economy context. *Frontiers in Nutrition*, *5*(December), 1–13. https://doi.org/10.3389/fnut.2018.00121

Guinee, J. B. (2002). Handbook on life cycle assessment operational guide to the ISO standards. *The International Journal of Life Cycle Assessment*, *7*(5), 311–313. https://doi.org/10.1007/BF02978897

Hernández-Muñoz, P., Cerisuelo, J. P., Domínguez, I., López-Carballo, G., Catalá, R., & Gavara, R. (2019). Nanotechnology in Food Packaging. In *Nanomaterials for Food Applications* (Issue August 2016 (pp. 205–232). Elsevier. https://doi.org/10.1016/B978-0-12-814130-4.00008-7

Hunt, R. G., Franklin, W. E., & Hunt, R. G. (1996). LCA—How it came about. *The International Journal of Life Cycle Assessment*, *1*(1), 4–7. https://doi.org/10.1007/BF02978624

ISO (2006). *Environmental Management: Life Cycle Assessment; Principles and Framework.* (ISO 14040:2006). International Organization for Standardization, Geneva.

Kuswandi, B. (2016). Nanotechnology in food packaging. *Nanoscience in Food and Agriculture*, *1*, 151–183.

Laura, F., Rolf, F., & Trisha, M. (2013). *An Analysis of Life Cycle Assessment in Packaging for Food & Beverage Applications. Life Cycle Initiative.*

Lorite, G. S., Rocha, J. M., Miilumäki, N., Saavalainen, P., Selkälä, T., Morales-Cid, G., Gonçalves, M. P., Pongrácz, E., Rocha, C. M. R., & Toth, G. (2017). Evaluation of physicochemical/microbial properties and life cycle assessment (LCA) of PLA-based nanocomposite active packaging. *LWT—Food Science and Technology*, *75*, 305–315. https://doi.org/10.1016/j.lwt.2016.09.004

Marsh, K., & Bugusu, B. (2007). Food packaging—roles, materials, and environmental issues. *Journal of Food Science*, *72*(3), R39–R55.

Papadaki, D., Foteinis, S., Mhlongo, G. H., Nkosi, S. S., Motaung, D. E., Ray, S. S., Tsoutsos, T., & Kiriakidis, G. (2017). Life cycle assessment of facile microwave-assisted zinc oxide (ZnO) nanostructures. *Science of The Total Environment*, *586*, 566–575. https://doi.org/10.1016/j.scitotenv.2017.02.019

Pardo, G., & Zufía, J. (2012). Life cycle assessment of food-preservation technologies. *Journal of Cleaner Production*, *28*, 198–207. https://doi.org/10.1016/j.jclepro.2011.10.016

Petrucci, R., Fortunati, E., Puglia, D., Luzi, F., Kenny, J. M., & Torre, L. (2018). Life cycle analysis of extruded films based on poly(lactic acid)/cellulose nanocrystal/limonene: a comparative study with ATBC plasticized PLA/OMMT systems. *Journal of Polymers and the Environment*, *26*(5), 1891–1902. https://doi.org/10.1007/s10924-017-1085-3

Pira, S. (2019). *The Future of Packaging: Long-term Strategic Forecasts to 2028. Recuperado de.* https://www.smithers.com/resources/2019/feb/future-packaging-long-term-strategic-forecasts-2030.

Ponnusamy, P. G., & Mani, S. (2022). Life cycle assessment of manufacturing cellulose nanofibril-reinforced chitosan composite films for packaging applications. *International Journal of Life Cycle Assessment*, *27*(3), 380–394. https://doi.org/10.1007/s11367-022-02035-y

Primožič, M., Knez, Ž., & Leitgeb, M. (2021). (Bio)nanotechnology in food science—food packaging. *Nanomaterials*, *11*(2), 292. https://doi.org/10.3390/nano11020292

Rai, M., Ingle, A. P., Gupta, I., Pandit, R., Paralikar, P., Gade, A., Chaud, M. V., & dos Santos, C. A. (2019). Smart nanopackaging for the enhancement of food shelf life. *Environmental Chemistry Letters*, *17*(1), 277–290. https://doi.org/10.1007/s10311-018-0794-8

Restuccia, D., Salomone, R., Spizzirri, U. G., Saija, G., Ioppolo, G., Parisi, O. I., & Picci, N. (2016). Chapter 16—Industrial Applications: Regulatory Issues and Life Cycle Assessment of Food Packaging. In J. Barros-Velázquez (Ed.), *Antimicrobial Food Packaging* (pp. 215–227). Academic Press. https://doi.org/10.1016/B978-0-12-800723-5.00016-4

Rhim, J.-W., Park, H.-M., & Ha, C.-S. (2013). Bio-nanocomposites for food packaging applications. *Progress in Polymer Science*, *38*(10), 1629–1652. https://doi.org/10.1016/j.progpolymsci.2013.05.008

Riaz, F., Rasul, I., Azeem, F., Zubair, M., Nadeem, H., Imran, M., Muzammil, A., Afzal, M., & Siddique, M. H. (2024). Chapter 17—Environmental Impact, Health Implications, and Life Cycle Assessment of Bionanocomposites. In B. Sharma, S. Thomas, P. Kumar Bajpai, K. Ghosal, & S. Shekhar (Eds.), *Advances in Bionanocomposites* (pp. 387–406). Elsevier. https://doi.org/10.1016/B978-0-323-91764-3.00009-7

Scaraboto, D., Joubert, A. M., & Gonzalez-Arcos, C. (2020). Using lots of plastic packaging during the coronavirus crisis? You're not alone. *The Conversation*, *668*, 1077–1093.

Shukla, P., Chaurasia, P., Younis, K., Qadri, O. S., Faridi, S. A., & Srivastava, G. (2019). Nanotechnology in sustainable agriculture: studies from seed priming to post-harvest management. *Nanotechnology for Environmental Engineering*, *4*(1), 11. https://doi.org/10.1007/s41204-019-0058-2

Silvestre, C., Duraccio, D., & Cimmino, S. (2011). Food packaging based on polymer nanomaterials. *Progress in Polymer Science*, *36*(12), 1766–1782. https://doi.org/10.1016/j.progpolymsci.2011.02.003

Siracusa, V., Dalla Rosa, M., Romani, S., Rocculi, P., & Tylewicz, U. (2011). Life cycle assessment of multilayer polymer film used on food packaging field. *Procedia Food Science*, *1*, 235–239.

Siracusa, V., Ingrao, C., Lo Giudice, A., Mbohwa, C., & Dalla Rosa, M. (2014). Environmental assessment of a multilayer polymer bag for food packaging and preservation: an LCA approach. *Food Research International*, *62*, 151–161. https://doi.org/10.1016/j.foodres.2014.02.010

Sundqvist-Andberg, H., & Åkerman, M. (2021). Sustainability governance and contested plastic food packaging—an integrative review. *Journal of Cleaner Production*, *306*, 127111. https://doi.org/10.1016/j.jclepro.2021.127111

Temizel-Sekeryan, S., & Hicks, A. L. (2020). Global environmental impacts of silver nanoparticle production methods supported by life cycle assessment. *Resources, Conservation and Recycling*, *156*, 104676. https://doi.org/10.1016/j.resconrec.2019.104676

Tripathi, A., Srivastava, S., Pandey, V. K., Singh, R., Panesar, P. S., Dar, A. H., Rustagi, S., Shams, R., & Pandiselvam, R. (2023). Substantial utilization of food wastes for existence of nanocomposite polymers in sustainable development: a review. *Environment, Development and Sustainability*, 1–27. https://doi.org/10.1007/s10668-023-03756-2

Tsagkaris, A. S., Tzegkas, S. G., & Danezis, G. P. (2018). Nanomaterials in food packaging: state of the art and analysis. *Journal of Food Science and Technology*, *55*(8), 2862–2870. https://doi.org/10.1007/s13197-018-3266-z

Tsouti, C., Papadaskalopoulou, C., Konsta, A., Andrikopoulos, P., Panagiotopoulou, M., Papadaki, S., Boukouvalas, C., Krokida, M., & Valta, K. (2023). Investigating the environmental benefits of novel films for the packaging of fresh tomatoes enriched with antimicrobial and antioxidant compounds through life cycle assessment. *Sustainability (Switzerland), 15*(10). https://doi.org/10.3390/su15107838

Turkcu, D., & Tura, N. (2023). The dark side of sustainable packaging: battling with sustainability tensions. *Sustainable Production and Consumption, 40*, 412–421. https://doi.org/10.1016/j.spc.2023.07.007

Vanapalli, K. R., Sharma, H. B., Ranjan, V. P., Samal, B., Bhattacharya, J., Dubey, B. K., & Goel, S. (2021). Challenges and strategies for effective plastic waste management during and post COVID-19 pandemic. *The Science of the Total Environment, 750*, 141514. https://doi.org/10.1016/j.scitotenv.2020.141514

Vigil, M., Pedrosa-Laza, M., Alvarez Cabal, J. V., & Ortega-Fernández, F. (2020). Sustainability analysis of active packaging for the fresh cut vegetable industry by means of attributional and consequential life cycle assessment. *Sustainability, 12*(17). https://doi.org/10.3390/su12177207

Vignali, G. (2016). Life-Cycle Assessment of Food-Packaging Systems. In S. S. Muthu (Ed.), *Environmental Footprints of Packaging* (pp. 1–22). Springer, Singapore. https://doi.org/10.1007/978-981-287-913-4_1

Wesley, S. J., Raja, P. R., Raj, A. A., & Tiroutchelvamae, D. (2014). Review on—nanotechnology applications in food packaging and safety. *International Journal of Engineering Research, 3*, 645–651.

Zhang, H., Hortal, M., Dobon, A., Jorda-Beneyto, M., & Bermudez, J. M. (2017). Selection of nanomaterial-based active agents for packaging application: using life cycle assessment (LCA) as a tool. *Packaging Technology and Science, 30*(9), 575–586. https://doi.org/10.1002/pts.2238

Zhang, B. Y., Tong, Y., Singh, S., Cai, H., & Huang, J.-Y. (2019). Assessment of carbon footprint of nano-packaging considering potential food waste reduction due to shelf life extension. *Resources, Conservation and Recycling, 149*, 322–331. https://doi.org/10.1016/j.resconrec.2019.05.030

15

Migratory Effects, Safety, and Concerns of Nanofood Packaging

Rahul Kumar and Kanishka Bhunia

Agricultural and Food Engineering Department, Indian Institute of Technology Kharagpur, Kharagpur, West Bengal, India

15.1 Introduction

In nanotechnology, materials have at least one dimension between 1 and 100 nm. Nanomaterials and nanocomposites are being used in food packaging to create active and intelligent packaging, improve physical and antibacterial properties, and lengthen the shelf life of food (Ashfaq et al., 2022). Montmorillonite, zinc oxide, and kaolinite are the three nanomaterials that are most frequently employed in food packaging (Chaudhary et al., 2020). Traditional food packaging materials are combined with nanoparticles to create nanocomposites, which can enhance barrier properties and have a favorable effect on the thermal and mechanical properties of the packaging material (Honarvar et al., 2016). However, nanocomposite-based food packaging raises concerns about the potential migration of nanomaterials into the food matrix, which requires further investigation (Honarvar et al., 2016). Enhanced packaging through nanocomposites, intelligent/smart packaging, and antimicrobial agents are a few examples of how nanotechnology is used in food packaging (Sharma et al., 2017). The development of biomaterials suitable for extensive industrial applications using nanotechnology in food packaging shows promise (Nilsen-Nygaard et al., 2021). A bottom-up approach, in which atoms and molecules are assembled upward to form nanoscale particles, or a top-down approach, in which larger materials are ground down to the nanoscale, can theoretically produce any particulate substance in nanoform (Chaudhry et al., 2017).

Due to their high volume-to-surface ratio, nanoparticles are a viable material for food packaging. Nanoparticles can, however, move from food contact material (FCM) into food. Nanomaterials have the potential to harm human cells by increasing membrane permeability, disrupting mitochondrial activity, producing active oxygen, and producing toxic effects and chronic illnesses like allergies, asthma, various inflammatory conditions, cardiovascular diseases, and cancer. Numerous pathways exist for nanomaterials to enter the body (Paidari et al., 2021). According to their kind, size, shape, aspect ratio, concentration, duration of exposure, and a person's sensitivity, nanoparticles' toxicity varies (Egbuna et al., 2021;

Nanotechnology for Sustainable Food Packaging, First Edition. Edited by C. Anandharamakrishnan, Jeyan A. Moses, and M. Maria Leena.

Khan et al., 2019). Organic nanoparticles made of starch, chitosan, lipids, proteins, and vitamins are typically regarded to be nontoxic (Divya et al., 2018). On the other hand, metallic compound-based inorganic ones are thought to be indigestible and considered to be toxic. Silver, titanium, and silica are some examples. Organic nanoparticles may, however, differ from their original forms in terms of bioavailability, digestion, and nutritional attributes. Organic nanoparticles might increase a molecule's bioavailability, which occasionally might be detrimental. According to Bumbudsanpharoke and Ko (2015), the majority of research on the migration of nanoparticles in food products is concentrated on the migration of silver nanoparticles. Overall, there are worries regarding the possible effects of nanomaterials on human health and the environment, even as their usage in food has the potential to improve food safety, quality, and shelf life. More research is needed for a comprehensive understanding of the potential risks and benefits of using nanomaterials in food and to develop appropriate regulations to ensure their safe use.

15.2 Nanomaterial Types in Nanofood Packaging

Inorganic metal particles come in a variety of forms and are utilized in food packaging. It has special physical and chemical qualities that improve the functionality of carriers for food packaging. Because of their potential antibacterial effects, a number of oxides of NP have attracted a lot of attention. Metal oxide NPs were discovered to have antibacterial activity in Ag_2O, TiO_2, SiO_2, CuO, ZnO, CaO, and MgO. Various nanoparticles used in nanofood packaging extend the packaged food's shelf life (Table 15.1).

15.2.1 Titanium Dioxide (TiO_2)

It is an inorganic nanoparticle that is used in chewing gum and other food items as a coloring agent. Due to their brilliant white coloring, TiO_2 nanoparticles are frequently used as pigments in a number of products, including meals and cosmetics (Garcia et al., 2018). The incorporation of TiO_2 in polymer improves the thermophysical and mechanical qualities.

15.2.2 Silver (AgNPs)

Silver nanoparticles (AgNPs) have strong antimicrobial properties against harmful microorganisms, including viruses, fungi, yeasts, and bacteria. According to studies, nanocomposite films containing AgNPs have a high antagonistic effect on both gram-positive and gram-negative bacteria (Cao et al., 2020; Lin et al., 2020; Ramos et al., 2020; Sonseca et al., 2019; Valipour Motlagh et al., 2021).

15.2.3 Gold (AuNPs)

Due to its pharmacological, oxidative catalytic, and antibacterial characteristics, gold nanoparticles (AuNPs) are used in food packaging. AuNPs are used to fix other NP flaws due to their harmless nature. AuNPs are expensive, which limits their application in food packaging. Size, shape, and concentration are the main factors that determine the antifungal effects of AuNPs (Paidari and Ibrahim, 2021).

Table 15.1 Nanomaterials in food packaging and their role in food quality attributes

Nanoparticles	Matrix	Concentration	Size (nm)	Food/Stimulant	Salient Findings	Reference
ZnO	Agar	2% (w/w) and 4% (w/w)	14–48	Green grape	Fresh up to 14 and 21 d in ambient conditions for 2% and 4% concentration.	Kumar et al. (2019)
Ag NPs	LDPE	1000, 500, 250, 125, 62.5 (ppm)	10–40	50% ethanol, 3% acetic acid, and distilled water	Under the same conditions, 3% acetic acid produced the maximum quantity of migration, followed by distilled water and 50% ethanol.	Deng et al. (2020)
TiO_2	PLA	3%	<100	Mangoes	Throughout the whole 15-d storage period, the PLA nanocomposite films were able to successfully postpone the deterioration of mango firmness. PLA nanocomposite films delayed the changes in color, total acidity, vitamin C content, and microbiological properties. Fresh mangoes might have a 15-d postharvest shelf life and retain their freshness by being wrapped with PLA nanocomposite films.	Chi et al. (2019)
Nano-silica	Chitosan	30%, w/v		Sweet cherries	Prevent reactive oxygen species (ROS) build-up in the fruit during cold storage and lipid peroxidation in the cell membrane.	Meng et al. (2022)
TiO_2, ZnO nanofillers	Thermoplastic starch (TPS)	1–5%	10–40	10% ethanol, 3% acetic acid, and distilled water are the ingredient	Migration of TiO_2 was maximum at 3–4% TiO_2, followed by a sharp decrease at higher TiO_2 levels. Films containing ZnO showed higher migration levels in acid simulant due to higher dissolution of ZnO particles.	Phothisarattana and Harnkarnsujarit, (2022)
ZnO	Thermoplastic starch (TPS)	1–5%	20	10% ethanol, 3% acetic acid, and distilled water are the ingredient	ZnO-containing films exhibited greater migration rates, especially in acid simulants.	Phothisarattana and Harnkarnsujarit, (2022)

(*Continued*)

Table 15.1 (Continued)

Nanoparticles	Matrix	Concentration	Size (nm)	Food/Stimulant	Salient Findings	Reference
AuNPs and graphene oxide (GO)	PVA			Banana	In comparison to GO-PVA crosslinked composite film, AuNPs included in PVA crosslinked composite film have a stronger capacity to avoid microbiological contamination due to the establishment of a bigger inhibitory zone. With PVA-glyoxal-AuNPs, banana shelf life has been increased. The best preservation effect on bananas was found by PVA-glyoxal-AuNPs film.	Chowdhury et al. (2020)
ZnO	Chitosan and gelatin	2% and 4%	20–40	Cassia fistul	The nanocomposite films demonstrated effective antibacterial action against *Escherichia coli*. ZnO NPs' antimicrobial efficacy increased with concentration.	Kumar et al. (2020)
Copper oxide	Sodium al-alginate (SA) and cellulose nanowhisker (CNW)	5 mM	–	Fresh cut pepper	Prevents microbial contamination in fresh-cut pepper for up to 7 d. The film showed antioxidant activity in terms of ABTS and DPPH scavenging.	Saravanakumar et al. (2020)
AgNPs	polyvinyl-pyrrolidone (PVP)-based glycosomes	0.1 M	37.01–123	Bell pepper	The G-PVP-AgNPs (green synthesized) showed better antibacterial activities compared to the C-PVP-AgNPs (chemically synthesized). G-PVP-AgNPs extended the shelf life of the fresh-cut bell pepper for 12 d at 4 °C.	Saravanakumar et al. (2020)

AgNPs	PVA	0.01–0.05 g/mL	7–22	Strawberries	AgNPs exhibited outstanding antibacterial activity against *E. coli*. Results showed that the Ag@CS/PVA films effectively preserved strawberries for up to 9 d.	Yang et al. (2023)
Nanoclay	LDPE	-	1–10 μm	Ethanol 10% and acetic acid 3%	Both samples showed aluminum migration, with the Aisaika bags showing the highest migration at 51.65 ng/cm^2 and the Debbie Meyer bags at 24.14 ng/cm^2. The conditions under which the migration values are higher for both types of samples are 3% acetic acid for 10 d at 40 °C.	Echegoyen et al. (2016)

LDPE: Low Density Polyethylene; HDPE: High Density Polyethylene; LLDPE: Linear Low Density Polyethylene; PLA: Poly Lactic Acid; PVA: Poly Vinyl Alcohol; ABTS: 2,2′-azinobis-3-ethylbenzothiazoline-6-sulfonic acid; NPLA: Nano Poly Lactic Acid; DPPH: 2,2-Diphenyl-1-picrylhydrazyl.

15.2.4 Zinc Oxide (ZnO)

Due to their distinctive optical, electrical, catalytic, and photochemical capabilities, zinc oxide nanoparticles (ZnONPs) have attracted significant interest from all over the world in industrial and biomedical applications. Nanocomposites food packaging containing ZnONP has shown improved gas barrier properties (Indumathi et al., 2019; Rukmanikrishnan et al., 2020; Shankar et al., 2018), mechanical properties (Jayakumar et al., 2019; Wang et al., 2020), antimicrobial activity (Abdollahzadeh et al., 2018) and thermal characteristics (Indumathi et al., 2019; Kumar et al., 2020; Rukmanikrishnan et al., 2020).

15.2.5 Silicon Oxide (SiO_2)

An inorganic nanoparticle is used as an anticaking agent in food products (Sarfraz et al., 2020).

15.2.6 Nanocapsules

Organic nanoparticles are used as nano food additives or nutritional supplements of magnesium oxide (MgO) (Sarfraz et al., 2020).

15.2.7 Liposomes

Organic nanoparticles are used as delivery systems for nutrients and bioactive compounds in food products (Ghorbanpour et al., 2020).

15.3 Migration of Nanomaterials from Packaging

Nanoparticles can migrate from packaging in several ways, including through diffusion, abrasion, or chemical processes. Migration, sometimes also known as the inadvertent transfer of packaging components into the food, is the release of a substance from one medium to another. Foods can be contaminated with nanomaterials and low-molecular–weight compounds (such as oligomers, plasticizers, or additives) (Alamri et al., 2021), causing the food's nutritional and organoleptic qualities to decline (Dey et al., 2021). The migration has three sub-processes. At first, migrants will diffuse through the polymer toward the direction of the food the virtue of concentration gradient. As the migrant reaches the food-plastic interface, the polymer will desorb it (desorption process), and the food will absorb it (sorption process). The migrant will further diffuse into the entire food matrix due to the concentration gradient (Figure 15.1). Additionally, the convection process will accelerate migration within the food matrix. Therefore, sorption and diffusion both control the migratory process.

To protect consumer safety, regulatory bodies from all over the world have set standards and laws governing the use of metal-based nanoparticles in food packaging (Table 15.2). To assure the safety of their products, manufacturers are also encouraged to carry out their own investigation and testing.

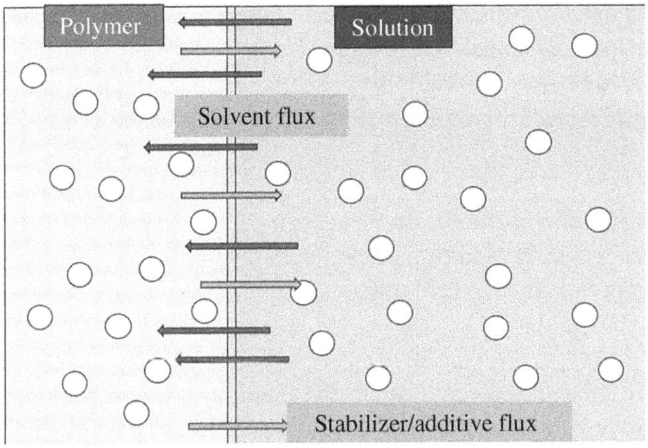

Figure 15.1 Diffusion of solvent and additives in food from polymer (*Source:* Adapted from Ferrara et al. (2001))

Table 15.2 Permissible limits of various metals into food and food simulantse

Metal	Permissible Limit (mg/kg Food or Food Simulant)	Metal	Permissible Limit (mg/kg Food or Food Simulant)
Barium	1	Cobalt	0.05
Copper	5	Iron	48
Lithium	0.6	Manganese	0.6
Zinc	25		

Source: Adapted from European Commission (2011)

15.3.1 Types of Migration

In general, migration is indicated by the specific migration (SM) and overall migration (OM). The word "OM" refers to the total amount of compounds that packaging releases per unit area of packaging for a specified condition. The SM indicates the migration of a specific component into food (Robertson, 2005).

15.3.1.1 Migration Related to the Nature of Foods

Three kinds of migration exist in relation to food systems: nonmigratory, volatile, and leaching systems. In contrast to the high molecular weight of packing polymers, relatively little mass transfer of pigments or some inorganic compounds occurs in a nonmigrating system. In a volatile migrating system, small volatile aromatic chemicals migrate to the package. When it comes to dry goods, where there is less direct touch between the food and the container, this form of migration is considered. The migration of volatile compounds occurs in three steps under these circumstances: desorption from a product, movement by evaporation or diffusion, and adsorption on the product surface. For a leaching process, the food must come into contact with the packaging for the migration to take place. In this system, a migration begins with the diffusion of the migrant from the packaging material, continues with its disintegration, and ends with its dispersion into the food product. Typically, the mass transfer of a substance from plastic packaging to fluid or semisolid foods occurs due to direct contact (Lee et al., 2008).

15.3.1.2 Contact Migration

As the name suggests, the migration of compounds from packaging to food occurs by contact only. For example, it occurs when the additives, monomers, or plasticizers migrate from a plastic tray, pouch, or wrapper to the food (Alamri et al., 2021).

15.3.1.3 Gas-Phase Migration

With this kind of migration, the substance seeps through the inner layer of the packing material and through the exterior coating or printed layer of the package. A specific substance's mass is transferred through the channel of gas (Barnes et al., 2006).

15.3.1.4 Penetration Migration

In penetration migration, a substance from the packaging material's outer coated or printed layer migrates through the packaging material itself to the inner layer or contacting side. As the material enters the package's interior, it may move to the food within by contact or gas-phase migration (Barnes et al., 2006).

15.3.1.5 Set-Off Migration

It occurs when packaging films are stacked (like printed cartons). Inks, varnishes, and coatings present in packaging are transferred in large quantities from the outer printed side to the interior side (e.g., winding printed wrappers into a reel). Depending on the substance, the set-off movement may be either visible or invisible. Food that is packaged or wrapped could become contaminated by substances that are clinging to the inner side through set-off migration. These substances could easily spread by direct contact or gas-phase migration (Barnes et al., 2006).

15.3.1.6 Condensation/Distillation Migration

Although heat treatment of foods is used to increase their shelf stability, substances may transfer during the boiling or sterilizing operations of in-package food or food in cartons. Volatile compounds released from packaging or the condensed moisture from the steam produced from an aqueous food can migrate from packaging material to food or vice versa (Barnes et al., 2006).

15.4 Factors Affecting the Migration

15.4.1 Nature of Foods

The kind and composition of the food are important considerations when assessing migration. The media's pH is a major element influencing the amount of migration (Gomez-Gomez et al., 2020). The difference in migration may not only be caused by the higher acidity but also by increased viscosity or, in the case of juice samples, by the presence of more carbohydrates

(Metak et al., 2015). It was reported that the migration of nanoparticles such as Ag/Cu nanoparticles is stronger in acidic environments (Ahari and Lahijani, 2021). It has also been reported that foods with an excess of fat have high migration levels (Triantafyllou et al., 2007).

15.4.2 Mode of Contact

Numerous studies have shown that the sort of interaction (direct or indirect) that the food content has with the package affects how much migration occurs. Particularly, there is growing evidence to suggest that applying direct contact with food considerably increases migratory extent (Anderson and Castle, 2003).

15.4.3 Temperature

Heat speeds up chemical migration, just like it does for practically all other chemical and physical processes. Therefore, increasing the temperature will hasten migration. Boiling, sterilizing, microwaving, and even baking in the pack are just a few of the temperature settings used for storage, in addition to deep freezing, refrigeration, and ambient temperature. It is obvious that a material that works well in one application may not necessarily work well in another (Barnes et al., 2006). Both the rate and the scope of migration appear to be directly influenced by the temperature component. Temperature increases cause increased migration rates and a quicker return to equilibrium (Triantafyllou et al., 2005).

15.4.4 Nature of Packaging Material

The characteristics of the package strongly influence how much migration occurs. As an example, thinner packages correlate with greater migration rates (Nerin et al., 2007), although the number of recycled ingredients showed no association with migration (de Fátima Poças et al., 2011). It has also been demonstrated that the thickness greatly influences the migration rate.

15.4.5 Migrant Characteristics

Highly volatile materials exhibit a higher rate and extent of migration, which is often influenced significantly by the migrant's features (Jickells et al., 2005). Additionally, compared to low molecular-weight materials, large molecular-weight components have lower migration levels (Triantafyllou et al., 2005). Established that the migration level is influenced by the substance's microstructure. If one were to examine the phenomena from a molecular perspective, it would be clear that complicated molecular configurations (such as molecules that are spherically orientated or include side chains) exhibit less migration.

15.4.6 Migrant Concentration in Packaging

After a given amount of time, food containing a migrant with low molecular weight will unavoidably contain increased amounts of the chemical, making the migration from package to food easier to detect Alamri et al. (2021).

15.4.7 Size and Shape of Nanoparticles

Smaller nanoparticles tend to migrate more easily than larger nanoparticles due to their increased surface area. Similarly, the shape of the nanoparticles can affect their migration, with nanoparticles that are more elongated or needle-like being more likely to migrate than spherical nanoparticles.

15.4.8 Surface Chemistry

The surface chemistry of the nanoparticles can also play a role in their migration. For example, nanoparticles with a higher surface charge may be more likely to migrate because they can interact more strongly with the packaging material and the food or product being package.

15.4.9 Properties of the Packaging Material

The properties of the packaging material, such as its permeability, thickness, and surface area, can also affect the migration of nanoparticles. For example, a more porous packaging material may allow for greater migration of nanoparticles than a less-porous material.

15.4.10 Contact Time

The amount of time the nanoparticles are in contact with the packaging material and the food or product being packaged can also influence their migration. Materials that are suitable for brief contact intervals might not be appropriate for prolonged service times. The kinetics of migration are, roughly speaking, first order because the amount of migration grows in proportion to the square root of the contact time. The time (duration) of contact for common packaging can vary enormously (Barnes et al., 2006). According to experimental results, the concentration of the migratory molecule is inversely related to the contact time squared. In addition, there is data that suggests a linear correlation between the inverse of temperature and the log of the time to equilibrium (de Fátima Poças et al., 2011).

15.4.11 Chemical Composition of the Food or Product

The chemical composition of the food or product being packaged can affect the migration of nanoparticles, as certain substances may interact more strongly with the nanoparticles than others.

It is important to consider these factors when designing and evaluating packaging materials to minimize the potential for nanoparticle migration and ensure the safety of consumers. The characteristics of the NPs (such as solubility, molecular weight, particle size, and diffusivity), food conditions (polarity, compositions, and pH), environmental factors (mechanical stress and temperature), packaging characteristics (structure and viscosity of polymer, and distribution of the NPs within the matrix), the interaction between the material and the NPs, and contact time are some of the factors that affect the migration of NPs to the food product.

15.5 Diffusion Models and Their Terminologies

15.5.1 Fickian Diffusion

Fickian diffusion is typically used to describe the transport of small molecules such as oxygen, carbon dioxide, and water vapor through packaging materials. For example, Fick's law can be used to predict the rate of oxygen diffusion through a plastic film as a function of the oxygen concentration gradient, the thickness of the film, and the diffusion coefficient of oxygen in the film. Fickian diffusion is often used to design packaging materials that provide specific levels of barrier properties to oxygen and other gases, which can help extend the packaged food's shelf life.

The migration of nanomaterials from nanofood packaging to food is described by Fick's second law, which is as follows (Sablani et al., 2020).

$$\frac{\partial C_P}{\partial t} = D_p \frac{\partial^2 C_P}{\partial x^2} \tag{15.1}$$

$$\frac{\partial C_F}{\partial t} = D_F \frac{\partial^2 C_F}{\partial x^2} \tag{15.2}$$

where C_p and C_F are the concentration (mg/g) of the migrant in the packaging material and food phase; D_p and D_F are the diffusion coefficient (cm^2/s) of the migrant in the packaging and food phase; x is the distance (cm) between food and packaging material, and t is the time. The amount of packaging components that may migrate into solid or liquid food depends on the physio-chemical properties of the food and packaging. Migration is influenced by various factors such as temperature, duration, migrant concentration, molecular weight, solubility, diffusivity, partitioning coefficient between polymer and food, chemical nature of the food (polar or nonpolar), and polymer structures (density, crystallinity, and chain branching) (Sablani et al., 2020).

The active substance is normally contained inside the active layer of the packaging system, which is made up of a barrier layer as well. A steady state Fick's first law of diffusion can explain the release of active chemicals from packaging. The thickness, area, and permeability of the barrier layer are three factors that affect how quickly the active component diffuses in a reservoir system. The following mathematical model can be used to apply Fick's first rule of diffusion to the release of active substances from reservoir systems under the right circumstances (Wang et al., 2022).

$$\frac{dM}{dt} = \frac{ADK_d \Delta C}{x} \tag{15.3}$$

where A and x represent the surface area and thickness of the layers, respectively, and dM/dt represents the release rate of migrants at time t. D is the diffusivity of the migrating component, K_d is the distribution coefficient of the migrant between the film and environment, and C is the concentration of the ingredient in the matrix. The dM/dt is stable when D, x, K_d, and ΔC don't fluctuate, which causes the release rate of the active components to remain constant.

15.5.2 Non-Fickian Diffusion

Non-Fickian diffusion, as the name indicates, does not follow Fick's law of diffusion. Non-Fickian diffusion can also occur in food packaging, especially in porous materials such as paper and cardboard. The transport of small molecules through these materials can be affected by various factors, such as the porosity, tortuosity, and surface chemistry of the material. Non-Fickian diffusion can result in more complex transport behavior, such as anomalous diffusion or hindered diffusion, which can affect the rate and extent of migration through the packaging material. In a study, non-Fickian diffusion of amyl acetate in polypropylene packaging was investigated (Zaki et al., 2009). The diffusion coefficient, surface mass transfer coefficient, and their dependency on amyl acetate concentration and aging temperature were calculated using a non-Fickian model (Zaki et al., 2009) as follows:

$$\frac{M_t}{M_\infty} = 1 - \sum_{n=1}^{\infty} \frac{2R^2}{\beta_n^2\left(\beta_n^2 + R^2 + R\right)} \exp\left(-\frac{\beta_n^2 Dt}{x^2}\right) \tag{15.4}$$

Where β_n are positive roots of $\beta \tan(\beta) = R$ and $R = xH/D$ is a dimensionless parameter. x is the thickness of the sample, H is the surface mass transfer coefficient at the interface, and D is the diffusion coefficient. The equation (15.4) is also known as the sigmoid model.

Overall, the understanding and control of diffusion processes in food packaging are important for ensuring the safety and quality of packaged food. Fickian and non-Fickian diffusion models can be used to design and optimize packaging materials and systems to provide the desired barrier properties and performance.

15.5.3 Partition Coefficient

The partition coefficient $K_{P/F}$, where the indexes P and F refer to the polymer or plastic and the food, respectively, can be used to quantitatively explain the sorption process. The diffusion process is further influenced by the function of time, temperature, material thickness, size of migrating substances, chemical nature of polymer and food, and chemical content. The chemical equilibrium of the migrating compounds in food and FCMs is expressed by the partition coefficient (K). The partition coefficient ($K_{P/F}$) of component 'a' between a polymer and a food is expressed by (Mahmud et al., 2022)

$$K_{P/F(a)} = \frac{C_{P,a,\infty}}{C_{F,a,\infty}} \tag{15.5}$$

where $C_{F,a,\infty}$ and $C_{P,a,\infty}$ are the equilibrium concentration of the component "a" in the food and polymer, respectively. At equilibrium ($t = \infty$), the migrant concentration in the food phase matches the migrant concentration in the polymeric phase when $K_{P/F(a)} = 1$. When more migrant is absorbed into the polymer than in the food, $K_{P/F(a)}$ it is greater. In terms of food safety, a high $K_{P/F(a)}$ prevents migrants from migrating from packaging to food; on the other hand, a low $K_{P/F(a)}$ shows that more migrant is absorbed into the food.

A low $K_{P/F(a)}$ is preferable, nevertheless, to reduce taste loss in a package. (de Fátima Poças et al., 2011) Sated that an increase in paper weight results in a higher partition coefficient between food and paper. The partition coefficient is influenced by factors such as the nature and composition of the packaging material, the type of food being packaged, and the storage conditions (Tehrany and Desobry, 2004). Understanding the partition coefficient is important in food packaging because it helps in predicting the amount and rate of migration of substances from the packaging material to the food.

15.5.4 Solubility and Diffusion Equations

In the context of food packaging, the solubility and diffusivity equations are important in understanding the behavior of food additives, such as antioxidants, antimicrobials, and flavorings, in the packaging materials.

The solubility of food additives in the packaging material is an important factor in determining the shelf life of the packaged food. If the solubility is too high, the additives may migrate from the packaging material into the food, potentially causing changes in the food's sensory characteristics, such as taste, odor, and color. On the other hand, if the solubility is too low, the additives may not be effective in protecting the food from oxidation or microbial spoilage. Therefore, it is important to balance the solubility of food additives in the packaging material to ensure the desired shelf life of the packaged food. The Arrhenius Equation 15.6 is used to explain how temperature and the diffusion coefficients of active substances relate to one another as follows (Samsudin et al., 2020)

$$D_P = D_0 \exp\left(-\frac{E_D}{RT}\right) \tag{15.6}$$

The diffusion rate of nanoparticles from various packing materials, including low-density polyethylene, linear low-density polyethylene, high-density polyethylene, polypropylene, polyethylene terephthalate, and polystyrene, may be calculated using a variety of models. The model, which is based on the assumption of Fickian diffusion, uses the Stokes–Einstein connection to calculate the diffusion coefficient (D), which describes the mobility of a spherical particle in a fluid (Duncan and Pillai, 2015).

$$D = \frac{K_B T}{6\pi\eta\alpha} \tag{15.7}$$

Also, semiempirical models can be used to estimate the diffusion of a substance from polymers into a liquid medium. The printer model is widely used for food contact applications. According to this model, we can predict diffusion constant at a 95% confidence limit (Duncan and Pillai, 2015).

$$D_P = 10^4 \exp\left(A_P - 0.1351 M_r^{2/3} + 0.003 M_r - \frac{10454}{T}\right) \tag{15.8}$$

Where in equations 15.6, 15.7, and 15.8, D_P stands for the active compound's film diffusion coefficient; E_D is the activation energy of diffusion (J/mol), R is the gas constant (8.314 J/mol/K), K_B is the Boltzmann constant, η is the dynamic viscosity of the amorphous phase of the polymer, α is the particle radius, T is the temperature (K), and D_0 is a pre-exponential factor (cm^2/s), A_P polymer-specific parameter, and M_r migrant's molecular weight.

Based on migration data collected for small molecule migrants of different molecular weights, tested at different temperatures in a wide range of polymers, the given A_P values and the coefficients in equation (15.8).

In a study, Duncan and Pillai (2015) proposed that if the diffusion constant is known and the direction of diffusion of any single particle is random, therefore the "radius of diffusion" (r), which describes the largest distance traveled by a particle through the host polymer away from its initial position for a given time (t), can be calculated by the expression as follows (Duncan and Pillai, 2015):

$$r = 2\left(\frac{Dt}{\pi}\right)^{1/2} = \left(\frac{2K_BTt}{3\pi^2\eta\alpha}\right)^{1/2} \tag{15.9}$$

In summary, the solubility and diffusivity equations are important tools in designing food packaging materials that can provide the desired shelf life and quality of packaged food. The diffusion coefficient of various migrants is listed in Table 15.3.

15.6 Migration Modeling

There are several methods for mathematically simulating the chemical transfer from food contact articles (FCAs) into food.

15.6.1 Deterministic/Mechanistic

Deterministic or mechanistic models provide a formal explanation of the physical and chemical mechanics underpinning the phenomena of migration. The benefit of deterministic models is that they are typically easier to analyze mathematically. The deterministic method predominates in the academic literature on migration modeling. Diffusion is a mass transfer process that frequently controls migration and can be described by Fick's law (Brandsch et al., 2002; Franz, 2005; Brandsch et al., 2015; Poças et al., 2008). To reasonably anticipate migration and dissemination, data for the following two fundamental constants are required:

1) The migration's partition coefficient between the FCM and the food stuff and
2) The substance's coefficient of diffusion in the FCM (Brandsch et al., 2002).

We will briefly provide further information about the gravity models and models related to replicator dynamics from the deterministic models. Finally, here is a highlight of a few urn models that are relevant to research. The gravity model of migration is a

Table 15.3 Diffusion coefficient of selected chemical substances

Migrant	Packaging Material	Food/Simulants	Temperature (°C)	Pressure (MPa)	D_P (cm^2/sec)	Reference
Cinnamaldehyde	PLA/$scCO_2$/ci	Ethanol (10%)	40		3.5×10^{-10}	Sepulveda et al. (2020)
	PLA/NP/$scCO_2$/ci		40		1.1×10^{-10}	
	PLA/NPLA/$scCO_2$/ci		40		0.9×10^{-10}	
TiO_2 nanoparticles	LDPE	95% ethanol	40		4.3×10^{-9}	Bott et al. (2014)
Natural astaxanthin	Monolayer LDPE	95% ethanol	23		3.54×10^{-11}	Colín-Chávez et al. (2013)
			30		7.40×10^{-11}	
			40		32.67×10^{-11}	
	Two-layer LDPE/HDPE		10		1.53×10^{-11}	
			23		4.12×10^{-11}	
			30		5.59×10^{-11}	
Thymol	LLDPE	10% ethanol	40	7	7.5×10^{-9}	Torres et al. (2014)
			40	9	1.4×10^{-8}	
			40	12	1.8×10^{-8}	
		10% ethanol	40	7	2.4×10^{-8}	
			40	9	2.3×10^{-8}	
			40	12	3.0×10^{-8}	
Nisin	CS/PLLA (3 : 1)	NaCl solution	5		1.04×10^{-14}	Wang and Rhim (2015)
			25		5.76×10^{-14}	
			45		26.06×10^{-14}	
Nisin	CS/PLLA (2 : 1)	NaCl solution	5		1.64×10^{-14}	
			25		7.69×10^{-14}	

(Continued)

Table 15.3 (Continued)

Migrant	Packaging Material	Food/Simulants	Temperature (°C)	Pressure (MPa)	D_P (cm^2/sec)	Reference
			45		29.74×10^{-14}	
Nisin	CS/PLLA (1 : 1)	NaCl solution	5		3.05×10^{-14}	
			25		11.8×10^{-14}	
			45		40.17×10^{-14}	
Carvacrol	Polypropylene	Ethanol	40		1.20×10^{-10}	Ramos et al. (2014)
		Acetic acid	40		1.70×10^{-10}	
		Isooctane	20		9.4×10^{-10}	
Thymol		Ethanol	40		1.75×10^{-10}	
		Acetic acid	40		2.51×10^{-10}	
		Isooctane	20		5.9×10^{-10}	

theory of place-to-place migration that holds that interregional movement is inversely linked to the distance between origin and destination areas and directly connected to their populations. The model output is a single value of migration since the model variables are assumed to have single, constant values. The gravity model can be written as (Vitanov and Vitanov, 2016).

$$\ln(M_{ij}) = \ln(G) + \beta_1 \ln(E_i) + \beta_2 \ln(E_j) - \alpha \ln(x_{ij}) \tag{15.10}$$

M_{ij} is the migration from region i to j; G is a constant; E_i and E_j are the population of region i and j.

Mass transfer from a packaging material into food or food-like substances may be very complicated (Huang et al., 2015). Fick's law of diffusion is generally used for the determination of the migration of food materials or mass transfer. The migration from a monolayer packing in isothermal contact with a limited amount of food may be predicted using a theoretical model developed by Crank (1979), but it is not used for two or more-layer packaging. To determine the extent of the migration process from films, computational migration modeling was done using the commercial finite element method program COMSOL Multiphysics (Chemical Engineering Module). Using a commercial simulation tool, mass transfer during migration was simulated, and diffusion coefficients of nanoparticles involved in the process were calculated. Following the governing equation is (Soleimani et al., 2017)

$$\frac{\partial C_i}{\partial t} = \nabla \cdot (D_i \nabla C_i) + Z_i \tag{15.11}$$

$$N_i = -D_i \nabla C_i \tag{15.12}$$

C_i is the concentration of nanoparticles in the simulant; D_i is the diffusion coefficient; N_i is the mass flux; Z_i is for source term due to reaction; and i for the simulant. Using the COMSOL Multiphysics 4.4 (Chemical Engineering Module) program, equations (15.10) and (15.11) were solved for nanocomposite films to determine the amount of the migration process from films. Authors have also reported various analytical models to predict the amount of migration from packaging material into food (Table 15.4).

15.6.2 Empirical

Mathematical equations that generate a good fit with experimental observations on migration are the foundation of empirical models. These models are entirely mathematical and have no interest in the underlying mechanics of migration or the physical significance of the model constants. One empirical model to explain migration from certain FCMs is the Weibull model, which is based on the Weibull distribution function (Poças et al., 2008).

15.6.3 Stochastic

Probability distributions' mathematical functions are used in stochastic models. The discrete method, which makes use of agent-based models, is typically stochastic and considers cell-scale features. The continuum method, on the other hand, offers a global

Table 15.4 Summary of the mathematical models reported in the literature in order to predict migration

Model	Reference
$\frac{M_{F,t}}{M_{P,0}} = \frac{2}{L_p}\left(\frac{Dt}{\pi}\right)^{0.5}$ and $\frac{M_{F,t}}{M_{F,\infty}} = \frac{2}{L_p}\left(\frac{Dt}{\pi}\right)^{0.5}$, Where is $M_{F,t}$ the amount of migrants in the food at time t; $M_{p,0}$ is the initial amount of migrants in the packaging film; $M_{F,\infty}$ is the number of migrants in the food at equilibrium; D is the diffusion coefficient of the migrant in the packaging film, and L_p is the thickness of the packaging film.	Chung et al. (2010)
$\frac{m_{F,t}}{A} = C_{P,0}\ \rho_P d_P\left(\frac{\alpha}{1+\alpha}\right)\times\left[1-\sum_{n=1}^{\alpha}\frac{2\alpha(1+\alpha)}{1+\alpha+\alpha^2 q_n^2}\exp\left(-\frac{D_P t q_n^2}{d_P^2}\right)\right]$, $\alpha = \frac{1}{K_{P/F}}\times\frac{V_F}{V_P}$ where $m_{F,t}$ Is the mass of the migrant transferred from P into F after time t, (μg); A is the area of P in contact with F (cm^2); $C_{P,0}$ is the initial concentration of the migrant in P (mg/kg); ρ_P is the density of P (g/cm^3); t is the migration time (s); d_p is the thickness of P (cm); V_P is the volume of P (cm^3); V_F is the volume of F (cm^3); q_n is the positive root of the equation $\tan q_n = -\alpha.q_n$; D_P is the diffusion coefficient for the migrant in the polymer (cm^2/sec); $K_{P/F}$ partition coefficient for the migrant between P and F.	Rodríguez-Martínez et al. (2016)
$\frac{M_t}{A} = \frac{2\sqrt{D_s t}\cdot K\cdot C_0}{1+K\cdot\alpha}\left[\frac{1}{\sqrt{\pi}} - i\cdot\text{erfc}\left(\frac{L}{2\sqrt{D_p t}}\right)\right]; \alpha = \sqrt{D_s / D_P}$; $i\cdot\text{erfc}(x) = \frac{1}{\sqrt{\pi}}\exp(-x^2) - x\cdot\text{erfc}(x)$ K = partitioning coefficient, D_P and D_s = diffusivity of migrant in polymer and food phase (m^2/s); C_p and C_s = concentration of migrant in the polymer and food phase (mg/g); L = thickness of polymer (m); t = time (s)	Sablani et al. (2020)
$\left[\frac{1}{\pi} - \frac{1}{\alpha}\frac{M_{F,t}}{M_{p,0}}\right]^{0.5} = -\frac{D_P^{0.5}}{\alpha}\frac{t^{0.5}}{L_p} + \frac{1}{\pi^{0.5}}$; $M_{F,t}$ is the amount of nano-TiO_2 particles migrated to food-simulated solutions at t time (mg), $M_{P,0}$ is the initial content of nano-TiO_2 (mg), and L_P is the thickness of the composite film (cm).	Tang et al. (2020)
$\frac{t}{M_t} = K_1 + K_2 t$; M_t is the content of AgNPs migrated into food simulant at different exposure times, where t is the migration time, k_1 is related to the migration rate, and k_2 is related to the asymptotic value of nanoparticle migration at equilibrium ($M_\infty = 1/K_2$).	Zhu et al. (2021)

description of the migration at the population level and frequently models cell density as a set of deterministic partial differential equations (Gavagnin and Yates, 2018). A probability distribution function that can be utilized in stochastic models is the normal distribution (Poças et al., 2008).

15.6.4 Probabilistic

Probabilistic models consider both the likelihood of the model variables' occurrence as well as its variability and uncertainty. This also includes mixed effect models, which combine deterministic models with parameter variability. As a result, rather than being represented by a single value, the model constants and output are represented by a distribution of values. Probabilistic models employ techniques that spread knowledge about variation and uncertainty across the model. A common numerical simulation technique used in probabilistic models is the Monte Carlo approach, which is based on simulated random sampling. A further illustration of a numerical simulation technique is Latin hypercube sampling (Poças et al., 2008). Alternative approaches include molecular thermodynamics or molecular modeling, as presented by the works of (Nguyen et al., 2016).

15.7 Modeling Approaches for Different Packaging Materials

15.7.1 Plastic

The majority of studies on migration models utilize a deterministic strategy based on Fick's law of diffusion and concentrate on plastic packaging. Actual laboratory migration studies, which are frequently time consuming and expensive, can be supplemented by or partially replaced by migration modeling. As a result, it has gained popularity as a tool for regulators, businesses, and scholars to predict migration. To determine whether plastic FCMs abide by regulatory-specific migration constraints (SMLs), there are acknowledged deterministic migration models in the EU and the United States (Fang and Vitrac, 2017). These models use assumptions and simplifications to estimate migration in the worst-case scenario (i.e. the whole concentration of a particular chemical in a plastic FCM will migrate into the food). The primary presumptions are:

1) The migrant in the FCM is originally uniformly distributed (concentration must be specified),
2) Initially, there is no migrant present in the food,
3) For the exchange between the FCM and the food, there is no border resistance.
4) After transfer, the migrant is evenly dispersed throughout the foodstuff.
5) The total number of migrants in the FCM and the food quantity remain constant (no loss, no response) and
6) In food, the migrant is highly soluble.

The usual "worst-case" deterministic models have recently been modified in an effort to better reflect more plausible migration scenarios. These methods use stochastic and probabilistic modeling and consider the unpredictability and uncertainty of the mass transport parameters. Hence, the use of migration/diffusion modeling for the assessment of risk that has a minimal migration risk or creating plastic additives with low diffusivity could become more prevalent (Gavriil et al., 2018).

For a large number of chemicals, other methods use migration modeling to forecast consumer exposure to chemical migrants from FCMs (Ernstoff et al., 2017). To assess dietary exposure to FCCs, food intake data can be integrated with mechanistic or empirical migration models. Such high-throughput (HT) techniques are compatible with other screening and prioritization techniques like life cycle assessment and high-throughput risk-based screening (HTRS) and can be used as screening-level tools for prioritizing compounds for closer examination (HTRS). But they are linked to a higher degree of scientific ambiguity.

15.7.2 Paper and Board

Understanding the variables affecting the migration of chemicals from paper and board FCMs and creating mathematical models for them have received far less attention. This is a result of these materials' inhomogeneity, which makes modeling challenging. Nonetheless, a number of scientific papers addressing this issue have appeared recently (Nguyen et al., 2021).

Because the mechanistic diffusion model cannot effectively represent the mass transfer processes taking place in these heterogeneous, porous, fiber-based materials, its relevance to paper and board is restricted. Nonetheless, attempts to successfully modify the traditional diffusion model to account for observed migration from certain paper and board FCMs into food have been attempted (Barnkob and Petersen, 2013).

It has been observed that the empirical Weibull model accurately captures the form of empirically obtained chemical migration curves in paper and board FCMs (Cai et al., 2017). The model can describe "a number of mass transfer processes within the food engineering domain and in food packaging systems, in circumstances where diffusion theories could not adequately explain the mass transfer process" due to its simplicity and flexibility.

15.7.3 Other Materials

The scientific literature on migration modeling strategies for other FCMs like ceramics or printing inks is much less extensive. When it comes to the former, the mechanistic diffusion model has been used to forecast harmful metal migration into acidic meals (Lago et al., 2019). The migration from ceramics at room temperature has been shown to be well described by or somewhat underestimated by the diffusion model. The latter has also seen the diffusion model applied and modified (Aparicio and Elizalde, 2015; Wang and Rhim, 2015). For compounds that moved from printing inks on paper and board packaging into food, there was good agreement between predictions from the diffusion model and the experimental evidence.

15.8 Migration Modeling Software

The scientific literature has mentioned user-friendly migration modeling software., for example MIGRATEST Lite 2000/2001 (Fabes GmbH, Germany), AKTS-SML (Advanced Kinetics Technology Solution AG, Switzerland), and (SFOPH), SMEWISE

(Simulation of Migration Experiments with Swelling Effect), MULTITEMP, MULTIWISE, the Safe Food Packaging Portal, and SFPP3 (INRA) by France and FMEC Aengine (Failure Mode Effects and Criticality Analysis) developed for migration analysis. The additives, flavors, and Food Contact Materials Exposure Task (FACET) model is a further tool for the probabilistic modeling of user exposure to FCCs from the primary package (Maia et al., 2016).

The migration of chemicals from FCMs into food can be predicted using mathematical modeling. It can, in some cases, take the role of lab migration experiments, saving time and money. To create accurate predictions and understand the results, migration models should only be utilized by qualified professionals with in-depth knowledge of chemical migration. Moreover, migration modeling makes assumptions and simplifies information about the FCM, the chemical migrant, and the migration process. As a result, it may overestimate or underestimate actual migration. To model its transmission from the FCM to the food, the migratory chemical and its initial concentration in the packaging must also be determined. As a result, these techniques are inadequate for evaluating the migration of unidentified nonintentionally added substances. Moreover, migration modeling can be useful to the chemical industry and plastic converters as beginning concentrations are frequently unavailable to package end-users.

Most of the migration modeling up to this point has been on plastic FCMs. To demonstrate compliance with regulatory SMLs, "worst-case" migration modeling—in which the complete amount of a given chemical will migrate from the FCM into food—can be utilized. The assumption of "worst-case" scenarios may not necessarily result in an overestimation of actual migration, and it does not reflect true migration scenarios (Ernstoff et al., 2017). Since probabilistic models are more complex and provide more precise migration estimates, they can be utilized for risk assessment and the creation of safe packaging (Fang and Vitrac, 2017). Only one chemical can be subjected to "worst-case" and probabilistic diffusion models at once. HT models provide a method to simultaneously screen a large number of chemicals, estimate exposure levels, and rank substances for further investigation. Unfortunately, these methodologies are inappropriate for the evaluation of single compounds due to the numerous uncertainties connected with them (Ernstoff et al., 2017).

To create reputable migration models, more research on paper, board, and other materials is required. The deterministic diffusion model created for plastics has been used and modified in the majority of studies on modeling migration from nonplastic materials to date. It has been demonstrated that the empirical Weibull model may predict chemical migration from paper and board FCMs with promising outcomes. Empirical models, however, fail to clarify the underlying migration mechanisms.

Nanomaterials are another area in particular where diffusion-based migration modeling has limited application. In contrast to experimental studies, which have also noted the migration of considerably bigger nanoparticles from FCMs, diffusion modeling only predicts migration for small nanoparticles (up to a few nm) (Jokar et al., 2017; Störmer et al., 2017). Diffusion is probably not as essential a migratory mechanism for nanoparticles as desorption, dissolution, and polymer degradation. Thus, it is necessary to create new mathematical models that take these mechanisms into account (Jokar et al., 2017).

15.9 Migratory Effects of Nanoparticles on Food

The migratory effects of nanoparticles on food depend on various factors such as the size, shape, surface charge, and chemical composition of the nanoparticles, as well as the type of food and the conditions of exposure.

Nanoparticles can potentially migrate from food packaging materials or food contact surfaces into the food itself and then be ingested by consumers. This could have both positive and negative effects on food quality and safety (Ashfaq et al., 2022).

On the positive side, nanoparticles can be used as antimicrobial agents, antioxidants, and nutrient delivery systems in food. For example, silver nanoparticles can be used to inhibit the growth of bacteria and fungi in food (Zorraquín-Peña et al., 2020), while iron nanoparticles can be used to fortify food with essential nutrients (Kumari and Chauhan, 2022). ZnO nanoparticles can be used as an antimicrobial material (Garcia et al., 2018).

On the negative side, some nanoparticles may have toxic effects on human health, especially when ingested in high doses or for prolonged periods. Nanoparticles can potentially damage DNA, disrupt cellular function, and cause inflammation in the body (Nile et al., 2020).

Therefore, it is important to carefully evaluate the potential risks and benefits of using nanoparticles in food and food packaging and to ensure that any potential migration of nanoparticles into food is kept at safe levels. Regulatory agencies around the world are working to establish guidelines and standards for the use of nanoparticles in food and to monitor their potential effects on human health and the environment.

15.10 Migration Tests

European Commission (2011) defines a migration test as the measurement of a substance's release from a material or object into food or a food substitute. Food simulants are test materials that replicate food behavior, including migration from items that come into touch with food. There are six distinct food imitators: vegetable oil, poly(2,6-diphenyl-p-phenylene oxide), ethanol 10% v/v, acetic acid 3% v/v, ethanol 20% v/v, and ethanol 50% v/v. After a contact period, the examination of migrating chemicals is done in the simulant media by quantification or identification using an analytical method like spectroscopy or chromatography.

The extraction capacity of the molecule under research has a significant impact on how compounds migrate from packaging materials to meals. This feature may be quantified by exposing the packing material to food at a certain temperature and time and measuring the amount of chemical that migrates to the food (Arvanitoyannis and Stratakos, 2011). However, figuring out migration in actual circumstances is rather challenging. The majority of the time, migration situations are simulated for this purpose. For instance, migration studies are conducted for a shorter period of time at a higher temperature for foods that are kept for a long period of time. Additionally, to facilitate the experiments, standard chemicals known as food simulants are utilized (instead of foods) (Stoffers, 2005).

15.10.1 Food Simulants

Migration experiments are typically carried out utilizing FSs because food chemical and physical structures are so complex. The latter are chemicals that are either liquid or solid and have a straightforward and well-known composition, making testing simpler and more easily read. Several other FSs have been employed sometimes, and some of them are currently recognized as official simulants by-laws that must be followed (Lee et al., 2008). The FS listed above is categorized in accordance with EU Directive 85/572 and Grob (2008) on the grounds that they are thought to share characteristics with one or more food categories. Even though the FS is divided into a number of groups (see the paragraphs that follow), there are some foods that combine fatty and aqueous foods.

15.10.1.1 Water

Water is widely used in the preparation of aqueous foods with a pH greater than 4.5. This category of foods also includes alcoholic beverages and nonalcoholic beverages with an alcohol content of less than 5% vol. Examples include water, cider, fruit or vegetable juices of average or concentrated strength, musts, fruit nectars, lemonades, and mineral waters, as well as syrups, liquid chocolate, bitters, coffee, tea, beers, pastry, cakes, and other freshly baked goods, moist, honey, molasses, and sugar syrups, fruits that have been processed, vegetables that have been pureed, and fish.

15.10.1.2 Acetic Acid

Acidic aqueous foods with a pH below 4.5, such as vinegar, alcoholic beverages, and nonalcoholic beverages with low alcohol content, such as yogurt, buttermilk, and similar foods with or without fruit and fruit products, are frequently treated with an acetic acid solution in water at a concentration of 3%.

15.10.1.3 Ethanol

Aqueous ethanol at a 10% concentration is frequently used to make alcoholic beverages. As long as the alcoholic grade of the food is greater than 10% v/v, the concentration of the solution is adjusted to that level. As an illustration, alcoholic beverages with an alcohol strength of at least 5% vol can be imitated using aqueous ethanol (liqueurs, wines, etc.).

15.10.1.4 Oil

Olive oil, other oils, like sunflower oil, and synthetic triglyceride combinations are frequently used to make fatty dishes. This category includes fats and oils originating from both animals and plants, a variety of sauces (such as mayonnaise and mustard), milk, processed meat products, and items in the oily medium.

The most recent EU rule (10/2011) for plastic packaging materials mentions three categories of FS (Table 15.5), with the exception of water and the addition of various ethanolic solution concentrations.

15.10.2 Specific Migration Testing

A sample of the substance under research (such as a polymer, metal, etc.) comes into touch with the food simulant under controlled settings (temperature, duration) after the food

Table 15.5 List of food simulants foode

Food Simulant	Abbreviation
Ethanol 10% (v/v)	Food simulant A
Acetic acid 3% (w/v)	Food simulant B
Ethanol 20% (v/v)	Food simulant C
Ethanol 50% (v/v)	Food simulant D_1
Vegetable oil[a]	Food simulant D_2
Poly (2,6-diphenyl-p-phenylene oxide), particle size 60–80 mesh, pore size 200 nm	Food simulant E

[a] This may be any vegetable oil with a fatty acid distribution of food
Source: European Commision (2011)

Table 15.6 Conditions of contact when using food simulants

Contact Time		Contact Temperature	
Contact Time in Worst Foreseeable Use	**Test Time**	**Conditions of Contact in Worst Foreseeable Use**	**Test Temperature**
$t \leq 5$ min	5 min	$T \leq 5$ °C	5 °C
5 min < $t \leq 0.5$ h	0.5 h	5 °C < $T \leq 20$ °C	20 °C
0.5 h < t ≤ 1 h	1 h	40 °C < $T \leq 70$ °C	40 °C
1 h < $t \leq 2$ h	2 h	20 °C < $T \leq 40$ °C	70 °C
2 h < $t \leq 6$ h	6 h	70 °C < $T \leq 100$ °C	100 °C or reflux temperature
6 h < $t \leq 24$ h	24 h	100 °C < $T \leq 121$ °C	121 °C[a]
1 d < $t \leq 3$ d	3 d	121 °C < $T \leq 130$ °C	130 °C[a]
Above 30 d	See specific conditions	150 °C < T < 175 °C	175 °C[a]
		T > 175 °C	Adjust the temperature to the real temperature at the interface with the food[a]

[a] Only the food simulants D_2 and E may be used at this temperature. Migration testing under pressure at the appropriate temperature may be done for applications heated under pressure. A test at 100 °C or at reflux temperature for a length of four times the time determined in accordance with the contact time criteria may be used to replace the test for food simulants A, B, C, or D_1
Source: EU (2011)

simulant has been chosen based on the food under inspection. For conducting the migration test, specific time and temperature requirements and testing conditions are recommended by the European Union (Tables 15.6 and 15.7). The gravimetric test is the technique that is used the most commonly to analytically evaluate migration. To calculate the percentage of solid component migration by difference, the simulant is specifically removed

Table 15.7 Standardized testing conditions

Test Number	Contact Time in Days (d) or Hours (h) at Contact Temperature in (°C)	Intended Food Contact Conditions
OM1	10 d at 20 °C	Any food contact at frozen and refrigerated conditions.
OM2	10 d at 40 °C	Any prolonged storage at room temperature or lower, as well as short-term heating of up to 100 °C for up to 15 min or up to 70 °C for up to 2 h.
OM3	2 h at 70 °C	Any contact conditions that do not involve long-term storage at room temperature or a refrigerator, such as heating to 70 °C for up to two hours or 100 °C for up to fifteen minutes.
OM4	1 h at 100 °C	All food simulants can be used at high temperatures of up to 100 °C.
OM5	2 h at 100 °C or at reflux or alternatively 1 h at 121 °C	High-temperature applications up to 121 °C.
OM6	4 h at 100 °C or at reflux	Any circumstances where food is in touch with food simulants *A*, *B*, or *C* at a temperature higher than 40 °C.
OM7	2 h at 175 °C	Applications involving fatty foods at elevated temperatures that go beyond OM5 conditions.

(in the case of volatile simulants) or evaporated (in the case of oils), and the remaining substance, or the sample material after the contact, is weighted. The total migration can be estimated using the following equation:

$$OM = W_1 - W_2 \tag{15.13}$$

Where W_1: weight of the sample before the contact
W_2: weight of the sample after the contact

15.11 Toxicological Effect of Nanoparticles on Human Health

Due to the potential migration of NPs from FP into food products and the potentially harmful consequences on consumer health that might result from ingesting and being exposed to NPs, its broad commercialization must overcome significant obstacles (Enescu et al., 2019). NP migration into food or food simulants from polymer composites for FP has been evaluated utilizing European and U.S. (FDA) standard migration assays, and European Regulation (European Commission, 2011) has established limits. Plastic materials and items are not permitted to discharge more than 10 mg/dm^2 of total components per dm^2 of food contact surface from plastic materials and products to food simulants. Plastic products and materials that are meant to come into contact with food intended for infants and young children—as defined by Commission Directives 2006/141/EC (1) and 2006/125/EC (2)—must not transfer constituents to food simulants in amounts greater than 60 mg of total constituents released per kg of food simulant (European Commision, 2011). The only alarming awareness and certainty regarding human exposure to NPs is that they are not typically consumed and metabolized. Ironically, while the use of nanotechnology in FP presents new opportunities for ingesting higher-quality and safer foods, it may also introduce new risks to both human health and the environment (Enescu et al., 2019). It is not yet entirely understood what health risks are involved with eating food that contains nanoscale chemicals that have leached from the packaging. The fact that smaller particles are often absorbed more quickly and easily means that they might possibly cause harm to cells and tissues by releasing reactive oxygen species (ROS) or by either direct or indirect toxicity (Han et al., 2011). Generally speaking, human and/or rodent cell lines from the gut, liver, lung, and skin, as well as plant cells, have been used to study basal cytotoxicity and its processes. The most often used cytotoxicity assays nowadays are the lactate dehydrogenase release assay, live-dead assay, cell counting, Alamar blue assay, neutral red absorption, protein content, and trypan blue dye exclusion. Through the observation of changes in many indicators, including ROS generation, glutathione levels, inflammatory response, DNA damage, and cell death, the toxicity processes have been examined. Ames test, Comet assay, and micronucleus essay are a few more genotoxicity tests that have been employed.

Multiwalled Carbon Nanotubes (CNTs), when injected into an animal's peritoneal cavity, induced peritoneal mesothelioma, according to studies done on mice and rats (Fukushima et al., 2018). Although research conducted in vitro and in vivo has shown that TiO_2NPs accumulate in mammalian tissues and are also removed extremely slowly, the data regarding their accumulation and toxicity, even when given in large dosages, are contradicting. To be able to draw conclusions about the danger of oral exposure to TiO_2NPs, it is necessary to have trustworthy in vivo toxicokinetic data (Enescu et al., 2019).

In the area of SiNPs, some studies on SiO_2NPs revealed that they may generate ROS and accumulate in the liver at levels that pose a risk to health, but other studies that administered SiO_2NPs to rats for 13 weeks found no accumulation or toxicity (Enescu et al., 2019). Based on the current data, no definitive conclusions can be made regarding SiO_2NPs. In an in vitro model, physiological dosages of SiO_2NPs were examined to assess their effects on gastrointestinal health and function. Brush border membrane damage and unfavorable effects in cells treated for 4 hours and 5 days were observed (Guo et al., 2018). Similar to SiNPs, FeO, and Fe_2O_3 have the potential to be hazardous due to their capacity to create ROS, although investigations, including their oral administration to male or female rats over a 13-week period, have shown little accumulation in tissues and no toxicity.

15.12 Safety Regulation

Both so-called horizontal law and vertical legislation are capable of regulating nanotechnologies. Despite not having a stated goal, the current horizontal law is comprehensive and includes aspects of nanotechnologies. The terminology employed in vertical legislation makes it more appropriate to problems faced by users of nanotechnologies since it is expressly intended to regulate nanotechnologies and sections of the industry likely to use them. Vertical law for nanotechnologies is very new compared to horizontal regulation and was nonexistent until a few years ago. Chemical safety regulations apply to the migration of nanoparticles from food packaging materials into food, which is a significant safety risk. Only carbon black, titanium dioxide, and silicon dioxide have been approved for use in plastic food packaging under current EU regulations. All substances used in food packaging that come into contact with food or have the potential to do so are governed by European rules (Paidari et al., 2021). There is continuing study into the possible health hazards brought on by the migration of nanoparticles into food. The issue of environmental effect, primarily on the packaging material's manufacture but more importantly on its use and disposal. The efficacy of current recycling systems and whether incorporating nanomaterials might jeopardize them is a special concern. These uncertainties in consumer safety and environmental safety limit regulation and market adoption. To ensure the health and safety of customers and the environment, regulations are in place for packaging materials and their ingredients. Consumer and environmental safety issues limit regulation and market adoption.

In the USA, The Food and Drug Administration (FDA) is the principal agency in charge of setting guidelines and enforcing regulations to guarantee food safety in the United States. FDA has not yet created any specifications or rules for goods based on nanotechnology. However, the FDA has released a set of recommendations for the sector on food safety, titled "Considering Whether an FDA-Regulated Product Involves the Application of Nanotechnology." The FDA's guidelines addressed the legal situation, safety procedures, and health impacts of using nanotechnology in food products (USFDA, 2014).

The FSSAI has made a significant contribution to the revolution in India's food safety standards. As the primary authority on food safety and regulations in India, the Food Safety and Standards Authority of India (FSSAI), it is essential to create stringent laws on the use of nanomaterials in foods in order to limit their usage. Despite the fact that the Indian Department of Science and Technology (DST) has published recommendations for using nanomaterials safely in research, businesses, and laboratories (Mishra et al., 2019).

To regulate and track food safety "from farm to table," the EU Commission produced the White Paper on Food Safety in January 2000. Regulation (EC) No. 178/2002, which created the European Food Safety Authority (EFSA) and the General Food Law, was subsequently enacted by the EU Council and the European Parliament in 2002.

15.13 Final Remarks

One of the many issues relating to food safety and preservation can be resolved using novel and cutting-edge techniques, such as nanofood packaging. Nanofood packaging, in contrast to other traditional packaging, has the ability to eradicate, suppress, or delay the growth of a number of food pathogen species without degrading the food's quality or appearance. Future-focused nano packaging includes nano preservatives for food, nano sensors, nanocoating, antibacterial nanoparticles, and nano-delivery systems. Nanomaterial migration from packaging poses a risk to people's health. Understanding the migration of nanoparticles from various packaging materials into food requires research. An appropriate mathematical model can predict migration and also can be employed to quantify the effect of various internal and external stimuli on the migration process. It will also assist in the decision-making process for migration-related food safety.

References

Abdollahzadeh, E., Mahmoodzadeh Hosseini, H., & Imani Fooladi, A. A. (2018). Antibacterial activity of agar-based films containing nisin, cinnamon EO, and ZnO nanoparticles. *Journal of Food Safety*, *38*(3). https://doi.org/10.1111/jfs.12440

Ahari, H., & Lahijani, L. K. (2021). Migration of silver and copper nanoparticles from food coating. *Coatings*, *11*(4), 380. https://doi.org/10.3390/coatings11040380

Alamri, M. S., Qasem, A. A. A., Mohamed, A. A., Hussain, S., Ibraheem, M. A., Shamlan, G., Alqah, H. A., & Qasha, A. S. (2021). Food packaging's materials: a food safety perspective. *Saudi Journal of Biological Sciences*, *28*(8), 4490–4499. https://doi.org/10.1016/j.sjbs.2021.04.047

Anderson, W. A. C., & Castle, L. (2003). Benzophenone in cartonboard packaging materials and the factors that influence its migration into food. *Food Additives and Contaminants*, *20*(6), 607–618. https://doi.org/10.1080/0265203031000109486

Aparicio, J. L., & Elizalde, M. (2015). Migration of photoinitiators in food packaging: a review. *Packaging Technology and Science*, *28*(3), 181–203. https://doi.org/10.1002/pts.2099

Arvanitoyannis, I., & Stratakos, A. (2011). Migration from food packaging materials in technologies of preservation of food and food pac. *kaging (in Greek)*. University Studio Press, Thessaloniki, Hellas (Greece).

Ashfaq, A., Khursheed, N., Fatima, S., Anjum, Z., & Younis, K. (2022). Application of nanotechnology in food packaging: pros and cons. *Journal of Agriculture and Food Research*, *7*, 100270. https://doi.org/10.1016/j.jafr.2022.100270

Barnes, K., Sinclair, R., & Watson, D. (2006). *Chemical Migration and Food Contact Materials*. Woodhead Publishing.

Barnkob, L. L., & Petersen, J. H. (2013). Effect of relative humidity on the migration of benzophenone from paperboard into the food simulant Tenax® and modelling hereof. *Food Additives & Contaminants: Part A*, *30*(2), 395–402. https://doi.org/10.1080/19440049.2012.741717

Bott, J., Störmer, A., & Franz, R. (2014). A model study into the migration potential of nanoparticles from plastics nanocomposites for food contact. *Food Packaging and Shelf Life*, *2*(2), 73–80. https://doi.org/10.1016/j.fpsl.2014.08.001

Brandsch, J., Mercea, P., Rüter, M., Tosa, V., & Piringer, O. (2002). Migration modelling as a tool for quality assurance of food packaging. *Food Additives & Contaminants*, *19*(sup 1), 29–41. https://doi.org/10.1080/02652030110058197

Brandsch, R., Dequatre, C., Mercea, P., Milana, M. R., Störmer, A., Trier, X., Vitrac, O., Schaefer, A., & Simoneau, C. (2015). In E. J. Hoekstra (Ed.), *Practical Guidelines on the Application of Migration Modelling for the Estimation of Specific Migration* (p. 40). Publications Office.

Bumbudsanpharoke, N., & Ko, S. (2015). Nano-food packaging: an overview of market, migration research, and safety regulations. *Journal of Food Science*, *80*(5). https://doi.org/10.1111/1750-3841.12861

Cai, H., Ji, S., Zhang, J., Tao, G., Peng, C., Hou, R., Zhang, L., Sun, Y., & Wan, X. (2017). Migration kinetics of four photo-initiators from paper food packaging to solid food simulants. *Food Additives & Contaminants: Part A*, *34*(9), 1632–1642. https://doi.org/10.1080/19440049.2017.1331470

Cao, W., Yan, J., Liu, C., Zhang, J., Wang, H., Gao, X., Yan, H., Niu, B., & Li, W. (2020). Preparation and characterization of catechol-grafted chitosan/gelatin/modified chitosan-AgNP blend films. *Carbohydrate Polymers, 247*, 116643. https://doi.org/10.1016/j.carbpol.2020.116643

Chaudhary, P., Fatima, F., & Kumar, A. (2020). Relevance of nanomaterials in food packaging and its advanced future prospects. *Journal of Inorganic and Organometallic Polymers and Materials, 30*(12), 5180–5192. https://doi.org/10.1007/s10904-020-01674-8

Chaudhry, Q., Watkins, R., & Castle, L. (2017). *Nanotechnologies in Food: What, Why and How?* (pp. 1–19). *Chemical Society Reviews.*

Chi, H., Song, S., Luo, M., Zhang, C., Li, W., Li, L., & Qin, Y. (2019). Effect of PLA nanocomposite films containing bergamot essential oil, TiO_2 nanoparticles, and Ag nanoparticles on shelf life of mangoes. *Scientia Horticulturae, 249*, 192–198. https://doi.org/10.1016/j.scienta.2019.01.059

Chowdhury, S., Teoh, Y. L., Ong, K. M., Rafflisman Zaidi, N. S., & Mah, S.-K. (2020). Poly(vinyl) alcohol crosslinked composite packaging film containing gold nanoparticles on shelf life extension of banana. *Food Packaging and Shelf Life, 24*, 100463. https://doi.org/10.1016/j.fpsl.2020.100463

Chung, D., Papadakis, S. E., & Yam, K. L. (2010). Simple models for assessing migration from food- packaging films. *Food Additives & Contaminants, 19*(6), 611–617. https://doi.org/10.1080/0265203021012638

Colín-Chávez, C., Soto-Valdez, H., Peralta, E., Lizardi-Mendoza, J., & Balandrán-Quintana, R. (2013). Diffusion of natural astaxanthin from polyethylene active packaging films into a fatty food simulant. *Food Research International, 54*(1), 873–880. https://doi.org/10.1016/j.foodres.2013.08.021

Crank, J. (1979). *The Mathematics of Diffusion*. Oxford University Press.

Deng, J., Ding, Q. M., Li, W., Wang, J. H., Liu, D. M., Zeng, X. X., Liu, X. Y., Ma, L., Deng, Y., Su, W., & Ye, B. (2020). Preparation of nano-silver-containing polyethylene cxomposite film and Ag ion migration into food-simulants. *Journal of Nanoscience and Nanotechnology, 20*(3), 1613–1621. https://doi.org/10.1166/jnn.2020.17346

Dey, A., Dhumal, C. V., Sengupta, P., Kumar, A., Pramanik, N. K., & Alam, T. (2021). Challenges and possible solutions to mitigate the problems of single-use plastics used for packaging food items: a review. *Journal of Food Science and Technology, 58*(9), 3251–3269. https://doi.org/10.1007/s13197-020-04885-6

Divya, K., Smitha, V., & Jisha, M. S. (2018). Antifungal, antioxidant and cytotoxic activities of chitosan nanoparticles and its use as an edible coating on vegetables. *International Journal of Biological Macromolecules, 114*, 572–577. https://doi.org/10.1016/j.ijbiomac.2018.03.130

Duncan, T. V., & Pillai, K. (2015). Release of engineered nanomaterials from polymer nanocomposites: diffusion, dissolution, and desorption. *ACS Applied Materials & Interfaces, 7*(1), 2–19. https://doi.org/10.1021/am5062745

Echegoyen, Y., Rodríguez, S., & Nerín, C. (2016). Nanoclay migration from food packaging materials. *Food Additives & Contaminants: Part A, 33*(3), 530–539. https://doi.org/10.1080/19440049.2015.1136844

Egbuna, C., Parmar, V. K., Jeevanandam, J., Ezzat, S. M., Patrick-Iwuanyanwu, K. C., Adetunji, C. O., Khan, J., Onyeike, E. N., Uche, C. Z., Akram, M., Ibrahim, M. S., El Mahdy, N. M., Awuchi, C. G., Saravanan, K., Tijjani, H., Odoh, U. E., Messaoudi, M., Ifemeje, J. C., Olisah, M. C., Ezeofor, N. J., Chikwendu, C. J., & Ibeabuchi, C. G. (2021). Toxicity of nanoparticles in biomedical application: nanotoxicology. *Journal of Toxicology, 2021*, 1–21. https://doi.org/10.1155/2021/9954443

Enescu, D., Cerqueira, M. A., Fucinos, P., & Pastrana, L. M. (2019). Recent advances and challenges on applications of nanotechnology in food packaging. A literature review. *Food and Chemical Toxicology, 134*(May), 110814. https://doi.org/10.1016/j.fct.2019.110814

Ernstoff, A. S., Fantke, P., Huang, L., & Jolliet, O. (2017). High-throughput migration modelling for estimating exposure to chemicals in food packaging in screening and prioritization tools. *Food and Chemical Toxicology, 109*, 428–438. https://doi.org/10.1016/j.fct.2017.09.024

EU (2011). Commission regulation (EU) No 10/2011 of 14 January 2011 on plastic materials and articles intended to come into contact with food. *Official Journal of the European Union, 22*(1).

European Commision (2011). Commission Regulation (EU) No 10/2011 on plastic materials and articles intended to come into contact with food. *Official Journal of the European Union, L(12)*, 1–89.

European Commission (2011). COMMISSION REGULATION (EU) No 10/2011 on plastic materials and articles intended to come into contact with food. *Official Journal of the European Union, L*(12), 1–89.

Fang, X., & Vitrac, O. (2017). Predicting diffusion coefficients of chemicals in and through packaging materials. *Critical Reviews in Food Science and Nutrition, 57*(2), 275–312. https://doi.org/10.1080/10408398.2013.849654

de Fátima Poças, M., Oliveira, J. C., Pereira, J. R., Brandsch, R., & Hogg, T. (2011). Modelling migration from paper into a food simulant. *Food Control, 22*(2), 303–312. https://doi.org/10.1016/j.foodcont.2010.07.028

Ferrara, G., Bertoldo, M., Scoponi, M., & Ciardelli, F. (2001). Diffusion coefficient and activation energy of Irganox 1010 in poly(propylene-co-ethylene) copolymers. *Polymer Degradation and Stability, 73*(3), 411–416. https://doi.org/10.1016/S0141-3910(01)00121-5

Franz, R. (2005). Migration modelling from food-contact plastics into foodstuffs as a new tool for consumer exposure estimation. *Food Additives and Contaminants, 22*(10), 920–937. https://doi.org/10.1080/02652030500157700

Fukushima, S., Kasai, T., Umeda, Y., Ohnishi, M., Sasaki, T., & Matsumoto, M. (2018). Carcinogenicity of multi-walled carbon nanotubes: challenging issue on hazard assessment. *Journal of Occupational Health, 60*(1), 10–30. https://doi.org/10.1539/joh.17-0102-RA

Garcia, C. V., Shin, G. H., & Kim, J. T. (2018). Metal oxide-based nanocomposites in food packaging: applications, migration, and regulations. *Trends in Food Science & Technology, 82*, 21–31. https://doi.org/10.1016/j.tifs.2018.09.021

Gavagnin, E., & Yates, C. A. (2018). Stochastic and Deterministic Modeling of Cell Migration. In S. R. S. Rao, & C. R. Rao (Eds.), *Handbook of Statistics* (Volume 39, pp. 37–91). Elsevier. https://doi.org/10.1016/bs.host.2018.06.002

Gavriil, G., Kanavouras, A., & Coutelieris, F. A. (2018). Food-packaging migration models:a critical discussion. *Critical Reviews in Food Science and Nutrition, 58*(13), 2262–2272. https://doi.org/10.1080/10408398.2017.1317630

Ghorbanpour, M., Bhargava, P., Varma, A., & Choudhary, D. K. (Eds.). (2020). *Biogenic Nano-Particles and their Use in Agro-ecosystems*. Springer, Singapore. https://doi.org/10.1007/978-981-15-2985-6

Gomez-Gomez, B., Perez-Corona, M. T., & Madrid, Y. (2020). Using single-particle ICP-MS for unravelling the effect of type of food on the physicochemical properties and gastrointestinal stability of ZnONPs released from packaging materials. *Analytica Chimica Acta*, *1100*, 12–21. https://doi.org/10.1016/j.aca.2019.11.063

Grob, K. (2008). The future of simulants in compliance testing regarding the migration from food contact materials into food. *Food Control*, *19*(3), 263–268. https://doi.org/10.1016/j.foodcont.2007.04.001

Guo, Z., Martucci, N. J., Liu, Y., Yoo, E., Tako, E., & Mahler, G. J. (2018). Silicon dioxide nanoparticle exposure affects small intestine function in an in vitro model. *Nanotoxicology*, *12*(5), 485–508. https://doi.org/10.1080/17435390.2018.1463407

Han, W., Yu, Y., Li, N., & Wang, L. (2011). Application and safety assessment for nano-composite materials in food packaging. *Chinese Science Bulletin*, *56*(12), 1216–1225. https://doi.org/10.1007/s11434-010-4326-6

Honarvar, Z., Hadian, Z., & Mashayekh, M. (2016). Nanocomposites in food packaging applications and their risk assessment for health. *Electronic Physician*, *8*(6), 2531–2538. https://doi.org/10.19082/2531

Huang, J.-Y., Li, X., & Zhou, W. (2015). Safety assessment of nanocomposite for food packaging application. *Trends in Food Science & Technology*, *45*(2), 187–199. https://doi.org/10.1016/j.tifs.2015.07.002

Indumathi, M. P., Saral Sarojini, K., & Rajarajeswari, G. R. (2019). Antimicrobial and biodegradable chitosan/cellulose acetate phthalate/ZnO nano composite films with optimal oxygen permeability and hydrophobicity for extending the shelf life of black grape fruits. *International Journal of Biological Macromolecules*, *132*, 1112–1120. https://doi.org/10.1016/j.ijbiomac.2019.03.171

Jayakumar, A., Heera, K. V., Sumi, T. S., Joseph, M., Mathew, S., Praveena, G., Nair, I. C., & Radhakrishnan, E. K. (2019). Starch-PVA composite films with zinc-oxide nanoparticles and phytochemicals as intelligent pH sensing wraps for food packaging application. *International Journal of Biological Macromolecules*, *136*, 395–403. https://doi.org/10.1016/j.ijbiomac.2019.06.018

Jickells, S. M., Poulin, J., Mountfort, K. A., & Fernàndez-ocaña, M. (2005). Migration of contaminants by gas phase transfer from carton board and corrugated board box secondary packaging into foods. *Food Additives and Contaminants*, *22*(8), 768–782. https://doi.org/10.1080/02652030500151992

Jokar, M., Pedersen, G. A., & Loeschner, K. (2017). Six open questions about the migration of engineered nano-objects from polymer-based food-contact materials: a review. *Food Additives & Contaminants: Part A*, *34*(3), 434–450. https://doi.org/10.1080/19440049.2016.1271462

Khan, I., Saeed, K., & Khan, I. (2019). Nanoparticles: properties, applications and toxicities. *Arabian Journal of Chemistry*, *12*(7), 908–931. https://doi.org/10.1016/j.arabjc.2017.05.011

Kumar, S., Boro, J. C., Ray, D., Mukherjee, A., & Dutta, J. (2019). Bionanocomposite films of agar incorporated with ZnO nanoparticles as an active packaging material for shelf life extension of green grape. *Heliyon*, *5*(6), e01867. https://doi.org/10.1016/j.heliyon.2019.e01867

Kumar, S., Mudai, A., Roy, B., Basumatary, I. B., Mukherjee, A., & Dutta, J. (2020). Biodegradable hybrid nanocomposite of chitosan/gelatin and green synthesized zinc oxide nanoparticles for food packaging. *Foods*, *9*(9), 1143. https://doi.org/10.3390/foods9091143

Kumari, A., & Chauhan, A. K. (2022). Iron nanoparticles as a promising compound for food fortification in iron deficiency anemia: a review. *Journal of Food Science and Technology, 59*(9), 3319–3335. https://doi.org/10.1007/s13197-021-05184-4

Lago, M. A., Sendón, R., Bustos, J., Nieto, M. T., Paseiro Losada, P., & Rodríguez-Bernaldo de Quirós, A. (2019). Migration studies of two common components of UV-curing inks into food simulants. *Molecules, 24*(19), 3607. https://doi.org/10.3390/molecules24193607

Lee, D. S., Yam, K. L., & Piergiovanni, L. (Eds.). (2008). *Food packaging science and technology*. CRC Press.

Lin, D., Yang, Y., Wang, J., Yan, W., Wu, Z., Chen, H., Zhang, Q., Wu, D., Qin, W., & Tu, Z. (2020). Preparation and characterization of TiO_2-Ag loaded fish gelatin-chitosan antibacterial composite film for food packaging. *International Journal of Biological Macromolecules, 154*, 123–133. https://doi.org/10.1016/j.ijbiomac.2020.03.070

Mahmud, J., Sarmast, E., Shankar, S., & Lacroix, M. (2022). Advantages of nanotechnology developments in active food packaging. *Food Research International, 154*, 111023. https://doi.org/10.1016/j.foodres.2022.111023

Maia, J., Rodríguez-Bernaldo de Quirós, A., Sendón, R., Cruz, J. M., Seiler, A., Franz, R., Simoneau, C., Castle, L., Driffield, M., Mercea, P., Oldring, P., Tosa, V., & Paseiro, P. (2016). Determination of key diffusion and partition parameters and their use in migration modelling of benzophenone from low-density polyethylene (LDPE) into different foodstuffs. *Food Additives & Contaminants: Part A, 33*(4), 715–724. https://doi.org/10.1080/19440049.2016.1156165

Meng, X., Chen, C., Song, T., Xu, J., Zhang, X., Wang, J., Pan, Z., Zhang, H., & Zhang, H. (2022). Effect of nano-silica coating combined with pressurized Ar treatment on postharvest quality and reactive oxygen species metabolism in sweet cherry fruit. *Food Chemistry, 374*, 131715. https://doi.org/10.1016/j.foodchem.2021.131715

Metak, A. M., Nabhani, F., & Connolly, S. N. (2015). Migration of engineered nanoparticles from packaging into food products. *LWT—Food Science and Technology, 64*(2), 781–787. https://doi.org/10.1016/j.lwt.2015.06.001

Mishra, M., Dashora, K., Srivastava, A., Fasake, V. D., & Nag, R. H. (2019). Prospects, challenges and need for regulation of nanotechnology with special reference to India. *Ecotoxicology and Environmental Safety, 171*, 677–682. https://doi.org/10.1016/j.ecoenv.2018.12.085

Nerin, C., Contin, E., & Asensio, E. (2007). Kinetic migration studies using Poropak as solid-food stimulant to assess the safety of paper and board as food packaging materials. *Analytical and Bioanalytical Chemistry, 387*, 2283–2288. https://doi.org/10.1007/s00216-006-1080-3

Nguyen, P.-M., Guiga, W., & Vitrac, O. (2016). Molecular thermodynamics for food science and engineering. *Food Research International, 88*, 91–104. https://doi.org/10.1016/j.foodres.2016.03.014

Nguyen, H.-L., Tran, T. H., Hao, L. T., Jeon, H., Koo, J. M., Shin, G., Hwang, D. S., Hwang, S. Y., Park, J., & Oh, D. X. (2021). Biorenewable, transparent, and oxygen/moisture barrier nanocellulose/nanochitin-based coating on polypropylene for food packaging applications. *Carbohydrate Polymers, 271*, 118421. https://doi.org/10.1016/j.carbpol.2021.118421

Nile, S. H., Baskar, V., Selvaraj, D., Nile, A., Xiao, J., & Kai, G. (2020). Nanotechnologies in food science: applications, recent trends, and future perspectives. *Nano-Micro Letters, 12*(1), 45. https://doi.org/10.1007/s40820-020-0383-9

Nilsen-Nygaard, J., Fernández, E. N., Radusin, T., Rotabakk, B. T., Sarfraz, J., Sharmin, N., Sivertsvik, M., Sone, I., & Pettersen, M. K. (2021). Current status of biobased and biodegradable food packaging materials: Impact on food quality and effect of innovative processing technologies. *Comprehensive Reviews in Food Science and Food Safety*, *20*(2), 1333–1380. https://doi.org/10.1111/1541-4337.12715

Paidari, S., & Ibrahim, S. A. (2021). Potential application of gold nanoparticles in food packaging: a mini review. *Gold Bulletin*, *54*(1), 31–36. https://doi.org/10.1007/s13404-021-00290-9

Paidari, S., Tahergorabi, R., Anari, E. S., Nafchi, A. M., Zamindar, N., & Goli, M. (2021). Migration of various nanoparticles into food samples: a review. *Foods*, *10*(9), 2114. https://doi.org/10.3390/foods10092114

Phothisarattana, D., & Harnkarnsujarit, N. (2022). Migration, aggregations and thermal degradation behaviors of TiO_2 and ZnO incorporated PBAT/TPS nanocomposite blown films. *Food Packaging and Shelf Life*, *33*, 100901. https://doi.org/10.1016/j.fpsl.2022.100901

Poças, M. F., Oliveira, J. C., Oliveira, F. A. R., & Hogg, T. (2008). A critical survey of predictive mathematical models for migration from packaging. *Critical Reviews in Food Science and Nutrition*, *48*(10), 913–928. https://doi.org/10.1080/10408390701761944

Ramos, M., Beltrán, A., Peltzer, M., Valente, A. J. M., & del Carmen Garrigós, M. (2014). Release and antioxidant activity of carvacrol and thymol from polypropylene active packaging films. *LWT—Food Science and Technology*, *58*(2), 470–477. https://doi.org/10.1016/j.lwt.2014.04.019

Ramos, M., Beltran, A., Fortunati, E., Peltzer, M., Cristofaro, F., Visai, L., Valente, A. J. M., Jiménez, A., Kenny, J. M., & Garrigós, M. C. (2020). Controlled release of thymol from poly(lactic acid)-based silver nanocomposite films with antibacterial and antioxidant activity. *Antioxidants*, *9*(5), 395. https://doi.org/10.3390/antiox9050395

Robertson, G. L. (2005). *Food Packaging: Principles and Practice*. CRC Press.

Rodríguez-Martínez, A. V., Sendón, R., Abad, M. J., González-Rodríguez, M. V., Barros-Velázquez, J., Aubourg, S. P., Paseiro-Losada, P., & Rodríguez-Bernaldo de Quiros, A. (2016). Migration kinetics of sorbic acid from polylactic acid and seaweed based films into food simulants. *LWT—Food Science and Technology*, *65*, 630–636. https://doi.org/10.1016/j.lwt.2015.08.029

Rukmanikrishnan, B., Ismail, F. R. M., Manoharan, R. K., Kim, S. S., & Lee, J. (2020). Blends of gellan gum/xanthan gum/zinc oxide based nanocomposites for packaging application: Rheological and antimicrobial properties. *International Journal of Biological Macromolecules*, *148*, 1182–1189. https://doi.org/10.1016/j.ijbiomac.2019.11.155

Sablani, S. S., Bhunia, K., & Rahman, M. S. (2020). Food–Packaging Interactions. In M. S. Rahman (Ed.), *Handbook of Food Preservation. CRC Press.*

Samsudin, H., Auras, R., Burgess, G., Dolan, K., & Soto-Valdez, H. (2020). Migration of antioxidants from polylactic acid films, a parameter estimation approach: reparameterization of the Arrhenius equation. *Food Control*, *113*, 107208. https://doi.org/10.1016/j.foodcont.2020.107208

Saravanakumar, K., Sathiyaseelan, A., Mariadoss, A. V. A., Xiaowen, H., & Wang, M.-H. (2020). Physical and bioactivities of biopolymeric films incorporated with cellulose, sodium alginate and copper oxide nanoparticles for food packaging application. *International Journal of Biological Macromolecules*, *153*, 207–214. https://doi.org/10.1016/j.ijbiomac.2020.02.250

Sarfraz, J., Gulin-Sarfraz, T., Nilsen-Nygaard, J., & Pettersen, M. K. (2020). Nanocomposites for food packaging applications: an overview. *Nanomaterials*, *11*(1), 10. https://doi.org/10.3390/nano11010010

Sepulveda, J., Villegas, C., Torres, A., Vargas, E., Rodriguez, F., Baltazar, S., Prada, A., Rojas, A., Romero, J., Faba, S., & Galotto, M. J. (2020). Effect of functionalized silica nanoparticles on the mass transfer process in active PLA nanocomposite films obtained by supercritical impregnation for sustainable food packaging. *The Journal of Supercritical Fluids*, *161*, 104844. https://doi.org/10.1016/j.supflu.2020.104844

Shankar, S., Wang, L.-F., & Rhim, J.-W. (2018). Incorporation of zinc oxide nanoparticles improved the mechanical, water vapor barrier, UV-light barrier, and antibacterial properties of PLA-based nanocomposite films. *Materials Science and Engineering: C*, *93*, 289–298. https://doi.org/10.1016/j.msec.2018.08.002

Sharma, C., Dhiman, R., Rokana, N., & Panwar, H. (2017). Nanotechnology: an untapped resource for food packaging. *Frontiers in Microbiology*, *8*. https://doi.org/10.3389/fmicb.2017.01735

Soleimani, J., Ghanbarzadeh, B., Dehghannya, J., Baheri Islami, S., & Sorouraddin, S. M. (2017). Comparative numerical study of titanium and silver nano-particles migration from nano-composite of polystyrene into simulants on experimental data basis. *International Journal of Food Engineering*, *13*(12). https://doi.org/10.1515/ijfe-2017-0091

Sonseca, A., Madani, S., Rodríguez, G., Hevilla, V., Echeverría, C., Fernández-García, M., Muñoz-Bonilla, A., Charef, N., & López, D. (2019). Multifunctional PLA blends containing chitosan mediated silver nanoparticles: thermal, mechanical, antibacterial, and degradation properties. *Nanomaterials*, *10*(1), 22. https://doi.org/10.3390/nano10010022

Stoffers, N. H. (2005). *Certified reference materials for food packaging specific migration tests: development, validation and modelling*. Wageningen University and Research.

Störmer, A., Bott, J., Kemmer, D., & Franz, R. (2017). Critical review of the migration potential of nanoparticles in food contact plastics. *Trends in Food Science & Technology*, *63*, 39–50. https://doi.org/10.1016/j.tifs.2017.01.011

Tang, Z., Fan, F., Fan, C., Jiang, K., & Qin, Y. (2020). The performance changes and migration behavior of PLA/nano-TiO_2 composite film by high-pressure treatment in ethanol solution. *Polymers*, *12*(2), 471. https://doi.org/10.3390/polym12020471

Tehrany, E. A., & Desobry, S. (2004). Partition coefficients in food/packaging systems: a review. *Food Additives and Contaminants*, *21*(12), 1186–1202. https://doi.org/10.1080/02652030400019380

Torres, A., Romero, J., Macan, A., Guarda, A., & Galotto, M. J. (2014). Near critical and supercritical impregnation and kinetic release of thymol in LLDPE films used for food packaging. *The Journal of Supercritical Fluids*, *85*, 41–48. https://doi.org/10.1016/j.supflu.2013.10.011

Triantafyllou, V. I., Akrida-Demertzi, K., & Demertzis, P. G. (2005). Determination of partition behavior of organic surrogates between paperboard packaging materials and air. *Journal of Chromatography A*, *1077*(1), 74–79. https://doi.org/10.1016/j.chroma.2005.04.061

Triantafyllou, V., Akridademertzi, K., & Demertzis, P. (2007). A study on the migration of organic pollutants from recycled paperboard packaging materials to solid food matrices. *Food Chemistry*, *101*(4), 1759–1768. https://doi.org/10.1016/j.foodchem.2006.02.023

USFDA (2014). Guidance for industry: Assessing the effects of significant manufacturing process changes, including emerging technologies, on the safety and regulatory status of food ingredients and food contact substances, including food ingredients that are color additives. https://www.fda.gov/regulatory information/search-fda-guidance-documents/considering-whether-fda-regulated-product-involves application-nanotechnology.

Valipour Motlagh, N., Aghazamani, J., & Gholami, R. (2021). Investigating the effect of nano-silver contained packaging on the olivier salad shelf-life. *BioNanoScience*, *11*(3), 838–847. https://doi.org/10.1007/s12668-021-00876-9

Vitanov, N. K., & Vitanov, K. N. (2016). Box model of migration channels. *Mathematical Social Sciences*, *80*, 108–114. https://doi.org/10.1016/j.mathsocsci.2016.02.001

Wang, L.-F., & Rhim, J.-W. (2015). Preparation and application of agar/alginate/collagen ternary blend functional food packaging films. *International Journal of Biological Macromolecules*, *80*, 460–468. https://doi.org/10.1016/j.ijbiomac.2015.07.007

Wang, C., Chang, T., Dong, S., Zhang, D., Ma, C., Chen, S., & Li, H. (2020). Biopolymer films based on chitosan/potato protein/linseed oil/ZnO NPs to maintain the storage quality of raw meat. *Food Chemistry*, *332*, 127375. https://doi.org/10.1016/j.foodchem.2020.127375

Wang, Q., Chen, W., Zhu, W., McClements, D. J., Liu, X., & Liu, F. (2022). A review of multilayer and composite films and coatings for active biodegradable packaging. *Npj Science of Food*, *6*(1). https://doi.org/10.1038/s41538-022-00132-8

Yang, D., Liu, Q., Gao, Y., Wan, S., Meng, F., Weng, W., & Zhang, Y. (2023). Characterization of silver nanoparticles loaded chitosan/polyvinyl alcohol antibacterial films for food packaging. *Food Hydrocolloids*, *136*, 108305. https://doi.org/10.1016/j.foodhyd.2022.108305

Zaki, O., Abbès, B., & Safa, L. (2009). Non-fickian diffusion of amyl acetate in polypropylene packaging: experiments and modelling. *Polymer Testing*, *28*(3), 315–323. https://doi.org/10.1016/j.polymertesting.2009.01.002

Zhu, B. et al. (2021). Migration kinetic of silver from polylactic acid nanocomposite film into acidic food simulant after different high-pressure food processing. *Journal of Food Science*, *86*(6), 2481–2490. https://doi.org/10.1111/1750-3841.15746

Zorraquín-Peña, I., Cueva, C., Bartolomé, B., & Moreno-Arribas, M. V. (2020). Silver nanoparticles against foodborne bacteria. Effects at intestinal level and health limitations. *Microorganisms*, *8*(1), 132. https://doi.org/10.3390/microorganisms8010132

16

Global Regulatory Frameworks for Nanomaterials in Food Packaging

Singam Suranjoy Singh[1], *Anns Annie Gigi*[2,3], *Prasanth K. S. Pillai*[4] *and K. V. Ragavan*[2,3]

[1] *Department of Food Science, University of Guelph, Guelph, Ontario, Canada*
[2] *Agro-Processing and Technology Division, CSIR-National Institute for Interdisciplinary Science and Technology, Thiruvananthapuram, India*
[3] *Academy of Scientific and Innovative Research (AcSIR), Ghaziabad, India*
[4] *Department of Food Science and Nutrition, University of Minnesota, St. Paul, MN, United States*

16.1 Introduction

Food production systems across the globe are facing the daunting task of overcoming the issue of food loss and waste (Xue et al., 2017). The main reasons for food spoilage are associated with (i) different kinds of food processing parameters and techniques, (ii) microbial contamination, (iii) phase changes upon storage, and (iv) ineffective food packaging systems. In addition, it is estimated that the food spoilage count is going to hit 200 million tons by 2050. Having appropriate food packaging systems could mitigate this problem to a greater extent (Wikström et al., 2019; Wohner et al., 2019).

Food packaging is one of the critical steps in food processing that has evolved from being just a barrier to external factors to being much more advanced as an interactive component, either communicating the quality of packaged material or interfering with the shelf-life of fresh produce (Amin et al., 2022; Cheng et al., 2022; Gundewadi et al., 2022). Food packaging materials are sourced from diverse raw materials, including agri-biomass, petrochemicals, microbial, glass, metal, etc. (Sid et al., 2021). In the past two to three decades, nanomaterials with fascinating properties have been blended with polymeric and other packaging materials to produce nanocomposite-based packaging materials (Sharma et al., 2017).

16.2 Scope of Nanomaterials in Food Packaging

Nanomaterials bring exciting features to the packaging material, such as antibacterial, antifungal, enhanced barrier, and mechanical properties. Some of them are superior to the commonly used food packaging materials (Mei and Wang, 2020). Another important

Nanotechnology for Sustainable Food Packaging, First Edition. Edited by C. Anandharamakrishnan, Jeyan A. Moses, and M. Maria Leena.

feature is the potential of nanomaterials to act as an indicator/sensor for spoilage/food quality through their inherent recognition/transducing properties (Kumar et al., 2021; Mustafa and Andreescu, 2020). Despite all these features exhibited by the nanomaterials in food packaging, there is a dearth of information related to the impact on food quality, migration into the food matrix, food safety, human health, and the environment. Regulatory agencies across the globe are actively involved in drafting and enforcing regulations to ensure the safety of consumers and the environment (Chau et al., 2007; Paidari et al., 2021; Ranjan et al., 2019; Souza and Fernando, 2016). In this conjecture, the regulations available across the geographies are reviewed in the following sections.

In recent years, research and development in nanomaterials for food packaging applications have gained substantial traction. Several nano- to submicron-scale materials (organic—carbohydrates, proteins, fats, etc.; inorganic—metals and metal oxides; and combinations of both) are being incorporated into polymer matrices to enhance the functional properties of packaging systems (Bumbudsanpharoke et al., 2015; He et al., 2019). Enhanced moisture or gas barrier, mechanical strength, active and interactive (intelligent) packaging materials, nanosensors-based smart packaging, impart functionalities such as antimicrobial activity, controlled release of active compounds that can prolong the shelf life of the packaged food are a multitude of ways by which nanomaterials can maintain the safety and quality of foods (Ashfaq et al., 2022). The primary intention of introducing nanomaterials in food packaging is to enhance the essential attributes of conventional packaging systems, i.e. (i) protection and preservation, (ii) convenience, (iii) containment, and (iv) communication and marketing (Robertson, 2005; Vanderroost et al., 2014).

Nanomaterials exhibit unique and better physicochemical features as compared to macroscale materials. Owing to their nanoscale size, nanomaterials possess a large surface-to-volume ratio and a high surface activity. Nanomaterials, when incorporated into suitable polymers, result in enhanced material properties such as mechanical strength, flexibility, barrier properties, electrical conductivity, and thermal stability. The developed nanomaterials augmented food packages will help reduce interactions between packaging and food matrices, affecting consumer health barriers against gases and light, while also providing active and intelligent packaging systems (Mihindukulasuriya and Lim, 2014). Based on their structural forms, nanomaterials are typically categorized into particulates (nanoparticles), platelets, nanoclays, and nanofibers. Additionally, they could also be broadly classified based on their sources as organic (natural materials including carbohydrate, protein, and fat), inorganic (metal and metal oxide), and their combination (nanoclay), which are typically integrated into a polymer matrix to enhance the functional properties of packaging materials (Ranjan et al., 2014). Silver nanoparticles (AgNP), nanoclay, nano-titanium dioxide (TiO_2), nano-zinc oxide (nano-ZnO), carbon nanotubes (CNTs), cellulose nanowhiskers, and starch nanocrystals are some of the prominent nanomaterials used in food packaging as functional components, as can be seen in Figure 16.1 (Ashfaq et al., 2022; Trinh et al., 2023).

Generally, the applications of nanomaterials in food packaging are predominantly centered on three major packaging functionalities such as enhanced properties, intelligent/smart packaging, and active packaging (Figure 16.2).

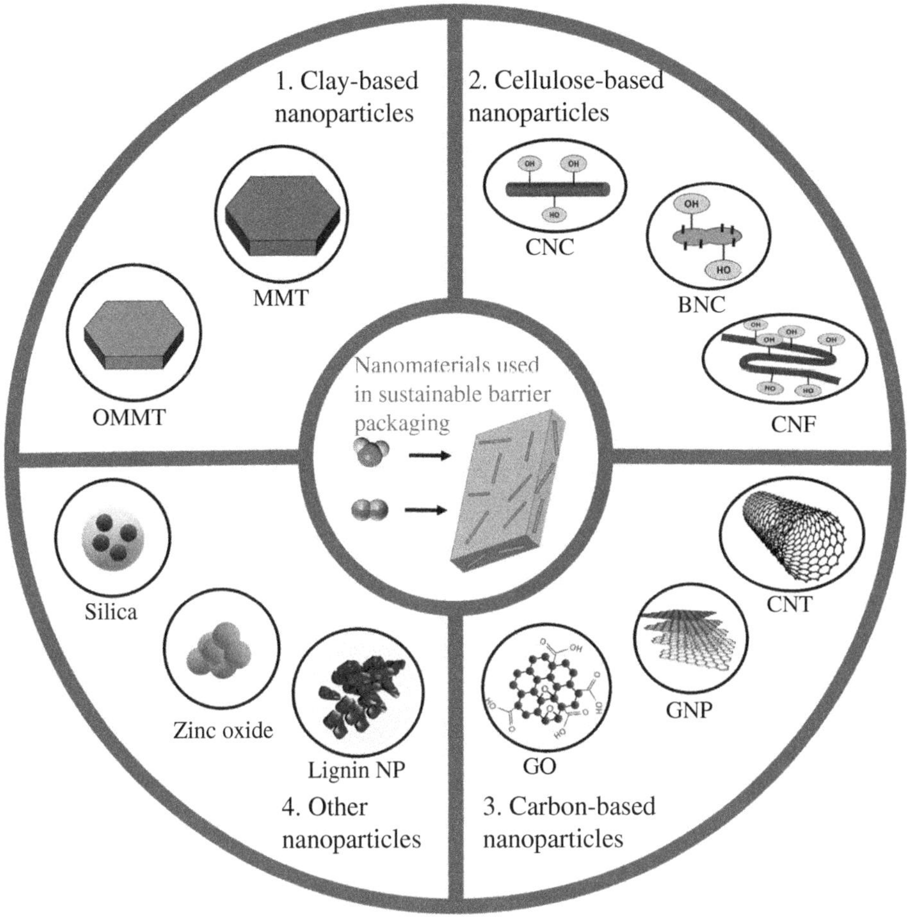

Figure 16.1 Common nanomaterials employed in sustainable food packaging applications (*Source:* Trinh et al. (2023)/with permission of Elsevier)

i) *Enhanced packaging properties*: the nano-enabled food packaging exhibits the ability to preserve color, flavor, texture, and consistency of food products by enhancing the overall properties of the packaging system, such as the mechanical, gas barrier (diffusion of O_2, CO_2, and volatile flavor compounds), and stability under a range of moisture and temperature conditions, (Barage et al., 2022; Silvestre et al., 2011). Nanocomposite is a material composed of multiple phases, created by the amalgamation of a matrix (continuous phase) and a nanoscale material (discontinuous phase). The addition of nano-sized components enhances the strength and flexibility of the polymer by transferring elastic stress to the reinforced nanomaterials. This property has made nanocomposite the benchmark for improving the mechanical and barrier properties of polymers (Othman, 2014). Furthermore, the applications of polymer

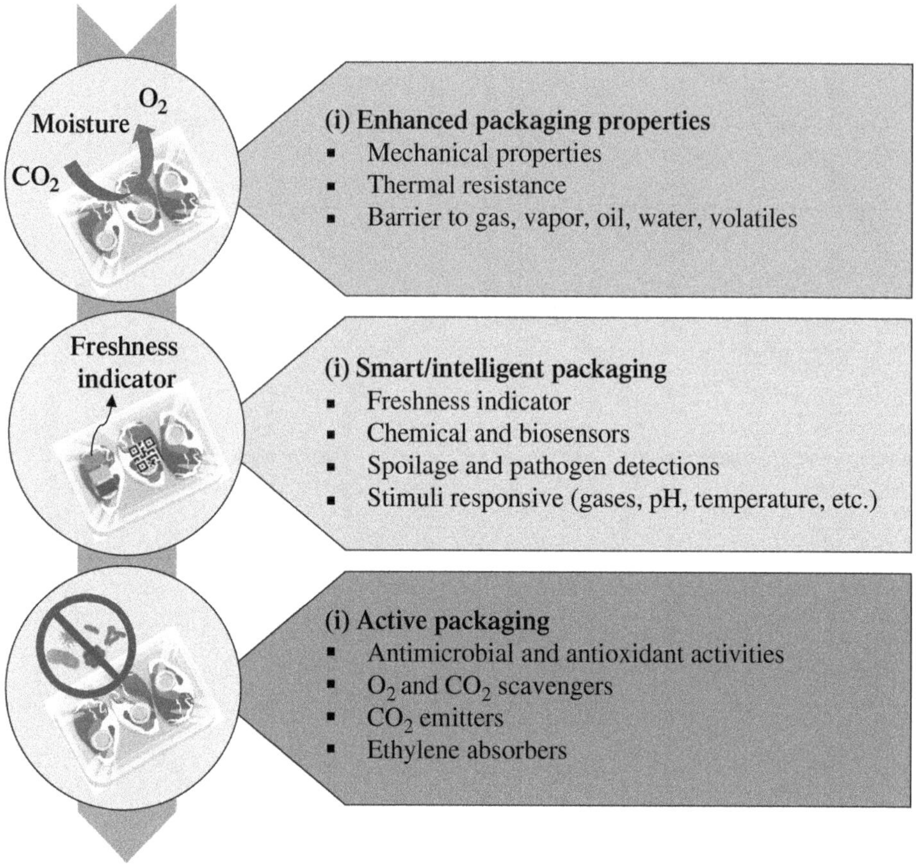

Figure 16.2 Three primary applications of nanomaterials in food packaging (*Source:* Reproduced from Trinh et al. (2023)/with permission of Elsevier)

nanocomposites as noble food packaging materials can lead to new opportunities for enhancing the balance between the qualities and cost-effectiveness of packaging materials. The aspect ratio (the proportion of the largest to the smallest dimension of the filler) in packaging filler material is crucial; fillers with higher aspect ratios have a greater specific surface area, resulting in enhanced reinforcing properties (Rafieian and Simonsen, 2014). The distinct chemical structure and characteristics of each nanomaterial impart unique capabilities to the substrate materials, leading to a wide range of functional packaging applications. For example, when nanoclay or CNTs are combined with traditional thermoplastics, they are evenly distributed throughout the polymer matrix. The molecular mobility and polymer relaxation of the plastic are altered by the nanoparticles, which substantially enhance the mechanical and thermal properties (Bumbudsanpharoke et al., 2015).

a) *Clay nanoparticles*

Clay nanoparticle-incorporated polymers are one of the first types of polymer nanomaterials to be introduced as advanced materials for food packaging. Nanoclays can be dispersed in a polymer matrix in three different patterns, including phase separated, intercalated, and exfoliated, to develop polymer-clay nanocomposites either through melt blending, solution blending, or *in situ* polymerization techniques. It is necessary to attain a homogenous dispersion of the clay nanoparticles in the polymer matrix to benefit from the addition of clay. Enthalpy and entropy parameters are primary factors that control how the clay nanoparticles are arranged morphologically within the polymer matrix (de Paiva et al., 2008). Clay dispersion in a polymer demands sufficient desirable enthalpic factors, which occur when the interactions between the polymer and clay are favorable. For the majority of polar polymers, alkyl-ammonium surfactants can provide enough additional enthalpy to create homogeneous nanocomposites (Mackay et al., 2006). A well-proportionally balanced polymer-clay nanocomposite structure displayed a substantial improvement in modulus, elasticity, and heat distortion temperature compared to pristine polymer structures. The dispersion of clay nanoparticles in the polymer matrix in a layer-by-layer pattern increases the distance (tortuous path) that gas or water molecules have to travel to pass through the packaging layer, resulting in improved barrier properties. Furthermore, the composites exhibit lower water sensitivity, reduced gas permeability, and a relatively similar coefficient of thermal expansion (Barage et al., 2022). Research has demonstrated that the decrease in gas permeability of clay nanocomposites is heavily influenced by factors such as the type of nanoclay, the compatibility between clay and polymer matrices, the aspect ratio of clay nanoparticles, and the nanocomposite structures. For instance, polymer nanocomposites containing fully exfoliated clay particles and high aspect ratios would yield the highest gas barrier qualities. The most extensively utilized nanoclay is montmorillonite (MMT) [$Mx(Al_4\text{-}xMgx)\ Si_8O_{20}(OH)_4$], consisting of an edge-shared aluminum hydroxide $Al(OH)_3$ octahedral sheet sandwiched between two layers of silica tetrahedrons, linked by weak electrostatic forces (de Paiva et al., 2008; Mittal, 2009). Its extensive utilization in polymer composites is attributed to its exceptional capacity for cation exchange, extensive surface area, and favorable swelling properties (de Azeredo, 2009). This clay is known for having a moderately negative surface charge that is balanced by exchangeable cations, usually Na^+ and Ca^{2+} (Tan et al., 2008).

The use of nanoclay in food packaging started in the 1990s. An example of a commercially available nanoclay composite is Durethan® KU2-2601, which is a nanoclay-reinforced polyamide (Nylon 6) film created by Bayer and sold by Lanxess Deutschland GmbH (Chun, 2009; Cushen et al., 2012; Hatzigrigoriou and Papaspyrides, 2011). Durethan® KU2-2601 is composed of clay platelets dispersed throughout the polymer matrix, acting as a barrier. The composite film can block the penetration of oxygen (O_2) and carbon dioxide (CO_2) from the outside while maintaining the moisture content of the food within. Voridian of Eastman Chemical and Nanocor of AMCOL (American Colloid Company) International developed Imperm®, a composite of nanoclay and MXD6 resin (a nylon 6 polymer). This newly created resin has been manufactured into single-layer, multilayer films, and multilayer coinjection PET bottles suitable for sensitive drinks like juices, beer, specialty water, and flavored alcoholic drinks (Bumbudsanpharoke and Ko, 2015; Peters et al., 2011).

b) Starch nanoparticles

Starch is a complex polysaccharide consisting of two glycosidic macromolecules called linear amylose and branched amylopectin. It is a type of polysaccharide that is naturally occurring, renewable, nontoxic, and biodegradable. Owing to its cost-effectiveness, nontoxic, and environmentally benign nature, it has been extensively researched and utilized as a desirable material in the food, pharmaceutical, bioplastic manufacturing, papermaking, and food packaging industries (Oladebeye et al., 2013; Yu et al., 2021). Some of the major sources of starch nanoparticles used in food packaging-associated applications are potato, pea, rice, wheat, corn, cassava, and tapioca. The preparation of starch nanoparticles involves the breaking down of starch granules followed by the isolation of crystalline starch from the amorphous-crystalline complexes (Yu et al., 2021). Several methods, including mechanical treatment (e.g., microfluidizer), precipitation, enzymes, and acid hydrolysis, are employed to produce amorphous and crystalline starch nanoparticles. Starch nanoparticles have dimensions ranging from 150 to 450 nm and a substantially large surface area to volume ratio (Klass, 2007). Starch nanoparticles, also referred to as starch nanocrystals, have been discovered to be effective as nanofillers in composite materials. Their use enhances the strength, flexibility, biodegradability, water resistance, thermal properties, and barrier properties of the composites (Campelo et al., 2020). Besides being used as a filler and reinforcement component, starch nanocrystals are also used as a natural and cost-effective binding agent, making them a potential substitute for synthetic emulsion latex. Cationic starch can be utilized to create the starch-based nano-biopolymer binder, which is typically employed in wet-end processes to substitute traditional dry-strength resins like glyoxalated polyacrylamide (Bumbudsanpharoke et al., 2015). In stark contrast to conventionally cooked starch, starch nanoparticles produce a flexible, transparent film rather than a brittle one, making them ideal for coating cardboard since they can resist cracking caused by folding and scoring. Furthermore, the adhesive properties of nano-starch latex are utilized for container boards and multi-ply paper boards.

ii) *Smart/intelligent packaging*: the incorporation of nanodevices or nanosensors in smart/intelligent packaging provides real-time information on the quality of packaged food items, serves as a deterrent to counterfeit products, and indicates any potential exposure to harmful factors such as extreme temperatures or high oxygen levels. Additionally, it can also be used for marketing and feedback purposes. Nanomaterial-based indicators/sensors incorporated into food packaging can interact with both internal factors, such as food components and headspace, and external environmental factors (Figure 16.3).

iii) *Active packaging*: the incorporation of nanomaterials in packaging materials enables them to interact with food and the packaging environment, thus playing a dynamic role in protecting and preserving the food. The carrier component (nanomaterials) can then interact with internal and/or external elements to trigger actions that extend shelf life, improve food quality, and promote product safety. Active packaging provides protection and preservation depending on the mechanism(s) that are triggered by intrinsic and/or extrinsic factors, Figure 16.4 (Sharma et al., 2017; Silvestre et al., 2011).

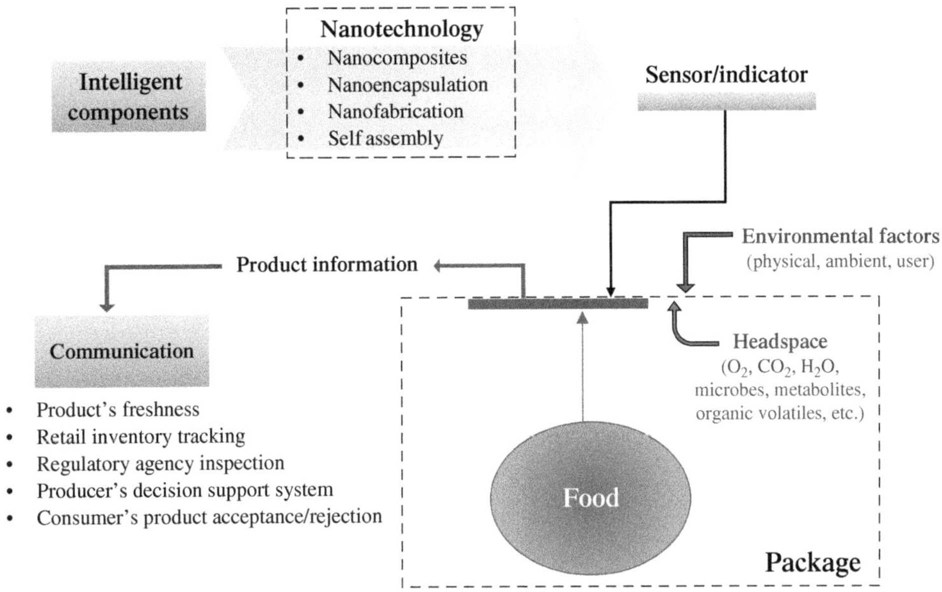

Figure 16.3 An illustration showing the involvement of nanotechnology in developing intelligent/smart food packaging (*Source:* Reproduced from Mihindukulasuriya and Lim (2014), Salgado et al. (2021))

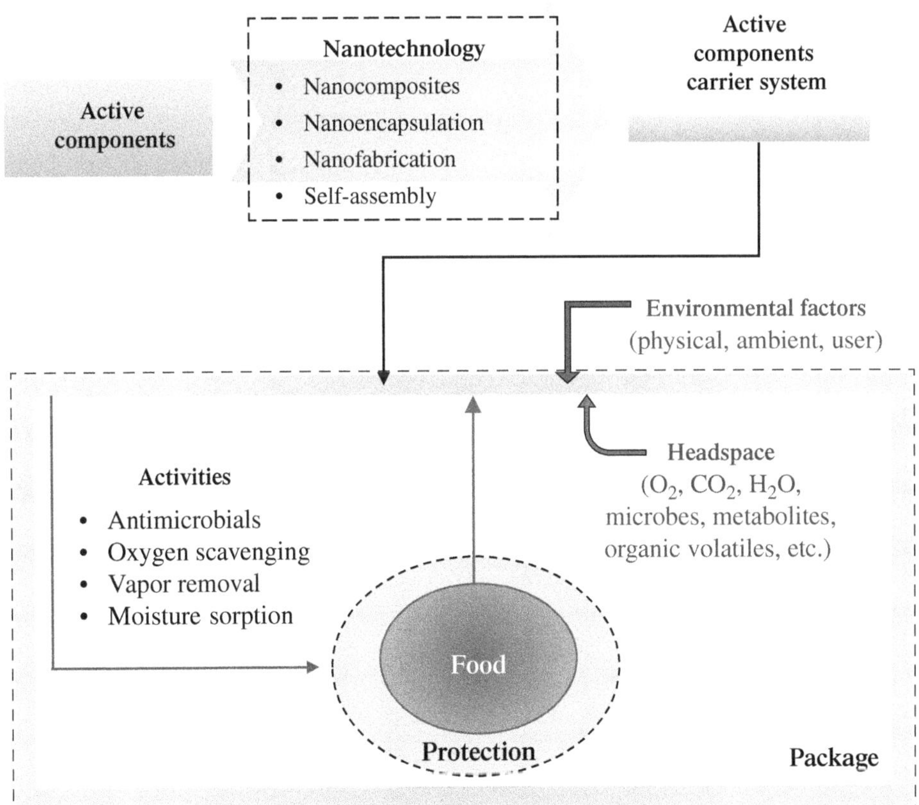

Figure 16.4 Application of nanotechnology in the development of active food packaging system (*Source:* Sharma et al. (2017)/Frontiers Media S.A./CC-BY-4.0)

Table 16.1 Nanomaterials and their composites reported for food packaging applications (Bumbudsanpharoke and Ko, 2015)

<table>
<tr><th>Composition of Nano Packaging Material</th><th>Product Name and Manufacturer</th><th>Country of Origin</th><th>Application</th><th>Reference</th></tr>
<tr><td>Silver Nanomaterials</td><td></td><td></td><td></td><td></td></tr>
<tr><td>Nanosilver/Tritan™</td><td>Incense Nano Silver Food Container
Dong Yang Chemical Co., Ltd.</td><td>South Korea</td><td>It can inhibit the growth of bacteria, viruses, or fungi on the surface of container.</td><td>Bumbudsanpharoke and Ko (2015)</td></tr>
<tr><td>Nanosilver/polyethylene (PE)</td><td>Fresh Box Nano Silver Food Container
Fine Polymer, Inc.</td><td>South Korea</td><td>Due to the effect of finely dispersed nanosilver particles excellent antimicrobial properties against various bacteria and fungus and hence it makes food fresh longer compared with conventional food containers.</td><td>Bumbudsanpharoke and Ko, (2015)</td></tr>
<tr><td>Nanosilver/polyethylene sulfonate (PES), polypropylene (PP)</td><td>Baby Dream®
Baby Dream Co., Ltd</td><td>South Korea</td><td>It is the first environmental hormone-free nursing bottles. It has 99.9% antimicrobial activity.</td><td>("BABYDREAM CO., LTD.—baby, bottle, silver," n.d.); ("Baby Dream Co. Ltd|Nanotechnology Company|NPD," n.d.)</td></tr>
<tr><td>A silver-base zeolite antimicrobial agent/PP, polystyrene (PS), acrylonitrile butadiene styrene (ABS)</td><td>Zeomic Sinanen Zeomic Co., Ltd.</td><td>Japan</td><td>It has excellent safety, sustainability, heat resistance.</td><td>("Silver-base inorganic antimicrobial agent 'Zeomic'|Sinanen Zeomic Co., Ltd.," n.d.)</td></tr>
<tr><td>Nanosilver/PP, PE</td><td>Anson Nano Freshness-Keeping film, storage bag, and fresh containers.
Anson Nano-Biotechnology (zhuhai) Co., Ltd.</td><td>China</td><td>Keeps the food fresh longer. It is used for storing vegetables and other foods.</td><td>Bumbudsanpharoke and Ko (2015)</td></tr>
</table>

Nanosilver/PP, silicon	Nanosilver food container. Cixi Mingxin Plastic & Rubber Factory	China	It has antibiotic and antibacterial property.	Bumbudsanpharoke and Ko (2015)
Nanosilver/PP, silicon	Double-handle nanosilver baby bottle. Shenzhen Ibecare Commodity Limited Company	China	It has antibacterial property and BPA free.	Bumbudsanpharoke and Ko (2015)
Nanosilver (particles size 25 nm)/PP	Fresher Longer™ Plastic Storage Bags Fresher Longer™ Miracle Food Storage. Sharper Image R © Company	USA	Keep the foods fresher three or even four times longer for fruits, vegetables, herbs, breads, cheeses, soups, sauces, and meats	Bumbudsanpharoke and Ko (2015)
Nanosilver/PP, silicon	Sina Antibacterial Food Storages Dai Dong Tien Corporation	Vietnam	Prevent the growth of fungus and keep the foods fresh for longer.	Bumbudsanpharoke and Ko (2015)
Nanosilver/PP, Copolyester (Tritan™)	Window R © Nano Silver Airtight Container	South Korea	It helps to sterilize food containers and reduce bad smells. Approved by USFDA.	Bumbudsanpharoke and Ko (2015)
Nanosilver	Everin Food Containers Nano Silver Airtight NewLife Co., Ltd.	South Korea	The silicone seal has antibacterial nanosilver particles that kill harmful bacteria, keeping food fresher for longer.	Bumbudsanpharoke and Ko (2015)
Metal Oxide Nanomaterials (Nanoclay)				
Nanoclay/nylon 6	Aegis R © OXCE Barrier Nylon Resin Honeywell International Inc.	USA	It provides an excellent barrier which is comparable to the glass bottles performance	Deshmukh et al. (2023)

(Continued)

Table 16.1 (Continued)

Composition of Nano Packaging Material	Product Name and Manufacturer	Country of Origin	Application	Reference
Nanoclay/nylon 6	Imperm R © Nylon nanocomposite. Mitsubishi Gas Chemical Company, Inc.	USA	It can replace the EVOH with a more cost-effective material that allows for easier processing and maintaining barrier properties.	Deshmukh et al. (2023)
Nanoclay/starch	Plantic R © Plastic Tray Plantic Technologies Limited	Australia	They are biodegradable. It has improved mechanical and rheological properties and reduced sensitivity to moisture	Deshmukh et al. (2023)
Nanoclay	Bread Bags™ Debbie Meyer Bread Bags	United States of America	Used in bakery products. It has high moisture barrier.	Deshmukh et al. (2023)
Nanoclay/montmorillonite-based polymer	NaturaSorb ® SN series FUJIGEL SANGYO LTD.	Japan	Used as a food package barrier film. It is chemical inert and economical.	Deshmukh et al. (2023)
Nanoclay/polyethylene terephthalate (PET) and nano clay particles N-coat	FOOTHILL FARMS ™ Multifilm Packaging Corporation	United States of America	Used in cereal-based snacks and ready-to-eat food. It has a high gas barrier.	Deshmukh et al. (2023)
Nanoclay/kaolin	BARRISURFTM^{FX} Imerys SA	France	Used in fast foods packaging. It has a gas barrier, i.e., oxygen, nitrogen, carbon dioxide, or microorganism barrier	Deshmukh et al. (2023)
Nanoclay/polyethylene, ethylene vinyl alcohol (EVA), polyamide (PA), PP, and organoclay	Poliplen® Mondi Uralplastic	Russia	Used in meat, fish, and dairy products (tea, coffee). It has a barrier to gases and odor.	Deshmukh et al. (2023)
Nanoclay/PP/PE/organoclay	Polyelf® TERM Mondi Uralplastic	Russia	Used in meat and cheese package. It has reduced moisture migration	Deshmukh et al. (2023)

Nanoclay/nylon resin/ organoclay	AegisBarrierPro2™ Honeywell	United States of America	Used in beverages. It is an excellent barrier to CO_2 and O_2 and good recyclability.	Deshmukh et al. (2023)
Nanoclay/organoclay	DELLITE™ Laviosa Chimica Mineraria	Italy	Used in fruits and vegetables. It has less weight, greater barrier to gases	Deshmukh et al. (2023)
Nanoclay/PE/PA/ organoclay	Suhou™ BYBK Materials Technology	China	It has antibacterial activity	Deshmukh et al., (2023)
Nanoclay/kaolinite	O_2 Block™ NanoBioMatters Industries S.L	Spain	Used in Meat. It has O_2 scavenging activity.	Deshmukh et al. (2023)
Miscellaneous Nano Materials (Graphene)				
Graphene-based polymer nanocomposites (graphene thickness = ~0.34 nm, graphene oxide = ~1.1 nm and reduced graphene oxide)	—	—	Incorporated in food packaging material, which provides good mechanical strength, barrier to light, water vapor, and gases, and also has good antimicrobial and antioxidant properties, which can prevent the food spoilage	Han Lyn and Nur Hanani (2022)

16.3 Regulations in Different Regions

Given that nanotechnology is a relatively new field, it is crucial to consider consumer attitudes toward nanomaterials in food packaging during the early stages of technological development. This is particularly important due to widespread public skepticism and opposition to technologies such as GMOs in the agri-food sector. The marketability of products derived from nanotechnology faces a significant obstacle in the form of social rejection (Faccio and Guiotto Nai Fovino 2019). Consumer perception varies depending on whether nanotechnology is applied "nano-inside" (where nanomaterials are incorporated into the food product itself) or "nano-outside" (where nanomaterials are used in processing or packaging but do not directly enter the body). Even when the safety of nanomaterials in food packaging has been thoroughly established, the public still has concerns regarding potential health effects. Consumer safety is the fundamental issue surrounding the use of nanomaterials, as studies have shown that these materials may migrate from packaging into food. The amount of migration may be linked to the percentage of nanomaterials present in the packaging. One potential danger is that nanomaterials in packaging could seep into the food, possibly resulting in consumers being exposed to harmful substances. Moreover, there are worries about the possible adverse health impacts of consuming food that has been packaged using nanomaterials over an extended period. Regulations for nanomaterials vary by country and are still evolving, creating uncertainty about their future use. Several countries, including the EU, Canada, China, Japan, India, and many others, have also invested their efforts in developing safety guidance or legislation for nanotechnology production (Amenta et al., 2015). Most governmental agencies have established a general framework for marketing and utilizing products derived from nanotechnology in the food and agricultural industries. In the United States, various governmental agencies, including the Environment Protection Agency (EPA), Food and Drug Administration (FDA), Occupational Health and Safety Administration (OSHA), and Consumer Product Safety Commission (CPSC), play a crucial role in setting up regulations for nanotechnology. The fundamental regulations pertaining to nanomaterials remain in effect, although the proposed guidance has undergone several amendments subsequently. Regarding regulations concerning food ingredients and food-contact substances, the FDA's latest guidance is FDA-2011-D-0490, which was finalized in June 2014 (He et al., 2019). Additionally, the European Food Safety Authority (EFSA) initially issued guidance in 2006, which was later reformed in 2009 and 2011. The most recent version, titled "Nanotechnology Applications in the Agricultural, Feed, and Food Sector," represents the most up-to-date guidelines (Peters et al., 2011).

16.3.1 North America

Regulating food packaging produced with nanomaterials is crucial for ensuring consumer safety. In North America, this responsibility falls under the purview of two main agencies: the FDA in the United States and Health Canada in Canada. The FDA recommends that products using nanotechnology undergo safety assessments and be approved on a case-by-case basis. To evaluate the safety of new products, two criteria are considered: whether the material or end product has at least one dimension in the nanoscale range and whether it displays properties related to its dimension up to one micrometer. The FDA is responsible for ensuring the safety of food packaging and ingredients, including those made with nanomaterials, by setting guidelines for pre-market testing and risk assessment. They also require appropriate labeling of all food packaging containing nanomaterials to inform consumers. Meanwhile, Canada evaluates the potential risks and benefits of nanomaterials in packaging within existing legislative and regulatory frameworks. Health Canada, responsible for this topic, recommends that manufacturers of health and food products containing nanomaterials consult with regulatory authorities before submitting the product for approval to discuss its safety. Health Canada has established regulations to ensure the safe use of nanomaterials in food packaging, with a focus on protecting human health and the environment. They require premarket testing and risk assessment, as well as proper labeling of all food packaging containing nanomaterials.

The FDA has not yet determined the safety of nanomaterials in food packaging and therefore, the responsibility for providing evidence of their safety lies with the industries involved. Prior to the report by The New York Times, an article by Mark Astley on FoodProductionDaily.com in December 2011 highlighted the absence of robust regulations from the FDA concerning the risks associated with nanomaterials in the food sector (confectionerynews.com, 2011). FoodProductionDaily.com, in April 2013, a study conducted by Cushen et al. (2012) investigated the transfer of nanosilver from PVC to chicken meat. The study indicated that the migration levels of nanosilver were within the permissible limits set by regulations. However, the specific impact of these migrated nanomaterials remained uncertain. Within the European Union (EU), there is insufficient toxicological data available for nanosilver, prompting ongoing risk assessments. The FDA played a pivotal role in overseeing nanomaterial regulation in the United States. Notably, certain nanomaterials such as carbon black, aluminum, nanoclay, and zinc oxide were classified under the Generally Recognized as Safe (GRAS) category. The FDA believes that nano-additives can be used if their levels are within specified limits, without the need for further notification.

16.3.2 European Union

In Europe, the use of nanoparticles is generally prohibited unless they are specifically authorized in their nanoform. The EFSA recommends that nanomaterial risk assessment be performed on a case-by-case basis, with a few exemptions. In cases where there are environmental, health, or safety concerns, the 50% nanoparticle threshold should be lowered to 10% for food and medical applications. According to EU regulations, carbon black, titanium nitride, and silicon dioxide are the three authorized nanomaterials for use in plastic food packaging. However, nanoclay and metal nanoparticles, including silver, zinc oxide, and aluminum, have not been authorized in the EU. In 2011, the EFSA published a guidance document called "Guidance on the Risk Assessment of the Application of Nanoscience and Nanotechnologies in the Food and Feed Chain." This document outlines the requirements for physicochemical information from manufacturers. In March 2013, during a parliamentary debate, the Plastics Europe Organization and the Parliament Magazine presented a unique perspective on applications of nanomaterials in food packaging. For the EU Parliament, they carried out tests and research on the migration of nanoparticles from plastic packaging to food. The results of their testing showed that the migration of nanomaterials in food packaging is extremely low, indicating that it is safe and convenient. Vittorio Prodi, an Italian deputy, supported this experimental work and proposal (Bumbudsanpharoke et al., 2015).

In the EU, only products that are subject to one of the following regulations are required to clearly state the presence of nanomaterials on the label of their product and product ingredients, and the rest do not have to do so.

- Regulation (EC) No 1223/2009 Cosmetic Products
- Regulation (EU) No 1169/2011 Provision Food Information
- Regulation (EU) No 10/2011 Plastic Materials and Articles intended to come into contact with food (FCM)

16.3.3 Asia

16.3.3.1 Japan

In Japan, regulatory decisions are made through cooperative efforts among various governmental organizations, such as the Ministry of Health, Labor, and Welfare; the Ministry of Education, Culture, Sports, Science, and Technology; the Ministry of the Environment; and the Ministry of Economic, Trade, and Industries (Chau et al., 2007). The regulations pertaining to the safety of nanomaterials or nanotechnology, nevertheless, remain somewhat ambiguous. During the 190th Conference for the Promotion of Food Import Facilitation Report in January 2016, the Ministry of Health, Labor, and Welfare introduced fresh safety standards for agriculture and food (He et al., 2019).

16.3.3.2 China

The responsibility for establishing national nanotechnology standards in China rests with the National Centre for Nanoscience and Technology (NCNST), which was established in March 2003. The NCNST then embarked on the development of protocols for characterizing nanomaterials and establishing safety requirements for manufacturing. In 2005, the Commission on Nanotechnology Standardization, affiliated with the NCNST, was established to oversee the regulation of nanomaterial production, engineering, and safety considerations (Bai, 2005)). Nonetheless, at that time, the prevailing regulations continued to depend on the China National Standard, and there was a lack of dedicated legislation pertaining to any nanomaterials or nanotechnology connected to the food and agricultural industries.

16.3.3.3 Australia and New Zealand

The Food Standards Australia New Zealand (FSANZ) is a self-governing organization under the Australian government's authority that assumes the responsibility of formulating regulatory measures for the food industry. FSANZ has adopted various approaches to assess and manage the potential risks associated with nano-ingredients in food and nanomaterials used in packaging. In collaboration with other regulatory agencies in Australia, revisions were made to the Application Handbook in 2008. These revisions included the inclusion of application information specifically related to the use and approval of food nanotechnology. An essential aspect of these revisions was the incorporation of comprehensive risk assessment as a mandatory component. The Application Handbook serves as a practical guide for applicants seeking approval from FSANZ, providing them with guidance on navigating the regulatory process effectively. These regulatory advancements were made to prioritize public health and safety in the production and use of new nanomaterials in the food industry. Additionally, FSANZ established a Scientific Nanotechnology Advisory Group, consisting of experts in nanotechnology, nanotoxicology, and nano-safety, to provide advice on the safe application of food nanomaterials (He et al., 2019).

India

The Food Safety and Standards Act, enforced by the Food Safety Standard Authority of India, presides over the existing food safety regulations in India. The introduction of the Nano Mission, a project under the Department of Science and Technology (DST) in India, has raised concerns regarding the possible hazards of nanotechnology on human health and the environment. To address these concerns and promote the safe use and disposal of nanomaterials in research and industry, the DST issued a guidance document in 2016. This document provided a comprehensive review of reports from various regulatory agencies, including the ISO, OECD, OSHA, and United States National Institute for Occupational Safety and Health, with the aim of providing information for the safe handling and disposal of manufactured nanomaterials.

16.4 Guidelines for the Evaluation of Nano-Based Agri-Inputs and Food Products in India

16.4.1 Nanomaterials

A material is considered a nanomaterial (NM) if at least one dimension of its size is between 1–100 nm or if it exhibits improved characteristics or phenomena as a result of dimensional effects, even if the dimensions are more than the nanoscale range up to 1,000 nm.

16.4.2 Nano-Agri Products

These are formulations used in the agricultural industry that contain nanomaterials. These preparations are designed for consumption or application in food, feed, supplements, or as a delivery system for nutraceuticals. Nano-Agri Products (NAPs) comprise materials that display dimensions of zero, one, or two within the nanoscale spectrum or possess a structure either internally or on the surface at the nanoscale level.

16.4.3 Nano-Agri-Input Product (NAIP)

It is a preparation of agricultural inputs (AI) containing nanomaterials intended for various applications in farming, including soil, seed, foliar, and drip applications, as well as other methods on crops. Nano-Agri-Input Products (NAIPs) consist of materials with zero, one, or two dimensions on the nanoscale or possess an internal or surface structure at the nanoscale.

Categorization of NAIPs and NAPs can be done based on their properties, synthesis methods, and functionalities, as follows:

i) Categorization based on the nanoform of the ingredient:
   - *Nanocarriers loaded with AI*: These are soft and hard nanomaterials (NMs) used as carrier systems for targeted agricultural input NMs. They possess the capability to deliver agri-inputs in a controlled and slow-release manner.
   - *AI converted to nanoform*: In order to enhance stability and effectiveness, active molecules or compounds can be transformed into nanoforms.

ii) Categorization based on synthesis methods:
   - *Biologically synthesized NMs*: synthesized using biological agents and their bioactive components.
   - *Chemically synthesized NMs*: produced through the utilization of synthetic chemicals as reducing agents, oxidizing agents, and templates.
   - *Physically synthesized NMs*: created employing physical procedures including ball milling, laser ablation, temperature- and microwave-assisted ultrasonication, pulsed laser deposition, plasma, and UV-assisted techniques.

iii) Based on their chemical composition, NMs can be categorized as organic, inorganic, and composite NMs:
   - *Organic NMs*: These NMs are composed of organic compounds such as lipids, proteins, and carbohydrates. They are predominantly used in agriculture due to their low toxicity.
   - *Inorganic NMs*: Inorganic NMs exhibit high stability and can be synthesized using simple bottom-up methods. They offer a diverse array of adjustable physicochemical properties, including shape, size, surface area, surface charge, composition, and crystallinity. Compared to organic NMs, inorganic NMs are more versatile for agri-inputs. They can be easily tailored to meet specific application requirements, such as optical properties (fluorescence and absorption), electrical properties (surface charge and conductivity), as well as thermal and magnetic properties.
   - *Composite NMs*: Composite NMs are composed of a mixture of materials from different categories.

iv) Based on their degradation characteristics, NMs can be categorized as biodegradable and nonbiodegradable:
   - *Biodegradable*: These NMs possess unique and beneficial properties, making them commonly used in nanocarrier systems and other agri-input applications. They can undergo degradation through natural processes.
   - *Non-biodegradable*: These NMs are frequently utilized in NAIPs and NAPs, particularly in the development of controlled-release fertilizers. They do not undergo degradation easily and provide sustained release of their active components.

Specific considerations need to be taken into account when evaluating NAIPs and NAPs within the regulatory frameworks of the Central Insecticide Board and Registration Committee (CIBRC), the Food Safety and Standard Authority of India (FSSAI), the Fertilizer Control Order (FCO), and the Bureau of Indian Standards (BIS). The approval status of AI and carrier NMs dictates the necessity for additional assessment and approval of NAIPs and NAPs based on the following criteria:

Table 16.2 Nanomaterial definition by several countries

Country and Regulation	Size	Solubility	Aggregates and Agglomerates	Distribution Threshold	Intentionally Manufactured or Engineered	Novel Properties
European Commission (EC)	1–100 nm	No	Yes	50% by number	No	No
EC for food products	1–100 nm and larger	No	Yes	--NA--	Yes	Yes
EC for cosmetics	1–100 nm	Ye	Yes	No specific mention	Yes	No
EC for biocides	1–100 nm	No	Yes	50% by number	No	No
United States Food and Drug Administration (USFDA)	1–100 nm and larger	No	--NA--	No specific mention	Yes	Yes
United States Environmental Protection Agency (US EPA)	1–100 nm	No	Yes	10% by weight	Yes	Yes
Australian Government Department of Health and Ageing	1–100 nm	No	Yes	10% by number	Yes	Yes
Health Canada	1–100 nm and larger	No	Yes	No specific mention	Yes	Yes
Taiwan Council of Labor Affairs	1–100 nm and larger	No	No specific mention	No specific mention	Yes	Yes
Swiss Federal Office of Public Health and Federal Office for the Environment	1–100 nm and larger	No	Yes	50% by number	No	No
Norwegian Environmental Agency	1–100 nm and larger	No	Yes	50% by number	No	No
Belgian Federal Public Service Health, Food Chain Safety and Environment (Belgium 2014)	1–100 nm	No	Yes	50% by number	Yes	No
Danish Ministry of the Environment (2014)	1–100 nm	No	Yes	50% by number	No	No

Source: Adapted from Boverhof et al. (2015) (CC BY-NC-ND 4.0)

- *AI not registered and new nanocarrier*: If the AI is not registered according to existing provisions and the nanocarrier is new and not yet approved in the country, further evaluation and approval of NAIPs or NAPs are required.
- *AI not registered but approved nanocarrier*: If the AI is not registered under existing provisions but the nanocarrier is already approved for other NAIPs or NAPs, further evaluation and approval of NAIPs or NAPs will depend on this scenario.
- *Conventional/traditional NAIPs or NAPs with a new nanocarrier*: If the conventional/traditional form of NAIPs or NAPs is approved for use in the country according to existing provisions, but the nanocarrier system is new and not approved, further evaluation and approval will be necessary.
- *Conventional/traditional NAIPs or NAPs with an approved nanocarrier*: If both the conventional/traditional form of NAIPs or NAPs and the nanocarrier system are approved for use in the country under existing provisions, additional evaluation and approval may not be required.

The assessment procedure should take into account the distinctive process environments of nano-formulations in relation to traditional agricultural inputs and products. Moreover, it is important to disclose the waste disposal method for nanomaterials and their potential environmental consequences.

Apart from the above, the BIS, the national standards body of India, has framed specifications (IS/ISO/TS 20660:2019) for evaluating the physicochemical properties of silver nanoparticles in either powder or colloidal form intended for antibacterial applications (BIS, 2019).

16.5 Global Standards

International Standards Organization (ISO) is a nongovernmental organization represented by most of the countries involved in framing global standards for technology and manufacturing. A technical committee on nanotechnology was established by ISO in 2005. Standardization in the field of nanotechnologies is the scope of this committee, which includes the following:

- Understanding and control of matter and processes at the nanoscale, typically, but not exclusively, below 100 nm in one or more dimensions where the onset of size-dependent phenomena usually enables novel applications,
- Utilizing the properties of nanoscale materials that differ from the properties of individual atoms, molecules, and bulk matter to create improved materials, devices, and systems that exploit these new properties"(ISO/TC 229—Nanotechnologies," 2021).

ISO has developed a standard for nanomaterials, including definitions, terminologies, vocabulary, and hierarchy ("ISO/TS 80004-11:2017(en), Nanotechnologies—Vocabulary—Part 11: Nanolayer, nanocoating, nanofilm, and related terms," n.d.). which are extensively covered in ISO/TC229 for nanotechnology (Soltani and Pouypouy, 2019). Some of the important ones include the following.

Table 16.3 Member countries and the concerned stakeholder involved in framing ISO/TC229 (Soltani and Pouypouy, 2019)

Country	Stakeholder
India	Bureau of Indian Standards (BIS)
Germany	German Institute for Standardization (DIN)
Australia	Standards Australia (SA)
Austria	Austrian Standards International (ASI)
Belgium	The Bureau of Normalization (NBN)
Brazil	The Brazilian National Standards Organization (ABNT)
Bulgaria	Bulgarian Institute for Standardization (BDS)
Canada	Standards Council of Canada (SCC)
China	Standardization Administration of China (SAC)
Colombia	The Colombian Institute of Technical Standards and Certification (ICONTEC)
Czech Republic	Czech Office for Standards, Metrology and Testing (UNMZ)
Denmark	Danish Standards (DS)
Finland	The Finnish Standards Association (SFS)
France (AFNOR)	French Standardization Association (AFNOR)
Indonesia (BSN)	National Standardization Agency of Indonesia (BSN)
Iran	Iran National Standards Organization (INSO)
Ireland (NSAI)	National Standards Authority of Ireland (NSAI)
Israel (SII)	The Standards Institution of Israel (SII)
Italy (UNI)	The Italian National Standards Institute (UNI)
Japan	The Japanese Industrial Standards Committee (JISC)
Korea	Korean Agency for Technology and Standards
Malaysia	The Department of Standards Malaysia (DSM)
Mexico (DGN)	The Directorate General for Standards (DGN)
Mongolia (MASM)	The Mongolian Agency for Standardization and Metrology (MASM)
Netherlands (NEN)	The National Ecological Network (NEN)
Norway (SN)	Standards Norway (SN)
Poland (PKN)	National Standards Body in Poland (PKN)
Portugal (IPQ)	The Portuguese Institute of Quality (IPQ)
Russian Federation (GOST R)	The Federal Agency on Technical Regulating and Metrology (GOST R)
Singapore (SPRING SG)	The Standards, Productivity and Innovation Board
South Africa (SABS)	South African Bureau of Standards (SABS)
Spain (UNE)	Spanish Association for Standardization (UNE)
Sweden (SIS)	Swedish Institute for Standards, SIS,
Switzerland (SNV)	The Swiss Association for Standardization (SNV)
United Kingdom (BSI)	The British Standards Institution (BSI)
United States (ANSI)	The American National Standards Institute (ANSI)

Source: Adapted from Soltani and Pouypouy (2019)

Nanotechnology: Application of scientific knowledge to manipulate and control matter predominantly in the nanoscale to make use of size- and structure-dependent properties and phenomena distinct from those associated with individual atoms, molecules, or extrapolations from larger sizes of the same material.

Nanomaterial: Material with any external dimension in the nanoscale or having internal structure or surface structure in the nanoscale.

Manufactured nanomaterial: Nanomaterial intentionally produced to have selected properties or composition.

Engineered nanomaterial: Nanomaterial designed for a specific purpose or function

ISO has also defined coatings, layers, films, freestanding films, and membranes (film, layer, multilayer, foil, membrane, coating, nanoscale, nanostructured, nanoplate, nanopore, nanoporous material), as well as the elements having thickness in the nanoscale (nanocoatings, nanolayers, nanofilms, and related terms) ("ISO/TS 80004-11:2017(en), Nanotechnologies—Vocabulary—Part 11: Nanolayer, nanocoating, nanofilm, and related terms," n.d.).

16.6 Market Trends and Consumer Preference

The level of consumer approval for nanomaterials employed in food packaging is greatly impacted by their understanding and access to relevant knowledge and information (Cushen et al., 2012). Increasing social trust and knowledge among consumers can enhance the acceptance of food packaged with nanomaterials (Siegrist et al., 2007). Enhancing consumer education is a recommended approach to improving the acceptance of both nano-packaged foods and nano-packaging (Chen et al., 2013). Studies on consumer perception of nanotechnology in the food industry have been conducted in various countries, including Australia (Evans et al., 2010), Brazil (Vidigal et al., 2015), Canada (Matin et al., 2012), Chile (Schnettler et al., 2013), Taiwan (Chang et al., 2017), and the United States (Zhou and Hu, 2018). Similarly, research has been carried out in European countries such as Austria (Joubert et al., 2020), Germany (Roosen et al., 2015), Italy(Sodano et al., 2016), Ireland (Handford et al., 2015), Switzerland (Siegrist et al., 2007), and the United Kingdom. These studies demonstrate regional variations in knowledge and public perception of nanotechnology, which evolve over time and are challenging to predict (Gómez-Llorente et al., 2022).

One of the leading global suppliers of nanoclays is Nanocor, a subsidiary of AMCOL International Incorporation, Illinois, USA. Nanocor specializes in plastic nanocomposites and has successfully commercialized resin and packaging products that utilize montmorillonite minerals. An example of their innovation is Durethan R© KU2-260, a transparent composite used for barrier films and coatings in packaging. This composite combines polyamide 6 (nylon 6), an engineered plastic, with nanoclay. The nanoclay particles are evenly distributed within a polymer matrix, resulting in exceptional properties such as gas and moisture barrier, strength, toughness, abrasion resistance, and chemical resistance. Durethan R© can be employed in various packaging applications, ranging from everyday food items to medical products (Sekhon, 2010; Chun, 2009; Durán and Marcato, 2013; Zeng et al., 2005; Cushen et al., 2012). This product is available from Lanxess Deutschland GmbH (Hatzigrigoriou and Papaspyrides, 2011). Another noteworthy development in the field is Aegis™OXCE by Honeywell Polymer, which is a nylon 6 nanoclay composite. It is specifically used for high-oxygen barrier packaging catering to beer and flavored alcoholic beverages. Unlike Nanocor's products, Aegis™OXCE is modified with an additional oxygen scavenger (Bumbudsanpharoke and Ko, 2015). This innovation has found commercial success, including its use in the 1.6-L Hite Pitcher beer bottle manufactured by Hite Brewery Co. in South Korea (; Peters et al., 2011; Picó and Blasco, 2012.

A collaboration between Nanocor, Voridian (a division of Eastman Chemical), and the use of MXD6 Imperm R© has led to the development of nylon nanocomposites. These nanocomposites serve as a barrier layer in thermoformed containers or multilayer PET bottles, making them suitable for beer, liquor, or small carbonated soft drink beverages (Bumbudsanpharoke and Ko, 2015). Miller Brewing in the United States currently commercializes this nanocomposite in their plastic beer bottles, specifically in brands such as Miller Lite, Miller Genuine Draft, and Icehouse (Chaudhry et al., 2008). Another nanoclay-based product, Nanoblend™, has been created by PolyOne Corp. It contains 40% nanoclays and is incorporated into a matrix of homo- and modified polypropylene, low-density polyethylene, linear low-density polyethylene, and high-density polyethylene. Nanoblend™ serves as an alternative to polyolefin-based nanoclay composites on the market (Bumbudsanpharoke and Ko, 2015). While it provides excellent oxygen and water barrier properties, it has slightly reduced clarity and a higher haze compared to the nylon-based composite. Plantic Technologies Ltd. has introduced a packaging solution called Plantic R© R1 Tray, which incorporates a blend of thermoformed starch and nanoclay. This innovative composite material is utilized in packaging various chocolate products, including Cadbury R© Dairy Milk™, Milk Tray™, and Marks & Spencer Swiss chocolates (Bumbudsanpharoke and Ko, 2015). Another notable application is the use of nanosilver, which possesses antimicrobial properties. Silver molecules in nanosilver can eliminate around 650 disease-causing pathogens within 6 minutes of contact, surpassing the effectiveness of general antibiotics that can only target five to six pathogens (Han and Li, 2008; Thanh and Phong, 2009). Nanosilver is commonly incorporated into food packaging to impart antimicrobial capabilities. A wide range of antibacterial food packaging products utilizing nanosilver can be found in global markets, particularly through e-commerce platforms. Examples of branded and unbranded nanosilver food containers available online include e.Window R©, Everin R© Food Containers Nano Silver Airtight, Dokdo Anti-Bacterial Food Storage Container, and FresherLonger™ Miracle Food Storage and Plastic Bag. According to the manufacturers of these products, nanosilver successfully lowers bacterial growth and prolongs the freshness of stored goods.

In China, Quan Zhou Hu Zheng Nano Technology Co., Ltd. introduced the Baoxianhe Nanosilver Storage Box for fresh food storage, while Anson Nano-Biotechnology Co., Ltd. marketed nano-packaging products under the Anson Nano Antimicrobial Storage Series, including plastic bags and film. While in Taiwan, SongSing Nano Technology Co. Ltd. utilizes nano-zinc oxide as an antimold plastic wrap (Bumbudsanpharoke and Ko, 2015). EcoSphere Biolatex R© is a novel technology produced by EcoSynthetix, a multinational business known as Nano Technology Co., Ltd. This technology involves modifying starch molecules derived from renewable raw materials, including potato or corn starch, to produce a natural base binder. The resulting biopolymer nanosphere exhibits high solid dispersion when immersed in water, with granule sizes ranging from 50 nm to 150 nm. Compared to natural starch granules, this biopolymer nanomaterial has a 400 times larger surface area, making EcoSphere

Biolatex R© superior to traditional starch and comparable to polyvinyl acetate adhesive (Klass, 2007). In the United States, nanostarch has been utilized as a replacement for traditional adhesive in McDonald's hamburger clamshells (Blasco and Pico, 2011). Another type of nanomaterial, TiN, is incorporated into plastic for processing purposes. When mixed with plastics like PET, it enhances the material's thermal properties and increases PET bottle production output (EFSA Panel on food contact materials and aids, 2012).

16.7 Perspectives and Conclusion

It is evident from the literature that there is an increasing trend in the incorporation of nanomaterials in food packaging applications for the superlative properties discussed at the beginning of the chapter (Ashfaq et al., 2022; Chaudhary et al., 2020; Garcia, 2022). In the current scenario, the outlook for the application of nanocomposite for food packaging is supposed to increase in the coming years. Hence, there is a need for elaborate regulations in place to ensure food quality, food safety, migration of nanomaterials, Human health, and the environment. Long-term risk assessment studies are necessary to ascertain the impact and role of nanomaterials in packaging materials on human health and ecotoxicity (Gallocchio et al., 2016; Garcia et al., 2018; de Oliveria Morais et al., 2020; Störmer et al., 2017). Most developed countries have drafted regulations for the application of nanomaterials in food packaging. It is prudent to undertake a holistic and structured approach to arrive at regulations related to nanomaterials in food packaging materials. It must include the byproducts formed during the synthesis of nanomaterials for the packaging application migration levels into different foods. Recycling nanocomposite will be a challenge with existing facilities; it might require additional facilities such as containers, awareness and training programs for handlers etc. It is crucial to calculate the additional cost of recycling. Another important aspect of the inclusion of nanomaterials in the packaging film is the education of consumers about the presence of nanomaterials (Joubert et al., 2020; Siddiqui et al., 2022). It might require new logos on the packaging materials along with disposal/recycling instructions. Regulatory agencies have a huge responsibility and task in front of them to frame or expand the regulations to cover all aspects of nanomaterial-based food packaging. A coordinated approach including concerned stakeholders such as industries, researchers, consumers, and others might be the best possible solution to move forward.

References

Amenta, V., Aschberger, K., Arena, M., Bouwmeester, H., Botelho Moniz, F., Brandhoff, P., Gottardo, S., Marvin, H. J. P., Mech, A., Quiros Pesudo, L., Rauscher, H., Schoonjans, R., Vettori, M. V., Weigel, S., & Peters, R. J. (2015). Regulatory aspects of nanotechnology in the agri/feed/food sector in EU and non-EU countries. *Regulatory Toxicology and Pharmacology, 73*, 463–476. https://doi.org/10.1016/j.yrtph.2015.06.016

Amin, U., Khan, M. K. I., Maan, A. A., Nazir, A., Riaz, S., Khan, M. U., Sultan, M., Munekata, P. E. S., & Lorenzo, J. M. (2022). Biodegradable active, intelligent, and smart packaging materials for food applications. *Food Packaging and Shelf Life*, *33*, 100903. https://doi.org/10.1016/j.fpsl.2022.100903

Ashfaq, A., Khursheed, N., Fatima, S., Anjum, Z., & Younis, K. (2022). Application of nanotechnology in food packaging: Pros and Cons. *Journal of Agriculture and Food Research*, *7*, 100270. https://doi.org/10.1016/j.jafr.2022.100270

de Azeredo, H. M. C. (2009). Nanocomposites for food packaging applications. *Food Research International*, *42*(9), 1240–1253. https://doi.org/10.1016/j.foodres.2009.03.019

Bai, C. (2005). Ascent of nanoscience in China. *Science*, *309*, 61–63. https://doi.org/10.1126/science.1115172

Barage, S., Lakkakula, J., Sharma, A., Roy, A., Alghamdi, S., Almehmadi, M., Hossain, M. J., Allahyani, M., & Abdulaziz, O. (2022). Nanomaterial in food packaging: a comprehensive review. *Journal of Nanomaterials*, *2022*, 1–12. https://doi.org/10.1155/2022/6053922

BIS. (2019). Nanotechnologies—Antibacterial Silver Nanoparticles—Specification of Characteristics and Measurement Methods. *IS/ISO/TS20660:2019*.

Blasco, C., & Pico, Y. (2011). Determining nanomaterials in food. *Trends in Analytical Chemistry*, *30*, 84–99.

Boverhof, D. R., Bramante, C. M., Butala, J. H., Clancy, S. F., Lafranconi, M., West, J., & Gordon, S. C. (2015). Comparative assessment of nanomaterial definitions and safety evaluation considerations. *Regulatory Toxicology and Pharmacology*, *73*(1), 137–150. https://doi.org/10.1016/j.yrtph.2015.06.001

Bumbudsanpharoke, N., & Ko, S. (2015). Nano-food packaging: an overview of market, migration research, and safety regulations. *Journal of Food Science*, *80*, R910–R923. https://doi.org/10.1111/1750-3841.12861

Bumbudsanpharoke, N., Choi, J., & Ko, S. (2015). Applications of nanomaterials in food packaging. *Journal of Nanoscience and Nanotechnology*, *15*(9), 6357–6372. https://doi.org/10.1166/jnn.2015.10847

Campelo, P. H., Sant'Ana, A. S., & Pedrosa Silva Clerici, M. T. (2020). Starch nanoparticles: production methods, structure, and properties for food applications. *Current Opinion in Food Science*, *33*, 136–140. https://doi.org/10.1016/j.cofs.2020.04.007

Chang, H. H., Huang, C. Y., Fu, C. S., & Hsu, M. T. (2017). The effects of innovative, consumer and social characteristics on willingness to try nano-foods: product uncertainty as a moderator. *Information Technology & People*, *30*, 653–690.

Chau, C.-F., Wu, S.-H., & Yen, G.-C. (2007). The development of regulations for food nanotechnology. *Trends in Food Science & Technology*, *18*(5), 269–280. https://doi.org/10.1016/j.tifs.2007.01.007

Chaudhary, P., Fatima, F., & Kumar, A. (2020). Relevance of nanomaterials in food packaging and its advanced future prospects. *Journal of Inorganic and Organometallic Polymers and Materials*, *30*(12), 5180–5192. https://doi.org/10.1007/s10904-020-01674-8

Chaudhry, Q., Scotter, M., Blackburn, J., Ross, B., Boxall, A., Castle, L., Aitken, R., & Watkins, R. (2008). Applications and implications of nanotechnologies for the food sector. *Food Additives & Contaminants: Part A*, *25*(3), 241–258. https://doi.org/10.1080/02652030701744538

Chen, M.-F., Lin, Y.-P., & Cheng, T.-J. (2013). Public attitudes toward nanotechnology applications in Taiwan. *Technovation*, *33*, 88–96.

Cheng, H., Xu, H., Julian McClements, D., Chen, L., Jiao, A., Tian, Y., Miao, M., & Jin, Z. (2022). Recent advances in intelligent food packaging materials: principles, preparation and applications. *Food Chemistry*, *375*, 131738. https://doi.org/10.1016/j.foodchem.2021.131738

Chun, A. L. (2009). Will the public swallow nanofood? *Nature Nanotechnology*, *4*, 790–791.

Cushen, M., Kerry, J., Morris, M., Cruz-Romero, M., & Cummins, E. (2012). Nanotechnologies in the food industry—Recent developments, risks and regulation. *Trends in Food Science & Technology*, *24*(1), 30–46. https://doi.org/10.1016/j.tifs.2011.10.006

Deshmukh, R. K., Hakim, L., Akhila, K., Ramakanth, D., & Gaikwad, K. K. (2023). Nano clays and its composites for food packaging applications. *International Nano Letters*, *13*(2), 131–153. https://doi.org/10.1007/s40089-022-00388-8

Durán, N., & Marcato, P. D. (2013). Nanobiotechnology perspectives. role of nanotechnology in the food industry: a review. *International Journal of Food Science and Technology*, *48*, 1127–1134. https://doi.org/10.1111/ijfs.12027

Evans, G., Kermarrec, C., Sable, T., & Cox, D. N. (2010). Reliability and predictive validity of the food technology neophobia scale. *Appetite*, *54*, 390–393.

Faccio, E., & Guiotto Nai Fovino, L. (2019). Food neophobia or distrust of novelties? Exploring consumers' attitudes toward GMOs, insects and cultured meat. *Applied Sciences*, *9*, 4440.

Gallocchio, F., Cibin, V., Biancotto, G., Roccato, A., Muzzolon, O., Carmen, L., Simone, B., Manodori, L., Fabrizi, A., Patuzzi, I., & Ricci, A. (2016). Testing nano-silver food packaging to evaluate silver migration and food spoilage bacteria on chicken meat. *Food Additives & Contaminants: Part A*, *33*(6), 1063–1071. https://doi.org/10.1080/19440049.2016.1179794

Garcia, C. V. (2022). Editorial: nanomaterials in food packaging. *Frontiers in Nutrition*, *9*. https://doi.org/10.3389/fnut.2022.1083185

Garcia, C. V., Shin, G. H., & Kim, J. T. (2018). Metal oxide-based nanocomposites in food packaging: applications, migration, and regulations. *Trends in Food Science & Technology*, *82*, 21–31. https://doi.org/10.1016/j.tifs.2018.09.021

Gómez-Llorente, H., Hervás, P., Pérez-Esteve, É., Barat, J. M., & Fernández-Segovia, I. (2022). Nanotechnology in the agri-food sector: consumer perceptions. *NanoImpact*, *26*, 100399.

Gundewadi, G., Rudra, S. G., Prasanna, R., Banerjee, T., Singh, S. K., Dhakate, S. R., Gupta, A., & Anand, A. (2022). Palladium encapsulated nanofibres for scavenging ethylene from sapota fruits. *Frontiers in Nutrition*, *9*. https://doi.org/10.3389/fnut.2022.994813

Han, M., & Li, S. (2008). Study on the medical application of silver material. In *Micro-and Nanotechnology: Materials, Processes, Packaging, and Systems IV* (pp. 299–304). SPIE.

Han Lyn, F., & Nur Hanani, Z. A. (2022). Graphene-based polymer nanocomposites in food packaging and factors affecting the behaviour of graphene-based materials: a review. *Journal of Nanoparticle Research*, *24*(9), 179. https://doi.org/10.1007/s11051-022-05558-5

Handford, C. E., Dean, M., Spence, M., Henchion, M., Elliott, C. T., & Campbell, K. (2015). Awareness and attitudes towards the emerging use of nanotechnology in the agri-food sector. *Food Control*, *57*, 24–34.

Hatzigrigoriou, N. B., & Papaspyrides, C. D. (2011). Nanotechnology in plastic food-contact materials. *Journal of Applied Polymer Science*, *122*(6), 3719–3738. https://doi.org/10.1002/app.34786

He, X., Deng, H., & Hwang, H. (2019). The current application of nanotechnology in food and agriculture. *Journal of Food and Drug Analysis*, *27*(1), 1–21. https://doi.org/10.1016/j.jfda.2018.12.002

Joubert, I. A., Geppert, M., Ess, S., Nestelbacher, R., Gadermaier, G., Duschl, A., Bathke, A. C., & Himly, M. (2020). Public perception and knowledge on nanotechnology: a study based on a citizen science approach. *NanoImpact*, *17*, 100201.

Klass, C. P. (2007). New nanoparticle latex offers natural advantage. *Paper360 Magazine*, *2*(1), 30–31.

Kumar, A., Choudhary, A., Kaur, H., Mehta, S., & Husen, A. (2021). Metal-based nanoparticles, sensors, and their multifaceted application in food packaging. *Journal of Nanobiotechnology*, *19*(1), 256. https://doi.org/10.1186/s12951-021-00996-0

Mackay, M. E., Tuteja, A., Duxbury, P. M., Hawker, C. J., Van Horn, B., Guan, Z., Chen, G., & Krishnan, R. S. (2006). General strategies for nanoparticle dispersion. *Science*, *311*(5768), 1740–1743. https://doi.org/10.1126/science.1122225

Matin, A. H., Goddard, E., Vandermoere, F., Blanchemanche, S., Bieberstein, A., Marette, S., & Roosen, J. (2012). Do environmental attitudes and food technology neophobia affect perceptions of the benefits of nanotechnology? *International Journal of Consumer Studies*, *36*, 149–157. https://doi.org/10.1111/j.1470-6431.2011.01090.x

Mei, L., & Wang, Q. (2020). Advances in using nanotechnology structuring approaches for improving food packaging. *Annual Review of Food Science and Technology*, *11*(1), 339–364. https://doi.org/10.1146/annurev-food-032519-051804

Mihindukulasuriya, S. D. F., & Lim, L.-T. (2014). Nanotechnology development in food packaging: a review. *Trends in Food Science & Technology*, *40*(2), 149–167. https://doi.org/10.1016/j.tifs.2014.09.009

Mittal, V. (2009). Polymer layered silicate nanocomposites: a review. *Materials*, *2*(3), 992–1057. https://doi.org/10.3390/ma2030992

Mustafa, F., & Andreescu, S. (2020). Nanotechnology-based approaches for food sensing and packaging applications. *RSC Advances*, *10*(33), 19309–19336. https://doi.org/10.1039/D0RA01084G

Oladebeye, A. O., Oshodi, A. A., Amoo, I. A., & Karim, A. A. (2013). Morphology, X-ray diffraction and solubility of underutilized legume starch nanocrystals. *International Journal of Science and Research*, *2*(3), 497–503.

de Oliveria Morais, L., Macedo, E. V., Granjeiro, J. M., & Delgado, I. F. (2020). Critical evaluation of migration studies of silver nanoparticles present in food packaging: a systematic review. *Critical Reviews in Food Science and Nutrition*, *60*(18), 3083–3102. https://doi.org/10.1080/10408398.2019.1676699

Othman, S. H. (2014). Bio-nanocomposite materials for food packaging applications: types of biopolymer and nano-sized filler. *Agriculture and Agricultural Science Procedia*, *2*, 296–303. https://doi.org/10.1016/j.aaspro.2014.11.042

Paidari, S., Tahergorabi, R., Anari, E. S., Nafchi, A. M., Zamindar, N., & Goli, M. (2021). Migration of various nanoparticles into food samples: a review. *Foods*, *10*(9), 2114. https://doi.org/10.3390/foods10092114

de Paiva, L. B., Morales, A. R., & Valenzuela Díaz, F. R. (2008). Organoclays: properties, preparation and applications. *Applied Clay Science*, *42*(1–2), 8–24. https://doi.org/10.1016/j.clay.2008.02.006

Peters, R., ten Dam, G., Bouwmeester, H., Helsper, H., Allmaier, G., vd Kammer, F., Ramsch, R., Solans, C., Tomaniová, M., Hajslova, J., & Weigel, S. (2011). Identification and characterization of organic nanoparticles in food. *TrAC Trends in Analytical Chemistry*, *30*(1), 100–112. https://doi.org/10.1016/j.trac.2010.10.004

Picó, Y., & Blasco, C. (2012). Nanomaterials in food, which way forward? In *Comprehensive Analytical Chemistry* (pp. 305–353). Elsevier.

Rafieian, F., & Simonsen, J. (2014). Fabrication and characterization of carboxylated cellulose nanocrystals reinforced glutenin nanocomposite. *Cellulose*, *21*(6), 4167–4180. https://doi.org/10.1007/s10570-014-0305-4

Ranjan, S., Dasgupta, N., Chakraborty, A. R., Melvin Samuel, S., Ramalingam, C., Shanker, R., & Kumar, A. (2014). Nanoscience and nanotechnologies in food industries: opportunities and research trends. *Journal of Nanoparticle Research*, *16*(6), 2464. https://doi.org/10.1007/s11051-014-2464-5

Ranjan, S., Dasgupta, N., Singh, S., & Gandhi, M. (2019). Toxicity and regulations of food nanomaterials. *Environmental Chemistry Letters*, *17*(2), 929–944. https://doi.org/10.1007/s10311-018-00851-z

Robertson, G. L. (2005). *Food Packaging: Principles and Practice*. CRC Press.

Roosen, J., Bieberstein, A., Blanchemanche, S., Goddard, E., Marette, S., & Vandermoere, F. (2015). Trust and willingness to pay for nanotechnology food. *Food Policy*, *52*, 75–83.

Salgado, P. R., Di Giorgio, L., Musso, Y. S., & Mauri, A. N. (2021). Recent developments in smart food packaging focused on biobased and biodegradable polymers. *Frontiers in Sustainable Food Systems*, *5*. https://doi.org/10.3389/fsufs.2021.630393

Schnettler, B., Crisóstomo, G., Sepúlveda, J., Mora, M., Lobos, G., Miranda, H., & Grunert, K. G. (2013). Food neophobia, nanotechnology and satisfaction with life. *Appetite*, *69*, 71–79.

Sekhon, B. S. (2010). Food nanotechnology–an overview. *Nanotechnology Science and Applications*, 1–15.

Sharma, C., Dhiman, R., Rokana, N., & Panwar, H. (2017). Nanotechnology: an untapped resource for food packaging. *Frontiers in Microbiology*, *8*. https://doi.org/10.3389/fmicb.2017.01735

Sid, S., Mor, R. S., Kishore, A., & Sharanagat, V. S. (2021). Bio-sourced polymers as alternatives to conventional food packaging materials: a review. *Trends in Food Science & Technology*, *115*, 87–104. https://doi.org/10.1016/j.tifs.2021.06.026

Siddiqui, S. A., Zannou, O., Bahmid, N. A., Fidan, H., Alamou, A.-F., Nagdalian, A. A., Hassoun, A., Fernando, I., Ibrahim, S. A., & Arsyad, M. (2022). Consumer behavior towards nanopackaging—a new trend in the food industry. *Future Foods*, *6*, 100191. https://doi.org/10.1016/j.fufo.2022.100191

Siegrist, M., Cousin, M.-E., Kastenholz, H., & Wiek, A. (2007). Public acceptance of nanotechnology foods and food packaging: the influence of affect and trust. *Appetite*, *49*(2), 459–466. https://doi.org/10.1016/j.appet.2007.03.002

Silvestre, C., Duraccio, D., & Cimmino, S. (2011). Food packaging based on polymer nanomaterials. *Progress in Polymer Science*, *36*(12), 1766–1782. https://doi.org/10.1016/j.progpolymsci.2011.02.003

Sodano, V., Gorgitano, M. T., Verneau, F., & Vitale, C. D. (2016). Consumer acceptance of food nanotechnology in Italy. *British Food Journal*, *118*, 714–733.

Soltani, A. M., & Pouypouy, H. (2019). Standardization and Regulations of Nanotechnology and Recent Government Policies Across the World on Nanomaterials. In *Advances in Phytonanotechnology* (pp. 419–446). Elsevier. https://doi.org/10.1016/B978-0-12-815322-2.00020-1

Souza, V. G. L., & Fernando, A. L. (2016). Nanoparticles in food packaging: Biodegradability and potential migration to food—a review. *Food Packaging and Shelf Life*, *8*, 63–70. https://doi.org/10.1016/j.fpsl.2016.04.001

Störmer, A., Bott, J., Kemmer, D., & Franz, R. (2017). Critical review of the migration potential of nanoparticles in food contact plastics. *Trends in Food Science & Technology*, *63*, 39–50. https://doi.org/10.1016/j.tifs.2017.01.011

Tan, W., Zhang, Y., Szeto, Y., & Liao, L. (2008). A novel method to prepare chitosan/montmorillonite nanocomposites in the presence of hydroxy-aluminum oligomeric cations. *Composites Science and Technology*, *68*(14), 2917–2921. https://doi.org/10.1016/j.compscitech.2007.10.007

Thanh, N. V. K., & Phong, N. T. P. (2009). Investigation of Antibacterial Activity of Cotton Fabric Incorporating Nano Silver Colloid. In *Journal of Physics: Conference Series* (p. 012072). IOP Publishing.

Trinh, B. M., Chang, B. P., & Mekonnen, T. H. (2023). The barrier properties of sustainable multiphase and multicomponent packaging materials: a review. *Progress in Materials Science*, *133*, 101071. https://doi.org/10.1016/j.pmatsci.2023.101071

Vanderroost, M., Ragaert, P., Devlieghere, F., & De Meulenaer, B. (2014). Intelligent food packaging: the next generation. *Trends in Food Science & Technology*, *39*(1), 47–62. https://doi.org/10.1016/j.tifs.2014.06.009

Vidigal, M. C., Minim, V. P., Simiqueli, A. A., Souza, P. H., Balbino, D. F., & Minim, L. A. (2015). Food technology neophobia and consumer attitudes toward foods produced by new and conventional technologies: A case study in Brazil. *LWT-Food Science and Technology*, *60*, 832–840.

Wikström, F., Verghese, K., Auras, R., Olsson, A., Williams, H., Wever, R., Grönman, K., Kvalvåg Pettersen, M., Møller, H., & Soukka, R. (2019). Packaging strategies that save food: a research agenda for 2030. *Journal of Industrial Ecology*, *23*(3), 532–540. https://doi.org/10.1111/jiec.12769

Wohner, B., Pauer, E., Heinrich, V., & Tacker, M. (2019). Packaging-related food losses and waste: an overview of drivers and issues. *Sustainability*, *11*(1), 264. https://doi.org/10.3390/su11010264

Xue, L., Liu, G., Parfitt, J., Liu, X., Van Herpen, E., Stenmarck, Å., O'Connor, C., Östergren, K., & Cheng, S. (2017). Missing food, missing data? A critical review of global food losses and food waste data. *Environmental Science & Technology*, *51*(12), 6618–6633. https://doi.org/10.1021/acs.est.7b00401

Yu, M., Ji, N., Wang, Y., Dai, L., Xiong, L., & Sun, Q. (2021). Starch-based nanoparticles: stimuli responsiveness, toxicity, and interactions with food components. *Comprehensive Reviews in Food Science and Food Safety*, *20*(1), 1075–1100. https://doi.org/10.1111/1541-4337.12677

Zeng, Q. H., Yu, A. B., Lu, G. Q., & Paul, D. R. (2005). Clay-based polymer nanocomposites: research and commercial development. *Journal of Nanoscience and Nanotechnology*, *5*, 1574–1592.

Zhou, G., & Hu, W. (2018). Public acceptance of and willingness-to-pay for nanofoods in the US. *Food Control*, *89*, 219–226.

17

Nanotechnology Solutions in Food Packaging

Present and Future

Navjot Kaur[1], *Hamid*[1], *and Pintu Choudhary*[2]

[1] *Department of Food Technology and Nutrition, School of Agriculture, Lovely Professional University, Phagwara, Punjab, India*
[2] *Department of Food Technology, Chaudhary Bansal Lal Government Polytechnic, Bhiwani, Haryana, India*

17.1 Introduction

Nanotechnology is a field that involves various techniques and tools for generating, controlling, or investigating matter in nanoscale dimensions (Primožič et al., 2021). The creation of innovative nanomaterials has made it feasible to enhance agricultural growth, quality of food, and environmental monitoring (He et al., 2019). Materials are improved in terms of their other physiochemical qualities, including color, diffusivity, solubility, strength, toxicity, and magnetic and optical properties, among others, and have unique properties, including a high surface-to-volume ratio (Gupta et al., 2016; Singh et al., 2017). When the particle size is reduced below the threshold, the material's physical and chemical properties differ significantly from those of macroscale materials consisting of the same component. Metal, glass, and paper are common ingredients in food packaging. While these materials remain in use, plastics are indeed a desirable option due to their light weight, ease of processing, exceptional versatility in physical qualities, and relatively inexpensive. Polymers that are most frequently used for food packaging are polystyrene, polyethylene terephthalate, polyvinyl chloride, polypropylene and various grades of polyethylene such as, low-density polyethylene (LDPE) and high-density polyethylene (HDPE). LDPE has a branched structure whereas HDPE has a linear structure with minimal branching, resulting in stronger intermolecular forces and thus greater tensile strength. Although polymers have transformed the food business and are superior to conventional materials in many ways, they also have certain drawbacks, including permeability to gases and other tiny molecules (Duncan, 2011).

The primary purpose of packaging is to transport and store food while shielding it from contaminants, chemicals, light, moisture, and oxygen in order to maintain its safety and quality, resulting in increased shelf life. These days, this is made feasible by the use of materials that are utilized in various forms, such as packaging films and trays, with materials based on plastic, and glass. Cellulose-based packaging is the most popular choice.

Nanotechnology for Sustainable Food Packaging, First Edition. Edited by C. Anandharamakrishnan, Jeyan A. Moses, and M. Maria Leena.

In addition to serving as a means of communication with consumers and the entire food supply chain, packaging also serves as a marketing tool for the product, with the appearance of the packaging being one of the most crucial aspects of consumer choice (Simmonds and Spence, 2017). With the advancement in research and development, innovation entered the packaging industry. Nanotechnology can help the packaging sector develop innovative packaging materials, enabling the development and manufacturing of packaging systems with new features. Utilizing NPs is appealing because of their special qualities, which are only possible at the nanoscale. When compared to their micro- and macroscopic counterparts, particles in the nanoscale range have drastically distinct physical and chemical properties, which can have considerable benefits for packaging materials. The effects of nanoscale technology on high surface area-to-volume ratio materials are typically related to their enhanced surface area, chemical stability, and low density. Specialized optical behavior and high mechanical characteristics (Gleiter, 2000). Thus, nanomaterials allow the fabrication of nanocomposites, which results in packaging material with improved properties like increased barrier capacity and strong mechanical characteristics. In addition to all of these possibilities, it is expected that nanotechnology-enabled packaging materials will utilize fewer raw materials, allow the use of bio-based and biodegradable materials, and retain the core qualities of a "conventional or traditional" packaging solution.

There is no such material that is totally impermeable to air, water vapor, natural compounds present in the food, or even the material being used for packaging, which makes migration and permeability a crucial concern in food packaging (Finnigan, 2009). High barriers to migration or diffusion of gases may not be desirable in some circumstances, such as when packaging fresh produce whose shelf life depends on accessibility to oxygen for maintaining continuous cellular respiration (Robertson, 2005). Boosting food security, enhancing flavor and nutrition delivery and extending shelf life are the three main connections between nanotechnology and the food industry. Therefore, nanotechnology transforms conventional forms of food and beverage packaging into innovative, active, and intelligent packaging that provides features like moisture control and antioxidant activity in addition to product protection and brand presentation. For example, tiny sachets decrease exposure to oxygen by scavenging or capturing leftover oxygen or ethylene from the environment or from the food itself. The food may develop microorganisms or undergo chemical changes as a result of oxygen exposure, among other things. An oxygen scavenger is designed to overcome such conditions in order to extend the food's shelf life. The use of oxygen scavengers can be done in the packaging of milk powder. Antimicrobial activity is seen in sachet systems, where the antimicrobial agents are released gradually to preserve the freshness of the product and also extend its shelf life (Enescu et al., 2019).

Almost every part of the food industry is likely to be impacted by nanotechnology in some way. Most promising applications include the development of nanosensors that can detect microbial or chemical contaminants at surprisingly low concentrations and food packaging materials with extremely high gas barriers and antimicrobial properties. Other potential applications of nanotechnology in food are nano encapsulants for nutrient and flavor or aroma delivery. Food packaging and its preservation are the major fields that have been affected by nanotechnology. The acceptance of this cutting-edge technology has led to improvements in food processing and storage strategies, ensuring the safety of food. Numerous traditional compounds used as food additives have also been discovered to partially exist at the nanoscale scale. For instance, food-grade TiO_2 NPs have recently been discovered to be up to about 40% in the nanoscale range (Dudefoi et al., 2017; Weir et al., 2012). The purpose of this article is to demonstrate to the reader the exciting advantages that nanomaterials can bring to the food business, including improved packaging materials and safer goods that are less likely to be contaminated by bacteria that could be fatal.

17.2 Current Status of Food Nanotechnology

Nanotechnology has significant opportunities for the creation of novel goods and applications, and it is anticipated that future production volumes and product offerings utilizing nanotechnology will rise. In contrast to bulk materials, which have a significantly smaller surface-to-mass ratio, nanoparticles, which are collectively referred to as nanomaterials, have very specific functional qualities that give rise to the advantages that nanotechnologies provide. Numerous industries, particularly those in the food and agricultural industries, are using these special properties of NPs. Nanomaterials are employed in food to boost the bioavailability of vitamins and minerals, improve colors, flavors, taste, and texture, and prevent microbial deterioration of packaged foods. The use of NPs has also made it possible to create novel packaging materials that can extend the shelf life of the product along with ensuring its safety. These opportunities have sparked creative advances in food, agriculture, and allied industries (Peters et al., 2017). The interest and trust of consumers is the point of major concern. Numerous nanotechnology initiatives, such as commissions or centers, have been formed to guarantee the quick development of nanotechnology and encourage economic progress. In addition, there are regulations that are being developed to better safeguard human health and the environment (Siegrist et al., 2007).

Nanotechnology has been widely used in the food processing industry for the creation of innovative functional nanomaterials for food applications, micro- and nanoscale processing, the development and production of foods with improved qualities, and varied instrument design and techniques for their potential application in food safety and biosafety (Augustin and Sanguansri, 2009). The food processing industry, from production to conservation, processing, packaging, and even waste treatment, can be catalyzed by incremental advancements employing nanotechnology to improve and enhance the amount and quality of food. While the nanofood industry suffers from public safety concerns, the food packaging sector is advancing with nanotechnology products at full pace. Leading the charge is "smart" or "active" packaging, which claims to increase food quality and safety while extending product shelf life. The development of nanotechnology for the food industry is heavily focused on food packaging and monitoring (R&D) (Ravichandran, 2010).

17.2.1 Food Packaging

Packaging enables the safe and secure transportation of processed food products across considerable distances from their source of origin while being wholesome and nutritious when consumed. Food packaging's primary functions are to enclose the food, protect it from external influences and harm, and provide consumers with information about the

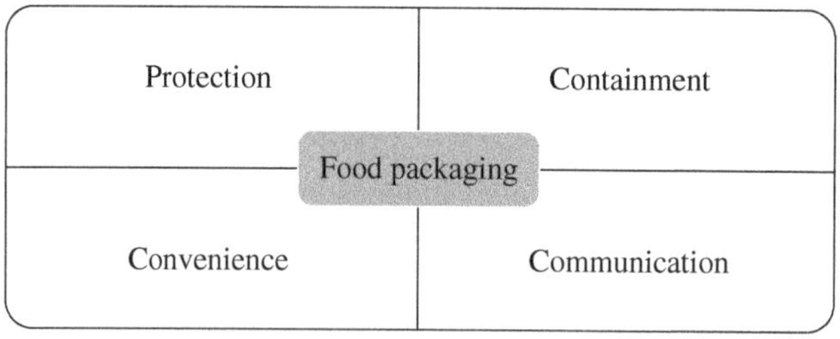

Figure 17.1 Conventional functions of food packaging

ingredient and their nutritional composition (Coles et al., 2003). The supplementary functionalities are usability, traceability, and tamper indication. Food packaging's goal is to contain food while fulfilling industry requirements and consumer preferences, maintaining the safety and quality of food, and minimizing environmental impact. Figure 17.1 represents the major motive behind the packaging of food.

With the nanotechnological interventions, various possibilities are investigated and explored in the field of food, including the direct inclusion of nanomaterials into food products, their integration with food packaging materials, and their use in the food processing sector. The perception and acceptance of newly presented technologies by the consumer and their application affect the commercialization, implementation, and reactions to different applications of nanotechnology (Kim et al., 2014). The fabrication of nanomaterial packaging by joining nanoparticles to polymers improves standard packaging by giving it better barrier qualities, flexibility, and thermal and mechanical stability. Using NPs for filling, i.e., nanofillers with reinforcing properties has improved food packaging in both physical and biochemical characteristics. The mechanical strength, barrier properties, stability, and flexibility are increased. Also, there is increased biocompatibility, biodegradability, and low-waste generation (Kuswandi, 2017). One of the more recent applications of nanotechnology in the food packaging sector is food quality monitoring using biosensors. Other examples include smart and active food packaging systems, nanoemulsions, and nanoencapsulation of bioactive compounds. Nanotechnology can be used to evaluate food quality and safety more effectively. Nanosensors are employed for detecting internal and external conditions in food packaging. Electrochemical sensors based on carbon nanotubes for the detection of organic compounds in food and different aptamers for detecting microbial cells, pesticides, and toxins are just a few of the nanosensors being developed to fulfill a variety of needs (Wegari Dera and Bogale Teseme, 2020).

17.3 Applications in Food Packaging Industry

Nanotechnology has the potential to be used in all stages of the food chain, from farm to fork. The food packaging industry sector is considered an area where the application of nanotechnology has started to deliver on its promises (Hills, 2008). The advancements in plastic material, the incorporation of bioactive constituents that can provide functional benefits beyond those provided by standard packaging techniques, and sensing and signaling of relevant information are the key areas of nanotechnology applications and

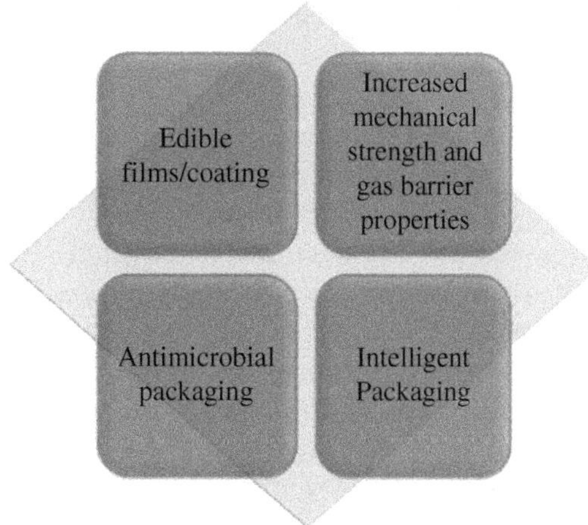

Figure 17.2 Applications of nanotechnology in food packaging

functionality that are being applied in food packaging development. These applications of nanotechnology can be used to generate more intense flavors and colors, increase the barrier and strength of the packaging material, and increase the level of safety by identifying microbial contamination in the packaging (Sekhon, 2010). Figure 17.2 represents the applications of nanotechnology in the food packaging industry.

17.3.1 Edible Films/Coating

Food quality is maintained, shelf life is increased, and waste is somewhat reduced using edible packaging. Edible materials can now be transformed into edible films due to technological advancements and the availability of cutting-edge converting techniques. Wraps and pouches made of these films are also used for food packaging. Food products, on the other hand, may also be covered with coatings. Because edible packaging is a component of the product, consumers don't need to open it, which is a distinct advantage. To avoid moisture and flavor loss, from the foods as well as between them, edible packaging can replace and possibly strengthen the layers at the outside of packaged goods. It also allows for the controlled exchange of vital gases such as oxygen, carbon dioxide, and ethylene, which are involved in the respiration of food (Galus and Kadzińska, 2019). Additionally, edible packaging can improve the organoleptic qualities of packaged meals by adding various colors and flavors and customizing the hydrophobic and hydrophilic properties. These can also carry functional components that could have further health advantages. A material's mechanical strength at the nanoscale typically differs from that at the macroscopic size. The three main factors that contribute most to the overall characteristic properties of nanocomposites are nanofiller chemistry, filler size, and filler shape. The interactions between fillers and between polymers are impacted by the chemistry of the nanofiller. While inter-filler interactions cause agglomerates at high volume fractions of fillers, polymer-nanofiller interactions are more significant for stress transfer at the interface of nanofiller-polymer. In an attempt to enhance the composite's stress transmission, homogenous dispersion of the fillers, breakage of the agglomerates, and high wettability of the fillers are necessary. Mechanical properties are thus greatly influenced by nanofiller chemistry (Jeevahan and Chandrasekaran, 2019). The impact of the origin of the waxy or regular maize starch nanocrystals on the amaranth edible films based on protein created using film casting was examined by Condes et al. In comparison to the protein control films, the nano edible films displayed better mechanical and physical qualities like reduced water uptake and water vapor transmission and increased surface hydrophobicity. Without changing the thickness or optical qualities, integrated films containing starch nanocrystals were more uniform and transparent than the clean protein film (Condés et al., 2015).

Multilayers may result in improved barrier properties and other features because edible packaging materials cannot meet all of the barrier requirements on their own. This is because the desired properties of a particular packaging material can be blended with another packaging material with specific properties. Additionally, nanotechnology has been used for developing edible packaging materials with enhanced barrier and mechanical qualities (Trajkovska Petkoska et al., 2021). Because no known pure polymer displays all of the requisite mechanical attributes for all conceivable applications in the food packaging sector, complex multilayer films or polymer mixtures are frequently considered. Even though blending of polymer and multilayer films results in the development of materials with respectable gas barrier properties that do not have the restrictions associated with certain monolayer films made of polymers with ultrahigh barrier properties, they come at a higher cost in terms of production and materials, additional additives are also required which complicate their regulation, and present additional challenges when focusing on recycling. As a result, the polymer sector continues to exert enormous pressure to produce monolayer films with enhanced strength and improved gas barrier properties, especially those made of biocompatible polymers. The most recent materials developed to address the issues are polymer nanocomposites (PNCs). An innocuous dispersion is formed by mixing nanoscale filler across a polymeric matrix in order to produce PNCs.

17.3.2 Increased Mechanical Strength and Gas Barrier Properties

Food contact materials are in direct contact with food products during the production process, delivery, and storage. The food sector has been studying and developing nanotechnology as an innovative alternative to food packaging (Pereda et al., 2019). Nanomaterials used in food packaging have several advantages over standard packaging materials. Due to its mechanical strength, enhanced barrier properties and thermal characteristics, and lower expenses, nano clay is one of the many new nanomaterials that has been explored and used extensively for food packaging purposes (Gabr et al., 2015). All the oxygen-induced degradation events result in discolorations, odd tastes, nutritional value, and occasionally poisonous substances that endanger consumer safety as well as the shelf life of food. In the end, oxygen can be seen as one of a number of elements that unintentionally increase food wastage and economic harm along the food supply chain. As a result, one of the primary goals of the food industry is to limit the accessibility of oxygen inside the packaging, i.e., oxygen in contact with the food (Rovera et al., 2020). The preservation of food goods involves keeping them in an inert, low-oxygen environment to prevent microbial development and spoiling; hence, the material should be gas-impermeable. Nanocomposites can enhance food packaging materials' barrier properties against moisture, UV radiation, carbon dioxide, oxygen, and volatiles while also increasing their strength, reduction in weight, and resistance to heat (de Azeredo, 2009).

Nanocomposites compete with other materials for certain uses, such as in food packaging, because they are significantly lighter than conventional composite materials. It will assist in lowering the amount of packaging waste generated by processed food along with food preservation, thus extending the shelf-life. Due to the wide surface area, interactions between filler and matrix are facilitated, thus improving performance; nanocomposites are inserted into the polymer matrix of the substances. Although biopolymers are environment friendly and have the ability to keep food better while maintaining its quality and safety, their inferior mechanical and barrier qualities in comparison to their petroleum-based counterparts are by far this class of materials' major drawback. This shortcoming inhibits numerous applications from using biopolymers for packaging foods. It has been demonstrated that nanocomposite technology is effective in improving the material properties of synthetic thermoplastics and biopolymers (Mihindukulasuriya and Lim, 2014). The biggest significant issue with using PHAs in food packaging is a lack of sufficient flexibility. Although blending might lessen brittleness, adding nanofillers to PHAs might increase Young's modulus. These nano-packaging materials significantly increase their flexibility without sacrificing their other qualities in order to compete with petrochemical-based polymers used in food packing (Fortunati et al., 2013).

In 2010, Arora and Padua identified several nanostructures, including montmorillonite (MMT), kaolinite clays, and graphite nanoplates, as the most promising nanoscale size fillers. Because improved permeability for significant components in food packaging is essential, all of these materials are utilized. The interactions between clay and polymers, as well as nanoplate dispersions, are areas of great interest for use in the food packaging sector. Additionally, it was established that the inclusion of kaolinite nanofillers, which is a nano clay, improved the thermal, mechanical, and gas barrier properties of the film formed of polylactic acid (PLA). A composite film containing thymol displayed excellent antibacterial properties when used as a packing medium for chicken breasts constructed of Gelidium corneum-gelatin with nano clay. This prevented microbial growth during storage, along with increased mechanical and gas barrier properties (Jang et al., 2010). In other applications, films based on starch could be used to package perishable foods. Clay has been utilized as a potential filler to enhance thermoplastic starch (TPS) characteristics. MMT nanoparticles have been dispersed using polymer melt processing procedures to create starch/clay nanocomposite films. It has been demonstrated that the presence of sodium MMT, even in smaller amounts (>5%), boosted TPS's tensile strength (Ranjan et al., 2016).

17.3.3 Antimicrobial Packaging

Antimicrobial packaging has significant potential in the food industry with the motive to reduce the growth of microorganisms, thus extending the shelf life of food products. When the release of antimicrobials slows or stops, they can spread across a gradient upon saturation. In other instances, an antibiotic is delivered all at once, and its impact ends when it is digested (Han, 2005). Additionally, it has been discovered that the metal NPs, even in their nanoscale form, exhibit promising antibacterial action. Due to its antibacterial properties, silver nanoparticles (NPs) are the one metal NP that is produced and used most commercially. In comparison to bulk, silver NPs have a greater antibacterial impact because they have a higher surface area and a smaller number of surface atoms. Silver NPs antimicrobial properties have been applied to inhibit the growth of a broad range of pathogens (He et al., 2021).

According to Pathakoti and his team, the release of nanoparticles through the packaging system prevents microbial development and prevents food spoiling (Pathakoti et al., 2017). When added to a polymer (He et al., 2021), nanoparticles with antibacterial capabilities can function as active ingredients, extending the protective function of the food packaging material. Bactericidal NPs have been immobilized and coated on surfaces, and their active release has been managed (Ruparelia et al., 2008). Many efforts have been made to enhance the qualities of PLA to make it significantly more competitive than other plastic films. One of the most important goals is the addition of functional agents such as polyphenols, chitosan, and other compounds that can enhance the antibacterial properties of nanofiber membranes made of PLA (Au et al., 2012). A few appropriate NPs added into the PLA matrix can also give PLA-based antimicrobial packaging materials new functions. A combination of essential oils with ZnO-NPs are widely known for their antibacterial properties, has also been investigated. As a result, covered slices of sea bass in gelatin films made of protein isolate and skin of fish that included 3% ZnO-NPs (20–40 nm) and essential oil of basil leaf (BEO). The fish samples wrapped in this active film showed the lowest growth of psychrophilic bacteria, lactic acid bacteria, H_2S-producing bacteria, and spoilage microorganisms, in comparison to the BEO-containing films, the ZnO-NP films, and the control fish protein isolate/fish skin gelatin films. Additionally, sensory analysis of the sea bass demonstrated that the ZnO-NP/BEO films had the greatest shelf life (12 days) in comparison to the control, which had six days. Consequently, the addition of essential oil offered antioxidant characteristics that reduced lipid oxidation (Arfat et al., 2015).

The antimicrobial activity of titanium oxide NPs is because of the generation of reactive oxygen species photo catalytically on exposure to UV radiation. Titanium dioxide is significant in many contemporary industries and application areas. In the food processing industry, titanium oxide's photocatalytic self-disinfecting characteristics are utilized to clean surfaces that come into touch with food and to stop the growth of biofilms. In order to prevent or lessen microbial development on the surface of solid meals, titanium oxide NPs are being used in active food packaging. They can be incorporated into the polymer matrix or can also be applied as a coating on the surface of the polymer film. Metal doping enhances titanium oxide's ability to absorb visible light and inhibit photolytic bacteria growth. For this, noble metals have been employed, particularly metallic silver and silver ions (Silvestre et al., 2011). Broad-spectrum antimicrobial, UV protection, and high photocatalytic activity are all properties of nano-TiO_2. The photocatalytic activity of nano-TiO_2 is largely explored in the field of food packaging. It has been suggested that one could create beneficial novel materials for packing applications through the combination of nano-TiO_2 particles with high-density polyethylene and other components (Gumiero et al., 2013). Because of the potential of Ag NPs to enter bacteria and bind to their cell membranes to kill them, they are also promising antibacterial substances (Jiang et al., 2013). Films based on chitosan nanocomposite were prepared and characterized by Rhim and co-workers, for demonstrating the antimicrobial activity of Ag. Strong antibacterial action against both Gram-positive and Gram-negative bacteria was observed by chitosan film composed of ag-zeolite (Rhim et al., 2006). In comparison to a single nanoparticle, the combinatorial usage of two or more nanoparticles shows a synergistic effect with powerful antibacterial action. *Escherichia coli* and *Bacillus cereus* spores can be successfully eliminated by combining silver NPs with titanium dioxide and carbon nanotubes, respectively (Krishna et al., 2005). The silver-doped TiO_2 NP eliminated *B. cereus* spores that were on aluminum and polyester surfaces. The silver-doped TiO_2 NPs kill the airborne bacteria and molds that are captured in the air filters (Vohra et al., 2005).

17.3.4 Intelligent Packaging

An effective interdisciplinary technique for creating intelligent packaging systems is nanotechnology. The potential benefits of using nanotechnology-based detection techniques over traditional laboratory procedures are numerous. The detection of pathogens and mycotoxins in food can also be done by nanosensors. Toxins and pathogens can be difficult to regulate conventionally, but the nanosensor can quickly find them in food, during processing, and during storage. When the nano-nano sensor comes into touch with certain food pathogens, it fluoresces in a variety of hues, making it useful for smart packaging. Nanowires and antibodies have been used in various devices to identify various poisons, diseases, and chemicals in food packaging (Mcclements, 2015). The ability to integrate wireless sensors with nano-based communication devices like barcodes and Radio Frequency Identification (RFID) tags in packaging materials could revolutionize packaging technology and the food supply chain (Biji et al., 2015). RFID is a technology that uses radio waves to identify and interact with specific targets without needing physical contact (Zuo et al., 2022). Fruits and vegetables inside a package use up oxygen and release carbon dioxide, which affects their freshness. RFID tag equipped with sensors that measure both temperature and carbon dioxide, can be used in food packaging industry, allowing for real-time monitoring of the fruits and vegetables freshness (Tao et al., 2024). The development of intelligent packaging can also help to improve the Hazard Analysis and Critical Control Points (HACCP) and Quality Assurance Critical Control Points (QACCP) systems (Heising et al., 2014), which are intended to detect unsafe food on-site, identify potential health risks, and develop measures to reduce or eliminate their occurrence.

Nanoscale particles are well suited for use in the creation of sensors due to their distinct electro-optical and chemical properties. The size, shape, and chemical environment of nanoparticles can all change how they behave. Conductivity, surface plasmon resonance, and redox potential are physicochemical properties that might change as a result of the interaction with the target molecules and produce a detectable signal. Additionally, the development of nanosensors to identify diseases, product tampering, and contamination has drawn attention to nanotechnology applications in this industry. In food microbiology, nanosensors are used to quantify the available food ingredients, detect infections in food material, and notify distributors and consumers of the food's safety status (Cheng et al., 2006). The nanosensor operates as an indicator, reacting to changes in ambient environmental conditions like temperature or humidity, microbial contamination, or the deterioration of items (Bouwmeester et al., 2009). Thin films, nanorods, nanoparticles, and nanofibers are a few of the nanostructures that have been examined for potential use in biosensors (Jianrong et al., 2004). Radiofrequency identification is a fascinating field in which nanotechnology can be used (RFID). Because of the increased demand for traceability in the management of agro-food products, the food packaging industry has already begun to employ this technology. The RFID tag system is a potential tracking tool. That could result in the creation of possible nanosensors using quantum dots, tiny barcodes, and magnetic NPs (Abad et al., 2009).

Several authors have contributed to the field of nanosensors. The detection of chemicals, gases, and microbes is done using these nanomaterials. Oxygen detection nanosensors in raw bacon packaging (Mills, 2005); pathogen and toxin detection sensors in food (Burris and Stewart, 2012); pesticide detection (organophosphate residues) in food (Liu et al., 2008); biosensor for the detection of *B. cereus* in food (Pal et al., 2007) and biosensor for detection of *Salmonella* in foods (Fu et al., 2008).

17.4 Future Prospective

The packaging industry has evolved over time through the use of modern technology, moving away from conventional types of packaging such as paper, wood, and glass and toward the majority of other environment-friendly biodegradable materials. An automated way of packing has replaced the traditional method of manual packaging, resulting in increased output with little risk of damage (Verma et al., 2021). The first packaging ever created was based on the principle of providing the product with a physical appearance with regard to the producer and its long-term storage (Risch, 2009). Later, the idea of packaging was expanded to include more than only storage and transportation of packaged goods; it was also used to guarantee product quality, provide information about copyright, and communicate information about products and marketing tactics. In the current scenario, packaging technology is always developing, moving from outdated conventional methods to cutting-edge automated systems. Paper, paperboard, polymers, metals, wood, and glass have all historically been used for the purpose of food packaging. The effects of environmental contamination and the need for nonrenewable resources in manufacturing are strongly correlated with these materials. The expanding food business, which continues to use conventional food packaging, is directly responsible for the rise in packaging waste. Consumers are becoming more and more conscious, as conventional packaging, particularly plastic products, causes significant harm to the environment and the overall ecosystem despite being extremely helpful. Therefore, it is now more important than ever to use alternative packaging materials that are eco-friendly along with other strategies for decreasing wastage due to food packaging, such as source reduction, recycling, and reusing (Ljungberg, 2007). Active and intelligent packaging materials are the subject of very dynamic research that advances in tandem with the hunt for environment-friendly packaging solutions. If properly regulated, the presence of nanoparticles in food will have a significant positive impact on human health as well as the environment and sustainability. However, the development of innovative nanomaterials for application in the food sector has made it increasingly difficult to evaluate the safety of nano foods and nanopackaging. Therefore, more cooperation is required to make sure that as new products are introduced, social and environmental interests are not compromised. The idea of nanocomposites has made it possible to create innovative and inventive polymer materials in technologically fascinating ways. This has made it possible to create different PNCs with a variety of fascinating and superior features, such as mechanical, thermal, and barrier capabilities.

Additionally, these materials show heat deflection, retardancy to flame, and cases of polymer matrix transparency (Idumah et al., 2019).

Edible coating is a food preservation technique in which various edible materials are employed to form a thin layer of barrier on desired food products. Additionally, contributing to the aesthetic improvement of food products involves the use of edible coating with the help of edible ingredients on food products' surfaces. In regard to coatings, edible food packaging is utilized to provide a semipermeable wall against hazardous environmental elements, including light, heat, gaseous elements, microbiological elements, etc. Another effective way to increase the shelf life of perishable foods is the addition of health-beneficial ingredients like antioxidants and antimicrobials to food products using edible food packaging. The packaging for food serves as a barrier against a variety of harms, including microbiological, thermal, chemical, mechanical damage, etc. (Falguera et al., 2011). Live microorganisms can also be used in edible coatings and films. The creation of edible films and coatings containing probiotics that act as efficient carriers for the bacteria by boosting their survival is one of the most recent developments in this sector. Direct application of these bacteria has certain drawbacks; to overcome these and improve the functional qualities of edible coatings, probiotics are introduced in the edible films and coatings. These may be used as stabilizing matrices for retaining probiotic microorganisms. This strategy of incorporation of probiotics enables them to enter the consumer's gut in sufficient quantities, exerting any potential therapeutic benefits on the consumer and adding functionality to the food item (Pandhi et al., 2019). In bioactive packaging, the bioactive substances in the packaging materials may eventually diffuse into the food product (Lopez-Rubio et al., 2006). But in the case of probiotic-containing edible coatings, the release isn't even necessary, as the food should be eaten with the coating. Additionally, innovative formulations that mix films and coatings with emulsions, encapsulation at micro- or nanoscale range, or liposomes are being developed. These formulations give a homogeneous distribution and improve the stability of the target microbial population.

The development of various active packaging solutions as a result of recent technology breakthroughs offers tremendous potential for preserving the quality and safety of goods. In order to ensure the safety of our food, additional research and regulations about the effects of these nanomaterials on human health and their impact on the environment must undoubtedly be done and implemented in the near future. Despite the fact that nanoparticles improve the biocidal efficacy of packing materials, there still remain concerns about migration and toxicity. The knowledge that is currently known about the traces in food packaging and their effects on the human body is still insufficient. It has been strongly advised to take into account some key aspects of the migration of nanomaterials from materials that are in contact with the food, also including the mechanism and rate of migration, how to assess the trace, and the prediction of associated risk (Jokar et al., 2017). Such characteristics often shed light on all potential migration scenarios and offer the chance to address or prevent any negative effects. In order to maintain the food quality and increase the shelf life of food, antimicrobial packaging and packaging nanocomposite incorporating antimicrobial compounds may be useful. However, various challenges need to be resolved in order to get the desired results and make antimicrobial packaging feasible. Due to their potent antibacterial properties, nanoparticles are frequently employed in a variety of applications. However, once they are incorporated into the packaging matrix, their antimicrobial efficiency decreases. Therefore, it could be explained for being used in a wider variety of food packaging.

Another method for developing fresh and inventive packaging materials is the use of bionanocomposites. Packaging films with superior mechanical, thermal, and barrier properties can be developed by adding the right nanoparticles, such as MMT, kaolinitetitanium dioxide (TiO_2-NPs), zinc oxide (ZnO-NPs), and gold-silver nanoparticles (Au-NPs and Ag-NPs). They can significantly reduce flammability while maintaining the transparency of the polymer matrix (Youssef and El-Sayed, 2018). In order to increase the quality, marketability, and shelf life of various packaging materials, research, and development are still required in the field of bionanocomposites as packaging materials. Designs can be made with desired functionalities, which can include antimicrobial, antibiotic, biodegradable material, and combination. Additionally, it might be possible to create bioinspired nanostructure PNCs with enhanced mechanical strength. Multilayer nanocomposites can be created with the motive to increase gas barrier properties.

Packaging design is evolving in response to customer concerns about convenience and safety. The use of carbon nanotubes in packaging materials for the detection of hazardous microorganisms, proteins, and deterioration of food is also gaining a lot of attention (Tully et al., 2006). Additionally, carbon nanotubes have the ability to convert materials we use today for food packaging into active, intelligent packaging systems. Although numerous potential dangers and toxicity problems linked to the use of nanomaterials have been found, these issues are required to be addressed (Baranwal et al., 2016). In terms of health and safety concerns, it is noteworthy that the lack of standardized and comprehensive toxicity evaluation methods, as well as the frequently poor stability of the sensors that have been created so far and may cause substance migration in food products, have produced confusing and frequently inconsistent data. This presents a problem for the production and application of intelligent packages on a broad scale (Adeyeye, 2019). Therefore, one of the key difficulties that have to be resolved before future intelligent packages can be effectively used on a broad scale is the development, standardization, and validation of tests for their safety. Additionally, it is critical to ensure that intelligent packaging is environment-friendly and sustainable in terms of creation and design. Combining further enhanced, active, intelligent, and bio-based packaging could offer some practical solutions to the issues related to each category. These innovative combination packages will create new possibilities and opportunities for assuring the safety and quality of food. Based on the literature and patents now in existence, it looks likely that optical indication systems utilizing nanoparticles will soon become more widely used and commercially viable. Because some metallic NPs are sensitive to oxygen and easily oxidized, their optical characteristics can be reduced. The development of sophisticated fabrication methods must be pursued in order to include colloidal nanoparticles in indicator labels. To expand their use in the real world, these issues must be resolved (Bumbudsanpharoke and Ko, 2019).

17.5 Conclusion

The food industry has benefited from advances in nanotechnology, though many applications are still in the works. In order to improve nutrition and health, encourage the development and production of novel food products, for boost the microbiological quality during manufacturing and processing, there are several chances to utilize the benefits of nanotechnologies at various stages along the food chain. In comparison to traditional and other ways of food processing, nanotechnology has an advantage in the field of food science. In addition to improving mechanical strength and barrier qualities, nano-based materials and particles also aid in the detection of pathogens and inform consumers about the status of their food. Its influence on the food business is growing on a global scale. Nanoparticles implanted in food packaging warn consumers about the condition of food when it is no longer suitable for consumption. Sensors alert the customers before food spoils or tell us the precise nutritional condition of the contents. In reality, the fabrication of the entire packaging sector will change as a result of nanotechnology. Since its application offers improvements in packaging materials and packaging methods, nanotechnology is anticipated to be one of the primary development areas in the food packaging industry. The existing use of conventional polymer architectures can be replaced by using nanocomposites to develop lightweight, stronger, and more flexible films with superior aroma and oxygen barrier properties. Active packaging technologies for healthier and more recent foods are also being developed with the use of nanotechnology. Antimicrobial nanocomposites can be utilized to make packaging that extends food safety and microbiological shelf life. Therefore, smart packaging makes use of nanosensors, which are built into food packages to provide information about the condition of food within and if it is still safe for consumption. Although there are difficulties in the fabrication, integration, and mass production of such devices, the future of nanotechnology-based technologies is predicted to be brighter in many fields, including food.

References

Abad, E., Palacio, F., Nuin, M., de Zárate, A. G., Juarros, A., Gómez, J. M., & Marco, S. (2009). RFID smart tag for traceability and cold chain monitoring of foods: demonstration in an intercontinental fresh fish logistic chain. *Journal of Food Engineering*, *93*, 394–399. https://doi.org/10.1016/j.jfoodeng.2009.02.004

Adeyeye, S. A. O. (2019). Food packaging and nanotechnology: safeguarding consumer health and safety. *Nutrition & Food Science*. https://doi.org/10.1108/NFS-01-2019-0020

Arfat, Y. A., Benjakul, S., Vongkamjan, K., Sumpavapol, P., & Yarnpakdee, S. (2015). Shelf-life extension of refrigerated sea bass slices wrapped with fish protein isolate/fish skin gelatin-ZnO nanocomposite film incorporated with basil leaf essential oil. *Journal of Food Science and Technology*, *52*, 6182–6193. https://doi.org/10.1007/s13197-014-1706-y

Arora, A., & Padua, G. W. (2010). Review: Nanocomposites in Food Packaging. *Journal of Food Science*, *75*(1), R43–R49. https://ift.onlinelibrary.wiley.com/doi/10.1111/j.1750-3841.2009.01456.x

Au, H. T., Pham, L. N., Vu, T. H. T., & Park, J. S. (2012). Fabrication of an antibacterial non-woven mat of a poly (lactic acid)/chitosan blend by electrospinning. *Macromolecular Research*, *20*, 51–58. https://doi.org/10.1007/s13233-012-0010-9

Augustin, M. A., & Sanguansri, P. (2009). Nanostructured materials in the food industry. *Advances in Food and Nutrition Research*, *58*, 183–213. https://doi.org/10.1016/S1043-4526(09)58005-9

de Azeredo, H. M. C. (2009). Nanocomposites for food packaging applications. *Food Research International*, *42*(9), 1240–1253. https://doi.org/10.1016/j.foodres.2009.03.019

Baranwal, A., Mahato, K., Srivastava, A., Maurya, P. K., & Chandra, P. (2016). Phytofabricated metallic nanoparticles and their clinical applications. *RSC Advances*, *6*, 105996–106010. https://doi.org/10.1039/c6ra23411a

Biji, K. B., Ravishankar, C. N., Mohan, C. O., & Srinivasa Gopal, T. K. (2015). Smart packaging systems for food applications: a review. *Journal of Food Science and Technology*, *52*(10), 6125–6135. https://doi.org/10.1007/s13197-015-1766-7

Bouwmeester, H., Dekkers, S., Noordam, M. Y., Hagens, W. I., Bulder, A. S., de Heer, C., ten Voorde, S. E. C. G., Wijnhoven, S. W. P., Marvin, H. J. P., & Sips, A. J. A. M. (2009). Review of health safety aspects of nanotechnologies in food production. *Regulatory Toxicology and Pharmacology*, *53*, 52–62. https://doi.org/10.1016/j.yrtph.2008.10.008

Bumbudsanpharoke, N., Ko, S., 2019. Nanomaterial-based optical indicators: Promise, opportunities, and challenges in the development of colorimetric systems for intelligent packaging. *Nano Research* 12, 489–500 doi:https://doi.org/10.1007/s12274-018-2237-z

Burris, K. P., & Stewart, C. N. (2012). Fluorescent nanoparticles: Sensing pathogens and toxins in foods and crops. *Trends in Food Science and Technology*, *28*(2), 143–152. https://doi.org/10.1016/j.tifs.2012.06.013

Cheng, Q., Li, C., Pavlinek, V., Saha, P., & Wang, H. (2006). Surface-modified antibacterial TiO_2/Ag + nanoparticles: preparation and properties. *Applied Surface Science*, *252*, 4154–4160. https://doi.org/10.1016/j.apsusc.2005.06.022

Coles, R., McDowell, D., & Kirwan, M. J. (2003). Food Packaging Technology. CRC press.

Condés, M. C., Añón, M. C., Mauri, A. N., & Dufresne, A. (2015). Amaranth protein films reinforced with maize starch nanocrystals. *Food Hydrocolloids*, *47*, 146–157. https://doi.org/10.1016/j.foodhyd.2015.01.026

Dudefoi, W., Terrisse, H., Richard-Plouet, M., Gautron, E., Popa, F., Humbert, B., & Ropers, M. H. (2017). Criteria to define a more relevant reference sample of titanium dioxide in the context of food: a multiscale approach. *Food Additives & Contaminants. Part A, Chemistry, Analysis, Control, Exposure & Risk Assessment*, *34*, 653–665. https://doi.org/10.1080/19440049.2017.1284346

Duncan, T. V. (2011). Applications of nanotechnology in food packaging and food safety: barrier materials, antimicrobials and sensors. *Journal of Colloid and Interface Science*, *363*, 1–24. https://doi.org/10.1016/j.jcis.2011.07.017

Enescu, D., Cerqueira, M. A., Fucinos, P., & Pastrana, L. M. (2019). Recent advances and challenges on applications of nanotechnology in food packaging. A literature review. *Food and Chemical Toxicology*, *134*, 110814. https://doi.org/10.1016/j.fct.2019.110814

Falguera, V., Quintero, J. P., Jiménez, A., Muñoz, J. A., & Ibarz, A. (2011). Edible films and coatings: structures, active functions and trends in their use. *Trends in Food Science and Technology*, *22*(6), 292–303. https://doi.org/10.1016/j.tifs.2011.02.004

Finnigan, B. (2009). Barrier Polymers. In K. L. Yam (Ed.), *The Wiley Encyclopedia of Packaging Technology* (pp. 103–109). Wiley.

Fortunati, E., Peltzer, M., Armentano, I., Jiménez, A., & Kenny, J. M. (2013). Combined effects of cellulose nanocrystals and silver nanoparticles on the barrier and migration properties of PLA nano-biocomposites. *Journal of Food Engineering*, *118*, 117–124. https://doi.org/10.1016/j.jfoodeng.2013.03.025

Fu, J., Park, B., Siragusa, G., Jones, L., Tripp, R., Zhao, Y., & Cho, Y. J. (2008). An Au/Si hetero-nanorod-based biosensor for Salmonella detection. *Nanotechnology*, *19*. https://doi.org/10.1088/0957-4484/19/15/155502

Gabr, M. H., Okumura, W., Ueda, H., Kuriyama, W., Uzawa, K., & Kimpara, I. (2015). Mechanical and thermal properties of carbon fiber/polypropylene composite filled with nano-clay. *Composites Part B: Engineering*, *69*, 94–100. https://doi.org/10.1016/j.compositesb.2014.09.033

Galus, S., & Kadzińska, J. (2019). Gas barrier and wetting properties of whey protein isolate-based emulsion films. *Polymer Engineering and Science*, *59*, E375–E383. https://doi.org/10.1002/pen.24992

Gleiter, H. (2000). Nanostructured materials: basic concepts and microstructure. *Acta Materialia*, *48*(1), 1–29.

Gumiero, M., Peressini, D., Pizzariello, A., Sensidoni, A., Iacumin, L., Comi, G., & Toniolo, R. (2013). Effect of TiO_2 photocatalytic activity in a HDPE-based food packaging on the structural and microbiological stability of a short-ripened cheese. *Food Chemistry*, *138*, 1633–1640. https://doi.org/10.1016/j.foodchem.2012.10.139

Gupta, A., Eral, H. B., Hatton, T. A., & Doyle, P. S. (2016). Nanoemulsions: formation, properties and applications. *Soft Matter*, *12*(11). https://doi.org/10.1039/c5sm02958a

Han, J. H. (2005). Innovations in Food Packaging. Elsevier Academic.

He, X., Deng, H., & Hwang, H. M. (2019). The current application of nanotechnology in food and agriculture. *Journal of Food and Drug Analysis*, *27*(1), 1–21. https://doi.org/10.1016/j.jfda.2018.12.002

He, Y., Li, H., Fei, X., & Peng, L. (2021). Carboxymethyl cellulose/cellulose nanocrystals immobilized silver nanoparticles as an effective coating to improve barrier and antibacterial properties of paper for food packaging applications. *Carbohydrate Polymers*, *252*. https://doi.org/10.1016/j.carbpol.2020.117156

Heising, J. K., Dekker, M., Bartels, P. V., & Van Boekel, M. A. (2014). Monitoring the quality of perishable foods: opportunities for intelligent packaging. *Critical Reviews in Food Science and Nutrition*, *54*(5), 645–654. https://doi.org/10.1080/10408398.2011.600477

Hills, F. (2008). Nanotechnology in packaging: a revolution in waiting. *Food Engineering Ingred*, *33*, 6–9.

Idumah, C.I., Hassan, A., Ihuoma, D.E., 2019. Recently emerging trends in polymer nanocomposites packaging materials. *Polymer-Plastics Technology and Materials* 58, 1054-1109. doi:https://doi.org/10.1080/03602559.2018.1542718

Jang, S. A., Lim, G. O., & Song, K. B. (2010). Use of nano-clay (Cloisite Na+) improves tensile strength and vapour permeability in agar rich red algae (Gelidium corneum)-gelatin composite films. *International Journal of Food Science and Technology*, *45*, 1883–1888. https://doi.org/10.1111/j.1365-2621.2010.02363.x

Jeevahan, J., Chandrasekaran, M., 2019. Nanoedible films for food packaging: a review. *Journal of Materials Science* 54, 12290–12318. doi:https://doi.org/10.1007/s10853-019-03742-y

Jiang, T., Feng, L., & Wang, Y. (2013). Effect of alginate/nano-Ag coating on microbial and physicochemical characteristics of shiitake mushroom (*Lentinus edodes*) during cold storage. *Food Chemistry*, *141*, 954–960. https://doi.org/10.1016/j.foodchem.2013.03.093

Jianrong, C., Yuqing, M., Nongyue, H., Xiaohua, W., & Sijiao, L. (2004). Nanotechnology and biosensors. *Biotechnology Advances*, *22*, 505–518. https://doi.org/10.1016/j.biotechadv.2004.03.004

Jokar, M., Pedersen, G. A., & Loeschner, K. (2017). Six open questions about the migration of engineered nano-objects from polymer-based food-contact materials: a review. *Food Additives & Contaminants. Part A, Chemistry, Analysis, Control, Exposure & Risk Assessment*, *34*, 434–450. https://doi.org/10.1080/19440049.2016.1271462

Kim, Y. R., Lee, E. J., Park, S. H., Kwon, H. J., An, S. S. A., Son, S. W., Seo, Y. R., Pie, J. E., Yoon, M., Kim, J. H., & Kim, M. K. (2014). Comparative analysis of nanotechnology awareness in consumers and experts in South Korea. *International Journal of Nanomedicine*, *9*, 21–27. https://doi.org/10.2147/IJN.S57921

Krishna, V., Pumprueg, S., Lee, S. H., Zhao, J., Sigmund, W., Koopman, B., & Moudgil, B. M. (2005). Photocatalytic disinfection with titanium dioxide coated multi-wall carbon nanotubes. *Process Safety and Environmental Protection*, *83*, 393–397. https://doi.org/10.1205/psep.04387

Kuswandi, B., 2017. Environmental friendly food nano-packaging. *Environmental Chemistry Letters* 15, 205–221. doi:https://doi.org/10.1007/s10311-017-0613-7

Liu, S., Yuan, L., Yue, X., Zheng, Z., & Tang, Z. (2008). Recent advances in nanosensors for organophosphate pesticide detection. *Advanced Powder Technology*, *19*, 419–441. https://doi.org/10.1163/156855208X336684

Ljungberg, L. Y. (2007). Materials selection and design for development of sustainable products. *Materials and Design*, *28*, 466–479. https://doi.org/10.1016/j.matdes.2005.09.006

Lopez-Rubio, A., Gavara, R., & Lagaron, J. M. (2006). Bioactive packaging: turning foods into healthier foods through biomaterials. *Trends in Food Science and Technology*, *17*(10), 567–575. https://doi.org/10.1016/j.tifs.2006.04.012

Mcclements, D. J. (2015). Nanoscale nutrient delivery systems for food applications: improving bioactive dispersibility, stability, and bioavailability. *Journal of Food Science*, *80*, N1602–N1611. https://doi.org/10.1111/1750-3841.12919

Mihindukulasuriya, S. D. F., & Lim, L. T. (2014). Nanotechnology development in food packaging: a review. *Trends in Food Science and Technology*, *40*, 149–167. https://doi.org/10.1016/j.tifs.2014.09.009

Mills, A. (2005). Oxygen indicators and intelligent inks for packaging food. *Chemical Society Reviews*, *34*, 1003–1011. https://doi.org/10.1039/b503997p

Pal, S., Alocilja, E. C., & Downes, F. P. (2007). Nanowire labeled direct-charge transfer biosensor for detecting Bacillus species. *Biosensors & Bioelectronics*, *22*, 2329–2336. https://doi.org/10.1016/j.bios.2007.01.013

Pandhi, S., Kumar, A., & Alam, T. (2019). Probiotic edible films and coatings: concerns, applications and future prospects. *Journal of Packaging Technology and Reseach*, *3*, 261–268. https://doi.org/10.1007/s41783-019-00069-6

Pathakoti, K., Manubolu, M., & Hwang, H. M. (2017). Nanostructures: current uses and future applications in food science. *Journal of Food and Drug Analysis*, *25*(2), 245–253. https://doi.org/10.1016/j.jfda.2017.02.004

Pereda, M., Marcovich, N. E., & Ansorena, M. R. (2019). Nanotechnology in Food Packaging Applications: Barrier Materials, Antimicrobial Agents, Sensors, and Safety Assessment. In L. M. T. Martínez, O. V. Kharissova, & B. I. Kharisov (Eds.), *Handbook of Ecomaterials* (pp. 2035–2056). Springer International Publishing. https://doi.org/10.1007/978-3-319-68255-6_30

Peters, R., Brandhoff, P., Weigel, S., Marvin, H., Bouwmeester, H., Aschberger, K., Rauscher, H., Amenta, V., Arena, M., Botelho Moniz, F., Gottardo, S., & Mech, A. (2017). Inventory of Nanotechnology applications in the agricultural, feed and food sector. *EFSA Supporting Publications, 11*. https://doi.org/10.2903/sp.efsa.2014.en-621

Primožič, M., Knez, Ž., & Leitgeb, M. (2021). (Bio) nanotechnology in food science—food packaging. *Nanomaterials, 11*(2), 292. https://doi.org/10.3390/nano11020292

Ranjan, S., Dasgupta, N., & Lichtfouse, E. (2016). Nanoscience in Food and Agriculture 2. Springer.

Ravichandran, R. (2010). Nanotechnology applications in food and food processing: innovative green approaches, opportunities and uncertainties for global market. *International Journal of Green Nanotechnology: Physics and Chemistry, 1*(2), 72–96. https://doi.org/10.1080/19430871003684440

Rhim, J. W., Hong, S. I., Park, H. M., & Ng, P. K. W. (2006). Preparation and characterization of chitosan-based nanocomposite films with antimicrobial activity. *Journal of Agricultural and Food Chemistry, 54*, 5814–5822. https://doi.org/10.1021/jf060658h

Risch, S. J. (2009). Food packaging history and innovations. *Journal of Agricultural and Food Chemistry, 57*, 8089–8092. https://doi.org/10.1021/jf900040r

Robertson, G. L. (2005). Food Packaging: Principles and Practice. CRC Press.

Rovera, C., Ghaani, M., & Farris, S. (2020). Nano-inspired oxygen barrier coatings for food packaging applications: an overview. *Trends in Food Science and Technology, 97*. https://doi.org/10.1016/j.tifs.2020.01.024

Ruparelia, J. P., Chatterjee, A. K., Duttagupta, S. P., & Mukherji, S. (2008). Strain specificity in antimicrobial activity of silver and copper nanoparticles. *Acta Biomaterialia, 4*, 707–716. https://doi.org/10.1016/j.actbio.2007.11.006

Sekhon, B. S. (2010). Food nanotechnology-an overview. *Nanotechnology, Science and Applications, 3*, 1–15. https://doi.org/10.2147/nsa.s12187498

Siegrist, M., Keller, C., Kastenholz, H., Frey, S., & Wiek, A. (2007). Laypeople's and experts' perception of nanotechnology hazards. *Risk Analysis, 27*, 59–69. https://doi.org/10.1111/j.1539-6924.2006.00859.x

Silvestre, C., Duraccio, D., & Cimmino, S. (2011). Food packaging based on polymer nanomaterials. *Progress in Polymer Science (Oxford), 36*(12), 1766–1782. https://doi.org/10.1016/j.progpolymsci.2011.02.003

Simmonds, G., & Spence, C. (2017). Thinking inside the box: how seeing products on, or through, the packaging influences consumer perceptions and purchase behaviour. *Food Quality and Preference, 62*, 340–351. https://doi.org/10.1016/j.foodqual.2016.11.010

Singh, T., Shukla, S., Kumar, P., Wahla, V., & Bajpai, V. K. (2017). Application of nanotechnology in food science: perception and overview. *Frontiers in Microbiology, 8*, 1501. https://doi.org/10.3389/fmicb.2017.01501

Tao, B., Bai, J., Wu, H., & Miao, F. (2024). Design and research of multi-information fusion RFID sensor for fruit and vegetable quality detection. *Microchimica Acta, 191*(9), 1–16.

Trajkovska Petkoska, A., Daniloski, D., D'Cunha, N. M., Naumovski, N., & Broach, A. T. (2021). Edible packaging: sustainable solutions and novel trends in food packaging. *Food Research International, 140*, 109981. https://doi.org/10.1016/j.foodres.2020.109981

Tully, E., Hearty, S., Leonard, P., & O'Kennedy, R. (2006). The development of rapid fluorescence-based immunoassays, using quantum dot-labelled antibodies for the detection of *Listeria monocytogenes* cell surface proteins. *International Journal of Biological Macromolecules, 39*, 127–134. https://doi.org/10.1016/j.ijbiomac.2006.02.023

Verma, M. K., Shakya, S., Kumar, P., Madhavi, J., Murugaiyan, J., & Rao, M. V. R. (2021). Trends in packaging material for food products: historical background, current scenario, and future prospects. *Journal of Food Science and Technology*, *58*(11), 4069–4082. https://doi.org/10.1007/s13197-021-04964-2

Vohra, A., Goswami, D. Y., Deshpande, D. A., & Block, S. S. (2005). Enhanced photocatalytic inactivation of bacterial spores on surfaces in air. *Journal of Industrial Microbiology & Biotechnology*, *32*, 364–370. https://doi.org/10.1007/s10295-005-0006-y

Wegari Dera, M., & Bogale Teseme, W. (2020). Review on the application of food nanotechnology in food processing. *American Journal of Engineering and Technology Management*, *5*, 41. https://doi.org/10.11648/j.ajetm.20200502.12

Weir, A., Westerhoff, P., Fabricius, L., Hristovski, K., & von Goetz, N. (2012). Titanium dioxide nanoparticles in food and personal care products. *Environmental Science & Technology*, *46*, 2242–2250. https://doi.org/10.1021/es204168d

Youssef, A. M., & El-Sayed, S. M. (2018). Bionanocomposites materials for food packaging applications: concepts and future outlook. *Carbohydrate Polymers*, *193*, 19–27. https://doi.org/10.1016/j.carbpol.2018.03.088

Zuo, J., Feng, J., Gameiro, M. G., Tian, Y., Liang, J., Wang, Y., Ding, J., & He, Q. (2022). RFID-based sensing in smart packaging for food applications: a review. *Future Foods*, *6*, 100198. https://doi.org/10.1016/j.fufo.2022.100198

18

Sustainability and Future of Nanofood Packaging

J. Cyril Robinson Azariah[1], D. Sungeetha[1], and Ashley George Thomas[2]

[1] Department of Nanotechnology, Institute of Electronics and Communication Engineering, Saveetha School of Engineering, Saveetha Institute of Medical and Technical Sciences, Saveetha University, Chennai, Tamil Nadu, India
[2] Department of Medical Biotechnology and Integrative Physiology, Institute of Biotechnology, Saveetha School of Engineering, Saveetha Institute of Medical and Technical Sciences, Saveetha University, Chennai, Tamil Nadu, India

18.1 An Introduction to Sustainable Packaging

In recent years, sustainability has evolved into a pressing concern globally, particularly within the field of food packaging. The packaging industry plays an important role in the sustainable living of human beings and all living things (Jestratijevic et al., 2022; Parameswaranpillai et al., 2021). However, the plastic packaging industry has led to the degradation of our sustainable living. With an increasing awareness of environmental degradation and its implications for future generations, there has been a significant shift toward the adoption of sustainable practices across various industries. In this context, the field of food packaging emerges as a critical focal point, where the balance between convenience, preservation, and environmental impact must be carefully navigated. Sustainable packaging must be designed as early as possible to reduce environmental impacts and achieve sustainability goals (Ibrahim et al., 2022; Muthu, 2021). Sustainable food packaging, characterized by its emphasis on minimizing resource consumption, reducing waste, and promoting recyclability or biodegradability, represents a pivotal avenue for addressing these challenges. Such packaging can benefit the manufacturers and the consumers in many ways (Breetz, 2014; Butler, 2008). Also, food packaging has become essential, and hence, packaging has to be done in a distinct manner that appeals to the consumer (Rangappa et al., 2020). Even though, if enforced in the wrong manner, it can result in many hindrances. Hence, it is important to understand sustainable food packaging and its broader implications for global food systems and environmental conservation. Figure 18.1 shows the various importance of food packaging, highlighting key steps to ensure safety, quality, and environmental sustainability.

Nanotechnology for Sustainable Food Packaging, First Edition. Edited by C. Anandharamakrishnan, Jeyan A. Moses, and M. Maria Leena.

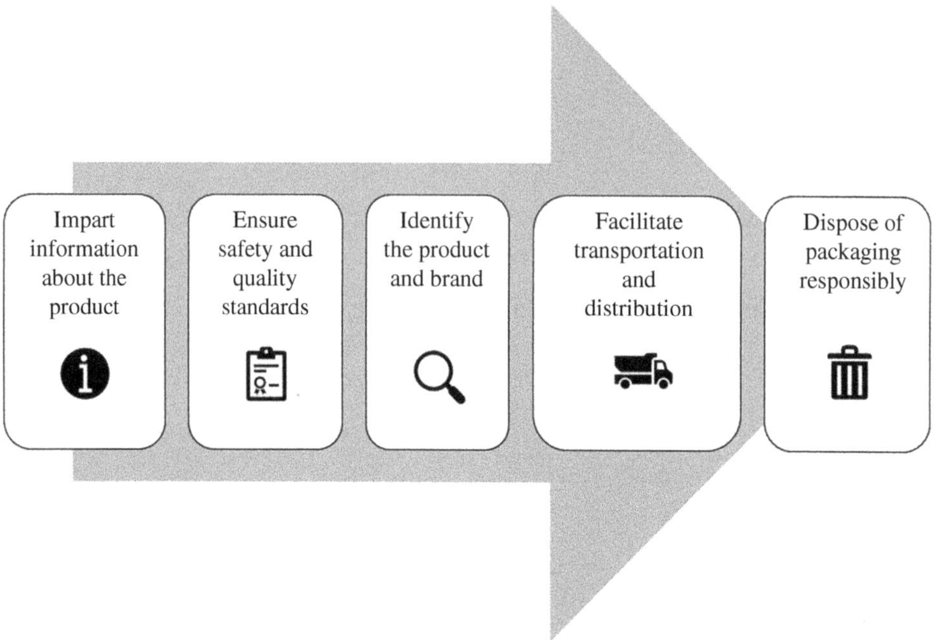

Figure 18.1 Flowchart illustrating the importance of food packaging, highlighting key steps to ensure safety, quality, and environmental sustainability

18.1.1 Merits of Packed Food

18.1.1.1 Safety Is Ensured

Across the world, food safety regulations exist to ensure safe, uncontaminated food packaging. This induces packaging to be an immensely secure option. This means the life of packaged food has been longer when compared to unpacked ones. Hence, it helps to remain fresh for a substantial period of time (Kasza et al., 2022).

18.1.1.2 Better and Hygiene

It has been proved that packaging definitely paves the path for maintaining hygiene. Hence, clear labeling and durability are factors for better packaging. After packaging, products undergo various processes and shift all over the world; at that time, there are high chances of getting spoiled. Therefore, packaging is the safest option for better hygiene (Barone et al., 2015).

18.1.1.3 Convenient Storage

The packaging concept has proven especially useful for people who are not near residential places. Food products can, therefore, be stored for a longer period of time without damaging their quality (Saegusa et al., 2005).

18.1.1.4 Food Quality Preserved

Packaging is one of the safe options for preserving food products and also ensures quality retained for a significant amount of time. It also protects the product from all environmental damage. Thus, glass or aluminum protects food from all possible damage and maintains the quality of food products (Galanakis, 2019; Vishwakarma et al., 2016).

18.1.2 Demerits of Packed Food

18.1.2.1 Adverse Effects of Preservatives

During the packaging process, various preservatives are loaded to increase the shelf life of products; such preservatives cause serious health issues. Thus, a wide range of preservatives are harmful to consumer and may lead to various health disorders (Ayalew and Paulos, 2019; Chandan, 2015; Ryan, 2015).

18.1.2.2 Presence of Harmful Bisphenol A (BPA)

Most food items are packed in plastic, which can be hazardous due to the presence of bisphenol, a compound that is harmful to consumers (Kawamura, 2014; Khan et al., 2021; Schug and Birnbaum, 2014).

18.1.2.3 Extremely Addictive

The main drawback of packaged food is that it can be highly addictive, posing significant risks to consumer health. As a result, packaged items have increasingly replaced natural foods, which remains the most sustainable and health-conscious option for consumer (Law, 1997).

18.1.2.4 Usage of Chemical Ingredients

In the packaging process, the usage of preservatives as artificial and chemical ingredients, such as color, artificial flavoring, texturing have been reported to cause few adverse heath. Health effects including hypertension, diabetes, unbalancing of internal organs, and dehydration have been reported (de Quirós et al., 2019).

18.2 Potential Future Applications for Smart Packaging with Various Food Products

Innovative food packaging is a key component of impression management and creating a brand among consumers about the value of the offer (Ginah and Akpotu, 2016; Singh et al., 2017; Vartanian, 2015). In addition, food packaging keeps the food at the proper temperature and preserves the food from contamination and further comfort to deliver fresh and luscious food consistently (Li, 2020; Passali et al., 2012). It creates a connection and builds trust with consumers; therefore, packaging serves as a powerful marketing tool (Keusseyan and Mobley, 2015). Smart Packaging improves the quality and safety of the product and also tracks thc pcrformancc of thc product throughout the supply chain (Sarkar, 2020). Smart packaging functions in both active and intelligent ways (Butler, 2008).

Its evolution has been driven by the integration of advanced electronics, wireless technologies, and cloud computing (Galanakis, 2019). Despite these advancements, the evolution of smart packaging has taken nearly 150 years. We will examine the timeline of packaging history, depicted in Figure 18.2, and explore the development of packaging (Zuo et al., 2022). Figure 18.3 illustrates the use of smartphone camera features for capturing and evaluating food freshness.

Before 19th century	• 1500 — Used natural materials tree leaves, bamboo, animal skin • 1800 — Invention of glass container in Europe • 1817 — Carboard production for cereal • 1830 — Production of boxes • 1840 — Production of paper bags • 1869 — Gelatine films to protect meat products in US
20th century	• 1920 — Frozen foods • 1930 — Waxing and lipid coating in fruits and vegetables • 1935 — Beverage container • 1946 — First plastic bottle • 1959 — Aluminum coating • 1977 — PET Bottles • 1980s: Microwave-safe packaging, barcode technology • 1990s: Vacuum packaging, biodegradable materials
Modern times (21st century)	• 2000s: Recyclable materials, nanotechnology, smart packaging • 2010s: Eco-friendly alternatives, blockchain technology, edible packaging • 2020s: Sustainable innovation, 3D printing, alternative formats

Figure 18.2 Evolution of food packaging: From ancient methods to modern innovations

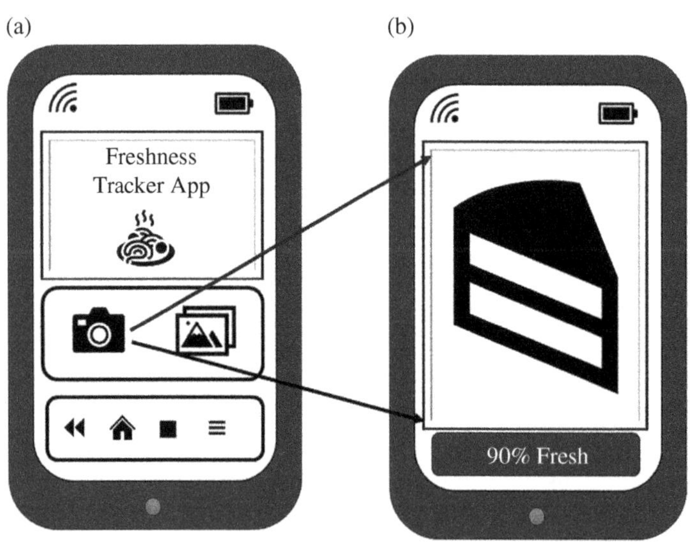

Figure 18.3 Empowering smartphone users: (a) accessing the intuitive interface of the user-centered freshness tracker app, with touch buttons for easy navigation, (b) utilizing the camera feature to capture and assess food freshness, promoting hygienic consumption with precision

18.2.1 Future Applications

- Two-dimensional quick response (QR) codes—which store more information, are easily traceable, and can also maintain a supply chain visibility (Brabazon et al., 2014)
- Augmented reality—videos, social media posts, holograms, etc. (Lee et al., 2020; de Souza Galdino Acioly and Soares, 2017)
- Time and temperature indicators—to know the freshness of the package (Dirpan et al., 2022; Meng et al., 2018)
- Labels with IoT integration (Trewhella, 2017; Tsai et al., 2018)

18.3 The Future of Antimicrobial Packaging Systems

The packaging industry is undergoing rapid change with the discovery of new materials that are safe and sustainable. Old methods are being rediscovered that are efficient and cost effective. However, food contamination and spoilage are a matter of serious concern in the food technology and packaging industry nowadays (Sperber and Doyle, 2009). Statistical data by the Food and Agriculture Organization indicates that almost one third of the food produced every year is wasted or lost every year because of microbial food spoilage and shelf life expiry (Food and Agriculture of the United Nations, 1994). During the different processes involved in food processing—from harvesting to being displayed in the market, food comes in contact with different elements, which can lead to food contamination. Microbial invasion and contamination are the main causes of food spoilage. This invasion of pathogens can lead to poor acceptance and can more seriously cause the spread of food-borne diseases (Blackburn, 2006). A lot of cases of food poisoning are reported every day, leading to an increased number of deaths annually. One of the major causes of food spoilage is incompetent food packaging. Conventional food packaging strategies are not efficient in controlling microbial growth. Hence a better packaging strategy can act as an obstacle against microbial contamination and prevent food spoilage (Taormina and Hardin, 2021).

The drawbacks associated with conventional food packaging can be overcome by imparting antimicrobial functionality in packaging. The development of antimicrobial packaging can minimize microbial contamination and food spoilage and can help meet the growing demand for food safety and preservation. This packaging can reduce microbial growth, improve food quality, and ensure a longer shelf life (Malhotra et al., 2015). It is crucial to evaluate a microbe for which antimicrobial is to be targeted. However, it is essential to select the right package of antimicrobial agents and environmental conditions for a particular food product. Although it might happen if the antimicrobial is too compatible, it may not get released or incompatible that will be released within minutes. Careful consideration should be necessary when choosing and integrating antimicrobial substances into food packaging materials, taking into account both the implications for food safety and the growing problem of microbial resistance. Therefore, it is crucial to adopt a multidisciplinary strategy by involving experts from various fields, such as biotechnology, specifically microbiology, food technology, engineering, and material science. This collaborative effort aims to pave the way for a resilient future in the food packaging industry regarding antimicrobials. Additionally, selecting appropriate materials capable of serving as effective antimicrobial agents against contaminations is of utmost importance (Malhotra et al., 2015).

Different materials belonging to the category of inorganic compounds, organic materials, and natural materials are widely used in antimicrobial packaging. Silver ions, zinc oxide, and copper oxide are different inorganic materials commonly used as antimicrobial packaging agents (Dar and Nayik, 2022; Egbuna et al., 2022). Silver ions are one of the most commonly used agents for antimicrobial packaging because of their excellent antibacterial activity. They are most commonly used in the form of nanoparticles. Usually, silver nanoparticles are functionalized with various organic ligands to alter their antibacterial activity. Zinc oxide particles also exhibit excellent antimicrobial activity. They can easily attach to bacterial cell surfaces, causing cell membrane disruption and causing bacterial cell death. Also, the small size of zinc oxide particles allows them to easily penetrate bacterial cell walls and inhibit gene expression. Copper oxide nanoparticles also show good antimicrobial activity against different bacterial species like *Escherichia coli* and methicillin-resistant *Staphylococcus aureus*. They can also be associated with silver nanoparticles and show excellent antimicrobial activity through the optimum release of ions (Dash et al., 2022; Malhotra et al., 2015).

Also, various organic and natural materials can be utilized in antimicrobial food packaging, offering various benefits such as biodegradability, renewability, and reduced environmental impact. Organic polymeric materials such as chitosan, lignin, starch, and polylactic acid (PLA) have been used as food packaging material (Muñoz-Bonilla et al., 2014). Chitosan is a biopolymer with inherent antimicrobial properties. It can be incorporated into packaging films or coatings to inhibit the growth of bacteria, fungi, and yeasts on the surface of food products, as listed in Figure 18.4. Lignin is a natural polymer found in plant cell walls and extracted from sources such as wood pulp or agricultural residues. Lignin-based films or coatings possess antimicrobial properties and can be used in food packaging applications. Starch-based materials can be modified to exhibit antimicrobial

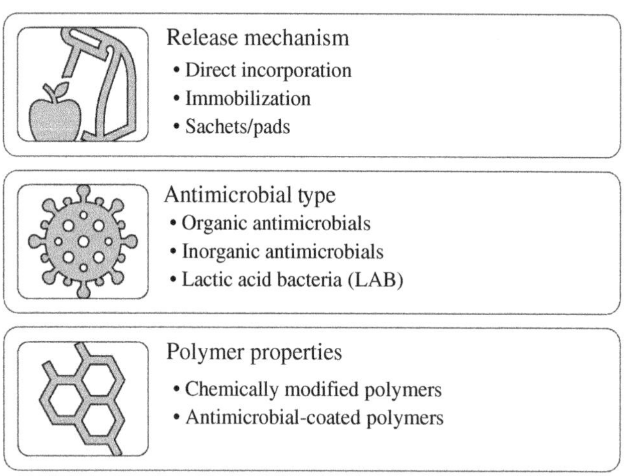

Figure 18.4 Combating food spoilage: exploring strategies for antimicrobial food packaging systems

activity. Starch films containing antimicrobial agents like essential oils or metal nanoparticles can be applied as coatings or packaging films. PLA is a biodegradable polymer derived from renewable resources such as corn starch or sugarcane. It can be modified or blended with antimicrobial agents to create packaging materials with enhanced functionality. Cellulose-based materials, derived from sources like wood pulp or cotton, can be used in antimicrobial packaging applications. Cellulose films or coatings can be functionalized with antimicrobial agents to prevent microbial growth on food surfaces.

Natural materials and compounds are widely used in antimicrobial packaging because of their attractive properties. Essential oils possess antimicrobial properties due to their bioactive compounds, such as terpenes and phenolics, and incorporating essential oils into packaging materials can help inhibit the growth of bacteria, molds, and yeasts (Muñoz-Bonilla et al., 2014; Thormar, 2011). In addition, certain plant extracts contain compounds with antimicrobial activity, such as polyphenols and flavonoids. These extracts can be incorporated into packaging materials to provide natural antimicrobial protection for food products (Makwana, 2013). Natural coatings made from materials like beeswax, shellac, or alginate can be applied directly onto food surfaces to provide a protective barrier against microbial contamination. Various natural bioactive compounds, such as carvacrol, thymol, and eugenol, exhibit antimicrobial properties and can be incorporated into packaging materials to enhance their functionality as listed in Figure 18.5.

However, there are different challenges associated with antimicrobial food packaging. There are very limited practical studies on the exact mechanism of action of different antimicrobial agents. Also, it is important to study the formation of resistant microbial strains. The study of the presence of these strains among the survivor microbial population is very important for the safety aspect and effectiveness of antimicrobial films. Another

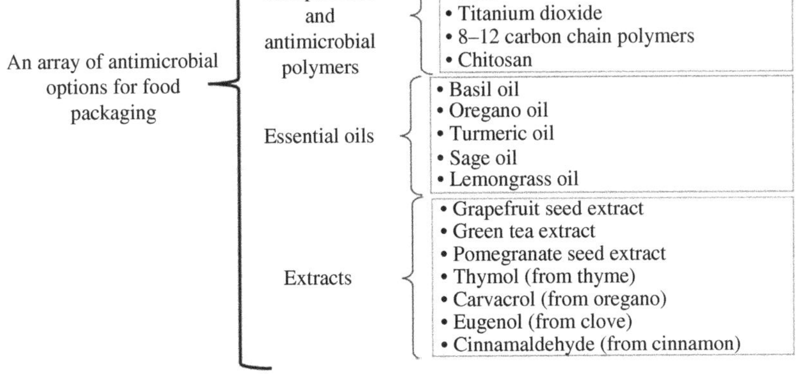

Figure 18.5 Diverse antimicrobial options for enhancing food packaging safety and shelf life

major challenge is the cost of antimicrobial packaging agents, although these agents can be affordable at the laboratory scale, scaling up them to an industrial level is expensive. This makes it inconvenient to produce them on an industrial scale.

Consumer acceptance and sensory effects need to be also considered for designing and commercialization of food packaging. Although antimicrobial packaging provides convenience and improves the shelf life, variations in odor, color, taste, and appearance will not be well received by the consumer and can result in product rejection. Many of these antimicrobial agents, like essential oils, contain phenolic compounds that come with strong odor, color, and flavor, which can interfere with the characteristics of the original packaged food, which is one of the biggest challenges of the food packaging industry. Also, various antimicrobial agents, such as essential oils and enzymes, can denature during the processing stage under high temperatures and pressure. This also is another major challenge that needs to be overcome (Malhotra et al., 2015).

Although the use of nanoparticles has several advantages, there are several limitations that have to be carefully looked into. It is established that the toxicity of nanoparticles increases with particle size. The migration of these nanomaterials to food from the package can result in several health problems, including carcinogenicity and genotoxicity. Therefore, it is important to devise effective techniques for the detection of risk and to analyze the interaction of the various nanoparticles with living organisms, study nanotoxicity and understand the required regulatory procedures and guidelines. Bionanocomposites are very attractive candidates because of their potential as future food packaging material. But there are a few concerns and challenges which need to be addressed. In the case of nanocomposite antimicrobial packaging films, poor compatibility between the polymer matrix and the incorporated antimicrobial packaging fillers can lead to inefficient antimicrobial activity. Also, the high processing and production costs and the low production rates limit their commercial application.

18.4 Nanotechnology in Sustainable Plastics for Food Packaging

Recent studies involving the use of nanotechnology in food packaging showed a great potential to improve packaging materials' main properties as well as add new functionalities, such as active (e.g., antimicrobial) and intelligent (controlled release) features. Nanotechnology promises many interesting changes for a better life. Nanotechnology can be used to improve health, wealth, products, and quality of life. Food nano packaging is still a rather unexplored field of nanoscience and food science. Biobased packaging is indeed an alternative to conventional packaging with nondegradable plastic polymers that are a threat to the environment. Biobased packaging reduces waste, extends the shelf life, and enhances food quality. Active packaging refers to the use of active materials such as antimicrobials and oxygen-scavenging reagents. Smart packaging is the use of nanosensors and nanodevices that detect freshness or contaminants in foods or monitor changes in packaging conditions or integrity.

Silver nanoparticles are already widely used in a number of applications, including food packaging, as a way to kill harmful bacteria because of their antimicrobial properties. Going one step further than just keeping food fresh, nanotechnology could also be used in intelligent packaging to indicate whether food is still good to eat. Researchers are exploring the idea of using carbon nanotubes in food packaging to create a sensor that could tell you whether the food is still fresh by detecting gases that are released when food spoils. The carbon nanotubes would be sprayed onto flexible plastic film used in the food packaging.

18.5 Sustainable Preparation Methods for Nanomaterials

In recent days, nanotechnology applications in food packaging have been utilized to detect bacteria within the packaging, thereby enhancing safety by improving barrier properties. Hence nanotechnology holds a great assurance to provide more benefits not only within the food packaging but surroundings of food products. In food packaging, nanomaterials are basically used for antimicrobial imparting and for increasing the barrier properties of packaging materials so that their life and freshness may extend. Mostly used nanomaterials in food packaging are titanium dioxide, silver, and zinc oxide hence, these materials act as coated films and make a barrier against oxygen and carbon dioxide. Thus, the specific factors for sustainable packaging include packaging ratio, minimal materials, compressed packaging, compressed layers of packaging, etc.

18.6 Future Research Areas Based on Societal Demand and Emerging Approaches

According to the research report, global smart packaging is anticipated to develop in the 20th century. Smart packaging and product packaging are anticipated to develop in bounds in the upcoming years. Thus, smart packaging includes environmental conditions, quality characteristics (indicators), and data carriers. The data carrier helps to give the information flow about the product in the supply chain more efficiently. The main functions of data carriers are to guarantee traceability and automation protection from theft. The most used data carriers are barcode and Radiofrequency identification (RFID). The barcodes are economical, simple, and broadly used to promote recording for stock, checkouts and for controlling inventory. A dimensional barcode is a parallel space pattern that needs a scanner and is associated with a system that can translate the coded information. Two-dimensional barcodes have more memory capacity, such as date of packaging, batch number, preparation instruction, weight, nutrition information etc., which are arranged in an array or a matrix of dots combination. The 2D barcodes are known as QR codes. RFID tags are advanced data carrier which has data storage of 1 MB noncontact, non-line-of-sight ability for real-time data. RFID tags are more expensive than barcodes and also need a powerful electronic information Network. The loaded information can be changed electronically. The improved version of RFID in the field is time temperature indicator (TTI), an intelligent packaging technology. This gives information about the product as the temperature progression that is used to track distribution and also used to reduce waste of food. The improvisation in TTI-RFID has a biosensor in which a specific pathogenic antibody is included in the membrane as a part, which indicates the presence of bacteria, which makes the barcode unreadable while scanning.

Freshness indicators help assess the quality of food products by signaling the loss of freshness, which indicates product conditions or exceeded durability. This have been achieved using nano integrated biosensors. Nanobased gas indicators monitor the quality of food by detecting atmospheric changes, with nano sensors used for this purpose.

Augmented reality videos, social media posts, and holograms in the food industry have increased the attention and recognition of consumers. Mixed reality is an upcoming research area in food packaging and is still under development. The increasing accessibility and wide usage of the internet have led to the development of cloud computing Internet of Things (IoT) and Blockchain technology in food packaging. Some of the features of these development are authentication, open sensors, track and trace, temperature sensors, best before warnings, reordering action, digital labels, real-time offers.

18.7 Nondestructive Quality Checks

Nondestructive is a form of quality check that involves different methods and techniques that destroy the food products being tested and do not damage the products. The NDT (Non-destructive testing) testing method includes acoustic emission testing, spectroscopy, ultrasound, nuclear magnetic resonance, liquid penetrant testing, and radiographic testing, as listed in Figure 18.6. The main advantage of NDT is it is cost-effective, identifying defects before becoming bigger.

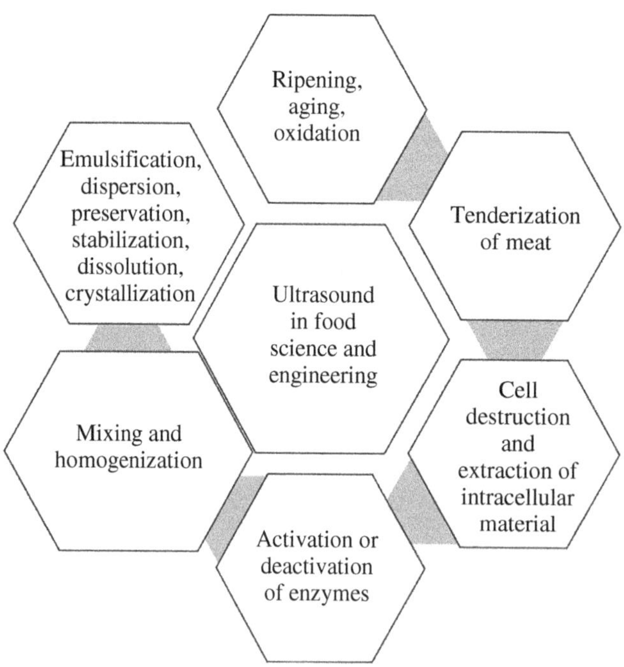

Figure 18.6 Flowgraph leveraging ultrasound in food science and engineering for enhanced processes from ripening to preservation and crystallization

18.8 Novel Sustainable Advanced Materials and Their Future Scope

Novel sustainable packaging mainly focuses on addressing food waste and loss reduction by preserving food quality as well as food safety issues by preventing food-borne diseases and food chemical contamination (Fink, 2023). It also addresses the long-term issue of persistent environmental plastic waste accumulation as well as the saving of oil and food material resources. Recent research has focused on the development of bio-packaging, involving the use of either bio-based packaging materials made from renewable resources or biodegradable materials. There are many technical issues faced with bio packaging materials. Hence, there is an urgent need to progress in the guidance of tools for packaging that will fast-track sustainable innovations into market penetration. The future scope of food packaging should be able to support the transition from a linear to a circular economy.

18.9 Conclusion

In conclusion, the past few decades have witnessed a significant rise in the prominence of sustainable packaging, as evidenced by the challenges, solutions, merits, and demerits discussed in this chapter. Through evaluating various food products, we have explored the potential future applications of smart packaging, shedding light on the evolving landscape of antimicrobial packaging systems and the associated materials. Furthermore, we have delved into the emerging field of nanotechnology in sustainable plastics for food packaging, highlighting its novelty and potential implications.

Moreover, this chapter has addressed sustainable preparation methods for nanomaterials and identified future research areas driven by societal demand and emerging approaches. Nondestructive quality checks for quality control have been emphasized, alongside discussions on the novelty of sustainable advanced materials and their future scope. Overall, this chapter emphasizes the importance of sustainable packaging solutions in addressing contemporary challenges while paving the way for innovative approaches that prioritize environmental responsibility and societal needs. As the field continues to evolve, there is ample opportunity for further exploration and advancement, guided by a commitment to sustainability and the pursuit of excellence in food packaging technology.

References

Ayalew, Y., & Paulos, M. (2019). Storability and maintaining quality of banana with pretreatments by preservatives and packaging materials. *International Journal of Postharvest Technology and Innovation*, *6*(2), 83. https://doi.org/10.1504/IJPTI.2019.10027675

Barone, C., Bolzoni, L., Caruso, G., Montanari, A., Parisi, S., & Steinka, I. (2015). *Food Packaging Hygiene*. Springer International Publishing. https://doi.org/10.1007/978-3-319-14827-4

Blackburn, C. W. (2006). *Food Spoilage Microorganisms*. Woodhead Publishing.

Brabazon, T., Winter, M., & Gandy, B. (2014). *Digital Wine*. Springer, Singapore. https://doi.org/10.1007/978-981-287-059-9

Breetz, C. (2014). *The Impact of Product Packaging on Consumers' Value Perception: Does Packaging Enable Retailers to Take Premium Pricing with Their Retailer Brand Premium Tier in the Food Segment in Germany?* GRIN Verlag.

Butler, P. (2008). Consumer Benefits and Convenience Aspects of Smart Packaging. In J. Kerry, & P. Butler (Eds.), *Smart Packaging Technologies for Fast Moving Consumer Goods* (pp. 233–245). Wiley. https://doi.org/10.1002/9780470753699.ch13

Chandan, R. C. (2015). Dairy Processing and Quality Assurance: An Overview. In A. Kilara, & N. P. Shah (Eds.), *Dairy Processing and Quality Assurance* (pp. 1–40). Wiley. https://doi.org/10.1002/9781118810279.ch01

Dar, A. H., & Nayik, G. A. (2022). *Nanotechnology Interventions in Food Packaging and Shelf Life*. CRC Press.

Dash, K. K., Deka, P., Bangar, S. P., Chaudhary, V., Trif, M., & Rusu, A. (2022). Applications of inorganic nanoparticles in food packaging: a comprehensive review. *Polymers, 14*(3), 521.

Dirpan, A., Djalal, M., & Ainani, A. F. (2022). A simple combination of active and intelligent packaging based on garlic extract and indicator solution in extending and monitoring the meat quality stored at cold temperature. *Foods, 11*(10), 1495. https://doi.org/10.3390/foods11101495

Egbuna, C., Jeevanandam, J., Patrick-Iwuanyanwu, C., & K., & N. Onyeike, E. (Eds.). (2022). *Application of Nanotechnology in Food Science, Processing and Packaging*. Springer International Publishing. https://doi.org/10.1007/978-3-030-98820-3

Fink, J. K. (2023). *Food Safety, Plastics and Sustainability: Materials, Chemicals, Recycling and the Circular Economy*. John Wiley & Sons.

Food and Agriculture Organization of the United Nations (1994). *The State of Food and Agriculture*. Food and Agriculture Organization of the United Nations.

Galanakis, C. M. (2019). *Food Quality and Shelf Life*. Academic Press.

Ginah, G. O., & Akpotu, C. (2016). Impression management and organizational performance in the fast food and restaurant sector in nigeria. *Archives of Business Research, 4*(1). https://doi.org/10.14738/abr.41.1815

Ibrahim, I. D., Hamam, Y., Sadiku, E. R., Ndambuki, J. M., Kupolati, W. K., Jamiru, T., Eze, A. A., & Snyman, J. (2022). Need for sustainable packaging: an overview. *Polymers, 14*(20), 4430. https://doi.org/10.3390/polym14204430

Jestratijevic, I., Maystorovich, I., & Vrabič-Brodnjak, U. (2022). The 7 Rs sustainable packaging framework: systematic review of sustainable packaging solutions in the apparel and footwear industry. *Sustainable Production and Consumption, 30*, 331–340. https://doi.org/10.1016/j.spc.2021.12.013

Kasza, G., Veflen, N., Scholderer, J., Münter, L., Fekete, L., Csenki, E. Z., Dorkó, A., Szakos, D., & Izsó, T. (2022). Conflicting issues of sustainable consumption and food safety: risky consumer behaviors in reducing food waste and plastic packaging. *Foods, 11*(21), 3520. https://doi.org/10.3390/foods11213520

Kawamura, Y. (2014). *Bisphenol A in Japanese Canned Foods* (pp. 155–166). National Institutes of Health Sciences, Tokyo, Japan. https://doi.org/10.1021/bk-2014-1162.ch015

Keusseyan, R., & Mobley, T. (2015). Material and Process Developments for Robust and High Reliability Glass Wafers for 2.5D Packaging. *Additional Conferences (Device Packaging, HiTEC, HiTEN, and CICMT), 2015*(DPC), 001364–001377. https://doi.org/10.4071/2015DPC-wp12

Khan, M. R., Ouladsmane, M., Alammari, A. M., & Azam, M. (2021). Bisphenol A leaches from packaging to fruit juice commercially available in markets. *Food Packaging and Shelf Life, 28*, 100678. https://doi.org/10.1016/j.fpsl.2021.100678

Law, F. (1997). Addictive behaviour: molecules to mankind-perspectives on the nature of addiction. *Addiction, 92*(8), 1041.

Lee, G. A., Ahn, S., Hoff, W., & Billinghurst, M. (2020). Enhancing First-Person View Task Instruction Videos with Augmented Reality Cues. *2020 IEEE International Symposium on Mixed and Augmented Reality (ISMAR)*, 498–508. https://doi.org/10.1109/ISMAR50242.2020.00078

Li, Y. (2020). Migration of metals from ceramic food contact materials. 1: effects of pH, temperature, food simulant, contact duration and repeated-use. *Food Packaging and Shelf Life, 24*, 100493. https://doi.org/10.1016/j.fpsl.2020.100493

Makwana, S. M. (2013). *Study of Antibacterial Property of Plant Based Phenolic Compounds and Food Contact Materials Coated with Functionalized Nanoparticles.* Southern Illinois University at Carbondale.

Malhotra, B., Keshwani, A., & Kharkwal, H. (2015). Antimicrobial food packaging: potential and pitfalls. *Frontiers in Microbiology, 6*. https://doi.org/10.3389/fmicb.2015.00611

Meng, J., Qian, J., & Tang, Y. (2018). A solid-state time-temperature indicator used in chilled fresh pork monitoring. *Packaging Technology and Science, 31*(5), 353–360. https://doi.org/10.1002/pts.2328

Muñoz-Bonilla, A., Cerrada, M. L., & Fernández-Garcia, M. (2014). Antimicrobial Activity of Chitosan in Food, Agriculture and Biomedicine. In Alexandra Muñoz-Bonilla; María Cerrada; Marta Fernández-García; *Polymeric Materials with Antimicrobial Activity. From Synthesis to Applications* (*RSC Polymer Chemistry Series*, Volume *10*, pp. 22–53). *The Royal Society of Chemistry.*

Muthu, S. S. (Ed.). (2021). *Sustainable Packaging*. Springer, Singapore. https://doi.org/10.1007/978-981-16-4609-6

Parameswaranpillai, J., Krishnankutty, R. E., Jayakumar, A., Rangappa, S. M., & Siengchin, S. (2021). *Nanotechnology-Enhanced Food Packaging*. John Wiley & Sons.

Passali, D., Gregori, D., & Foltran, F. (2012). Proper packaging for food and no-food products to avoid injuries. *International Journal of Pediatric Otorhinolaryngology, 76*, S53–S56. https://doi.org/10.1016/j.ijporl.2012.02.013

de Quirós, A. R. B., Cardama, A. L., Sendón, R., & Ibarra, V. G. (2019). *Food Contamination by Packaging: Migration of Chemicals from Food Contact Materials.* Walter de Gruyter GmbH & Co KG.

Rangappa, S. M., Parameswaranpillai, J., Thiagamani, S. M. K., Krishnasamy, S., & Siengchin, S. (2020). *Food Packaging: Advanced Materials, Technologies, and Innovations.* CRC Press.

Ryan, J. M. (2015). *Food Fraud*. Academic Press.

Saegusa, T., Ito, C., Yamakawa, H., & Shirai, K. (2005). Japanese perspectives and research on packaging, transport and storage of spent fuel. *Packaging, Transport, Storage & Security of Radioactive Material, 16*(3), 169–178. https://doi.org/10.1179/174650905775295332

Sarkar, D. K. (2020). *Packaging Technology and Engineering: Pharmaceutical, Medical and Food Applications.* John Wiley & Sons.

Schug, T. T., & Birnbaum, L. S. (2014). *Human Health Effects of Bisphenol A*. In S. M. Snedeker (Ed.), *Toxicants in Food Packaging and Household Plastics* (pp. 1–29). Springer-Verlag, London. https://doi.org/10.1007/978-1-4471-6500-2_1

Singh, P., Wani, A. A., & Langowski, H.-C. (2017). *Food Packaging Materials: Testing & Quality Assurance*. CRC Press.

de Souza Galdino Acioly, A., & Soares, M. M. (2017). Mobile Augmented Reality Systems Applied to Food Packaging - A Heuristic Evaluation. *2017 19th Symposium on Virtual and Augmented Reality (SVR)*, 243–252. https://doi.org/10.1109/SVR.2017.39

Sperber, W. H., & Doyle, M. P. (Eds.). (2009). *Compendium of the Microbiological Spoilage of Foods and Beverages*. Springer, New York. https://doi.org/10.1007/978-1-4419-0826-1

Taormina, P. J., & Hardin, M. D. (Eds.). (2021). *Food Safety and Quality-Based Shelf Life of Perishable Foods*. Springer International Publishing. https://doi.org/10.1007/978-3-030-54375-4

Thormar, H. (Ed.). (2011). *Lipids and Essential Oils as Antimicrobial Agents*. Wiley. https://doi.org/10.1002/9780470976623

Trewhella, J. (2017). Heterogeneous integration and iot collaborative packaging innovations. *Journal of Japan Institute of Electronics Packaging*, *20*(1), P1–P1. https://doi.org/10.5104/jiep.20.P1

Tsai, M., Chiu, R., He, E., Chen, J. Y., Chen, R., Tsai, J., & Wang, Y.-P. (2018). Innovative Packaging Solutions of 3D System in Package with Antenna Integration for IoT and 5G Application. *2018 IEEE 20th Electronics Packaging Technology Conference (EPTC)*, 1–7. https://doi.org/10.1109/EPTC.2018.8654330

Vartanian, L. R. (2015). Impression management and food intake. Current directions in research. *Appetite*, *86*, 74–80. https://doi.org/10.1016/j.appet.2014.08.021

Vishwakarma, G. S., Gautam, N., Babu, J. N., Mittal, S., & Jaitak, V. (2016). Polymeric encapsulates of essential oils and their constituents: a review of preparation techniques, characterization, and sustainable release mechanisms. *Polymer Reviews*, *56*(4), 668–701. https://doi.org/10.1080/15583724.2015.1123725

Zuo, J., Feng, J., Gameiro, M. G., Tian, Y., Liang, J., Wang, Y., Ding, J., & He, Q. (2022). RFID-based sensing in smart packaging for food applications: a review. *Future Foods*, *6*, 100198. https://doi.org/10.1016/j.fufo.2022.100198

Index

Nanotechnology for Sustainable Food Packaging, First Edition. Edited by C. Anandharamakrishnan, Jeyan A. Moses, and M. Maria Leena.

d

e

f

x

y

z